Wind Energy Revolution

Tarleton State University Southwestern Studies in the Humanities, No. 30

T. LINDSAY BAKER, GENERAL EDITOR

A list of other books in this series may be found at the back of this book.

Wind Energy Revolution

How the 1970s Energy Crisis Fostered Renewed
Interest in Electric-Generating Technology

Christopher C. Gillis Sr.   *Foreword by Michael Bergey*

TEXAS A&M UNIVERSITY PRESS   COLLEGE STATION

♾ This paper meets the requirements of ANSI/NISO Z39.48–1992
(Permanence of Paper).

Binding materials have been chosen for durability.

Manufactured in the United States of America

Library of Congress Cataloging-in-Publication Data

Names: Gillis, Christopher C., author.

Title: Wind energy revolution: how the 1970s energy crisis fostered
  renewed interest in electric-generating technology / Christopher C.
  Gillis Sr.; foreword by Michael Bergey.

Other titles: Tarleton State University southwestern studies in the
  humanities; no. 30.

Description: First edition. | College Station: Texas A&M University Press,
  [2023] | Series: Tarleton State University southwestern studies in the
  humanities; no. 30 | Includes bibliographical references and index.

Identifiers: LCCN 2023017651 (print) | LCCN 2023017652 (ebook) | ISBN
  9781648430626 (hardcover) | ISBN 9781648430633 (ebook)

Subjects: LCSH: Wind power—United States—History. | Wind turbines—United
  States—Design and construction—History—20th century. | Small power
  production facilities—United States—Technological
  innovations—History—20th century. | BISAC: TECHNOLOGY & ENGINEERING /
  History | HISTORY / United States / 20th Century

Classification: LCC TJ828 .G525 2023 (print) | LCC TJ828 (ebook) | DDC
  621.31/21360973—dc23/eng/20230420

LC record available at https://lccn.loc.gov/2023017651

LC ebook record available at https://lccn.loc.gov/2023017652

In memory of Karl H. Bergey Jr. (1922–2019), who not only founded a successful American small wind-electric turbine manufacturing enterprise but spent nearly fifty years unselfishly sharing his knowledge and passion for aerodynamics and wind power as a professor and scholar.

Contents

Foreword

It has been my good fortune to be exposed to the seductive challenges of wind energy design and engineering. I was trained as an engineer and had a father willing to apply his considerable aircraft design experience to my student projects and, in 1977, to a new small wind business venture. That led to four decades on the front lines of consumer power generation and off-grid energy supply. Through my involvement in the wind energy trade associations, I likewise have been fortunate to witness and to participate in the emergence of the wind power industry that we see today making important contributions to the environment, climate change, and the economy.

My father, Karl Bergey, and I loved the engineering challenge of wind power. The design of a wind turbine requires understanding and some level of mastery, as well as the integration of aerodynamics, structures, vibration, fatigue, electrical machines, electronics, controls, materials science, corrosion, soils, and civil works. Wind turbines can easily operate over 80 percent of the time, which has been likened to putting 150,000 miles a year on a car, so wear and tear are major issues as well. For me, and for many others I know, it is still magical to watch a wind turbine dance through gusts and eddies on a breezy day.

Nearly everyone who has entered the wind energy engineering and business realm, from backyard inventors to NASA scientists, has underestimated the difficulty in designing rugged and reliable wind turbines. All have been humbled by the challenge. I certainly have. An early example for Bergey Windpower was our use of aluminum blades on our first product, a 10-foot-diameter, 1-kilowatt turbine we introduced in 1980. In spite of careful design and an elaborate testing program, the blades started cracking after just a few months in the field. Our tests had not yet adequately duplicated the fatigue loads from wind gusts that the blades experienced in the real world. That led to a crash development project for a new airfoil and fiberglass blades, plus a costly retrofit program for over a hundred customers. Though we today are certainly more experienced now, we are still learning lessons in wind turbine engineering.

The industry has also faced significant policy and market access barriers. With the recent ascendance of clean energy technologies, it is easy to forget how powerful and assertive the utilities and the coal and nuclear industries were as wind energy struggled to gain a foothold in the market in the 1980s and 1990s. Bergey Windpower experienced the market chilling effects of utility interconnection requirements that included isolation transformers and expensive liability insurance. Neither was justified, and they just served to raise the costs and discourage potential consumers. Utility interconnection costs and delays remain a common barrier today for larger wind systems in distributed applications.

In Washington the wind industry often found itself outgunned. I vividly remember a meeting that AWEA had in the 1980s with the powerful chairman of the House Energy Appropriations Subcommittee. We had about ten minutes with him regarding the promotion of increased R&D funding for wind before one member of his staff whisked him away for an important meeting. We later learned that he actually left us to catch a ride to his district on a private jet owned by a coal company. The requested boost in funding was not forthcoming.

The future of large wind is assured, thanks to the tremendous strides the industry has made in developing ever larger land-based and offshore wind turbines with ever lower costs. Most people fail to realize how much utilities have been able to save their customers by buying energy from wind farms. The future of small wind is less certain due to severe competition from imported solar equipment. But I believe that small wind will contribute to our energy future by applying technology lessons learned by the large wind industry and finding innovative ways to provide new services that are valued by consumers and their utilities. The US Department of Energy is playing a critical role in supporting innovation and focusing on US manufacturing.

I am grateful to Chris Gillis for his well-researched and insightful book on the history of the US wind industry. I particularly appreciate the depth of his coverage of small wind because until now it has been largely overshadowed by the much larger, more successful and more impactful large wind and wind farm industries. The characters and stories that he has chronicled here provide a glimpse of the much broader struggles by hundreds of entrepreneurial firms that entered and did not survive the small wind business. Among them were some of the most altruistic, clever, and dedicated individuals I have had the pleasure to cooperate and compete with. Chris, thank you for your hard work, your ever-curious mind, and your dedication to ferret out the defining stories of our industry. I hope your readers enjoy this work as much as I have.

—Michael Bergey
Norman, Oklahoma

Preface

It sounds simple: Fashion a set of blades, attach them to a generator, set the machine on top of a tower, and let the wind do the work of creating electricity. Not so. Most of these attempts fall far short of expectations, even with the availability of the latest tools and materials. In fact, it would be naïve to suggest otherwise. When American windmill historian Dr. T. Lindsay Baker approached me five years ago about researching and writing a scholarly history of small wind-electric power in the United States, I realized from the outset that it would be a challenge to encapsulate in one book the breadth of an industry that has evolved over the past 150 years. My focus in this volume is on the efforts of entrepreneurs to develop "small" wind generators for use at homes, farms, and ranches following the 1973 Arab oil embargo. Dr. Baker and Texas A&M University Press editor Jay Dew put their faith in me to accomplish this awesome task, which I wholeheartedly embraced.

My interest in small wind-electric power started nearly two decades ago. In 2007, I encountered Craig Toepfer of Michigan while researching a book on the general history of electric-generating wind turbines. Toepfer introduced me to the world of small wind-electric plant manufacturers in the United States prior to 1950, such as Wincharger Corporation, Jacobs Wind Electric Company, and Winpower Company. Our numerous telephone calls, emails and in-person discussions heightened my awareness of the importance that these small wind-electric plants had for so many rural Americans, particularly across the Great Plains and Southwest, where during the first half the twentieth century they lacked access to "high-line"

electricity from faraway power plants. Toepfer explained to me how this onsite-generated power from small wind-electric plants afforded rural people the opportunity to enjoy the use of electrical appliances and work tools that were readily available to their brethren living in cities and towns. I also learned from Toepfer how the energy crisis and environmental concerns of the early 1970s sparked a renewed interest in the United States for small wind-electric turbines. As a young engineer during the 1970s, Toepfer enthusiastically traveled the back roads of the Midwest searching for wind turbines built before 1950, refurbished them in his workshop, and returned the units to service for numerous interested homeowners. He also had the good fortune to work alongside Marcellus Jacobs at the reestablished Jacobs Wind Electric Company in Minneapolis during the early 1980s. Through an evolving career path in the decades that followed, Toepfer has remained advocate, consultant, and lecturer for wind and solar energy.

To launch my book research, Dr. Baker generously opened his voluminous files in Rio Vista, Texas—the result of nearly six decades of research into the American windmill industry—in the summer of 2015. His files contained innumerable pertinent materials related to the wind-electric business, such as scholarly and trade journal articles, letters, brochures, sales literature, and various forms of advertising. I was already familiar with the extent of his archive from my research four years earlier for the book, *Still Turning: A History of Aermotor Windmills*. This extraction of information from hundreds of files provided the foundation for what would become a multiyear research endeavor for me. Throughout this process, Dr. Baker was never too busy to answer my questions and offer advice. He has been my steadfast mentor and editor since 1994, and for that I remain most grateful to him.

Another important research moment for this book occurred in late October 2015, when I traveled to Norman, Oklahoma, to visit with Michael Bergey and his father, Karl Bergey, to whom this book is dedicated. Both men from the start embraced the scholarly premise of this book and set aside time for me to interview them extensively about the history of their successful enterprise, Bergey Windpower, which today is considered the longest operating small wind turbine manufacturer in the United States. I have dedicated an entire chapter of this book to the founding of Bergey Windpower and its contributions to the development of modern small wind turbines. The chapter does not shy away from explaining the struggles the Bergeys endured during down periods in the small wind-electric industry. While other companies have come and gone in the small wind-electric industry, Bergey Windpower has remained a relevant participant in the national effort to reduce our reliance on electricity produced by fossil fuels.

Early on in the process of researching this book, I also determined that it was important to devote a separate chapter to the return of the Jacobs Wind Electric Company to the small wind-electric industry during the late 1970s. In late September 2016, I had the privilege of spending a breakfast meeting with Paul Jacobs near his home in Corcoran, Minnesota, during which he introduced to me to his family's rich legacy in the small wind turbine manufacturing business. He explained to me how, during the late 1920s, his father, Marcellus Jacobs, and his uncle, Joseph Jacobs, set out to build a wind-electric machine that would become the recognized leader for its high efficiency, reliability and durability. Many young engineers and entrepreneurs during the 1970s and early 1980s attempted to replicate the success of the Jacobs Wind Electric Company in their own machine designs but often to no avail. After our meeting, Paul Jacobs spent many hours corresponding with me by email during the winter of 2017 to answer my many questions, as well as share with me pertinent documentation about how he and his father reestablished the Jacobs Wind Electric Company and manufactured a new line of wind turbines from 1978 to 1985. Through Paul Jacobs's generosity and support of this book, I believe that my chapter about the Jacobs Wind Electric Company's return to the industry is most thorough and accurate.

Since I am not an engineer, I leaned heavily on the technical expertise of Herman Drees of Thousand Oaks, California. Drees spent his entire career in the wind-electric industry, including designing and manufacturing a residential-scale, vertical-axis wind turbine during the late 1970s and early 1980s. He would also become an authority on the operation of large wind turbines, even operating a small wind farm in Southern California from the mid-1980s to 2019. I was introduced to Drees by longtime wind energy consultant and author Paul Gipe at the start of researching this book. From the beginning, I could always count on Drees to be my advisor on technical matters involving wind turbine designs and operations. He patiently guided me through the writing of Chapter 8, which covers wind turbine technology developments during the 1970s and 1980s. Drees also shared with me numerous photographs from his personal files to help illustrate several chapters of this book.

In regard to the history of the US government wind energy research and development programs that started in the early 1970s, I am indebted to Warren Bollmeier, who now resides in Southern California. He was one of the first engineers to arrive at the Rocky Flats test site in Colorado, and he helped propel the wind-swept site into becoming the epicenter for US wind energy research in the decades that followed. Bollmeier supported the purpose of my book from the start. He even shared his research files and photographs which document his many years of work at Rocky Flats. These materials proved invaluable to my writing about the early years of the US government's wind energy research programs. R. Nolan Clark and Brian Vick, both of Amarillo, Texas, also shared with me details about the former wind energy research program operated by the US Department of Agriculture in Bushland, Texas. For the most recent view of the federal government's small wind-electric turbine research programs, I had the pleasure of interviewing Patrick Gilman of the US Department of Energy's Wind Energy Technologies Office and Ian Baring-Gould of the National Renewable Energy Laboratory's National Wind Technology Center, both in Golden, Colorado. Both of them provided valuable insights.

Other individuals who kindly shared with me their private collections of wind-electric industry materials, company memories, and technical knowledge include Jay Carter Jr., Carter Aviation Technologies, Wichita Falls, Texas; Jim and Jan Erdman, Custer, Wisconsin; Carlos Fernandez-Bueno, Dickerson, Maryland; Trudy Forsyth, Wind Advisors, Broomfield, Colorado; Coy Harris, former executive director, American Windmill Museum, Lubbock, Texas; Jonathan Hodgkin, Essex Junction, Vermont; Jennifer Jenkins, Flagstaff, Arizona; Sherry Jones, Gemini Controls, Cedarberg, Wisconsin; Richard Katzenberg, Amherst, New Hampshire; Ken Kotalik, Primus Windpower, Lakewood, Colorado; Andy Kruse, HOMER Energy, Boulder, Colorado; Bob McBroom, Kansas Wind Power, Holton, Kansas; Donald Mayer, Waitsfield, Vermont; Philip Metcalfe, Brimfield, Illinois; Leander Nichols, Anacortes, Washington; Allan O'Shea, CBSi Solar, Copemish, Michigan; Jack Park, Palo Alto, California; Phil Perry, Hepler, Kansas; Travis Price, Travis Price Architects, Washington, DC; Earle Rich, Amherst, New Hampshire; Hagen Ruff, Chava Wind, Miami; Larry Sherwood, Interstate Renewable Energy Council, Latham, New York; Paul Shulins, Boston, Massachusetts; Holly Spaulding, Buffalo, New York; Brent Summerville, Boone, North Carolina; Glen Swanson, Grand Rapids, Michigan; Randall Swisher, Silver Spring, Maryland; Russell, Tencer, United Wind, Brooklyn, New York; Walter Thompson, Manteca, California; Mike Werst, Manor, Texas; and Dan Whitehead, St. Johns, Florida.

I also received research assistance from a number of overseas windmill historians, including Frans Brouwers, *Levende Molens*, Ekeren, Belgium; the staff of the Poul la Cour Foundation, Middelfart, Denmark; Etienne Rogier, Toulouse, France; and Helen Walter, *The Windmill Journal*, Morawa, Australia. Their efforts to answer my many questions helped me present a clear picture of the early developments of wind-electric turbine technologies in Europe and Australia.

Preparation of the photos and line drawings used to illustrate this book would not have been possible without the kind technical assistance of my friend and photographer, Walter J. Leskuski Jr. Many hours during the evenings and weekends were spent at his kitchen table discussing and reviewing which images provide the reader the best visual presentation for this history of small wind-electric power in the United States. I would also like to thank Stephen James Govier of Suffolk, England, for his outstanding hand-drawn illustrations, which were used within the pages of this book.

Finally, I owe a tremendous debt of gratitude to my wife, Theresa, and children, Christopher Jr. and Elizabeth Ann. Without their support, this book would not have been possible. They tolerated my numerous hours spent behind closed doors in my home office during evenings after work and on weekends. During these times, I conducted phone interviews, corresponded with sources, reviewed research materials, and wrote my book chapters. The breadth and tediousness of this project required my utmost attention for four years.

Overall, it is my hope that this book will prove to be most enlightening and enjoyable for the reader with an interest in this country's innovative legacy in developing small wind-electric energy systems during the past 150 years. Volumes could surely be written about the historic and technical details presented in each of the nine chapters of this book. I am most humbled to be a participant among the ranks of other notable American wind power historians, such as Alfred R. Wolss, A. Clyde Eide, Dr. T. Lindsay Baker, Dr. Robert Righter, and Paul Gipe in the research, writing, and publication of this book. I hope that these historians and those in the future will find this book worthy to stand in their company.

Wind Energy Revolution

1 Electricity from the Wind

By 1900, the industrialized world, concentrated mostly in Europe and North America, had already burned vast amounts of coal and oil to power its factories and electric power plants during the preceding thirty years. City populations, which had grown exponentially during the late 1800s, benefited from access to industrial jobs, new products, and services, as well as electric power mostly in the form of lighting. They also suffered, however, from crowded, often filthy living conditions and dangerous work environments. While less concerned about the impacts of pollution on the earth, economists and scientists by the end of the nineteenth century were increasingly concerned that the coal and oil being taken from the ground would be quickly exhausted in the years ahead. There was also a growing awareness that the existing means of generating power—namely, through large steam engines—was highly inefficient. Charles R. Van Hise, an American geologist, explained at that time how the average steam engine used only 10 percent of the heat generated from burning coal to produce useful power; that loss reached as much as 99 percent when heat energy generated by this process was turned into electric power and then into light.[1] Scientists increasingly believed the answer to the world's future energy demand could be substantively extracted from the wind. "There are mighty rivers of air sweeping incessantly over the face of continent and ocean. The winds have neither source, shore, nor destination; they cannot be separately named and set down upon the map; so that we easily forget their vast extent and immeasurable force," a British authority stated in 1898.[2]

For hundreds of years, the Chinese used vertical-axis windmills, like the one shown in this postcard image (ca. 1910), for irrigation. From the author's collection.

For centuries wind power had been used for mechanical applications, such as lifting water, grinding grain, sawing wood, extracting oils, and crushing ores. The earliest documented mechanical windmills date to between AD 850 and 870 and were located in a rugged region of ancient Persia, what is now the modern-day border of Afghanistan and Iran. These windmills used lightweight, rectangular wood-framed sails filled in with woven reeds and spaced in a cylindrical manner around a vertical shaft, which turned a set of millstones to grind grain into flour.[3] The Chinese were likewise early users of vertical-axis windmills, which were coupled with a rudimentary system of gears and rope along with a series of evenly attached small buckets to lift water into fields surrounded by earthen berms.[4] Windmills are believed to have first appeared in Europe around AD 950, perhaps as a result of early European Christians encountering this technology during pilgrimages to Palestine and bringing it back with them. The evidence of windmills in Europe, however, appeared more prominently in medieval literature after the Third Crusade (1189–1192).[5] Unlike their counterparts in the Middle East and Far East, European windmill builders, known as millwrights, used a series of rectangular wooden-framed sails that were attached around horizontal wooden shafts. Often several stories tall, European windmills were constructed in three primary styles. Post mills—structures manually rotated into the wind on centered timber posts—were mostly found in northern Europe. Brick and stone tower mills, with rotating tops that turned the sails into the wind, could be found in northern Europe and along the Mediterranean Sea. Smock mills, which operated much like tower mills, had wood-framed polygonal sides that sloped outward at the bottom and were located mostly in the Netherlands, northern Germany, and Scandinavia.[6]

Millwrights and windmill operators have for centuries considered different sail designs and positions to gain the most power from the wind. It wasn't until the eighteenth century, however, that scientists and engineers gained a better understanding of how wind interacts with windmill sails. In 1759, English physicist and engineer John Smeaton presented groundbreaking research to the Royal Society of London in which he explained that the power available from the wind is proportional to the cube of the wind speed, meaning that as the wind speed doubles, the power from the wind

Traditional post mills, resting on vertical wooden posts, were manually turned into the wind. From *Le Spectacle de la Nature*, Paris (1751).

increases eight times. While English and Dutch windmills at the time generally used a consistent angle along the length of their sails, Smeaton's research concluded that the most efficient windmill sails should be twisted at about 18 degrees where they attach to the rotor shaft and contour to 7 degrees at the tip, giving them a varied pitch and a more swept appearance.[7] Smeaton's research was based on the standard cloth-covered wooden lattice sails predominantly used on northern European windmills. In the 1770s work was underway to refine the general appearance of windmill sails. Instead of the miller having to roll up part of the sailcloth in high winds, Scottish millwright Andrew Meikle in 1772 introduced the variable-spring sail, which replaced the cloth sail with a series of small wooden shutters connected to a single rod. The miller could manually move the rod to adjust the pitch of the shutters, depending on the wind velocity. In time, a system of springs and weights was introduced essentially to allow for the self-regulation of the shutter position and thus the windmill operating speed. When the wind

increased, the shutters would open, and conversely they would close when the velocity decreased.[8] This speed regulation prevented the machinery from running too fast and overheating. Meikle's basic design was refined by other millwrights in England and northern France through the early 1800s, but its widespread deployment was minimal. At the same time, the coal-fired steam engine, which was introduced in the mid-1700s, quickly marginalized the windmill for most industrial activities by the mid-1800s.[9] However, windmills remained important to Europe's rural communities for grinding grain into flour and animal feed and for draining soggy land.

By the early seventeenth century, windmill technology was carried throughout the world by European explorers and quickly became part of many settlements where there was plenty of timber, stone, and wind. The Dutch settlement, in what today is New York City's Lower Manhattan, included a windmill by the early 1600s, and it is recorded that a windmill was erected in 1621 on the Flowerdew Hundred estate 20 miles outside the

English physicist John Smeaton used this experimental device to determine the relationships between wind speed, peripheral rotational speed of sails, and achievable maximum power. He presented his groundbreaking research to the Royal Society in 1759. From *An Experimental Enquiry Concerning the Natural Powers of Water and Wind to Turn Mills and Other Machines Depending on Circular Motion*, London (1759).

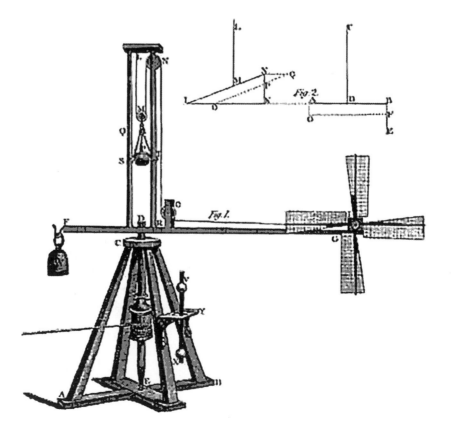

Jamestown settlement in Virginia. These windmills were built along the East Coast, from Canada and Maine to Florida and the Caribbean islands, and eventually deep into the North American interior with westward migration. The Russians, who settled Fort Ross in northern California, constructed a windmill by 1814.[10] Many windmills built along North America's East and West coasts became important tools in the production of salt, which was essential in colonial times for preserving food, drying leather, and making medicines. These rudimentary windmills pumped seawater into wooden vats or shallow ponds, where the sun evaporated the water, leaving behind the salt crystals.[11]

By the early nineteenth century, large-scale migration of people from the coasts of North America to the interior began in earnest. Land with access to surface water, such as springs and streams, was quickly settled and fiercely guarded, leaving those who secured property without readily accessible water to dig wells in search of a sustainable water supply. Water from these wells was obtained by a rope and bucket, or through a hand pump mechanism in the case of drilled or bored wells. While

this method met the water requirements of a rural household, garden, or small farm, it was far from efficient to quench the thirst of ever-larger cattle herds growing in numbers on the Great Plains.[12] By the 1850s, the railroads also began penetrating deeper into the country's interior from the East and West coasts. Their steam engines required lots of clean water to operate, and a dense network of storage tanks had to be erected along these rail lines to ensure a readily available water supply.[13] These two trends led to the emergence of a new industry in North America that was focused on providing mass-produced commercial windmills for pumping water. These windmills, however, needed to be relatively easy to erect, durable, and able to operate with minimal maintenance. Most importantly, they needed to "self-govern" in the wind. Without this ability, windmills could easily be destroyed by high winds.[14]

Several significant manufacturers emerged in the 1850s that offered self-governing windmills. These early designs generally consisted of a combination of wooden parts held together by iron hardware. The wood blades of the wind wheels

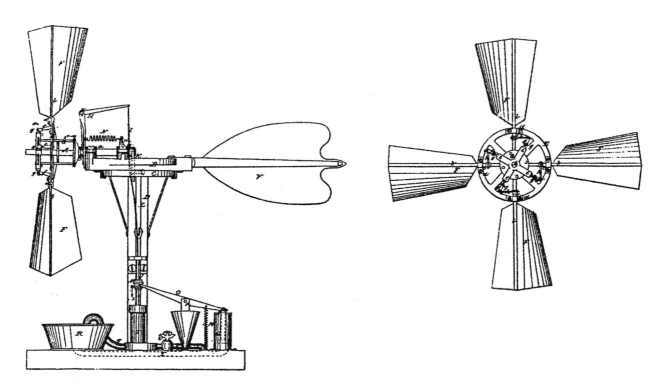

Patent drawing for the invention that became known as the Halladay Standard self-regulating windmill, 1854. Courtesy US Patent and Trademark Office, Washington, DC.

were trapezoidal-shaped and held in place by large metal hoops. These wind wheels were then attached to the ends of horizontal metal shafts at the top of timber-framed towers. Various geared mechanisms, driven by the turning wind wheel, provided a vertical, reciprocating motion to operate a piston-type pump at the water table. The wind wheels typically were turned into the wind by a large wooden vane. One of the early success stories of the American self-regulating windmill was sparked by the partnership between Connecticut mechanic Daniel Halladay and pump repairman John Burnham. In 1854, Halladay received a patent for his paddle-bladed, self-governing windmill. The Halladay Wind Mill Company manufactured its windmills from South Coventry, Connecticut, until 1862, when Burnham convinced Halladay to relocate the operation to Batavia, Illinois, closer to markets on the prairies. The U.S. Wind Engine and Pump Company, which took over the Halladay Wind Mill Company, benefited from closer proximity to its customer base—the western railroads, farmers, and cattlemen. The company also soon thereafter replaced the paddle-shaped blades with a wind wheel that consisted of narrowly spaced sections of thin wooden blades.[15] Another important self-governing windmill manufacturer among the railroads and cattlemen to emerge in the 1860s was the Eclipse Wind Mill Company of Beloit, Wisconsin, which was founded by Leonard H. Wheeler and his son of the same name. The Wheelers' windmill also started with four large angled wooden paddle-shaped blades but quickly shifted to a wind wheel similar to that of the U.S. Wind Engine and Pump Company.[16] There were also a number of windmill manufacturers that chose to focus on regions, such as William Isaac Tustin who dominated the California market from the 1850s to 1870s, and the Continental Windmill Company and Empire Wind Mill Manufacturing Company in New York, which focused much of their windmill sales on the northeastern states.[17] By the 1870s, dozens more windmill manufacturers scattered throughout the United States and Canada entered the industry. Their windmills were not only sold in North America but were also exported throughout the world.

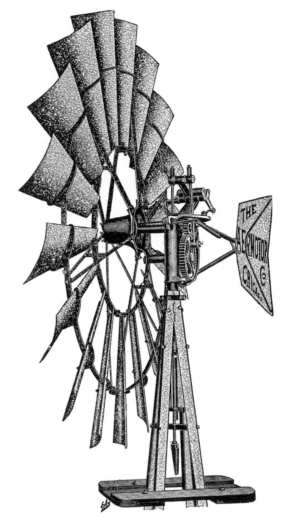

By the 1890s, the all-steel, back-geared Aermotor became one of the most recognized water-pumping windmills in the United States. From Aermotor Company, *7th Annual Catalogue* (Chicago, 1895), 24.

With advances in steel manufacturing and resulting price reductions at the start of 1870s, some designers turned their attention to all-metal windmill construction, promoting their durability over their wood competitors. The first commercially successful all-metal windmill, called the Iron Turbine, was manufactured by Mast, Foos and Company in Springfield, Ohio.[18] Another Springfield-based manufacturer, the Springfield Machine Company, followed with its Leffel Iron Wind Engine, and an Ellicott City, Maryland, company in the early 1880s introduced the Kirkwood Iron Wind Engine. Other producers of all-metal windmills included Indianapolis Machine and Bolt

Works, the Plymouth Iron Windmill Company of Michigan, and the Columbus Windmill Company in Ohio.[19] However, it was an ambitious pair of young engineers in the late 1880s who developed the most successful all-metal American windmill. Thomas Osborne Perry and La Verne W. Noyes, both of whom had previously been associated with the U.S. Wind Engine and Pump Company and formed the Chicago-based Aermotor Company in 1888, paid careful attention not only to the wind wheel's design, but also to the geared pumping mechanism behind the wheel. They designed a system of "back-gearing" to give their windmill a more consistent and smoother pumping action. Their gear arrangement allowed for one long pump stroke for approximately every three turns of the wind wheel, contrasted to the "jerky" single-stroke pumps at the time, which completed a single stroke for every 360-degree turn of the wind wheel.[20]

A number of manufacturers also offered windmills known as "power" mills that performed mechanical tasks. Instead of producing an up-and-down motion for a light pump rod, these windmills used a spur gear to drive a pinion gear and a pair of bevel gears at the top of the tower to spin a vertical steel shaft. Power mills were generally placed on short towers fixed to the roofs of barns and sheds. At the bottom of the power mill's shaft was a foot gear or fly wheel that drove farm machinery, such as feed grinders and choppers, corn shellers, circular buzz saws, and other devices offered for sale by the windmill manufacturers.[21]

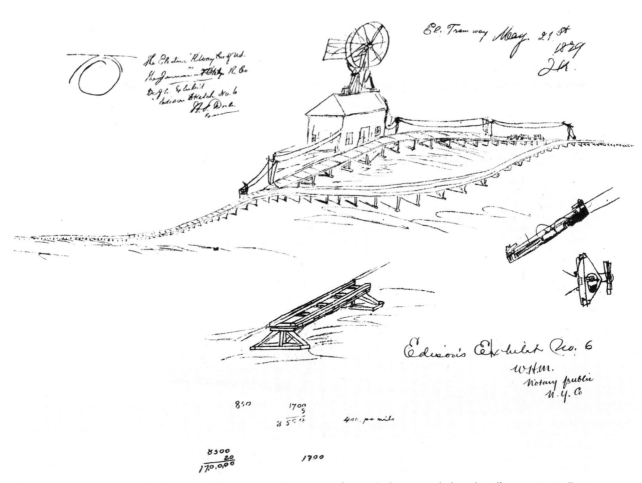

This original sketch made by Thomas Edison shows his concept for a wind-powered electric railway system. From a Technical Note by Thomas Alva Edison, March 21, 1879, Edison's Exhibit No. 6, *Electric Railway Company v. Jamaica and Brooklyn Road Company*. Courtesy Thomas A. Edison Papers, Rutgers University, Piscataway, NJ.

Perhaps the American windmill's biggest advantage over the traditional European windmill was its ability to be mass produced. These machines had interchangeable parts and were easy to ship and erect, not to mention requiring minimum attendance and maintenance. These attributes also made American windmills cost effective to own and operate in remote locations, not just in North America but throughout the world. The water-pumping windmill was emblematic of other technological breakthroughs that were occurring during the 1870s and 1880s, which not only saved labor and improved living conditions but conveyed a sense of modernity. The internal combustion engine replaced human and animal muscle and led to increased mechanization and mobility. The telegraph permitted people to send messages through electrical impulses over long distances via copper lines instead of relying on hand-delivered letters. The electric lighting apparatus allowed people to work more efficiently into the night and permitted retailers and restaurants to stay open longer. The dynamo generated electricity, which when coupled

with batteries could be stored for later use. Since the dynamo's armature shaft needed to rotate to generate electricity, it compelled some scientists and engineers at the time to consider the freely available wind as its future driving force. The first notable documented consideration of wind being used as a force to turn a dynamo was referenced in a January 6, 1841, Belgian patent issued to Professor François Nollet of the Military School in Brussels for his electromagnetic machine. This machine used a form of electrolysis to create light, heat, and power.[22] However, Nollet never acted on the patent, as the steam engine rapidly asserted itself in the industrialized world as the dominant force in electric power production. Nearly forty years later, wind-electric power sources were considered by two prolific American inventors of their day, Thomas A. Edison and Moses G. Farmer.

While passing through Iowa on a cross-country train trip in 1878 to observe a solar eclipse, Edison conceived the idea of building a network of small railroads throughout the midwestern states, which would be used to feed carloads of wheat into the

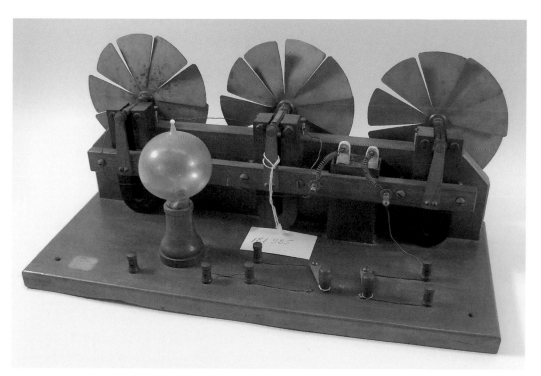

Inventor Moses Farmer in 1881 constructed this model as part of his patent to use wind power for charging batteries to power light bulbs. The patent model is part of the Smithsonian Institution's National Museum of American History collection. Courtesy Glen Swanson, Grand Rapids, MI.

mainline railroads for distribution throughout the country. These small railroads would use electricity for their power, instead of the coal-fired steam boilers used in locomotives on the main rail lines. Considering the strong winds that swept unhindered across the Great Plains, Edison said, "I thought that I could utilize those winds to run windmills which, in their turn, would give motion to dynamo machines and furnish electricity to operate the small motors and cars which I proposed to deploy." He explained this idea in testimony for an 1892 patent infringement lawsuit. Rough sketches used in the deposition showed Edison's May 1879 depiction of "a station for an electric railway with a turn-out track . . . the track on trestles and a station with windmill for obtaining power with wires running from the station out to the track." Edison extensively studied the idea at his Menlo Park, New Jersey, laboratory but determined that wind power at the time "was not sufficiently advanced to permit us to see our way clear to displace steam on a large railroad."[23]

Farmer, whose inventive prowess involving electricity dated to 1845, led him to develop an electromagnetic engine, the first fire alarm system, and improvements to telegraphy by the late 1840s. He also invented a battery-powered electric lighting system, which he used in the parlor of his Salem, Massachusetts, home in 1859.[24] In the late 1870s, Farmer had become fascinated by the force of the wind and its potential as an electric-generating energy source, and in 1881, he applied for a patent that incorporated wind power for charging batteries to power light bulbs. For reasons unknown, the US Patent Office never approved his patent application. A related ca. 1880 patent model was later donated to the Smithsonian Institute by his daughter, Sarah Farmer. The model consists of three solid brass wind wheels that drive the armatures of three dynamos, which connect to a small storage battery, an incandescent light bulb, and switches.[25]

However, public discussion of wind's potential as a facilitator of electric power did not actually begin until September 1, 1881, when Sir William Thomson, also known as Lord Kelvin, gave his introductory president's speech, "On the Sources

Alfred R. Wolff in 1885 published *The Windmill as a Prime Mover,* a compilation of nine years of research into the "economy and capacity" of American windmills. From "Death of Alfred R. Wolff," *Stevens Institute Indicator* (Hoboken, NJ), 26, no. 1 (January 1909).

of Energy in Nature Available to Man for the Production of Mechanical Effect," before Section A of the British Association for the Advancement of Science. Thomson was among a number of leading scientists of the day who believed that the industrialized nations were burning through the world's available coal and oil resources at such a rapid rate that they would run out by the early twentieth century. He stated: "Even now it is not utterly chimerical to think of wind superseding coal in some places for a very important part of its present duty—that of giving light. Indeed, now that we have dynamos and [Camille Alphonse] Faure's accumulator, the little want to let the thing be done is cheap windmills."[26] However, Thomson lamented the problems associated with prolonged periods when there was not sufficient wind to drive a windmill in order to maintain the electric charge of batteries. In addition, he expressed concern that windmills, when contrasted to steam engines, were not yet economical as prime drivers for electric power. "[W]indmills as hither to made, are very costly machines, and it does not seem probable that, without inventions not yet made, wind can be economically used, to give light in any considerable class of cases, or to put energy into store for work of other kinds."[27]

Challenging the position of this revered scientist's conclusions about wind power seemed almost blasphemy. Yet, the young mechanical engineer and associate editor of *The American Engineer*, named Alfred R. Wolff, immediately took exception to Thomson's conclusions about the economics of windmills. As he stated in an editorial, "In calculating the cost of our power, we should calculate all the expenses of producing that power. Such expenses when they relate to prime movers usually include the cost of fuel, oil and attendance, and the *interest*, repairs, depreciation and insurance of [the] plant. Thus considered, the windmill will be found to be the most economical prime mover."[28] In 1885, Wolff published *The Windmill as a Prime Mover*, a compilation of his nine years of research into the "economy and capacity" of American windmills. In his analysis, he noted that the windmill's propensity "[f]or driving dynamo-machines to charge electric accumulators" was near. However, three years later, he continued to battle naysayers within the engineering field who blamed the mechanics of the windmill for its lack of prolific use in electric power generation. As he stated in 1888,

> [T]he reason windmills are not used in this way is not that the windmills are not sufficiently economical or reliable, but that the electrical accumulators are not yet a satisfactory and assured success. When they are, windmills will come into extended use as prime movers for the generation of electricity, and electricians will be glad to avail themselves of the most economical motor, utilising the force of wind, otherwise going to waste, for this purpose.[29]

In the summer of 1887, a former student of Thomson and professor at the Glasgow and West Scotland Technical College, James Blyth, suc-cessfully built a windmill to drive a dynamo and charge batteries. First, he experimented with an English-style wind wheel that had four arms with cloth sails, and then he tried an "American-type" windmill with multiple blades. In a paper presented in 1892, he spoke of the difficulty of devising the most effective windmill:

> [A]ll these [windmill types] I found to answer very well so long as the wind had a moderate speed, but, like all other experimenters with windmills, I soon found that they had either to be made self-reefing or stopped altogether when a breeze came. This is obviously unsatisfactory, as the best of the wind for storage purposes is lost; and hence this problem presented itself—how to construct a windmill that would satisfy the following requisites:—1. It must be always ready to go. 2. It must go without attendance for lengthened periods. 3. It must go through the wildest gale, and be able to take full advantage of it.[30]

Blyth then studied the Robinson anemometer, which consisted of four hemispherical cups attached to the end of four arms that rotated horizontally around a vertical axis. He ultimately decided to use a similar design for his windmill, which he constructed in the garden at his summer cottage in the village of Marykirk in Kincardineshire. His machine consisted of four 26-foot-long horizontal wooden arms to which he attached two sets of semicylindrical boxes separated by a small space. The largest outer boxes measured 6 feet wide by 10 feet tall, and the smaller inner ones were 3 feet wide by 10 feet tall. Each wooden arm was fixed to metal sockets that were attached to the top of a vertical round iron shaft measuring 5 inches in diameter. The bottom of the shaft carried a pit wheel that turned a set of gears that drove a 6-foot-diameter flywheel. A belt connected this flywheel to a dynamo shaft. The electricity from the dynamo charged a set of thirteen cells. "With a good wind I reckon that it gives about four horse-power. I have also tested it in a strong gale by allowing it to run with no load, and the result was perfectly satisfactory, as a safe terminal speed was attained and all racing avoided," he said.[31] Blyth also addressed the issue of the electrical connection between the dynamo and batteries when the wind was low. "The only thing needed," he explained, "is that the circuit be broken when the dynamo is running at less than the storing speed. I have also tried a form of governor which throws a greater or less number of cells into the charging circuit as the wind varies, and in this way the machine is always doing some work."[32] Blyth used the electric power from the batteries to power up to ten 25-volt lamps at his summer cottage, but he predicted that one day light and power across Great Britain might be predominantly generated by windmills of his type.[33]

On the northern coast of France at Cotes-du-Nord, in July 1886, wealthy landowner Charles Marie Michel de Goyon proposed constructing a "self-directing wind motor" to charge batteries, which he would in turn rent to small landowners and farmers as sources of electric power. The idea was deemed impractical, however, since the batteries at the time were on one hand a "heavy apparatus, of inconvenient carriage, and, on another hand, very delicate."[34] Not to be dissuaded, he proposed in 1887 that a windmill and dynamo system could be used to charge batteries that could be used for lighting the lighthouse at Cape de la Hève, near the mouth of the Seine River. The idea

Drawing of an electric-generating windmill, based on the design of Scottish engineer James Blyth, which was erected at the Sunnyside Royal Hospital in Montrose, Scotland (1895). Courtesy Stephen James Govier, Suffolk, England.

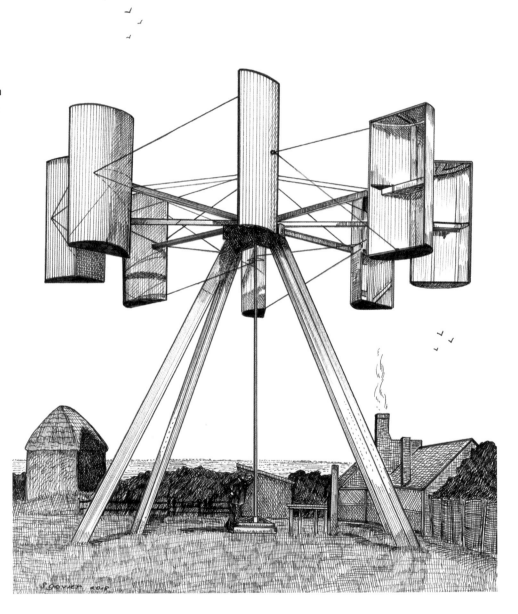

received the backing of France's minister of public works and the superintendent of the lighthouse service.[35] It was decided that the best self-regulating windmill to withstand the heavy wind conditions of France's northern coast would be the Halladay No. 12, imported from the US Wind Engine and Pump Company. Two Brush "Victoria" dynamos were shipped from London. The dynamos, each of which was capable of generating 75 volts, were set up in tandem and connected to the windmill drive shaft by an arrangement of pulleys. "The two machines work alternatively, according to the motive power furnished by the mill."[36] The builders of this system also developed an "interrupter" or "circuit cutter" between the dynamo and batteries when the power falls short. "The interrupter automatically re-establishes the communication between the dynamo and the accumulators as soon as the electro-motive power of the machine returns to its normal value."[37] In January 1890, a heavy storm damaged the windmill beyond repair.[38]

On the other side of the world in Auckland, New Zealand, in 1884 a matrimonial and employment agency proprietor, Thomas Brown Hannaford, presented to New Zealand's House of Representatives the design of an iron-constructed lighthouse, complete with its own windmill for turning a dynamo to charge batteries. The batteries would in turn be used to power the lighthouse's beacon. The British colony's Marine Department rejected the idea. However, a year later, the government approved a lighthouse Hannaford proposed, but without the electric-generating windmill, erecting it in October 1888 on Cuvier Island. Hannaford spent the next five years until his sudden death in late 1890 fighting the New Zealand government for what he viewed as a blatant theft of his idea.[39]

In the summer of 1887, the first electric-generating windmill in the United States was being constructed in the backyard of the estate of wealthy inventor and industrialist Charles F. Brush in Cleveland, Ohio. Brush made his name in the electric power generation and lighting industry starting in the mid-1870s. His inventions often improved upon existing technologies. For example, after studying the Gramme dynamo, Brush proceeded to design and build a more compact and efficient dynamo,

which won a competition in 1877 at the Franklin Institute in Philadelphia. Department store owner John Wanamaker was so impressed by Brush's dynamo that he purchased five of them to power up to twenty lights in his Philadelphia store. Between 1878 and 1880, Brush improved the electric arc light to make it useful for lighting interiors, as well as streets, at night. With both inventions, he and his partner George Stockley in 1880 launched the Brush Electric Company of Cleveland. That year Brush also invented a new storage battery that used a combination of lead plates and sulfuric acid. Meanwhile, the company's arc lights, dynamos, and batteries were sold throughout the United States and Europe, earning the company more than $1 million by 1885. In 1889 the Brush Electric Company was acquired by Thomson-Houston Company in Lynn, Massachusetts. Both companies became part of General Electric in 1892. The sale of the Brush Electric Company left Brush financially independent and enabled him to pursue scientific experiments without the pressures of running a manufacturing plant. By this time, he had already built a mansion on Euclid Avenue in Cleveland and decided to light his home and laboratory with electricity produced by a windmill of his own design.[40]

Brush first considered the use of a windmill for generating electricity to heat buildings in 1879. In his laboratory notes on June 3, 1880, he wrote that the "heating machines" could be designed to use the abundant winter wind to produce electrically heated air, which could then be circulated throughout a building using a system of flues.[41] In 1886, he drafted a patent application describing in detail "A System and Apparatus for Charging Secondary Batteries"; he referred to the windmill as his "prime variable source of power."[42] In early 1887, Brush hired a work crew to begin constructing his windmill. While the structure was nearing completion in early July, it was damaged by a wind storm, but Brush pressed forward, making the repairs and planning to have the windmill operational before the end of the month. With his typical reserve, he noted to the *Cleveland Plain-Dealer* in a July 11, 1887, newspaper article that the windmill was "intended purely as an experiment."[43] Unbeknownst

Illustration of a windmill designed and constructed by Cleveland inventor and industrialist, Charles F. Brush, to furnish electricity to his estate and laboratory. From "Charles Francis Brush," *Harper's Weekly* 34, no. 1753 (July 26, 1890).

to each other at the time, Brush and Blyth built and operated the world's first wind-electric plants by the summer of 1887.[44]

Brush's box-shaped, wood-framed windmill tower, somewhat resembling the traditional European post mill style, stood about 60 feet tall. The tower was mounted on an iron gudgeon, measuring 20 feet long with a 14-inch diameter. Eight feet of the gudgeon was secured in concrete, while the remaining 12 feet above ground supported the 80,000-pound structure. The wind wheel, which had a 58-foot diameter and was made of 144 wooden rotor blades in the style of the large commercially built mechanical power windmills of the period, was attached to a 20-foot-long horizontal wood shaft with a 6.5-inch diameter. The wind wheel was turned into the wind by a large tail vane. This vane, however, could be folded against the tower parallel with the wind wheel to keep it still when not in use. Inside the tower, the turning wheel shaft connected a series of shafts, pulleys, and belts, which turned a dynamo. The dynamo made about fifty revolutions to one turn of the wind wheel and reached full load at 500 revolutions a minute with a capacity of 12 kilowatts. Brush, however, designed the dynamo to start producing electric power at 330 revolutions per minute. He regulated the voltage at about 75 volts. This power was transmitted over a wire to a bank of batteries, consisting of 408 secondary cells, in the basement of his house. The batteries provided power for up to 350 incandescent lights throughout the house, in addition to two arc lights and three electric motors.[45] Brush continuously operated the windmill until 1907, when he took down the wind wheel and fully connected his house and laboratory to public utility electric power. He quipped a few years before his death in 1929 that his windmill was "built to go for twenty years and it never fail[ed] to keep the batteries charged until I took the sails down."[46] The total cost of Brush's windmill is unknown. However, the *Scientific American* in an 1890 article stated: "The reader must not suppose that electric lighting by means of power supplied by this way is cheap because the wind costs nothing. On the contrary, the cost of the plant is so great as to more than offset the cheapness of the motive of power."[47]

At the Danish village of Askov in 1891, a high school physics and mathematics teacher and inventor named Poul la Cour set out to construct a windmill to generate electricity for entire individual villages. His purpose was to find a way to deliver the benefits of electricity to the country's rural population. While the Brush windmill had already been in service for three years, Danish wind energy historians believe that la Cour had no prior knowledge of it when he started building his own machine. La Cour also had little personal funds available to construct his windmill and managed to secure 4000 kroners from the Danish government. Ever frugal, he participated in each aspect of the windmill's construction. With the assistance of millwright N. J. Poulsen, la Cour settled on a rotor design of four sails, which were self-regulated by a system of nine hinged shutters that opened and closed depending on the strength of the wind. The wind wheel was mounted on top of a 36-foot-tall wood timber-framed trussed tower.[48] La Cour invented the "kratostat," a device consisting of a system of rope drives and counterweights to stabilize the wind wheel's continuously fluctuating rotational speed and produce a more consistent electrical output from the dynamo. Since he could not afford lead-acid batteries, la Cour used a simpler electrical storage system based on electrolysis. With the dynamo's direct current, he broke down water into hydrogen and oxygen gases, which he then stored in separate tanks. He used the hydrogen gas to create a flame to heat ceramic containers of zirconium that gave off a bright white light. The Askov Folk High School used this system of lighting from 1895 to 1902.[49]

In 1897, la Cour built an even larger experimental windmill on the school grounds, capable of generating 110 volts. This second machine, which had a tower resembling a traditional smock mill, originally consisted of six experimental sails. However, the wind wheel design proved inefficient, and after three years of wind tunnel tests, la Cour concluded that a four-bladed rotor was most optimal. The six sails were then removed and replaced with a four-blade rotor by 1900.[50] He further analyzed the efficiency of a sixteen-blade wind wheel versus a four-bladed setup. In 1898, the

Inventor and high school teacher Poul la Cour designed and built his first experimental windmill to generate electricity at Askov, Denmark, in 1891. Courtesy Poul la Cour Foundation, Vejen, Denmark.

Scientific American relayed details from a lecture that la Cour had delivered in Copenhagen:

> In measuring the percentage of the power striking the wings, he arrived at the somewhat startling result of 143.7 per cent. This unlooked-for conclusion was owing to the . . . suction on the lee side of the wind passing between the wings. That the wings should not be plane, but have a bent or concave shape, was an old established truism; and the shape of the wings has in reality much influence upon the suction caused more especially by the wind, which just passes the edges of the wing. In measuring the percentage of the wind power utilized, the wind passing between the wings was taken into account, and instead of 143.7 per cent, the result was 21 percent.[51]

However, la Cour was unable to determine precisely the best aerodynamic design for his blades.[52]

By the start of the twentieth century, la Cour's experiments led to the commercial manufacture and installation of wind-electric generators with his four-blade rotor design across rural Denmark and into neighboring Germany. In 1903, la Cour helped create the Danish Wind Electric Society (Dansk Vind Elektrisitets Selskab or DVES), which published a technical journal about wind-electric power generation. A year later he oversaw the establishment of a rural electricians training program at the Askov High School, which included three months of classroom topics, such as chemistry, physics, mathematics, engineering, and technical drawing, followed by hands-on instruction involving actual wind-electric machine installations. It is

3. Hæfte.   Mai 1904.

TIDSSKRIFT FOR VIND-ELEKTRISITET

UDGIVET AF POUL LA COUR

GYLDENDALSKE BOGHANDEL · NORDISK FORLAG

Poul la Cour in 1903 established the Danish Wind Electric Society, which published a technical journal about wind-electric power generation. Courtesy Poul la Cour Foundation, Vejen, Denmark.

In 1904, Poul la Cour (seated first from the left) established a rural electricians training program at the Askov High School, Denmark. Courtesy Poul la Cour Foundation, Vejen, Denmark.

estimated that between 1906 and 1919, when the program ended, 230 rural electricians had been trained by the Askov program. It was estimated that by 1917–1918, when petroleum was in short supply because of World War I, about 250 wind-electric machines of la Cour's design were operating across Denmark as both power supply for farms and factories and supplemental power for village diesel-powered plants.[53] La Cour calculated that one of his wind-electric machines cost $4320. Starting in 1902, the electricity from his Askov machine, for instance, was sold to the village residents at a rate of 13.5 cents per kilowatt-hour for lighting and 4.5 cents per kilowatt hour for power purposes, in line with the cost of electricity in Copenhagen. La Cour noted that he earned receipts of $756 for the sold power, and after expenses of $216 were deducted, his wind-electric machine earned an annual profit of $540.[54]

On May 19, 1891, James M. Mitchell, of Lawrenceville, Georgia, became the first American to receive a patent for a "wind apparatus for generating electricity and charging secondary batteries."[55] All other patents issued before pertained solely to mechanical windmill designs and applications. Although it is unknown if an example of Mitchell's machine was ever built—he did not submit a model with his patent application—his design was ahead of its time. Mitchell proposed placing the dynamo on top of the tower and attaching the wind wheel on the end of the extended armature shaft. His design did away with the traditional setup of a wind-electric system of gears at the top of the tower to turn a vertical shaft, which then turned another set of gears, belts, or pulleys near ground level to turn a dynamo. Mitchell proposed what would later be referred to as a direct-drive wind plant. However, Mitchell retained the multibladed wind wheel commonly used on American pumping and power mills and incorporated a traditional tail vane to keep the wheel facing the wind. The dynamo pivoted on a center bearing at the top of the tower. Wires for carrying the electricity from the dynamo were lowered through the center of the tower to the ground-level storage batteries.[56]

A second patent, also obtained in 1891, included a circuit interrupter to disengage the dynamo from the batteries when the wind was low.[57] Mitchell wrote that "[t]his invention will be of high value in towns both large and small, as well as in cities where regular plants are not found. It can be used where suburbs are not easily reached with perfect success in all cases and with a nominal expense, as the motive power is without cost and the initial expense is comparatively small."[58] Early American wind-electric plant builders, however, continued using converted power mills with the dynamos at ground level.

The Farm Implement News in late 1894 estimated that more than one million water-pumping and power windmills had been constructed throughout the United States, adding that "wind mills or wind engines are seen upon almost every farm."[59] The publication believed that the American windmill industry and its technology were now ripe to drive an abundance of wind-electric power installations throughout the country and revolutionize farming operations: According to *The Farm Implement News*:

> In the near future the simplification of the machinery and apparatus for electric lighting now in progress will have made it practicable for the ordinary geared wind mill to provide the power necessary for lighting farm houses and farm buildings, all at a cost within the means of the ordinary well-to-do western farmer; and for this purpose the demand for the higher grades of wind mills or wind engines will be largely increased.[60]

A year earlier, in 1893, mechanical engineer G. D. Hiscock had calculated the cost of an American windmill with an electric-generating setup (see Table 1).

Many questions remained in the mid-1890s about the practicality of the wind to charge batteries due to its intermittent nature. It was suggested that two sets of batteries might be required for a windmill electric-generating outfit—while one set was in use the other could be charging—but this option

Table 1.

Considering the volume of power as the output of a windmill, allowing an average of eight hours' work of the mill and four hours' lighting, then a 20-foot mill, with an average of 1 horsepower, will charge a storage battery for a half-time lighting of 2 horsepower, or at least 16 candle lights per horsepower, will produce a lighting capacity of 16 candle lights or twenty 12 candle lights in a dwelling, enough for ordinary country houses. Such a plant with a supplementary water supply for a dwelling only cost approximately for:

One 20-foot wind mill	$700
A 40-foot frame tower	300
One 8-light dynamo and regulatory	250
Appliances, wiring, etc	25
A 16-light, 5-hour storage battery	200
Twenty 16-light electric lamps, fixtures, and wiring	40
Pump, piping, and water tank	200
Total	**$1,715**

A 30-foot windmill will be equal to the supply of 32 16 candle-power lamps and supply the water required for a large country house, and may be scheduled as follows:

One 30-foot windmill	$900
A 50-foot frame tower	500
One 3 ½ horsepower dynamo and regulator	300
Appliances, wiring, etc.	50
Thirty 2-light, 6-hour storage battery	300
Thirty 16-light electric lamps, fixtures, and wiring	60
Pump, piping, and water tank	225
Total	**$2,335***

* G. D. Hiscock, "Possibilities in Utilizing the Power of the Wind," The Farm Implement News (Chicago) 14, no. 9 (March 2, 1893): 20.

This 20-foot "Railroad Eclipse" windmill manufactured by the Fairbanks, Morse and Company to charge batteries for powering light bulbs on the estate of George E. McQuesten of Marblehead Neck, Massachusetts, in 1894. From Charles J. Jager Company, *Illustrated Catalogue of Windmills, Tanks and Pumps as Applied to Water Supply Systems, also Windmills Adapted for Power* (ca. 1895).

was too expensive for the average farmer. Others continued to propose using the windmill dynamo's power to produce hydrogen and oxygen gases by applying electrolysis to water and using them later for lighting and heating. Another proposed option was to use a water-pumping windmill to fill a raised tank and then release the water onto a water wheel to turn to a dynamo's armature and use the power as it was produced. This method, too, faced prolonged downtime for the dynamo, as the water tank would need to be refilled after it emptied.[61]

Yet, in the last years of the nineteenth century and into the first decade of the twentieth, engineers and entrepreneurs in both the United States and Europe continued to explore commercial opportunities for wind-electric power generation. One of those individuals was Isaac N. Lewis, a former US Army lieutenant and president of the Lewis Electric Company in New York City, who sought a new application for slow-speed dynamos that he had successfully coupled to the wheels of railcars of the New York Central Railroad in order to charge batteries and provide electric power to light bulbs in-side the passenger cabins. To do so, Lewis's dynamo used a series coil on the field, wound differentially to the shunt, so that it delivered constant current at variable speeds.[62] In late 1893, Lewis approached Andrew J. Corcoran, proprietor of A. J. Corcoran, a Jersey City, New Jersey, windmill manufacturer, to test one of his dynamos on a windmill, which like a railcar's wheels experienced fluctuating speeds. The test windmill at the A. J. Corcoran factory had an 18-foot-diameter wind wheel and a set of gears that drove a belt attached to the Lewis dynamo's shaft. The dynamo had a maximum current capacity of 35 amperes at 35 volts and cut into action when it reached a speed of 600 revolutions per minute. This was attained when the wind speed reached 20 miles per hour. A writer for *The Electrical Engineer* magazine, who visited the A. J. Corcoran factory in January 1894, noted: "The plant, though merely an experimental one, has operated without a single mishap from the start and the storage cells furnish current for 24 incandescent lamps distributed through the work shops. . . . The smoothness of working and the evident reliability of the entire arrangement leave little room for doubt that we shall see a wide application of this system."[63]

Lewis was delighted to witness one of his 2-kilowatt dynamos being used in a successful windmill-based electric lighting plant installation at Marblehead Neck, Massachusetts, in May 1894. Instead of an A. J. Corcoran windmill, the property owner, George E. McQuesten, used a 20-foot "Railroad Eclipse" windmill manufactured by the Fairbanks, Morse and Company of Chicago installed upon a 75-foot wood tower. Power was transmitted by bevel gears and shafting to an 18-foot 6-inch square shed at the base of the windmill tower which contained the Lewis dynamo and

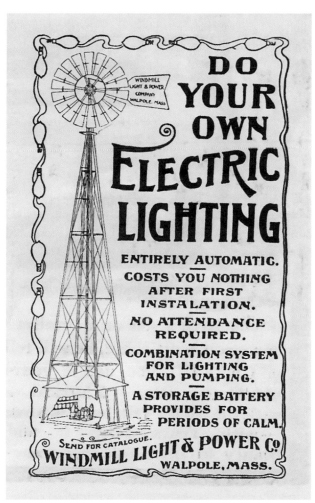

Joseph J. Freely, who owned and operated the Windmill Light & Power Company of Walpole, Massachusetts, during the late 1890s, offered for sale an all-metal American windmill that charged batteries for electric lighting and pumped water. From *Harper's Monthly*, June 1897. Courtesy of the *Windmillers' Gazette*, Rio Vista, TX.

forty-six Bradbury-Stone storage cells with capacity for 200 ampere-hours at 110 volts. The cost of the plant, including the windmill and tower, dynamo, cut-out circuit, storage battery, wiring to the house, and other buildings, as well as labor, was $1800. The windmill plant powered about sixty incandescent light bulbs in the house, with another fifteen in the stable and twelve in the workshop. It also provided electricity for another fifty light bulbs in the home of a neighbor, Elbert Wheeler. McQuesten operated the windmill annually from May 1 to November 5, and then shut it down during the winter. It is uncertain how many years the Marblehead Neck windmill was in service.[64] In the mid-1890s Charles J. Jager Company of Boston, which dealt in the sale and installation of Fairbanks, Morse and Company windmills, began offering Eclipse "Windmill Electrical Lighting Plants," highlighting the success of the McQuesten wind-electric plant.[65] Another wind-electric experimenter, Joseph J. Feely, of Walpole, Massachusetts, in the late 1890s, using an all-metal American power mill, developed a "self-regulating," constant-voltage dynamo and a "speed equalizer" device to counter the variations in wind speed. Electricity from the dynamo was stored in battery cells and used for lighting throughout his farm. Feely also calculated that he could build his electric-generating windmills for $700 to $800 apiece, the cheapest by far of other similar outfits, and by 1897 he offered these windmills for sale to others as the Windmill Light & Power Co. It is uncertain if his business was successful.[66]

Similar use of American-style or imported windmills for generating standalone electric power was underway in the United Kingdom during the 1890s. It was reported in 1892 that an electric-lighting plant in London was "being successfully operated by an American windmill" to charge a battery of twenty-eight cells, which lighted two arc lamps and "a sufficient number of incandescent lamps" to illuminate a large flour mill.[67] British manufacturers of windmills throughout the closing decade of the nineteenth century also began to expand beyond the water-pumping and mechanical applications of their windmills to offering them as electric generators. In June 1895, John Wallis Titt of Woodcock Ironworks in Warminster, Wiltshire, erected one of his 35-foot diameter, variable-pitch, multibladed Simplex Wind Engines on the English estate of cocoa magnate George Cadbury to provide electric power for lighting.[68] The variable blades of Titt's windmills were different from the American all-steel wind wheels in that they consisted of light wooden and round steel rod frameworks covered in canvas. While this made them lightweight and easy to regulate, the canvas coverings often required regular replacement in the humid British climate.[69] However, it was the machine Titt constructed in 1899 at Boyle Hall in West Ardsley, Yorkshire, an estate owned by the Colbeck family, that garnered him significant attention in the field of wind-generated electric power. The windmill, which was placed on top of a 35-foot-tall wrought-iron lattice tower, dynamo, and storage batteries provided power to 109 lights to the main house and outbuildings on the estate. The windmill was admired by many passengers riding on board the Great Northern Railway as they passed through the nearby Wood-kirk Station.[70]

By the end of the nineteenth century, scientists and engineers had proved that the kinetic energy of the wind through mechanical manipulation could be used as an alternative source to petroleum and coal-fired power plants for generating electricity, albeit on a much smaller level of output and usage. The complexity and cost of owning and operating wind-electric generating systems at the time, however, made them impractical for the majority of farmers and rural homeowners worldwide, who relied on their windmills simply to pump water and drive small farm machines. Electric lights and motors were still viewed as novelties to most rural people and not yet an affordable replacement to kerosene lamps and manual labor, although the promise of electricity was surely noted in the numerous stories they heard and read about from large towns and cities where its application was rapidly proliferating. It would take nearly two more decades of technological advancement and increased commercial manufacturing to make the wind-electric generator a practical and impactful feature on the rural landscape.

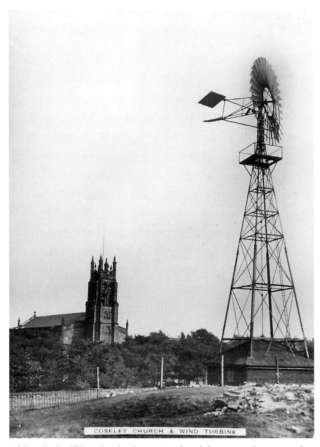

COSELEY CHURCH & WIND TURBINE

This windmill in Birmingham, England (postcard, ca. 1910), turned a dynamo located inside the building at the base of the tower. A storage battery provided enough electricity to Croseley Church for sixty light bulbs and a motor to pump the pipe organ. From the author's collection.

2 Wind-Electric Generators

By the start of the 1900s, the potential for electricity as a power source appeared boundless. In cities and towns across the United States, electricity manifested itself most prominently in the form of streetlamps along commercial thoroughfares, called "great white ways." It not only illuminated streets and sidewalks but also allowed businesses to showcase their wares in brightly lit storefront windows and to stay open later in the evenings. The proliferation of less expensive, longer-lasting incandescent lamps over earlier arc lamps, soon spread street lighting beyond the main arteries and into the residential areas.[1] Electrical lighting for homes and businesses also conveyed a sense of progress, as well as prosperity, when contrasted with gloomier oil- and gas-fueled lamps. Electricity for sizeable towns and cities originated from large central power-generating plants and was supplied over long stretches of copper wires suspended from trussed steel towers at up to 33,000 volts to reduce transmission losses. As this electricity reached its destinations, the voltage was often stepped down at a substation to perhaps 2200 volts, and by the time it reached its point of usage over smaller copper lines, the voltage was further reduced through smaller transformers to 110–220 volts suitable for powering electric lights, motors, and heaters.[2] By 1908, Chicago was the biggest consumer of electric power in the world at 323 million kilowatt-hours annually. As one magazine at the time put it, "If the connected commercial load of the Commonwealth Edison Company of Chicago, equivalent to 3,817,113 16-candlepower lamps, were strung out in a single line of lights, 35 feet apart, they would completely encircle the earth."[3]

By the first decade of the twentieth century, electricity had started to alter how the urban population in the United States lived and worked. Edwin J. Houston, a renowned electricity expert, wrote in 1905 that "To-day one can not converse intelligently on almost any topic without either, eventually, employing some electric term, or without referring to some electric fact or principle. . . . In our households we talk of dynamos, motors, trolleys, electric lamps, telephones, and batteries, quite as freely as we do of bread, butter, butcher's meats, milk, ice, coal and carpets."[4] Electric power companies, as well as numerous other manufacturers, produced household appliances powered by electricity, such as kitchen stoves, space heaters, refrigerators, fans, washing machines, sewing machines, clothes irons, toasters, waffle irons, coffee percolators, and tea pots. Companies marketed these appliances heavily toward women, the majority of whom ran the households at the time, as safe, labor-saving, and adding comfort to the home.[5] Electricity was playfully referred to as the "new servant girl" to the housewife, who "never gossips, never gets homesick, never is impertinent and is cleanly, quiet, and ever ready for duty."[6] Homebuilders in cities and towns began offering fully electrified dwellings, ushering in what was called "the age of the chimneyless home free from smoke, ashes and their attending dust."[7]

An American family at the start of the 1900s enjoys household lighting and appliances provided by electricity. From the author's collection.

It wasn't easy at first, however, to sell electric power and related appliances to urban homeowners at the start of the 1900s. Many people simply thought it was too expensive, especially when one added in the cost to wire a home for electric service. As the central power plants improved in operation and gained larger customers, such as industrial users, they were able quickly to reduce the cost per kilowatt-hour to household users in the vicinity. For example, by August 1908, the electric power station in Cadillac, Michigan, had an estimated 300 customers whose wages did not exceed $2 a day, but these homeowners were able to afford to pay $9 to $15 a year for electricity to light their homes. By 1909, there were more than 12,000 residential customers with Detroit's electric power company who had average monthly electric bills of $2.32, whereas Chicago's Commonwealth Edison Company served more than 30,000 small homes and apartments with electricity, whose occupants paid electric bills ranging from $1.15 to $2.50 a month. At 10 cents per kilowatt-hour near the end of the decade, household electricity had become as cheap as gas for lighting. If water power, or "white coal," was used in place of burning coal, then the per-kilowatt-hour cost could be even cheaper. And with the relentless marketing of electricity's benefits, as well as appliances, by the central power companies, it became increasingly difficult for urban homeowners to look away.[8]

The advent of large-scale electric power production and use in the United States and Europe at the start of the twentieth century also ushered in a new era of career opportunity and invention. Hundreds of thousands of young men in the United States attended newfound electrical schools where they prepared through classroom and hands-on instruction for occupations in construction, manufacturing, and maintenance.[9] "Electricians say rightly enough that they have just started in to make this the electrical century and also to make everything electrical," wrote the editors of the magazine *Popular Electricity* in June 1908. For rugged and adventurous individuals, or for those without an education, thousands of jobs were available in the erection of wooden utility poles for stringing telephone and electric power lines in cities and towns and across vast open areas. These linemen often spent many weeks at a time working and sleeping in mobile camps. Skilled linemen could earn as much as $11 to $12 a week. Many young African American men in the South and lower

Midwest found employment as linemen in the early 1900s. It's estimated that about four million wooden poles, the majority of which were used by telephone, telegraph, and railroad companies, were being erected annually in the United States by 1910.[10] For those with an interest in theoretical and design matters involving electricity, many universities and technical colleges began expanding their degree programs in electrical engineering. So many electrical-based patent applications were being filed to the US Patent Office by 1910 that the government agency set up four divisions of staff to analyze these inventions, "whereas no other industry in the whole range of human activities is represented by more than a single division in the Patent Office."[11]

None of these developments involving electricity was lost on the American farmer. They read about technological advances involving electricity in newspapers and magazines, as well as seeing electric lights for themselves during visits to town. Many farmers surely wondered how they could use electricity themselves in their own homes and day-to-day operations. A 1912 article in *Popular Electricity* magazine stated:

> The farmer of today has increased his standard of living. He demands many of the conveniences enjoyed by his brothers of the city. Most of the conveniences, which 20 years ago were looked upon as luxuries, are today considered

bare necessities. The proper and efficient lighting of the farm home and farm buildings is one of them. We have but to note the introduction of the gasoline engine and the telephone in farming communities. Fifteen years ago both of these modern conveniences were looked upon by farmers as luxuries. The modern electric light system marks another epoch of progress.[12]

While some farmers living close to towns were able to secure electricity from municipal power plants at reasonable cost through power lines run to their properties, the vast majority of rural American properties at the start of the 1900s were located far away from cities and towns, which made it too expensive for municipal power plants to provide them with cost-effective electricity.[13] Some farmers who had access to ample flowing water on their land, such as rivers and large streams, followed the path of numerous flour and feed mill operators at the time by using water wheels or turbines outfitted with flywheels and belt systems to drive electric dynamos for charging batteries to power light bulbs and small electric motors throughout their homes and farm buildings.[14] Dairy farmer Eli Crosiar of Utica, Illinois, used the former site of a mid-nineteenth century grist mill on his property as the foundation for erecting a hydroelectric plant in 1910. He was able to use a turbine wheel to turn a direct current dynamo rated at 12 kilowatts and

The rural community of Stoddard, Wisconsin, in 1910 constructed this water-powered electric light plant. From the author's collection.

220 volts which he used to electrify his entire farm. Crosiar estimated that the plant and equipment cost him $2500, a small fortune to many of his neighbors at the time, but it was worth it to him in terms of the efficiency in labor on his farm and comfort in his home.[15]

Most farmers with an interest in electricity, however, opted for a less expensive arrangement that included a stationary internal combustion engine to drive a dynamo, in some cases delivering electricity at the moment of need or for charging sets of batteries for later use.[16] A representative of the Southern California Edison Company of Los Angeles, speaking before the National Electric Light Association in 1913, called the internal combustion engine the "greatest competitor" to the municipal power companies attempting to reach new customers in the countryside.[17] Usually erected in small outbuildings due to the noise and exhaust fumes from the engines, it was estimated in the mid-teens that a lighting-only plant, including the wiring for an eight-room house and barn, cost between $350 and $750. If the plant was used to supply current to electric motors, in addition to lights, then the farmer might expect to spend $1200 or more for the system. The cost of operating a gas engine solely for lighting was about $15 a year, while the cost of operating these plants to supply electricity for powering equipment depended on the size of the motors and the amount of work performed.[18] Electric motors were used to run myriad farm equipment, such as cream separators, feed choppers, grinders, and irrigation pumps, while inside the home electricity powered small, motorized appliances. Electrical systems also raised the overall value of farm properties and country homes in the eyes of perspective buyers.[19]

The earliest farm electric light plants consisted of four basic parts: an internal combustion engine, dynamo, an electric control panel, and batteries. Farmers had the option of purchasing these components separately from different suppliers or as complete packages. The more frugal farmers sought out second-hand generators and gas-fueled engines, and they even learned how to install their own wiring. Kerosene and other internal combustion engine fuels became more readily available with

Charles Franklin Kettering, shown here at age fifty-two in 1928, set out fifteen years earlier to develop a quiet and efficient farm-electric light plant. He succeeded with the Delco-Light plant. From the author's collection.

the increase in automobile production at the start of the century. Harry T. Candee wrote to *Popular Electricity* in 1911 how he built his own inexpensive farm electric plant by purchasing a second-hand 2.5-kilowatt generator for $50, a new 3 horsepower gasoline engine for $90, a three-quarter horsepower motor with shunt for $35, and wires and lamps for $35, for a total outlay of $210. The system, which did not use battery storage, provided Candee with ample light for his home and pumped water from a well. "The engine is started each evening at dark and averages about four and one-half hours run per night, costing $2.50 a month, and using naphtha at 11 cents per gallon, which I find as satisfactory as gasoline and cheaper . . . The engine is stopped by pulling a wire from the house," Candee said.[20]

In 1913, Charles F. Kettering, a young electrical engineer who had successfully developed the first starting, lighting, and ignition system for automobiles, set out to provide electric lighting for the Loudonville, Ohio, farmhouse in which his parents lived. He used a single-cylinder gasoline engine belted to a dynamo to charge a 32-volt DC battery bank. The plant was not automatic, so when the engine quit running, the batteries were quickly drawn down and his parents returned to lighting

The first Delco-Light plants included a one-cylinder, air-cooled engine directly tied to an electric generator, which charged a set of sixteen, lead-plate storage batteries in glass jars for an electric output of 32 volts DC. (Postcard ca. 1918) From the author's collection.

with kerosene lamps. At the same time, Kettering's Dayton Engineering Laboratories Company (Delco) in Dayton, Ohio, received a telegraph from a cottage owner in Florida who requested a Delco starter system. It was discovered, however, that the customer wanted the starter for lighting his cottage, which he did by connecting his house wires to his running Cadillac car parked in the driveway. This gave Kettering and his engineers the idea for developing an electric light plant that would automatically start the engine whenever a light was turned on and shut off by itself when the batteries were fully charged. Over the next three years, the company spent a quarter of a million dollars in research and development on its "Delco-Light" plant, which consisted of a small, one-cylinder, air-cooled engine directly tied to an electric generator in a single compact unit. These first Delco-Light plants were designed to charge a set of sixteen lead-plate storage batteries in glass jars for a total output of 32 volts DC. Complete with batteries, these electric light plants retailed for $275 apiece in the United States and $375 each in Canada. The company, which initially operated under the Domestic Engineering Company before shortly changing over to Delco, sold about $2.5 million worth of Delco-Light plants in 1916, its first year. Delco's first advertisement for the electric light plant, "Electricity for Every Farm," appeared in the September 9, 1916, *Saturday Evening Post*,

stating: "It has a capacity of 40 to 50 lights, and is so simple that anyone can operate it. The turning of a switch starts it and it stops automatically when the batteries are full." Larger models, including 110-volt Delco-Light plants, soon followed.[21]

The internal combustion electric light plants of Delco quickly spawned an entire industry competing for the attention of farm and rural homeowners. A number of these companies developed the electric light plants as an extension of their other businesses, such as Westinghouse Electric and Manufacturing Company of East Pittsburgh, Pennsylvania; Allis-Chalmers Manufacturing Company of Milwaukee, Wisconsin; Fairbanks, Morse & Co. of Chicago; Phelps Light & Power Co. of Rock Island, Illinois; and Western Electric Company of New York.[22] Most of these companies bought existing electric light plants, tested them in their laboratories, and then developed plants of their own.[23] The Kohler Company of Kohler, Wisconsin, for example, entered the electric light plant business primarily as a way to expand its sales of plumbing equipment. "Perhaps the greatest contribution the Kohler Automatic is capable of making to the health and happiness is its ability to operate a *water system*. . . . Modern plumbing throughout the house, running water always available, in bath room, kitchen, laundry, anywhere—the Kohler Automatic makes all this a reality," the company said.[24] Many dealers for the various electric light plant

Using Poul la Cour's foundational designs, the Danish Wind Electricity Association installed 132 wind-electric plants throughout Denmark from 1903 to 1915. Courtesy Poul la Cour Foundation, Vejen, Denmark.

manufacturers set up stores in small towns, which not only sold the plants but also the numerous 32- and 110-volt DC appliances. They also exhibited at state and county fairs where rural dwellers gathered, and they made door-to-door sales.[25] The biggest selling point of a gas-powered electric light plant, as one journalist pointed out in 1920, is that "It is so simple to operate and maintain that most anyone without any experience in electrical matters can take care of it."[26] In the early 1920s, salesmen often encountered farmers who were hesitant to purchase an electric light plant due to the upfront cost. They had to work hard to convince these farmers that they needed the electric-light plant for their farms.[27] Some salesmen cleverly bypassed the farmer by talking to his wife, "as the farm woman is the one who is most directly interested in what the farm light and power plant will do for the farm home."[28]

The effort to electrify rural farms and communities was not lost on the windmill manufacturers at the start of the twentieth century, but the ability to partake in this market initially remained small and mostly centered in Europe. In Britain, a commercial enterprise that found some success at selling electric-generating windmills to rural cottage owners during the first decade was J. G. Childs & Co. of Willesden-green in northwest London. The company's steel windmill consisted of a 24-foot wind wheel with forty blades, with the generator located at the base of the tower offering 3 to 4 kilowatts of power at a voltage of about 70. J. G. Childs marketed its windmills through demonstrations at fairs. At a 1909 show in Gloucester, the company had an operational display called "Cooking by Wind," which used its 3-kilowatt windmill to supply electricity for cooking and lighting in a cottage erected by the Country Gentlemen's Association about 220 yards away. J. G. Childs also shipped some of its windmills to overseas markets, such as South Africa and Argentina.[29] In Germany, the Royal Technical Institute at Dresden successfully demonstrated the ability to generate electricity from the wind in 1913 by erecting a large windmill with a multibladed wind wheel on its campus. Through a vertical shaft connected to a bevel gear that turned a ground-mounted dynamo, the school offered that its test machine could supply 100 people with electric water pumping and lighting for an operating cost of $125 a year.[30] It was estimated that the Danish Wind Electricity Association oversaw the installation of 132 wind-electric plants for villages, farms, factories, and workshops, most of which produced 110 volts DC, in locations scattered throughout the country between 1903 and 1915.[31]

The prospects for wind power in the United States remained promising in the early 1900s. For the first two decades of the twentieth century, however, building those windmills to generate electric power remained in the hands of innovative tinkerers.[32] One of those individuals, R. W. Wilson of Noblesville, Indiana, in 1907 used a windmill to pump water into a cylinder fitted with a weighted plunger in the basement of his house. As the cylinder filled with water, the rising plunger eventually triggered a relief valve that allowed the water under the cylinder's downward pressure to drive a water motor that turned a dynamo. Wilson could use the electricity immediately or store it in a set of batteries for later consumption.[33] J. F. Forrest of Poynette, Wisconsin, attracted the interest of electrical engineer Putnam A. Bates, who in 1912 wrote an article about this wind-electric plant in the *Scientific American* and also presented a paper at the annual convention of the American Institute of Electrical Engineers in Boston in June 1912. Forrest converted a geared windmill with a 12-foot wind wheel that was used to operate light machinery on his farm to turn a series of gears and pulleys to rotate the armature of a dynamo on the second floor of his barn, which also housed a fourteen-cell storage battery unit. Forrest used the windmill to power fourteen tungsten lamps throughout his farm, as well as a drill press, corn sheller, feed grinder, saw and grindstone, washing machine, and grain elevator. He estimated that the unit cost him only $250 to build. Other farm and electrical trade journals picked up Forrest's story during 1912 and 1913.[34]

A setback for wind remained its intermittency—days on end it blew consistently, and then there were periods of calm, especially in the middle of summer. In 1911, P. C. Day, chief climatologist of the US Weather Bureau, recommended supplementing the electricity-producing windmill with a combustion engine-driven electric generator that automatically started when the wind became calm and then shut off when the wind was once again sufficient to turn the windmill and its generator.[35] That same year. inventor Caryl D. Haskins of Schenectady, New York, developed and patented a system in which when a wind-electric generator stopped for lack of wind, an ignition switch started a gas engine to continue turning the dynamo. When the windmill began turning at a sufficient speed, the ignition switch closed and shut down the gas engine automatically.[36]

American windmill manufacturers were aware of the success that some farmers and inventors had with converting their windmills into electric-generating plants, but for myriad reasons they weren't convinced enough to enter commercial production of these machines during the early 1900s. In 1922, the publication, *Farm Light and Power* surmised that "home-made" electric-generating windmills were highly inefficient due to "the great amount of friction caused by poor bearings, aggravated by inaccurately cut gears plus lack of alignment in the bearings themselves and the fact that the power had to go through so many steps before reaching the generator." The author concluded that "[i]t was highly discouraging to provide so much plant—that is, a windmill, a transmission, dynamo, and storage battery, involving a great outlay at that time, and get so little power in return for it."[37] The publication recommended the production of a new wind wheel

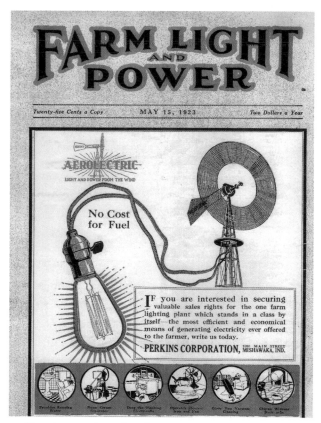

Perkins Corporation of Mishawaka, Indiana, became one of the first commercial manufacturers of electricity-generating windmills. It debuted the Aerolectric in 1921, as advertised in *Farm Light and Power* (May 15, 1923). From the author's collection.

specifically designed for the efficiencies required by electric-generating windmills.[38] Several windmill manufacturers understood this and began designing purpose-built wind-electric generators for the commercial market starting in 1920.

One of the foremost windmill manufacturers to offer a wind-electric generator for sale was the sixty-year-old Perkins Corporation of Mishawaka, Indiana. The company began developing its machine, called the Aerolectric, in 1919 and commercially debuted it in 1921. The Aerolectric incorporated a variable-speed, single-kilowatt dynamo sourced from Westinghouse. Sitting on top of a minimum 50-foot tower, the dynamo's armature was connected to the 14-foot steel wind wheel through stepped-up gearing that allowed it to turn forty revolutions per one turn of the wind wheel. The dynamo's current was carried through a collector brush and into wires that ran down the tower, through an automatic switchboard and into a battery storage unit. The wind wheel was turned into the wind by a traditional sheet metal vane tail.[39] The Aerolectric, which came in both 32-volt and 110-volt models, sold for a minimum of $695 in 1922.[40] The company promoted the Aerolectric as safer to operate than engine-based farm electric-light plants, citing that on average there were 93,400 fires on farms each year, of which 7.5 percent were caused by gasoline or kerosene.[41] Although the number of sales of these early Perkins electric-generating machines is unknown, the company not only sold them in the United States but also exported them worldwide by 1923. In Bombay (now Mumbai), India, for example, Perkins's sales representative, Cama Norton & Company, used clever billboard-style advertisements stating, "Why Pay for Light When Nature Provides It Free?"[42] Perkins's greatest marketing coup, however, was convincing Henry Ford in April 1923 to order one of its Aerolectrics to light the home and barn of his farm near Dearborn, Michigan. *Farm Light and Power* commented at the time that "Henry Ford doesn't buy anything extravagantly, and everything must be rigidly accounted for. This in itself is undoubtedly one of the highest tributes that could be paid to the Perkins Aerolectric, because if it could not stand the searching investigation of

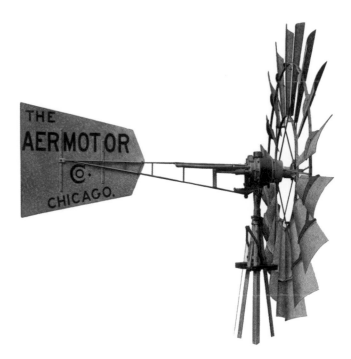

The Electric Aermotor was a steel mechanical windmill modified for electric power generation for farm and household use. From Aermotor Company, *Electric Aermotor*, Chicago (ca. 1920).

his engineers, the purchase order . . . would never [have] been sent out."[43]

One of the most recognized water-pumping windmill manufacturers, Chicago-based Aermotor Company, began selling a wind-electric generator in 1922. The company was no stranger to the concept, having used a windmill on the roof of its building at Park Place in New York City to charge batteries for office lighting in 1895, and its founder, La Verne W. Noyes, in 1911 filing a patent application for a battery-charging windmill consisting of six wind wheels that collectively turned into the wind on a tower by a large tail vane. Neither machine was built for commercial use, since Noyes was not convinced that such a device would have marketable success. That changed shortly after Noyes's death in 1919, when Aermotor's senior management directed Daniel R. Scholes, chief engineer, to design an electricity-generating windmill.[44] Like the Perkins Aerolectric, the Electric Aermotor consisted of a dynamo at the top of the tower. The dynamo's armature rotated with a set of step-up gears turned by the wind wheel's vertical shaft. The wind wheel was turned into the wind by a traditional Aermotor water-pumping windmill tail vane. The Electric Aermotor was sold only with a 12-foot wind wheel and could be ordered in two voltages of electric output—32 volts and 110 volts. The company said its 32-volt machine provided enough power for lighting a small country home, while the 110-volt unit could, in addition to lighting, power a range of farm machinery and household appliances with small electric motors.[45]

Brothers George and Wallace Manikowske of Wahpeton, North Dakota, began experimenting with windmill-generated electric power on their 1200-acre wheat farm in about 1910. At first, they built a plant using a traditional American power windmill using a system of gears, pulleys, and belts to turn a ground-based dynamo. They even exhibited a patent model of their wind-electric plant at the North Dakota Agricultural College electrical show in 1913.[46] Before the end of the decade, the Manikowskes had designed and manufactured an electric-generating windmill, called the Aerolite. The 14-foot multibladed wind wheel was designed in the fashion of a bicycle wheel held together with a channeled outer steel rim. The eighteen individual sheet metal blades that made up the wheel were designed to feather in varying winds. The 1500-watt dynamo, which was mounted just below the wind wheel on top of the tower, was driven by a specially designed Goodyear Rubber Company belt passing around the channel rim of the wind wheel, over an idler and then to the dynamo's armature. The entire unit rotated around a reinforced vertical steel mast pipe on Timken roller bearings and was kept in the wind by a small sheet metal vane placed behind the wind wheel, just below the dynamo. Sold in both 32-volt and 110-volt outputs, the Aerolite was manufactured by the Wind Electric Company of Minneapolis by the start of the 1920s. The wind machine mechanically could pump water while at the same time generating electricity. By the end of 1921, the company claimed already to have as many as 100 in service, of which forty were operating in North Dakota.[47]

Oliver P. Fritchle of Denver, who in 1904 had manufactured an electric-powered automobile, took an interest in using wind power to charge batteries for home and farm lighting in 1917. However,

Fritchle chose a different route to this objective than his contemporaries in that his wind-electric dynamo could be attached to the wind wheels of existing water-pumping mills. The 32-volt, three-quarter kilowatt Fritchle Wind-Electric System was mounted on a steel plate, which could then be bolted to the shaft box of the windmill. A large sprocket was bolted to the hub of the wind wheel, which was lined up with the countershaft sprocket attached to the steel plate. An inner sprocket on the same countershaft then drove another chain to a combination smaller sprocket and gear train that turned the armature on the dynamo. One turn of a 12-foot wind wheel, for example, resulted in forty revolutions of the dynamo. Electricity from the dynamo was carried by wires through the mast pipe of the windmill to a set of collector rings and then to the automatic switchboard, which

The Woodmanse Mfg. Co. of Freeport, Illinois, in the early 1920s sold a wind-electric plant based on the design of Oliver P. Fritchle, a pioneer in battery-powered automobiles at the start of the twentieth century. From Woodmanse Mfg. Co., *The Fritchle Wind-Electric System* (ca. 1922). Courtesy *Windmillers' Gazette*, Rio Vista, TX.

managed the electricity flow to the battery storage unit.[48] Fritchle erected his first plant on a 10-foot Dempster wooden water-pumping windmill on the Swanson farm at Derby, Colorado, in the spring of 1918, and another at his Denver factory on a 12-foot Dempster wooden wheel mill for conducting experiments.[49] In 1922, Fritchle struck a deal with Freeport, Illinois, windmill maker, Woodmanse Manufacturing Company, to market and sell its Fritchle Wind-Electric System with the 12 1/3-foot steel Woodmanse Mogul windmill, as well as the larger 14- and 16-foot models. A 12 1/3-foot Mogul with a 40-foot tower, Fritchle Wind-Electric System, and batteries in 1922 cost about $650.[50] However, by the early 1920s, engineers were already considering alternatives to multibladed windmills. The problem with the closely set blades of these traditional windmills was the fact that too much of the useful power from the wind became absorbed by the blades, a phenomenon known as "wind congestion," and reduced the rotational speed of the wind wheel.[51]

Extensive studies and tests were also made with various types of vertical-axis wind wheels. These horizontal-spinning wind wheels had been used for centuries in the Middle East for turning millstones and in China for lifting water. Examples of horizontal windmills could be found throughout the American colonies after the Revolutionary War. A handful of American windmill manufacturers in the 1880s, such as D. H. Bausman of Bausman, Pennsylvania; George L. Squier Manufacturing Company of Buffalo, New York; Cook Brothers of Southboro, Massachusetts; and Hercules Wind Engine Company of New York City, offered horizontal pumping and power mills for sale. Simpler homemade variants could be found on farms in the US Midwest. The wind wheels of these machines came in two basic designs. One included a partial hood or series of shutters around the wind wheel that opened to allow wind to strike half of the blades at a time for a more uniform rate of turn, while another design used a system of exposed vertical blades that pivoted on a rotating frame, allowing half the blades at a time to catch the wind while the other half swung edgewise away from it.[52] Examples of horizontal windmills being used to generate electricity

sprung up in the early 1900s, with their designers being heavily influenced by the work of those pioneering windmill manufacturers. Engineer W. E. C. Eustis in 1912 erected a horizontal windmill on his estate south of Boston. On top of a 100-foot steel lattice tower, the wind entered a 50-foot-tall outer stationary cylinder through curved pine board slats and struck an inner rotating cylinder. The windmill's shaft was configured to turn a dynamo at ground level and charge a set of batteries sufficient to light about 200 incandescent lamps, power a sawmill, and pump water into a holding tank inside the house.[53] Chicago millionaire Dorr Eugene Felt, inventor of the "comptometer" calculating machines, in the early 1920s erected a horizontal windmill with three wind wheels stacked one above

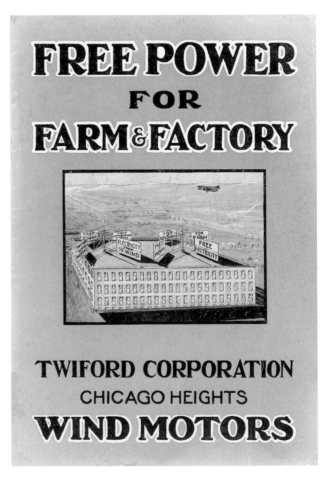

The Twiford Corporation of Chicago Heights, Illinois, in 1923 offered for sale a unique horizontal wheel windmill that could generate both electricity and serve as a rotating advertising billboard. From Twiford Corporation, *Free Power for Farm & Factory* (ca. 1924).

the other on his lakeside estate near Saugatuck, Michigan, to generate electricity.[54] *Popular Science* in 1923 described how an East Cleveland, Ohio, resident installed a 15-foot-tall horizontal windmill with "eight perpendicular 'sails' attached at top and bottom to circular, revolving frames" on top of the roof of his home to turn a 1-kilowatt dynamo.[55]

There were examples in the United States of experimenters who attempted commercial manufacture and sale of horizontal wheel windmills for electric power generation during the early 1920s. Manning Engineering and Sales Company of Elkhart, Indiana, in 1923 began offering a vertical-axis rotor-style wind machine called the "Cyclone Turbine" to generate electricity and pump water. The company's success with selling these

windmills is unknown.[56] Chicago Heights, Illinois-based Twiford Corporation offered two types of horizontal wheel windmills, including one with four large vertical rectangular sails that it dubbed the "Ad Motor" since companies could use them as advertising billboards, and another called the "Powerbowl," which had four half-cylinder sails each of which was composed of six perpendicular blades that opened and closed according to the wind and their position in the wheel's rotation. The company's founder, William R. Twiford, a young Nebraska farm boy, in 1922–1923 erected a test Ad Motor with a 75-foot-diameter wheel in Miami, Florida. Many people came to view the machine, which generated both electricity and pumped water. When Twiford launched his company in

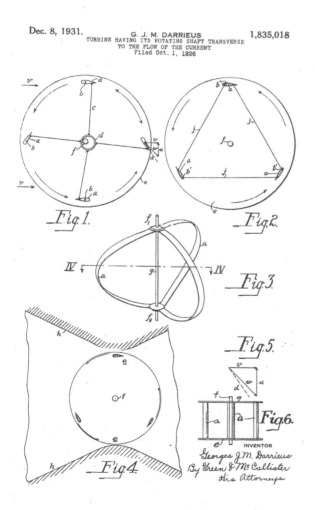

US patent drawing of Georges J. M. Darrieus's vertical-axis rotor design, 1931. Courtesy US Patent and Trademark Office, Washington, DC.

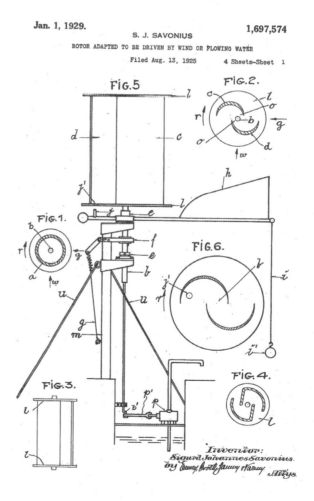

US patent drawing of Sigurd Johannes Savonius's "S"-shaped vertical-axis rotor design, 1929. Courtesy US Patent and Trademark Office, Washington, DC.

1923, he claimed to be capitalized with $1 million and had the backing of about a half-dozen wealthy businessmen. The machines' wind wheels ranged in size from 26 and 30 feet in diameter to 70 and 75 feet in diameter. Thirty-two volts of power were generated by ground-mounted dynamos ranging in output from 3/4 to 1.5 kilowatts. Because of their size and cost, it is uncertain if Twiford sold many of his windmills, since a complete unit could cost well over several thousand dollars and required a large amount of space to operate.[57]

The experimental work of German physicist Heinrich Gustav Magnus, who in 1852 studied the effect of a spinning cylinder in an air stream that became known as the Magnus effect, inspired Finnish engineer Sigurd J. Savonius between 1924 and 1925 to conduct numerous tests of a cylindrical "wing-rotor" design. The vertical-axis wind wheel developed by Savonius consisted of two offset sheet-metal cylindrical halves held in place at the ends by circular pieces of wood or sheet metal. When the wind struck the inside of the inner surface of the one-half cylinder, it streamed through the center opening between the two cylinders, striking the other, causing a rotational force. The wing-rotor remained upright and rotated on a central axis. Savonius tested his wing rotor against a number of different wind-wheel designs, including a tradi-tional eighteen-bladed steel water-pumping wind wheel. According to his tests, Savonius declared that his wing-rotor generated 30 percent more power than the multibladed wind wheel.[58] "Several months of experimental work and tests conducted under actual working conditions have shown that the Wing-rotor offers a new and simple means to utilise the windpower efficiently in various ways," Savonius wrote.[59] He quickly patented his wing-rotor design in both Europe and the United States.[60]

Another intriguing horizontal wind-wheel design from the 1920s was developed by French aeronauti-cal engineer Georges Jean Marie Darrieus, which consisted of three to four airfoil blades "curved in the form of a skipping rope" or "eggbeater" shape and attached to a vertical axis. The structure of the wind wheel was quite simple. It required no tail vane to turn it into the direction of the oncoming

wind, since it could rotate from any direction from which the wind blew. While the Darrieus wheel could turn at considerable speed, making it ideal for turning a dynamo, it did not easily start rotating on its own. Darrieus applied for a US patent for his wind-wheel design in 1926. However, the concept was not seriously pursued in commercial wind-electric power generation for another fifty years.[61]

The heavy multibladed steel wind wheels of the early American wind-electric plant manufacturers underwent a major simplification by the mid-1920s. Each of these companies, with the exception of Aermotor and Woodmanse, turned to aircraft-style propellers to drive their machines. Propellers had already proved themselves essential to fostering mechanical flight. However, while an aviation propeller was turned by an engine at high speeds to pull an aircraft through the air at velocities of up to 600 miles per hour, a propeller for a wind-electric generator was rotated by the power of the wind with velocities of 10 to 25 miles per hour.[62] Elisha N. Fales, who worked alongside aeronauti-cal engineer Frank W. Caldwell after World War I on propeller designs at Mount Cook Field (later Wright-Patterson Air Force Base) in Ohio, became intrigued by the idea of transferring the aircraft propeller's design to driving wind-electric genera-tors in the early 1920s. On July 14, 1924, Fales submitted a patent application to the US Patent Office for a propeller-driven, electric-generating "turbine."[63] Fales's concept was a single blade, most likely shaped from a lightweight wood, and includ-ed a counterweight for balance. The squared center of the blade bolted to a base plate which held fast to a generator shaft. He conducted numerous tests on the blade's shape, writing in his patent application:

> The blade of the propeller is designed with a very small blade pitch whereby a relatively small wind velocity will cause a rapid or high-speed revolution of the propeller and the shaft . . . The best results thus far obtained have resulted from a propeller having a pitch of 3 degrees at the tip increasing to 10 degrees at the hub, while good results are obtainable from propellers having a pitch at the blade tip which vary from 0 to 10 degrees, the corresponding pitch at the hub varying from 10 to 30 degrees. It will

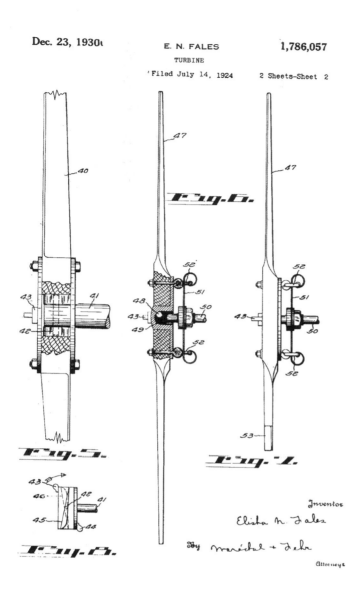

Dec. 23, 1930t
E. N. FALES
1,786,057
TURBINE
Filed July 14, 1924
2 Sheets—Sheet 2

Fig. 6.

Fig. 5.

Fig. 7.

Fig. 8.

Inventor
Elisha N. Fales
By Marechal + Dehr
Attorneys

A patent drawing showing the wind-electric generator blade developed by Elisha N. Fales during the early 1920s. Courtesy US Patent and Trademark Office, Washington, DC.

be understood that these dimensions may be varied as desired to meet different conditions. It has been found . . . that the blade should extend from the tip to the hub of the propeller, and further it has been found that the best results are obtainable in a blade in which the radius or length of the blade bears the ratio to the blade width of approximately six to one, although again it is obvious that this ratio is susceptible of some variation in the hands of a person skilled in the art.[64]

Fales surmised that his blade design could drive a generator requiring a shaft speed of 500 to 1000 rpm in a wind of about 15 mph.[65] He received his patent on December 23, 1930, six years after he initially applied, but by then a handful of wind-electric companies had already been manufacturing wind-electric generators with varying pitch or twisted propeller designs for four to five years.

It is unclear which of the wind-electric generator manufacturers with the multibladed wind wheels was the first actually to market a propeller-driven unit. However, if by a matter of months, some historians point to the Perkins Corporation, by then located at South Bend, Indiana, with its introduction of the "New Perkins Aerolectric" in mid-1924 as being the first.[66] The company said its shorter 18-pound "aeroplane type" wooden propeller was capable of generating more electric current than its former 14-foot-diameter, 400-pound multibladed steel wind wheel-based machine. Perkins also reduced the material weight of its 32-volt wind-electric plants from about 1500 pounds to less than 300 pounds, which made them cheaper to

Perkins Corporation of Mishawaka, Indiana, was one of the first wind-electric generator manufacturers to discard the multibladed wind wheel for a propeller-driven unit with its "New Perkins Aerolectric." From *Audels Plumbers and Steam Fitters Guide #3,* New York (1925).

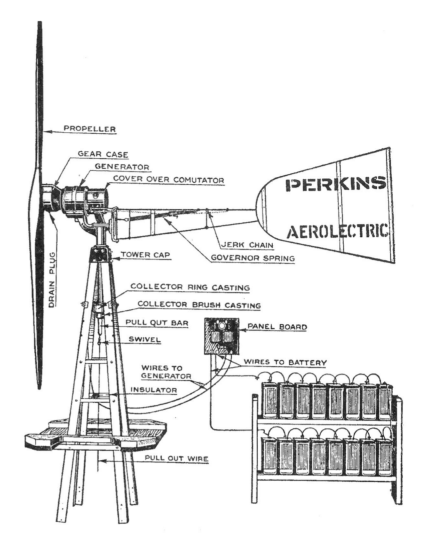

manufacture, ship, and erect in the field. By the spring of 1925, the company claimed it shipped the New Perkins Aerolectric machines not only throughout the United States and Canada but also as far away as England and Australia.[67] George Manikowske, who manufactured the multibladed Aerolite in Minneapolis, also claimed to shift to a wood propeller blade design by 1924, at which time he changed the name of his company and wind-electric machines to Aerodyne. While offering a propeller-driven combination electric and oil-bath water-pumping machine, it was Aerodyne's unique double-propeller unit that made it stand out. The front propeller of this machine was attached to the generator's forward armature and turned clockwise in the wind, while the back propeller, behind the armature and carrying the "field," turned counterclockwise. The unit, which was sized to produce either 32 volts or 110 volts, was kept in the wind by

a combination of wood-slatted main tail vane and a smaller side governing vane. The side vane, which would experience an increased drag force upon it as the wind speed increased, was used to fold the tail in high winds and turn the wheel out of the wind to reduce rotational speed and structural loading in high winds. While more complex than the Perkins Aerolectric, the double-propeller Aerodyne's performance was praised by E. A. Stewart, associate professor of the University of Minnesota's Division of Agricultural Engineering, after thorough testing, and by Simon Kruse, proprietor of the Hotel Radisson in Minneapolis, who had one of these machines installed at his Radisson Farm in Blaine, Minnesota, during the spring of 1926.[68]

Another prominent player to emerge with a propeller-drive wind-electric machine during the mid-1920s was the Herbert E. Bucklen Corporation (HEBCO) of Elkhart, Indiana. The company's

The Herbert E. Bucklen Corporation of Elkhart, Indiana, sold a propeller-driven wind-electric generator in 32-volt and 110-volt sizes during the late 1920s. From *General Specifications of HEBCO Wind-Electrics* (1928).

Brothers John and Gerhard Albers of Cherokee, Iowa, established Wincharger Corporation in the mid-1920s using a two-blade wind-electric generator design. From *There's Power in the Air with Your Wincharger 32-Volt*, Sioux City, IA (1938).

Wincharger Corporation sold inexpensive, battery-charging wind-electric generators ranging in size from 6-volt to 32-volt outputs. (Postcard ca. 1935). From the author's collection.

namesake, Herbert E. Bucklen, was the son of a turn-of-the-century millionaire manufacturer of medicines. The young Bucklen, however, developed an interest in wind-electric power and spent about $150,000 to research and develop his own propeller-driven unit, which entered the market in early 1926.[69] Bucklen, along with his chief engineer, Harlie O. Putt, developed what they called an "impeller" in which the air stream flowing past the blades "creates pressure on the wind side and a suction on the lee side whereby torque is produced and the impeller rotated to deliver power." The impeller's design, according to Bucklen and Putt, allowed it to start turning in winds of about 5 mph. The single blades, measuring from 11 feet, 6 inches to 14 feet in length, were made with layers of wood, which "permits proper contour, pitch, and camber to be secured without excessive machine work or carving, and it provides an impeller of light weight, the outer layers of wood being preferably of tough durable structure and requiring little hand finishing."[70] HEBCO offered its machine in both 32-volt and 110-volt sizes. The units were turned into the wind, or could be shut off, by a traditional hinged windmill tail vane assembly.[71] By 1928, HEBCO's machines sold for $680 for the smallest 32-volt Model 40G to $1060 for the 110-volt Model X.[72]

In the mid-1920s, two brothers from Cherokee, Iowa—John and Gerhard Albers—as an often-told story goes, were tired of having to take the family's 6-volt radio battery to the town auto garage once a month to have it charged at the cost of a dollar and then wait a couple of days to bring it back home for use. It was also said that Gerhard Albers was fascinated by the Norwegian polar explorer Fridtjof Nansen's book, *Farthest North*, in which he recounted the use of a deck-side wind-generator to charge batteries for lighting on the vessel, *Fram*. The brothers thus figured why not build their own wind-electric generator? Like their predecessors, they first took an old multibladed wind wheel, secured a neighbor's old windmill tower, obtained a junkyard automobile 6-volt generator, and removed gears from a discarded cream separator. The windmill worked, and no longer did they need to take the radio battery to town for charging. However, the Albers weren't satisfied with the performance of the wind wheel. Reading magazine and newspaper articles about early aircraft, they set out to make an aerodynamic wooden propeller to turn their generator. The prototype two-bladed propeller, about 6 feet in overall length, was attached to the front end of the family's automobile that was driven at a constant 10 mph to simulate the wind striking the face of the blade, causing it to turn by the pull of a partial vacuum against the backside of the blade. The brothers calculated that the blade rotated thirteen times as fast as the wind turning it. The Albers reasoned they could sell their 6-foot blades to area farmers, along with printed instructions on how to connect them to an old 6-volt automobile generator, fashion a tail vane, and erect them on a pole, roof, or windmill tower. With their newly found Albers Propeller Company, the brothers placed small advertisements in Iowa farm newspapers, which offered their blades and instructions for $3.50. They received more orders than their manual processes could keep up with. The Albers then built a machine to cut and carve their blades automatically. In 1927, the Albers moved their operation to Sioux City, Iowa, changed the name to Wincharger

Mrs. A. J. Jacobs of Dawson County, Montana, enjoys a radio program while ironing clothes. Her sons Marcellus and Joseph Jacobs in 1920 wired the family home and installed an electric-light plant, which spurred the brothers' interest in wind-electric machines. From "The Electrically Equipped Farm Home," *The Dakota Farmer* (Aberdeen, SD) 43, no. 5 (March 1, 1923).

Corporation, and built complete wind-electric generator systems for 6-volt and 32-volt outputs. Wincharger rapidly became one of the most recognized providers of inexpensive wind-electric machines, with a 6-volt unit costing less than $50 and 32-volt units selling for less than $100 apiece.[73]

While the Albers brothers may have developed an inexpensive, mass-produced wind-electric machine by the late 1920s, two other brothers, these from Vida, Montana, Marcellus L. and Joseph H. Jacobs, at the same time were building what would undoubtedly become known as one of the highest performing and most durable small wind-electric plants ever perfected. Like most farmers of the day, the Jacobs brothers learned at an early age how to be resourceful. They were also fascinated by the prospects of electricity for improving the living and working conditions of rural people. While formal education for most children in rural Montana ended with the eighth grade, the Jacobs brothers continued to educate themselves in the immediate years following school, with Joseph taking correspondence classes and Marcellus spending two winters in the early 1920s enrolled in high school and college-level classes. In the winter of 1924, Marcellus Jacobs participated in a five-month training program at the Sweeney Auto School in Kansas City, Missouri, to repair automotive electrical systems. He returned to Montana and worked as a shop foreman at a Chevrolet garage in Miles City from 1924 to 1929.[74] Meanwhile, the brothers were already known throughout the community for their electrical achievements, having successfully electrified their parents' farm in the spring of 1920 using

The first-generation Jacobs wind-electric plant prototype as shown at the Jacobs family ranch in Vida, Montana, in 1925. Courtesy Paul R. Jacobs, Corcoran, MN.

a 32-volt Delco-Light plant and battery set, and adding modern conveniences to the home, such as lighting, indoor plumbing, an electric sewing machine, clothes washer, and radio. The *Dakota Farmer*, in March 1923, published a letter written by the boys' parents, A. J. and Ida Jacobs, about their successful electrification of the family farm, which included a cover photograph of Ida Jacobs listening to a radio headset while ironing clothes.[75] The Jacobs brothers, however, were not satisfied with the gasoline generator for charging batteries. In addition to the noise and exhaust odor, they were most annoyed by the cost and logistics associated with picking up cans of fuel from the nearest town, which was about 40 miles away. In 1922, they turned their attention to developing a wind-electric generator.[76]

The Jacobs brothers' first wind-electric generator used a multibladed steel wind wheel, which they soon discarded for a propeller design. Marcellus Jacobs had prior knowledge of propellers since he had already learned how to fly an airplane.[77] The brothers first tested a two-bladed propeller in 1925 but noted the design suffered from excessive vibration. Marcellus Jacobs explained that the vibration with the two-bladed propeller occurred "every time the tail vane shifted, to follow the changes in wind direction" and "were found to be caused by the fact that the two-blade propeller, when in a vertical position, offers no centrifugal force resistance to the horizontal movement of the tail vane in following changes in wind direction. However, when the two-blade propeller is in the horizontal position, its maximum centrifugal force is applied to resist horizontal movement of the tail vane; thus the tail vane is forced to follow wind direction changes by a series of jerks, causing considerable serious vibration to the plant."[78] To correct this problem, the Jacobs brothers in 1927 created a three-blade propeller system with a 14-foot diameter made of lightweight Sitka spruce from Washington state. "When in operation the three-blade propeller creates a steady centrifugal force resistance, against which the tail vane reacts with constant pressure and produces a smooth shifting horizontal of the plant facing direction," Marcellus Jacobs said.[79] The brothers also developed a unique, compact fly-ball governing hub to which the blades were attached, allowing them to automatically and uniformly adjust to variable wind pressures and protecting the blades from overspinning the generator or splintering during storms.[80] For the first twenty wind plants, the Jacobs used 32-volt, 1000-watt generators and Ford truck rear-end gearboxes, with ring gears, and mounted the assemblies vertically in the top of the towers.[81] Their machines quickly gained a local customer base, and in 1928 the brothers decided to form the Jacobs Wind Electric Company by selling stocks to neighboring Montana ranchers.[82]

By 1931, the Jacobs brothers realized that to meet increased demand for their wind plants and to have wider and more efficient access to production materials and freight transportation channels, they needed to move their operation to Minneapolis.[83] Changes to their wind plants continued. For instance, they went to work on refining their gen-

erators, first increasing their output to 2500 watts and placing them in a horizontal position at the top of the tower and then connecting the propeller hub directly to the armature shaft for a direct-drive unit. They also remedied the problem of electric static buildup caused by the revolving armature and from nearby lightning strikes that over time harmed the integrity of their steel towers. "To correct this we installed dual sets of heavy grounding brushes on the armature shaft which completely eliminated any trouble from this cause. With the additional use of a large capacity oil filled condenser connected across the generator brushes and frame we practically eliminated any damage to the generators from lightning," Marcellus Jacobs explained.[84] They also applied aluminum paint and a copper leading edge to each of their propeller blades, which reduced frost and ice buildup on the wood during winter.[85] In short order, the brothers secured more than two dozen patents related to their wind-electric

Marcellus Jacobs shown with a second-generation Jacobs wind-electric plant employing a unique flyball governor and 13-foot-diameter rotor. Courtesy Paul R. Jacobs, Corcoran, MN.

plants and even offered a larger 3000-watt, 110-volt generator. The most noteworthy early wind plant sale made by the company was to Admiral Richard Byrd for his second Little America expedition to the South Pole in 1933–1935. According to the expedition's records, the wind plant operated from the top of a 70-foot-tall tower flawlessly in some of the harshest wintry conditions on earth, supplying the facility a steady supply of electricity for lighting, power, and radio broadcasting.[86] In 1938, Jacobs offered a three-bladed wind-electric plant

with two parallel 750-watt, 32-volt generators for a collective output of 1500 watts. "Either generator will continue to supply the electricity needed if the other [generator] is removed for inspection or adjustment. This exclusive feature is not found in any other plant. It insures uninterrupted service," the company touted in its trade literature.[87]

The Jacobs wind plants, however, were not cheap. The factory price for a 2500-watt, 32-volt plant alone was $490, while a complete package, which included Jacobs-branded storage batteries and a

Brothers Marcellus (left) and Joseph Jacobs are credited with manufacturing one of the most durable and efficient wind-electric plants during the first half of the twentieth century. From the author's collection.

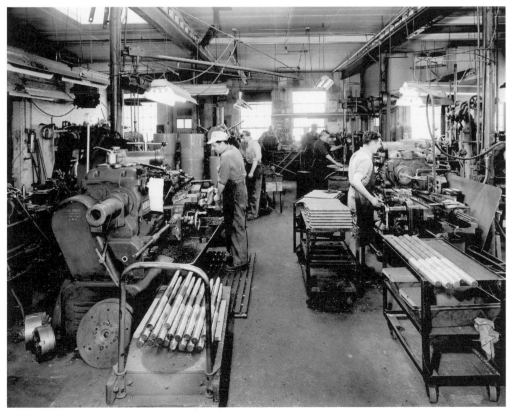

Machinists at the Jacobs Wind Electric Company factory in 1950 shown making generator shafts. Courtesy Paul R. Jacobs, Corcoran, MN.

50-foot steel tower sourced from the Challenge Company in Batavia, Illinois, cost $1025. Shipping and labor costs were additional. The Jacobs brothers were so confident in the quality of their wind plants that they offered an unconditional five-year warranty with every generator.[88] The Jacobs Wind Electric Company developed a network of more than 300 dealers, mostly in the United States and Canada, but thousands of its wind-electric plants were also sold by dealers throughout the world. Marcellus Jacobs often drove a pickup truck with a demonstration unit attached to the bed to display at county and state fairs throughout the country, as well as piloted his own plane to remote locations to cultivate dealer relationships.[89]

The wind-electric generator manufacturers that emerged during the mid-1920s not only sold their machines to farmers and ranchers, but many of their 32-volt units were purchased by the federal government for various remote electrical applications. In the mid-1920s, the US airmail service had begun. The pilots carrying the mail flew in and out of landing strips in rural locations scattered throughout the country which lacked access to electricity supplied by the large power providers in the cities and large towns. To ensure sufficient electricity for beacons used for nighttime flying, the US Department of Commerce, which operated the landing strips on behalf of the Post Office Department, purchased and installed multiple wind plants. Perkins, one of the early suppliers of wind-electric machines to the airmail service, explained how "a sunmeter, which automatically governs the turning on and off the lights" of the landing strips, essentially made its Perkins Aerolectric automatic for this operation. The company also said where its wind plants

A 32-volt HEBCO wind plant was purchased by the US Lighthouse Service in early 1930 to provide electricity to the lighthouse at Point Lookout, Maryland. From the author's collection.

only needed inspecting about once a month, "[e]ngine plants used for the purposes of generating light entail considerable expense as they require an attendant, besides the cost of gasoline and oil, which are difficult to deliver in these remote localities."[90] HEBCO wind plants were also prominent on remote landing strips, where they illuminated beacon lights placed in the upper portions of the towers. The company referred to these applications as "Lighthouses of the Air."[91] The US Lighthouse Service also found wind-electric plants ideal for charging battery banks that, in turn, ensured their remote lighthouse lamps burned brightly at night for safe coastal navigation. The first lighthouse supported by a wind-electric plant was erected by the Lighthouse Service at Arecibo Light Station on the northwestern coast of Puerto Rico in September 1926. The goal with the 32-volt wind-electric plant was to replace the aging, fire-prone white oil vapor light of the beacon with an electric lamp and improve conditions for the lighthouse attendant by adding electric lighting and running water.[92] The agency erected a similar wind-electric plant at Ka Lae at the lighthouse on the southern point of Hawaii's big island in November 1928 to supplement the auxiliary engine-based generator. A third plant was installed at the Point Lookout Light Station in Maryland, a year later. It was reported in the spring of 1930 that "[t]his wind-electric system [at Point Lookout] is one feature of the complete modernization of the light and quarters of this station, which was established in 1830."[93]

These early wind-electric plant manufacturers also spurred a host of indirect commercial enterprises to support the burgeoning industry. In addition to the numerous suppliers hired to make cast iron and steel parts used to build these machines, thousands of jobs were generated in sales and in the erection of wind plants for customers. Young men during the 1920s were flocking to the electrical fields, receiving training in universities, colleges, and trade schools, with the promise of compensatory paychecks and steady employment.[94] Some wind-electric companies partnered with electrical appliance manufacturers to sell packages of 32-volt DC products to customers and, in some cases, made arrangements to have certain products branded with their own labels. The Jacobs Wind Electric Company, for example, sold under its brand third-party manufactured food mixers, toasters, clothes irons, vacuum cleaners, radios, refrigerators, freezers, and fans, as well as farm equipment such as motors, pumps, milking machines, shearing equipment, and welders. The companies that made these products for Jacobs included Hamilton Beach, Eureka, Gould, and Marquette.[95] There were also the independent sellers of glass storage batteries, such as the Peerless Battery Manufacturing Company, National Batteries, Universal Batteries, the Edison Storage Battery Company, and the Electric Storage Battery Company (Exide), which catered to wind-electric plant operators.[96] In addition, a few firms specialized in the sale of replacement wind-electric plant blades. By the early 1930s, the

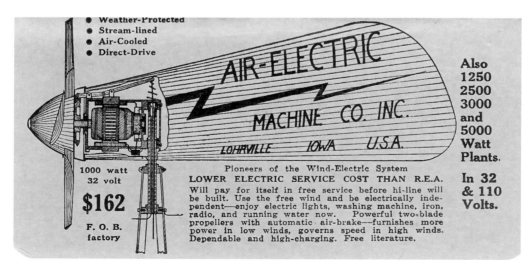

Air-Electric Machine Co. Inc., of Lohrville, Iowa, during the 1930s manufactured a wind-electric machine with a "streamlined" design. From *Air-Electric Machine Co. Inc.*[,] *Lohrville*[,] *Iowa*[,] *U. S. A.* (ca. 1935).

National Manufacturing Company of Lincoln, Nebraska, and Sheldon & Sartor Wind Electric Manufacturing Company in Nehawka, Nebraska, sold aluminum alloy wind plant propeller blades and promoted their benefits over wooden blades, which they said tended to be quickly damaged by storm winds and airborne debris, as well as bending and twisting with changes in humidity.[97]

By the early 1930s, the number of wind-electric plant manufacturers in the United States had grown to several dozen. In addition to the Wincharger Corporation and Jacobs Wind Electric Company, they included the likes of Air Electric Machine Company, Aerodyne, C. H. Charlson Manufacturing Company, Charles E. Miller Company, Nelson Electric, Parker-McCrory, Parris-Dunn Company, and Wind Power Light Company. Iowa was home

to seven of these firms, followed by Minnesota with a handful and the rest scattered among the states of Illinois, Kansas, Montana, and Indiana. But many of these companies watched their wind-electric plant sales shrivel as the United States found itself in the throes of the Great Depression at the start of the 1930s. Miller Motor Company of Newton, Iowa, which started selling three-bladed wind-electric generators in 1925, was in serious financial trouble by 1933, and its founder, Basil Miller, was eager to sell when the young out-of-work washing machine company executive, Edmund A. McCardell, approached him about purchasing the company, which he did for $5000.[98] Mergers among wind-electric firms were also taking place, such as the case in 1930 when the Herbert E. Bucklen Corporation and Perkins Corporation formed

This photograph (ca. 1935) shows the Webb Bros. Pty. Ltd.'s factory in South Melbourne, Australia. The company manufactured wind-electric plants for the Australian market from 1930 to 1948. Courtesy State Library Victoria, Melbourne, Australia.

Bucklen-Perkins Aerolectric, Inc. This new company was further absorbed into the Universal Battery Company of Chicago by 1935.[99] Wincharger became the market leader by volume shortly after radio manufacturer Zenith Corporation purchased a majority share of the company in 1935, resulting in the increased production of its two-bladed 6-volt machines for charging radio and automobile batteries to "several thousand" a month by the start of 1938.[100] During this time, Wincharger also made deals with nationwide catalog retailers Sears, Roebuck and Company and Montgomery Ward & Company to allow them the right to brand and sell its wind-electric plants. Sears sold its 32-volt Wincharger-built machines under the Powermaster name, while Montgomery Ward used the Powerlite label.[101] In addition, it was not uncommon for businessmen and engineers in the wind-electric industry to leave their employers to strike out on their own. Ernest A. Arndt, who in the early 1930s served as an engineer at Wincharger, formed a new wind-electric manufacturer, the Ruralite Engineering Company of Sioux City, Iowa, in 1937.[102] C. L. Parris, also formerly of Wincharger, joined with William Dunn of the Dunn Manufacturing Company in Clarinda, Iowa, in 1936 to form the Parris-Dunn Corporation, which became a prominent wind-electric plant manufacturer.[103]

Many US-built wind-electric plants were exported to remote, electricity-lacking locations overseas, such as small islands in the Caribbean and the Pacific and the expansive frontiers of Africa, South America, and Australia. Not to be outdone, a number of European and Australian manufacturers entered the small wind-electric plant business after World War I in an effort not only to satisfy their home markets but also to export these machines abroad. During the mid-1920s, a handful of these companies had been established in Britain. The Institute for Research in Agricultural Engineering of the University of Oxford from 1924 to 1925 conducted tests on seven wind-electric plants from five British manufacturers. These machines included wind wheels of the "American-style" multibladed design and those with four to six large aerodynamic blades. The university's report did not favor one machine over another.[104] A revised edition of the report published by the university in 1933 stated that "a wheel of blades made to a strictly aerofoil section is capable of giving a higher efficiency than the multi-bladed turbine wheel."[105] Yet, this did not dissuade some European companies from continuing to manufacture wind-electric plants with dated multibladed wind wheels. Inventors Robert Fis and François Pinaud, who operated the firm Etablissement des Aéromoteurs Cyclone at Margny-lès-Compiègne in France, constructed wind-electric plants with large multibladed wind wheels from 1923 to 1930.[106] Similarly, a handful of Australian companies produced wind-electric generators with steel multibladed wind wheels or wood rotors in the late 1920s and 1930s. James Alston and Sons Pty. Ltd. of South Melbourne, Australia, offered a 12-volt and 20-volt wind-electric generator patterned off its water-pumping windmill.[107] More successful 12-volt and 32-volt DC propeller-driven machines were produced and sold throughout Australia in significant numbers by firms such as Hannan Bros. Ltd., Dunlite Electric Co. Ltd., Webb Bros. Pty. Ltd., and Saunders Engineering Co. Ltd.[108] With the spread of large utility-produced electricity into rural areas, wind-electric plant manufacturers with smaller domestic markets, such as those in Europe, were already under pressure by the early 1930s to export their machines to other markets.[109] Some of these companies went so far as to try to tap into the US market with their machines, including applying for US patents. Agrico Manufacturing Company Ltd. of Copenhagen, Denmark, approached the Kregel Windmill Company in Nebraska City, Nebraska, in February 1925, and presumably other windmill manufacturers, about having its pumping and electric-generating windmills manufactured in the United States, "as we have not facilities to cover the amount of business which it seems is going to develop in the different countries of the world." Agrico noted in this correspondence that it had already sold manufacturing rights for its windmills to companies in the Netherlands, Argentina and Brazil. One of those concerns, Werkspoor in Amsterdam, paid Agrico "a considerable sum of cash and will further pay . . . on each mill made a royalty of 10 and 15% in the different countries covered by the agreement respectively," the company told Kregel.[110]

Winpower Corporation of Newton, Iowa, manufactured a unique three-bladed, downwind turbine (ca. 1940). From *Winpower 3500[,] Making the Dream of Cost Efficient Wind Energy A Reality* (1982).

Throughout the 1930s, wind-electric plant manufacturers continued to refine their equipment, with particular attention paid to their governor designs. Wincharger historian Mike Werst explained, "The governor was the primary design feature that defined a wind generator's longevity. A wind company's prized intellectual property was, in many cases, its speed regulating governor patent."[111] Without blade speed control, these wind-electric plants would destroy themselves in high winds. Governor designs varied from company to company. Wincharger used an airbrake mechanism on a majority of its wind-electric plants, which consisted of opposing sheet metal louvers with springs that produced a drag whenever the centrifugal force of the turning propeller superseded the restraining spring force. The Wincharger airbrakes, which attached perpendicular to the front of the hub and blade, measured 22 inches in length for the 6- to 12-volt units and up to 60 inches for the 32- and 110-volt models.[112] A more sophisticated method to control generator speed, which was perfected by the Jacobs Wind Electric Company, was through propeller blade pitch control. The Jacobs governing system used spring-loaded fly-ball weights to regulate the angles of the three propeller-type blades. As the speed of the propellers accelerated, the fly-ball weights would "swing out," causing each blade to turn more edgewise to the wind. This

action effectively reduced the propellers' lift and limited their speed. This governor design allowed the Jacobs wind-plant generators to operate on average 120–200 rotations per minute, with a maximum of 250 rotations per minute.[113] The Wind Power Light Company also developed a fly-ball governing system, but instead of operating in an upwind position like most wind-electric plants at the time, the company in 1935 developed a "tailless" unit. The company's chief designer, Merle Vincent, mounted fly balls behind each of the three propellers, which responded to the increased wind by folding back the blades in unison and making the wind plant "act as a huge dart."[114] One of the more interesting wind-plant governing systems, however, was developed by the Parris-Dunn Corporation, which let the power of the wind take it out of service. As the wind surpassed a desirable speed, the propeller and generator both began to pivot into a vertical position. Once in the full vertical position, what the company dubbed "slip the wind," the propeller remained immobile, pointing into the wind much like a weathervane, during severe winds and then dropping back into a horizontal operating position when the winds moderated. The Parris-Dunn machine could also be manually locked in the vertical position by a pull cable.[115]

Not everyone could afford a commercial wind-electric plant. Many farm families continued to

Many rural Americans before World War II still lacked electricity. The girl pictured here (ca. 1940) had the chore of cleaning the kerosene lamps, while her mother handwashed clothes. From author's collection.

experience the ill-effects of the Great Depression through the early to mid-1930s. "More farm homes have not enjoyed the benefits of electric lights and electrically operated equipment because they have not had the means with which to purchase the common electric light plant," wrote H. F. McColly and Foster Buck. These engineers at the Agricultural Experiment Station of the North Dakota Agricultural College at Fargo in 1935 provided detailed instructions on how to build a practical two-bladed, 6-volt wind-electric plant for powering a few light bulbs or a radio. "While the small low voltage system is not practical except for small loads, it will fulfill a long-felt desire by many rural families to enjoy some of the benefits of electricity and acquaint them with the possibilities of such service," McColly and Buck stated.[116] Other universities throughout the Midwest provided similar instructions for constructing wind-electric plants. In 1937, the Oklahoma Agricultural and Mechanical College at Stillwater provided farmers detailed instructions on how to build a three-bladed, 12-volt generator wind plant from scrap materials found

either around the farm or in a local junkyard at a cost of no more than $130.[117] Hobbyist magazines in the 1930s routinely published instructive articles on how to rebuild discarded automobile generators for use as 6-volt wind-electric plants to charge radio batteries.[118] Wind-electric plant parts, from the propellers and generators to the stub towers and control panels, could be purchased through mail-order suppliers like LeJay Manufacturing Company of Minneapolis.[119] Klinsick Mechanical Shop of Optima, Oklahoma, another wind-plant parts supplier, offered its customers a free 12-inch sample propeller with the purchase of a set of plans and building instructions for $2.50.[120]

People relying on wind-electric plants learned to live with their benefits, as well as the limitations. Farmer and author Fred W. Hawthorn and his wife, Nell, recalled in the 1970s how their lives on an Iowa farm with a 1500-watt, 32-volt Jacobs wind plant during the early 1930s "became quite wind oriented."[121] The young couple used their high-current consuming electric appliances, such as the washing machine and clothes iron, on the windiest days. Fred Hawthorn further described how the family coped during days without wind:

> During periods of prolonged calm, we carefully watched the three hydrometer balls on the battery. When all three balls floated, the cells were fully charged and we could use electric power extravagantly. The green ball would drop first. This caused us to conserve slightly. When the middle white ball dropped, only the "essential" appliances were used. When the final red ball dropped, all electric use was curtailed. The refrigerator was turned off and the kerosene lamps were brought out, since the expensive battery would be harmed by deep discharging. Seldom did we have to take such drastic measures for long because usually the wind would come up soon after the red ball dropped.[122]

Most wind-electric plant users were quite satisfied with the overall performance of their machines. According to a 1941 survey of ninety-three Wincharger users across Iowa, the average annual

Installation of wind-electric machine on an unknown rural property during the late 1930s. From the author's collection.

cost of owning and operating a wind-electric plant was $86.16. The survey further concluded that "[t]he success of wind electric plants is demonstrated by the fact that 92% of the owners reported that they would purchase wind driven plants if making a second selection of farm lighting plants."[123]

While wind-electric plants proved to be effective alternatives to gasoline-powered generators for lighting and other power applications on farms and ranches, they also became a partner to the oil and gas industry during the mid-1930s. By that time, high volumes of oil and gas were being shuttled around the country via buried steel pipelines. However, the pipes were at risk of rapid corrosion due to the natural propensity for small electric currents to run through them. It was estimated that 1 ampere of current could carry off as much as 20 pounds of steel a year, often in concentrated areas of pipe. To stop the corrosion of the pipes, companies buried scrap iron nearby. The scrap iron was then connected with a positive, low-voltage direct current source, and the negative pole was connected to the pipeline. Thus, a circuit was formed through which current flowed from the positive pole of the electrical source through a copper cable to the scrap iron—the anode—which corroded the cathode instead of the steel pipeline. Since pipelines crossed desolate parts of the country, it was determined that the source of direct current for the cathodic protection scheme could be supplied by 32-volt wind-electric generators.[124] Many large oil and gas pipeline operators, such as the Shell Pipe Line Company, Texas Company, Sinclair Oil Company, Phillips Oil Company, and Union Oil Company, turned to the Jacobs Wind Electric Company for their wind-electric cathodic protection plants. These plants first entered service in 1936. While other wind-electric plant manufacturers were also involved in this business, such as the Wind Power Light Company, Jacobs was by far the dominant player, installing "many hundreds" of its 1250- and 2000-watt units for this service over the next 15 years.[125] Marcellus Jacobs even helped to organize the National Association of Corrosion Engineers in 1935–1936, served on its first board of directors for several years, and remained an active member for the next 25 years.[126]

Since the mid-1920s, however, wind-electric plant manufacturers have been under constant competitive threat from the increasing delivery of electric power by large power companies. High voltages of alternating current could be carried over hundreds of miles via suspended copper wires to transformers where it could be stepped down for household use. Many farms and ranches throughout the United States remained disconnected from large-scale electric power during the early 1900s, since the utilities did not believe that the economics existed to run the hundreds of thousands of miles of copper lines to these mostly remote locations where electricity usage was minimal. Electrical engineer and author Ernest Greenwood lamented in 1928 that "there are still about 27,000,000 people living on 6,000,000 farms, still reading by obsolete oil lamps, washing clothes in old-fashioned tubs, carrying water from the well, and cooking in kitchens differing little from the kitchens of a hundred years ago."[127] At the start of the 1930s, the federal government estimated that eight out of nine of the estimated six million farms across the United States were without utility power.[128] Electric-light plants offered some electrical conveniences, but even then many farmers went without electricity. Rural Americans who sought electric power, even those located within several miles of existing power lines, were often denied connections due to the utilities' cost considerations. There were examples of rural homeowners who had their houses wired for utility power and waited 10 to 20 years for the service.[129] Frustrated by the antiquated operations of American farms still recovering from the Great Depression and the hesitation of the power companies to serve rural populations, President Franklin D. Roosevelt on May 11, 1935, issued an executive order to form the Rural Electrification Administration (REA), which was followed by passage of the Rural Electrification Act on May 20, 1936. As contrasted with certain other New Deal programs, the REA was generally considered successful. Within three years, the agency had set up 380 electric cooperatives in forty-four states, erecting more than 80,000 miles of rural power lines that served more than 250,000 rural customers, or an estimated 1.25 million people. Each co-op member paid a monthly

A lineman for the Texas Electric Service Company shown with his pole-climbing gear in 1936. From the author's collection.

service fee, in addition to a per-unit cost of electricity to the provider.[130] Once the power lines reached a farm or ranch, existing gasoline and wind-electric light plants on the property were disconnected and often removed. Some of these machines, along with the 32-volt appliances, ended up on the second-hand market to be used by farmers and ranchers in more remote locations.[131] But for many wind-electric plant operators, receiving power through the "high line" was still years, if not decades, away.

While rural electrification presented a long-term threat to the wind-electric plant manufacturers, a more immediate and crashing blow to the industry came with the United States' entry into World War II, following the Japanese surprise attack on the US Pacific fleet at Pearl Harbor, Hawaii, on December 7, 1941. Legislation and executive orders were issued that shaped the way manufacturing and commercial sales would be conducted during the subsequent war years. These included President Roosevelt's

War Production Board, which was put in charge of all war production and supply; passage of the Emergency Price Control Act to curb rising prices to avoid inflation; the president's wartime freeze on wages, salaries, and prices; and rationing on everything from meat and sugar to rubber and metals. Companies added women to their workforces as more men were either drafted or volunteered for military service. Upon entering the war, US companies were ordered to contribute to the production of war materiel; wind-electric plant manufacturers were no exception. The Wind Power Light Company, which had changed its name to the Winpower

Manufacturing Company, at the start of the war worked as a subcontractor to D. W. Onan and Sons of Minneapolis to assemble and wind armatures for 1.5-kilowatt DC plants for the military. This was also the time that the company hired its first female crew for the shop floor. Ed McCardell later recalled that within a few weeks of training the women and adding the proper shop equipment that they were producing more than a hundred armatures a day for more than a year.[132] The Jacobs Wind Electric Company obtained a significant contract from the US Navy to build 5-kilowatt gas-powered generators, which were installed aboard ships for "degaussing,"

Inventor and businessman Palmer Cosslett Putnam inspects his 1250-kilowatt, grid-tied wind turbine, which was installed at Grandpa's Knob near Castleton, Vermont, in the late summer of 1941. From the author's collection.

a process of changing a ship hull's polarity for the purpose of repelling magnetic mines placed in the English Channel and harbors by the German navy and air force. To keep up with this wartime production, the Jacobs Wind Electric Company ramped up its workforce to more than 100 employees and ran the operation day and night with three separate work shifts.[133] Wincharger Corporation, also using the commercial name Winco, secured a large contract from the War Department in late 1941 to manufacture dynamotors, which facilitated wireless two-way communications between aircraft and troops on the ground.[134] Interestingly, the Parris-Dunn Manufacturing Company manufactured nonfiring, bolt-action training rifles for the Army and Navy, churning out an estimated two million during the war.[135]

Before and during the war, American and European engineers considered various designs for large wind-electric generators to power entire communities or for use in industrial-scale applications. In the 1920s and early 1930s, cylindrical wind wheels were developed and tested by engineers such as Anton Flettner and Julius D. Mádaras for the purposes of electric-power generation, but they proved less efficient than propeller-type designs. In 1931, a wind generator with two automatically pitching blades measuring 100 feet in diameter was put into operation on a bluff at Balaclava near Yalta in what then was part of the Soviet Union to provide supplemental power to a peat-burning 20,000-kilowatt steam plant about 20 miles away at Savastopol. In 1925, German engineer Herman Honnef proposed building a "skyscraper windmill" that consisted of five 250-foot to 500-foot-diameter wind wheels on a 1000-foot-tall steel tower to generate up to 50 megawatts of power or enough electricity for a city with 100,000 inhabitants. Honnef's monstrous machine never got off the drawing board. However, a much smaller German machine, known as the Kumme wind turbine, was operational during the early 1920s; it consisted of a six-blade, 60-foot-diameter wind wheel that turned into the wind by a small rear-mounted fantail. Compagnie Electro-Mécanique in 1929 erected a two-blade wind turbine at Bourget, France, with a 65-foot diameter.[136] In the late 1930s, Boston

engineer Palmer Cosslett Putnam, with assistance from various corporate backers, set out to design and build a 1250-kilowatt wind generator with two 65-foot-long blades with a generator mounted on top of a 110-foot steel-girded tower. The machine was erected on the summit of Grandpa's Knob in Vermont in late summer 1941. The electricity was fed into the grid of the Central Vermont Public Service Corporation, which received most of its power from hydroelectric dams. The Putnam turbine remained in operation until March 26, 1945, when it threw a blade and sustained other structural damage. Although the project was stopped, the Putnam design became the foundation for future megawatt, alternating current wind turbine designs.[137] Even while fighting raged across the European continent during World War II, Honnef led a small wind turbine design program for Germany, and the Danish engineering firm F. L. Smidth erected both two- and three-bladed wind turbines rated between 60 and 70 kilowatts.[138] Interest in large-scale wind power's potential continued after the war, mostly in Europe due to the high cost of operating steam-driven power plants. Various designs of two- and three-bladed designs were tested by British, Danish, and German companies throughout the 1950s and early 1960s. British wind power expert, Edward W. Golding, in the mid-1950s envisioned two methods for deploying these wind turbines: (1) the mass use of "medium-size" machines with individual output of 100–200 kilowatts to supplement traditional fossil fuel-burning or hydropowered generating plants among the "hundreds of islands in many parts of the world," and (2) the use of large 1000-kilowatt, AC plants that fed directly into power grids and served as "fuel savers."[139]

The small wind-electric plant manufacturing industry in the United States did not fare so well after the war. Besides the high cost of retooling after four years of wartime production, the US government immediately turned its attention to bringing utility-produced electricity to the remaining 55 percent, or about 3.8 million, nonelectrified farms.[140] This move would be swift as materials previously used for the war, such as copper and steel, once again became available for commercial use, and young men returning home from the war were ready to resume civilian life. For example, farmers Charles and Mabel Bass of Torrington, Wyoming, received REA power at the end of 1949 after relying on a 6-volt Wincharger for ten years to power a few lamps in the farmhouse and a string of lights for the Christmas tree. The Bass house was wired for 110-volt AC electric power in November 1949 and received its REA connection just before Christmas that year. The Bass's son, Delwin, who was ten years old at the time the family received central station power, recalled:

> It was after dark and we could see the car lights of the [REA] crew about ¾ miles south along the main line. We were all set to go. My sister and I had a new toaster with the bread in it and the handle down and plugged in. We saw the flash when they connected our trunk line and we had lights instantly. The smell of the toast was wonderful. I looked at mother and she was crying. She said, "I have never seen such a dirty ceiling."[141]

That Christmas, Mabel Bass received an electric iron, and by the spring of 1950 she was using a new electric sewing machine and the family washing machine was outfitted with an electric motor.[142]

Some manufacturers had already stopped producing wind-electric plants altogether by the war's end and had started new product lines. Parris-Dunn Corporation sold the production of its 6- and 12-volt wind-electric plants to Winpower in 1949. While continuing to manufacture some wind-electric plants, Winpower began relying more on sales of cultivators, fertilizer spreaders, farm wagons, hay stackers, and post-hole diggers that it manufactured at its Newton, Iowa, factory. Air-Electric Machine Company of Lohrville, Iowa, became a manufacturer of gasoline-powered lawn mowers, while the Parker-McCrory Manufacturing Company of Kansas City, Missouri, pursued its long-standing electric-fencing business.[143] Yet a few wind-electric wind plant manufacturers persevered for a period after the war. The Jacobs Wind Electric Company, for example, used the war years to make further refinements to its wind-electric plants, including developing a simpler spring-based, auto- matic variable-pitch centrifugal governor to replace

Small wind plants still found use in remote application, like this 12-volt Wincharger being erected in 1963 to charge batteries at the US government's McMurdo Station in Antarctica. National Science Foundation photograph from the author's collection.

its former governor composed of fly-ball weights. The new Jacobs governor cost $238 from the factory. Marcellus Jacobs explained to a customer in 1953 that the new governor would "insure greatly improved performance of your plant, charging more in light winds, and is a much better acting and more positive governor speed control. It will not increase and race its speed during high winds as does your present governor."[144] As inventors and skilled marketers, the Jacobs brothers continued to hold out hope for their wind-electric plant business, even participating in television game shows such as "Queen for a Day" and "Truth or Consequences," in the early 1950s. During these shows, Marcellus Jacobs gave away the company's patented cork-in- sulated Wonder freezers to winning contestants.[145] However, the Jacobs Wind Electric Company finally closed its Minneapolis factory in 1956, the location from which it manufactured and sold an estimated 20,000 1.5- and 3-kilowatt wind-electric plants over a period of nearly twenty-five years.[146]

By 1960, many wind-electric plants in North America had already been left idle on top of their towers or were taken down. REA and the central station power providers had reached their goal of providing standard AC power to even the remotest parts of the country, as evidenced by the hundreds of thousands of miles of power lines carried on large steel towers and wooden poles that criss-crossed the landscape. Most Americans who had reached adulthood in the decades following the 1950s no longer considered what it was like to live without electric power. Electricity had become an expected and relatively inexpensive energy source that worked upon the flick of a switch on the wall or through the myriad electrical appliances available in the home and workplace. However, by the end of the 1960s and the start of the 1970s, a renewed interest in small wind-electric power emerged as utility-based electric power became increasingly expensive and a new generation of young people rebelled against the lifestyle norms and excesses of the preceding decade.

3 Back to the Wind

In the two decades following World War II, the standard of living across the United States rose, and with it came an increase in fossil fuel-based energy consumption in the form of utility-generated electricity for household appliances and lighting, oil for heating, and gasoline for transportation. Americans at the time benefited from a cheap and seemingly abundant supply of energy, with minimum regard for the pollution generated by power plants, factories, and vehicles to maintain this way of life. Energy conservation in the home appeared unnecessary, since one could simply change the temperature on the thermostat with little concern about cost. With gasoline averaging 25 cents a gallon, it was also more convenient to drive the car to a grocery store to purchase food than taking the time to grow it oneself.[1] By the start of the 1970s, unbridled fossil fuel consumption began taking a toll on the environment and human health in the form of choking smog and traffic congestion in urban areas.[2] However, it wasn't until the Arab nations, angry at the United States and Europe for supporting Israel, placed a temporary embargo on their oil exports in October 1973 that the majority of Americans took stock of their energy usage. The embargo caused gasoline to climb to 40 cents or more per gallon within two months, exacerbating concerns that gasoline prices could reach as much as a dollar per gallon. A number of states implemented "odd-even" rationing of gasoline, and newspapers printed photographs showing long lines of frustrated drivers waiting their turn at a gasoline pump and hastily written service station signs saying "sold out."[3] To reduce fuel consumption, people reduced weekend drives, speed limits were lowered to 55 miles per hour, and auto efficiency standards were legislated into law.[4] Home heating oil prices also increased from 20 and 30 cents per gallon to 30 and 35 cents per gallon. Power companies similarly raised their rates for residential electricity by 10 percent. By late 1973, it was estimated that the average American household spent approximately

By the mid-1970s, many Americans considered energy conservation a national priority, as depicted in this US Postal Service first day stamp issue in October 1977 espousing the benefits of wind energy. From the author's collection.

$743 of its annual income of $7150 on energy, with half of that for gasoline purchases, a quarter for heating, and the remainder for electricity.[5] At the start of the decade, clean and affordable alternative energy sources to fossil fuels were virtually nonexistent, and the public generally resisted proposals to construct additional nuclear power plants.[6] Combined with the realization that fossil fuels, solar photosynthetic energy stored underground for 350 million years, were finite, diminishing rapidly, and environmentally destructive led many Americans to embrace energy conservation. "Cutting back on our use of fossil fuel will have the incidental effect of forcing us inside ourselves for human resources long dormant, and on each other," wrote Ralph Keyes, a community living researcher and author, in a December 3, 1973, *Newsweek* article, titled "Learning to Love the Energy Crisis."[7]

As part of the energy conservation effort, a movement of going back to the land, also known as homesteading, began to take root in the early 1970s and ultimately drew tens of thousands of young people away from their middle-class lifestyles in the suburbs, towns, and cities to remote parts of the country where they physically detached themselves from modern conveniences, such as indoor plumbing, telephones, and televisions, and lived off what they could produce from the earth or with their hands. "Back to the land" movements were nothing new and had periodically arisen in the United States throughout much of the twentieth century, starting as far back as 1907 with the bank panic and reemerging during the Great Depression of 1929. Economist Ralph Borsodi's books, *This Ugly Civilization* and *Flight from the City*, published in the 1920s and 1930s, influenced many young people who came to embrace the back-to-the-land movement at the start of the 1970s. In *This Ugly Civilization*, Borsodi wrote, "For when we take the places in which we dwell away from the country; deprive our homes of intimate contact with the growth of the soil; shut off our access to sun and light on all sides, we do not merely deprive ourselves of fresh air and sunlight, green grass and majestic trees—we deprive ourselves of what is an elemental need of mankind: the inner discipline which comes from communion with the land."[8] In contrast to the 1862 Homestead

Act, under which the federal government gave title to 160-acre plots of public land to settlers if they lived on the property continuously and developed it over five years, Borsodi's definition of homesteading was to create small, self-sufficient communities. The center of these communities was a so-called School of Living in which all the inhabitants—children and adults—studied the philosophy of life. To sustain the communities, land trusts were established, and families were charged small amounts of money for rent. Several of these communities were established west of New York City during the 1930s. Communal-style living, however, was not for everyone, with some viewing homesteading as an individualistic lifestyle. This was the case of Scott and Helen Nearing, who became homesteaders in Vermont during the early 1930s and later in Maine. They published the influential book, *Living the Good Life*, in 1954, which described their desire to find "simplicity, freedom from anxiety or tension, an opportunity to be useful and to live harmoniously."[9] These books became highly sought after during the early 1970s as people once again considered moving back to the land. Eighty-eight-year-old Borsodi told *The Mother Earth News* in 1974 that "unrest usually spawns a 'back to the land' movement that catches fire for a while . . . and then times get better and we repeat the cycle all over again."[10]

Remote locations throughout New England, the Ozarks, Pacific Northwest, and Canada became ideal for homesteading during the early 1970s. While the lifestyle might have sounded appealing, it required a lot of hard work and commitment just to get started. First, one had to acquire the land and then clear away any trees and underbrush to erect a living space and other outbuildings, as well as prepare room for growing fruits and vegetables and raising small livestock. Producing one's own food to sustain an individual or family beyond the summer months and through the winter until new crops were in full production once again the following year took hard work and careful planning. Many homesteaders came from nonagricultural backgrounds and had to learn at first-hand how to live off the land. One of those young homesteading families, Eliot and Sue Coleman, along with their two-year-old daughter Melissa, in Bucksport,

Maine, was documented in a detailed article by *The Wall Street Journal* in 1971. At the time the article was written, the Colemans had already been living in an 18-by-22-foot one-room cabin on about 60 acres for about two and a half years. They had no indoor plumbing or electricity, and they grew 80 percent of their food. Eliot and Sue Coleman, 31 and 26 years old, respectively, were both college-educated and hailed from financially secure families. After marrying, they were inspired by the Helen and Scott Nearing book, *Living the Good Life*, to leave behind the American consumer economy; they even purchased the property for their homestead from the Nearings.[11] Even after establishing their homestead, the Colemans found it unrealistic to disassociate themselves completely from the outside world. In 1970, they estimated that they had $2000 for expenditures on food and other items that they could not produce themselves. They also earned extra money through local sales of surplus vegetables, in addition, Eliot Coleman earned some income by gardening and taking on

odd jobs for local residents several days a week during the spring and summer months, and Sue Coleman worked occasionally as a part-time secretary. While the family was satisfied with their existence, Eliot Coleman explained it bluntly: "I'm working 16 hours a day for survival. This isn't any game I'm playing. If I don't grow enough [during the summer], it's that much less to eat this winter."[12] The Colemans were successful, but many of those who attempted homesteading ultimately abandoned the endeavor owing to the associated hardships.

In 1971, a group of environmentally concerned scientists and peace activists, calling themselves the New Alchemy Institute, leased a 12-acre property near Falmouth on Cape Cod, Massachusetts, to pursue self-sufficient farming techniques. As part of this initiative, the institute in the summer 1974 designed and constructed a "sailwing" windmill to pump water, which was largely based on Cretan windmills. The resultant machine's downwind rotor consisted of three wood-framed, cloth-covered sails with an 18-foot diameter and used a simple

The New Alchemy Institute in 1971 purchased a 12-acre tract of land near Falmouth, Massachusetts, to pursue self-sufficient farming techniques, including homemade water-pumping windmills. Courtesy Bob McBroom, Holton, KS.

direct-drive crank mechanism to operate a vertical pump rod. The windmill, which rotated on a table bearing at top of a wood lattice tower, could pump an estimated 250 gallons of water per hour in a 6-mph wind. The institute continued to refine its sailwing windmill design through the decade and offered construction plans to do-it-yourselfers, as well as farmers in developing countries. The institute estimated the machine cost about $875 in materials and took twenty-five hours to install. It was four times cheaper than commercially manufactured water-pumping windmills at the time.[13] In the mid-1970s, John Todd, co-founder of the New Alchemy Institute, wrote that the purpose of its windmill research was "to participate in creating a body of knowledge that could lead to the replacement of fuel consuming engines and the hardware of present societies with equivalent support processes which could be derived from living systems coupled to sensitive technologies powered by the wind and sun."[14]

Mother Earth News (first issue cover shown) sparked deep interest in the prospects of wind-electric power among the new generation of homesteaders in the early 1970s. Courtesy Ogden Publications, Inc., Topeka, KS.

Homesteading, or going back to the land, didn't necessarily mean cutting oneself off from electric power to operate lights, a radio, or motor-based appliance. However, the logistics challenge of supplying electricity from a power company to remote sites proved costly. Gasoline-powered generators were widely available but were noisy, smelly, and temperamental to operate, not to mention required fuel. For many homesteaders the thought of relying on either of these two sources for electric power was counter to their vision of cutting ties to wasteful, polluting, big corporate power generation. In the November 1970 issue of *The Mother Earth News*, publisher John Shuttleworth reprinted the plans and diagram from a 1945 LeJay Manufacturing Company brochure for how to construct a direct-drive 32-volt wind-electric plant. "If you're heading off into the bush somewhere, have a grudge against the electric company or just want to know how to set up a home generating plant, this is a darn good practical handbook of basic information. I know: I was raised with wind generators and welders right out of the pages of LEJAY," Shuttleworth wrote.[15] At the start of the 1970s, Minneapolis-based LeJay, while it had left the wind-electric business at the end of World War II, still sold its do-it-yourself mechanical and electrical projects handbook to customers for $1.50 each. Constructing a Lejay wind plant, however, required one to scrounge for parts and have some mechanical acumen to put it together. Then there was the requirement of batteries to store the direct current (DC) power and knowledge of how to properly use the machine. With the addition of other articles, *The Mother Earth News* sparked deep interest in the prospects of wind-electric power among the new generation of homesteaders. Even the forbearers of the back-to-the-land movement supported the use of wind-electric power. "The oil is running out. Even the coal, which we still have a lot of, won't last forever. But the wind!" Borsodi said in the early 1970s. "You can use the wind to drive a motor and produce power and you can do so as much as you want. It doesn't lessen the quality of wind in the world a particle and it doesn't pollute anything. We ought to have literally thousands of windmills all over this country."[16]

Two of those early 1970s homesteaders who decided they could not do without electricity were Henry and Retta Clews. The young couple purchased 50 acres of property in North Orland, Maine, on which to build their farmstead. The property was about 5 miles from the closest paved road and equally far from access to utility power lines. Henry Clews, who had studied to become an engineer, inquired with the local power company about running a line to the home and was told it would cost more than $3000, plus a minimum electric bill of $15 per month for a period of five years. That didn't sit well with the Clews, and so they decided to look to the wind for their electric power. But options for available commercial wind-electric plants were extremely limited at the start of the decade. In the United States, only Dyna Technology of Sioux City, Iowa, manufactured the former Wincharger Corporation's small 200-watt, 12-volt wind plant to charge batteries for a small cottage and camping equipment on an intermittent basis. This was not a suitable wind plant for the Clews, who required sustained and heavy enough power to operate lights, household appliances, and shop tools. Upon further research, they located a wind-electric plant manufactured halfway around the world in Adelaide, South Australia, by the Davey-Dunlite Company. Through Dunlite's longtime distributor Quirk's, the Clews arranged for the import of a three-blade 2000-watt, 115-volt wind plant for about $1800. The unassembled plant arrived in Maine during the summer of 1972, and with raw experience and determination the couple, along with a brother-in-law, erected the Dunlite on top of a 50-foot steel tower in August that year. To make the intermittent wind system compatible with their continuous household electric requirements, the Clews set up a bank of twenty 6-volt batteries, an automatic "transistorized" voltage controller, and a system of rectifiers and inverters to provide 120 volts AC or DC. Henry Clews wrote in the November 1972 *The Mother Earth News* that within the first month of the wind-electric system's operation, "we've had uninterrupted power for lights, shop tools, water pump, hi-fi and—yes—even television . . . which is quite a change from candles and kerosene lamps!" He also announced that he had

Henry Clews of North Orland, Maine, climbs a tower to inspect a Dunlite wind-electric turbine on his property in July 1973. He operated Solar Wind Company, which imported Dunlite machines from Australia, as well as Elektros from Switzerland. Courtesy Stephen James Govier, Suffolk, England.

formed the Solar Wind Company to import Dunlite wind plants for the New England and New York area.[17] *The Mother Earth News* article resulted in the Clews receiving hundreds of letters inquiring about the Dunlite wind plants and their setup. They began publishing a thirty-page booklet, *Electric Power from the Wind*, in March 1973, and sold them for $2 each. By November that year, the Clews sold about 4000 copies, resulting in a second printing.[18]

After receiving numerous inquiries for wind-electric generators of larger than 2000 watts, the Solar Wind Company found a commercially available 6000-watt, 115-volt wind plant manufactured by Elektro GmbH of Winterthur, Switzerland. In addition to becoming a US dealer for Elektro, the Clews erected one of these wind plants on their property. The larger Elektro, with its 16.5-foot-diameter, three-blade propeller, was placed on top of a 50-foot guyed tower. For the wind plant, they added another fifty 6-volt battery set rated at 270 ampere-hours and purchased a DC-to-AC inverter rated at 3000 watts to produce 115 volts AC current for larger shop tools and appliances.[19] With just the Dunlite, the Clews were limited to about 300 watts of AC power, which they used for small electronics, such as their television and stereo but converted all larger appliances, like their water pump and shop table saw, to operate directly on DC. With the two wind plants, Henry Clews declared in November 1973 that "we are now powering two houses, which includes some ten 75-watt lights, ten small fluorescent lights in the workshop, a 1/3 hp deep-well water pump, a small refrigerator, radio, stereo, T. V. (color, no less!), miscellaneous appliances such as a blender, toaster, waffle iron, vacuum cleaner, sewing machine, etc. We have just acquired an electric hot water heater which we will hook up in conjunction with our present gas unit. This will be wired up so that it will switch on automatically whenever there is excess power available from the windmills."[20] Henry Clews was also invited that year to give a presentation about his home-based wind power system at the National Science Foundation's Wind Energy Conversion Systems Workshop, during which he stated: "Obviously, one wind powered home in Maine has little significance on the national energy crisis. But I feel that one operating wind power system, small though it may be, can demonstrate to many people that the wind is a viable and even practical source of energy for the future."[21]

By the end of its first year in business, the Solar Wind Company sold an estimated fifteen wind plants across New England and New York, as well as in Colorado, Alaska, and Nova Scotia, and was experiencing a six-month backlog for the larger units.[22] The Clews investigated other European wind plant manufacturers in 1973, including Aerowatt of Paris, which made a 4000-watt unit, and the Lubing Maschinenfabrik of Barnstorf, Germany, which built a 400-watt, vaneless genera-

tor. However, they determined that the Aerowatt, at more than $10,000 each, excluding controls or accessories, was too expensive, while the Lubing was aimed at water-pumping operations.[23] By January 1974, the Solar Wind Company had become a distributor of the Dyna Technology's 200-watt, 12-volt Winchargers; six different Elektro units ranging from a small 600-watt, 12-volt unit to a 12,000-watt, 115/230 AC volt system; and the Dunlite 1000-watt and 2000-watt, 115-volt DC or AC systems. The machines were not cheap. The Solar Wind Company's prices—delivered in Maine—ranged from $565 for the 200-watt, 12-volt Wincharger and $4395 for the 2000-watt, 115-volt Dunlite to $9785 for the Elektro 6,000-watt, 115-volt AC unit and $17,500 for the Elektro 12,000-watt, 115/230-volt AC system, placing significant limits on the US

In 1975 Earle Rich's home-built, 30-kilowatt wind generator in Scarborough, Maine, was considered the largest on the US East Coast. Courtesy Earle Rich, Amherst, NH.

company's sales.[24] For many homesteaders, the price tag of these commercial wind-electric plants, excluding labor and hookup, was cost prohibitive.

A philosophical underpinning of homesteading was figuring out how to create the things one needs by hand, and wind-electric plants in the early 1970s were no exception. "It's hard to recall when a single technical project—grabbing electricity out of the air—has attracted such an unlikely combination of experimenters," wrote Len Buckwalter of do-it-yourself magazine, *Mechanix Illustrated*, at the time. "Ordinary folk in Maine and elsewhere are scavenging through prop-pitch motors, old army surplus, automobile alternators and anything else to convert the turn of a windmill into current."[25] Yet the process of building one's wind-electric generator was

Windworks, which was formed by Hans Meyer with backing from architect Richard Buckminster Fuller, developed this cloth sail, wind-electric generator in September 1973. Courtesy Stephen James Govier, Suffolk, England.

no easy feat. It required the individual to have some basic knowledge of electricity, an array of shop tools, and plenty of determination. Earle Rich, a Maine native, set out to build his own wind-electric plant in the early 1970s. After graduating high school in 1958, he joined the Navy, during which time he received electronics training, and from 1962 to 1964, he worked in Vallejo, California, as an electronics instructor and technical writer, publishing five books. In 1965, he relocated back to Maine to take a job with Sylvania Central Devices, where he got into electronic design and control systems. The 1973 energy crisis steered Rich's interest toward wind power, and in 1974–1975 he set out to build a wind-electric plant of his own design. He purchased about a ton of steel to make a 60-foot lattice tower, a surplus aircraft 30-kilowatt generator, and a gearbox, and his wife stitched together a set of cloth sails. Once assembled, Rich said "it all worked," with an output of 240 volts, three-phase output. He used the system to provide electric lighting and heat for his home. At the time, Rich's wind generator was considered the largest on the US East Coast.[26]

In the early 1970s, some wind-electric plant experimenters began offering their construction plans for sale. In November 1972, Hans Meyer, a graduate aeronautical engineer and founder of the enterprise Windworks in Mukwonago, Wisconsin, led the research and design of simple wind-electric plants based largely on the architectural principles of the enterprise's lead backer, Richard Buckminster Fuller. Fuller was best known for his futuristic geodesic dome houses, which were designed to be inexpensive and efficient, yet withstand the world's harshest climates. Windworks experimented with a variety of sail designs, including cloth and wood. One of the more revolutionary blade designs developed in the early 1970s included a lightweight core of expandable, honeycombed paper glued to a 3/8-inch piece of plywood. The material was given an airfoil shape and covered with fiberglass. Windworks incorporated this design into a down-wind wind-electric plant with a three-bladed rotor of 10-foot diameter that sat on top of a 12-foot wooden tower. Because of its primary materials of wood, cloth, and fiberglass, the machine, plus tower, weighed only about 285 pounds. Like most

Windworks of Mukwonago, Wisconsin, which was led by Hans Meyer, attracted many young engineers and entrepreneurs interested in wind-electric turbine designs in the early 1970s. Courtesy Bob McBroom, Holton, KS.

of its early wind-electric plant designs, the first unit was erected on Fuller's island off the coast of Maine. Windworks calculated that the total cost of materials and parts, not including the used automobile alternator and batteries, was $182. To make the three blades, Windworks offered for sale three blocks of the honeycombed paper for $11. This simple, inexpensive Windworks wind-electric plant gained public attention when Meyer published a detailed article on how to build one in the November 1972 issue of *Popular Science*. However, the article was unclear on the design's actual electricity output, except to state that it ranged from one-quarter horsepower in 10 mph winds to 6 horsepower in winds of 30 mph.[27]

Another hallmark article for the do-it-yourself wind-electric generator was published in *The Mother Earth News* in March 1973. The article, with the attention-grabbing headline, "I built a wind charger for $400!," was authored by a young college graduate from northern California, named Jim Sencenbaugh. He spent about a year researching, designing, and constructing his first unit for a friend who lived on a remote property in Humbolt County, California, about 10 miles from the nearest power line. For his machine, Sencenbaugh said he chose a wooden, three-bladed variable-pitch propeller design with a 10-foot diameter for the best balance and efficiency. He shaped his blades from white pine boards and covered them in

an epoxy finish. To test his prototype rotors, he attached them to an A-frame bracket mounted to the front of his Porsche and drove up and down a residential street. To generate electricity, he chose a standard automobile alternator that could be obtained cheaply from a junkyard. Sencenbaugh ran the unregulated AC from the alternator through a diode bridge to convert it into a battery-storable DC. However, he had to overcome the problem of electricity in the batteries being slowly drained off through the alternator whenever the wind was idle. The alternator generates AC only when the armature rotates, cutting across the field current's electromagnetic lines, but if the field current remains on when the armature is stopped, the alternator will slowly drain the batteries to which it is connected. To remedy this problem, Sencenbaugh developed what he dubbed the "wind sensor," which consisted of a small vane and spring that, when the wind dropped below 8 mph, tripped a relay that switched off the alternators 3 amperes of field current. Subsequently, when the wind speed returned to 8 mph, the relay switched on to allow the field current to return to the alternator, creating electricity for the batteries. To drive the alternator, the propeller rotor was attached to two sets of chains and sprocket gears to reach a ratio of alternator to rotor rotations per minute of 7:1 to 9:1. Thus, when the wind reached 12 mph, the propeller should turn about 100 rotations per minute and the generator would turn 700 to 800 revolutions per minute, generating a substantive charging rate. To further control the wind-electric plant's operation in the wind, Sencenbaugh used a combination main sheet metal tail vane and smaller pilot vane, which he patterned off a 1908 Kenwood mechanical water-pumping windmill that was sold through the Sears, Roebuck & Company catalogs. He placed his machine on top of a 22-foot-tall tower made of 2-by-2-inch Douglas fir timbers. For the batteries, Sencenbaugh used 70 ampere-hour nickel-cadmium aircraft batteries, because they held up better than lead-acid storage batteries. The batteries were housed in a cabinet inside the base of the tower. To convert the batteries' 12-volt DC into a useful 110 AC for home lighting and appliances, he obtained a Heathkit Model MP-14 solid-state inverter.[28]

Sencenbaugh used *The Mother Earth News* article to launch a business in which he would offer the plans for constructing his wind-electric plant, called the Sencenbaugh O_2 Powered Delight, for $15 apiece. While he spent a little more than $400 to make his unit, Sencenbaugh believed that others could duplicate it for $300 to $350, depending on which types of batteries one used. He wrote:

> I'm also reasonably confident that you should be able to construct a wind charger like mine in any ordinary home woodworking shop. Everything but the blades can be made with hand tools and an electric drill. If you're exceedingly clever and determined, I suppose you might even fabricate the propeller with a 14 [inch] band saw and hand sand them . . . but I recommend that you figure on using a tilting table saw and belt sander for building the three airfoils.[29]

The following year, Sencenbaugh developed a new set of plans for his wind-electric generator which offered a blade design consisting of two pieces of three-quarter-inch plywood glued together, alternator rewinding instructions, and a heavier tower.[30] Within three years of starting his business, Sencenbaugh had shifted from just selling the plans for his machines to fabricating and selling the components and then to offering for sale completed units. In a 1976 interview with the *Wind Power Digest*, Sencenbaugh lamented that "out of the thousands of

Bob McBroom of Kansas Wind Power tested three water-pumping windmills—an Earthmind "S"-rotor, a two-blade Sparco, and a Dempster—at his property in Holton, Kansas, in the late 1970s. Courtesy Bob McBroom, Holton, KS.

sets [of plans] I sold, maybe five mills were actually built. It certainly wasn't the easiest thing to build; it took a lot of work, rewinding a car alternator, handcarving blades, using go cart parts—stuff that wouldn't make an attractive project."[31]

Numerous other experimenters constructed wind-electric generators of their own design, with varying degrees of sophistication and output. Newly minted publications in the early 1970s, such as *The Mother Earth News* and *Alternative Sources of Energy*, as well as longtime magazines like *Popular Science*, *Popular Mechanics*, and *Mechanix Illustrated*, heightened the promise and excitement surrounding the prospects of producing one's own electricity through their numerous published articles about how to construct inexpensive homebuilt wind-electric machines.[32] In a May 1974 article, Ken Smith, who worked in the School of Architecture and Planning at the University of Texas in Austin, explained first-hand how he constructed a 750-watt wind-electric plant, called the Windcycle, using a cloth sail-based rotor and gears from a 10-speed bike to drive an automotive alternator. "Our philosophy is designing funk-tech machines," which can be put together simply with discarded materials, Smith wrote.[33] Another homebuilt wind generator enthusiast, George Helmholz of Covelo, California, outlined in an article how he constructed an "easy to build," three-bladed wind-electric plant with a 4-kilowatt output in wind speeds reaching 23 mph for about $450.[34] In 1974, Michael Hackleman, who founded the organization Earthmind in Mariposa, California, published a popular book on how to build a wind-electric plant using a vertical-axis, triple-stacked "S-rotor" design, which traced its roots to research conducted by Finnish engineer Sigurd J. Savonius in the mid-1920s.

Earthmind's first rotor consisted of split 55-gallon drums, which were later replaced with half-cylinders made of smooth aluminum sheeting. The Earthmind wind wheel, which was mounted within a tower frame on top of a building roof, drove a gear train and alternator. The cost of constructing this outfit, as well as its output, varied widely. A man near Twenty-Nine Palms, California, in the mid-1970s claimed to build a machine based on

Earthmind's design that with batteries and wiring cost him $1200, with the expectation of generating 3000 watts in a strong wind.[35] A *Popular Science* editor, who wrote about the state of the reemerging use of wind-generated electricity in 1974, commented, "I couldn't help but marvel at the number of novel wind machines spawned by inventors: The devices range from an oil drum split lengthwise and welded to a center axle to complex assemblies of gears, chains, cogs, and flaps to 'swim' in the wind."[36] In the spirit of those seeking to generate their own clean, nonutility electric power, Don Marier, publisher of the *Alternate Sources of Energy*, supported the momentum behind homebuilt wind-electric plants. "My feeling is that by doing the work ourselves; organizing small design and production groups—then we can do the things we think should be done," he wrote in early 1974.[37]

There was another option, however, to either purchasing a high-output, but expensive new wind-electric plant or building a low-output, cheaply constructed machine, and that was to find and resurrect an earlier manufactured wind-electric plant. As more people began to explore the possibilities of wind-electric power, they encountered articles and trade literature about these earlier wind-electric plants and learned about how prolific the technology once was across the US Southwest and Midwest before power lines reached most rural locations by the mid-1950s. In the early 1970s, a group of mostly young men in their twenties set out in pickup trucks to explore the backroads of rural America in search of old wind-electric plants still left on top of their towers. They even knocked on farmers' doors to inquire if they had these machines sitting in their barns or sheds. Many farmers at first were perplexed why anyone would want to offer them cash to remove something that appeared to be useless junk.[38] "Some farmers who bought wind electric systems prior to the advent of the Rural Electrification Act are reselling their old machines for prices varying from scrap iron rates to the original price," wrote Jack Park, a wind-electric power expert, in a report for the US Department of Energy in the 1970s.[39] Craig Toepfer, a mechanical engineer who got his start in the small wind-electric business during the early 1970s, found most

Craig Toepfer returns to Michigan in 1975 with a pickup truck load of pre-1950 wind-electric machines which he gathered from farms in South Dakota. Courtesy Craig Toepfer, Chelsea, MI.

farmers to be agreeable on prices, often letting the units go for $100 to $200 each. He recalled how on a number of occasions he was invited to join the farmer's family for supper. "They were genuinely interested in what you were doing," he said.[40]

Taking down a 300- to 450-pound wind-electric plant from a 35- to 45-foot tower after thirty or more years of inactivity without damaging it or harming oneself in the process was no easy feat. An inactive generator often became stuck or damaged as it weathered on top of a tower over the years. Many individuals took down their first wind-electric plants with little to no experience on how to do it correctly, and the results were often disastrous. Some attempted to lower towers with their top-heavy wind-electric plants to the ground with block and tackle, only to helplessly watch as they went crashing to the ground, destroying the machine. They soon developed gin poles or small crane-like devices that clamped to the towers and extended just above the generators in order to unbolt them from the tower, lift them with a winch, and then slowly lower them to the ground. James DeKorne described the experience of taking down a wind-electric plant from its tower as "roughly analogous to removing an engine from an automobile at an altitude of 45 feet!"[41] The tower, once relieved of the generator's weight, could then be taken down safely.

Before leaving the site, it was advised to ask the farmer if he had any other related equipment for the wind plant, such as control panels and wiring. The glass batteries, if they weren't already scrapped, usually deteriorated after years of neglect. New batteries could still be purchased from manufacturers such as Gould, Exide, and Delco.

Many types of wind-electric plants were available in the field during the early 1970s in terms of their manufacture and output, ranging from the smaller 6- and 12-volt DC plants to the larger 32- and 110-volt DC plants produced by firms such as the Wincharger Corporation, Jacobs Wind Electric Company, Parris-Dunn, Wind King, and Winpower. The Wincharger was often referred to as the "Chevrolet" of wind-electric plants due to its wide availability and modest quality, whereas the highly coveted Jacobs machines were considered the "Cadillacs" for their durability and historic reputation. Many farmers soon realized that they could command a significantly higher price for their used Jacobs wind-electric machines. "Perhaps five years ago I could have got as many 110-volt Jacobs as I wanted, but now they are scarce and the price is high," said Paul Biorn, who sought to recover a Jacobs machine in 1975 for his personal use. "The last rancher I talked with had a 110-volt Jacobs on which the tail had broken loose and had swung around into the props; the props as well as the tail were shot, but the generator looked okay. The rancher still wanted $850 for the machine. Not being a rich man, I haven't conversed with him since," he said.[42] No matter the manufacturer or whether the wind-electric plant spent decades on top of a tower or stored in a shed, all of them required some overhauling before they could be returned to service.

The rush to acquire pre-rural electrification wind-electric plants led to the development of commercial enterprises that specialized in rebuilding these machines. Some of these firms sold a variety of early wind-electric plants. One of those was Coulson Wind Electric of Polk City, Iowa, which

Bob McBroom shown with a restored Winpower wind-electric plant that he installed on his Holton, Kansas, property in the mid-1970s. Courtesy Bob McBroom, Holton, KS.

specialized in rebuilding older Wincharger, Air-Electric, Parris-Dunn, and Winpower machines. Richard Coulson, the company's owner, even powered the 32-volt DC equipment in his workshop with a 2.5-kilowatt Winpower on a 65-foot tower and a 1.2-kilowatt Wincharger on an 80-foot tower.[43] Coulson was considered a meticulous rebuilder of Winchargers. "The armature will be growler tested, commutator turned, new bearings and seal, field coil checked for shorts, grounds, and proper resistance, as an overall condition indicator. The gear case will be complete, gears checked for soundness, shaft for trueness, and new bearings and brushes installed. The control panels will be tested for proper opening and closing voltages, and will be in good working order. The Wincharger unit will be painted original color and tail lettered to correspond with each model," Coulson's trade literature stated.[44] Coulson's father had been a Wincharger and Jacobs dealer in the 1930s and 1940s, and he reportedly had the original wood-working machines for carving Wincharger blades.[45] Likewise, Toepfer started Windependence Electric in Ann Arbor, Michigan, in the mid-1970s to mostly rebuild Winchargers, Jacobs, and Winpower plants. To maintain an inventory of these machines, he made contact through a Canadian farm magazine advertisement with Albert Stahl, a retired farmer in Medicine Hat, who offered to buy up the wind-electric plants in his area for Toepfer, including taking down the towers. In addition, Toepfer spent two weeks a summer in the Canadian provinces of Saskatchewan and Manitoba taking down these machines himself. "I would camp out and typically take down one or two machines each day," he said.[46] The founders of North Wind Power Company in Warren, Vermont—Donald Mayer and David Sellers—also started their business by selling reconditioned Jacobs wind plants. They collected these machines from around the country, bringing them back to their workshop for disassembly, cleaning, and reconditioning of the armature and field coils, as well as replacing all bearings, brushes, and bushings. They also fitted each machine with a new blade-actuated governor and a new set of "matched and balanced" blades made of Sitka spruce. While the company maintained an

Craig Toepfer, shown working on a Jacobs wind-electric generator, operated Windependence Electric at Ann Arbor, Michigan, from October 1975 to February 1982, during which he acquired and sold about eighty-five pre-1950 wind chargers. Courtesy Craig Toepfer, Chelsea, MI.

inventory of original Jacobs wind-plant parts, it occasionally manufactured new components that were either in short supply or had run out. North Wind sold its 2-kilowatt, 32-volt, and 110-volt Jacobs plants for $3000 to $3500 each, respectively, while its 3-kilowatt, 32-volt Jacobs plants retailed for $4000 and the 3-kilowatt, 110-volt machines sold for $4600 apiece.[47] Rebuilders enjoyed a brisk business in the mid-1970s. "An old Jacobs may have been purchased for $900, sold 25 years later for between $100 and $1000, repaired and restored to its original condition, and resold for $2000 to $3000. History shows that wind turbines, if properly maintained, can sell for their original cost plus an addition for inflation."[48]

One of the most prolific rebuilders, who gained sage-like status among young wind-electric power enthusiasts during the mid-1970s, was Martin Jopp of Princeton, Minnesota. By 1975, the nearly seventy-year-old Jopp, who was often photographed

with disheveled gray hair and wearing bib coveralls with long-sleeved plaid shirts, had already spent a lifetime involved with wind-electric power. As the story goes, he first became interested in wind-electric power in 1917 while living on his family's farm, and from there he began wiring many homes and farm buildings in the Princeton area for 6-, 12-, and 32-volt DC power as the Jopp Manufacturing Company. He also built and sold his own line of 12-volt and 32-volt wind plants during the 1930s and 1940s. During the 1960s, he rebuilt and sold Jacobs wind-electric plants, having accumulated many of these machines and related parts from farmers that had abandoned them after rural electrification took hold.[49] Jopp was known to be generous with his understanding of wind-electric plants, often allowing individuals to use his workshop to repair their own machines. "Knowledge is the one thing you can give away and still keep it for yourself," he was known to quip.[50] Don Marier, publisher of *Alternative Sources of Energy*, asked Jopp to write a wind-electric power column based on mailed-in questions from the readers.[51] In recognition of his contributions to development of alternative energy, Governor Rudy Perpich declared August 27, 1977, Martin Jopp Day in the state of Minnesota.[52] However, not everyone held Jopp in similar regard, citing misrepresentations and misleading information related to his machines and those of the Jacobs Wind Electric Company. Frustrated by what he had read about Jopp in the Spring 1977 issue of the *Wind Power Digest*, Marcellus Jacobs, president of the Jacobs Wind Electric Company, wrote a letter to the editor pointing out factual errors and embellishments in the article. "What engineering improvements or new patented designs did Mr. Jopp ever produce to improve wind electric plant performance?" Jacobs asked.[53] Yet, the *Alternative Sources of Energy* column, "Martin Answers," remained one of the magazine's "most popular and well-read" features; letters from around the country and even the world poured into Jopp's mailbox until his death on July 1, 1980, at the age of seventy-four, from the effects of Parkinson's disease.[54]

Utilizing the wind to fulfill the average homeowner's electricity requirements—either in the United States or Canada—during the 1970s was nearly impossible to achieve. At the time, many American homes were equipped with electric current-hungry kitchen stoves and refrigerators, water heaters, washing machines and clothes driers, and televisions and stereos. One observer wrote at the time that "Many Americans had no idea how many kilowatts their families routinely use until the utility bills began to go sky high."[55] The US Energy Information Administration estimated that in 1975 Americans paid on average 10.4 cents per kilowatt-hour for residential electricity, up from 9 cents per kilowatt-hour in 1970.[56] An all-electric home could easily consume more than 20,000 kilowatt-hours a year.[57] Even the largest commercially available wind-electric plants for homeowners in the mid-1970s would at best supply only a portion of this level of electricity consumption. "Most of these machines are still too small to meet the monthly electrical needs of a typical rural household or to be a significant replacement for an energy consum-

ing farm application," wrote Robert Meroney, a professor of civil engineering at Colorado State University, in 1976.

> For example, a household utilizing some 1000 kW-hrs/month, in an area with an annual average wind velocity of 15 m.p.h. assuming a 70% windmill generation efficiency and a 30% load factor, would require a 30 ft diameter mill. This mill[,] if its rated speed was 25 m.p.h.[,] would require a 20 kW generator! Since typical systems appear to cost over $1,000/kW, one would require a capital investment of some $20,000! If annual wind is only 10 m.p.h., this increases by a factor of three.[58]

There were those individuals, however, who did manage to sever their connection to utilities with

Woodward, Iowa farmer John Lorenzen became a hero among young wind-electric power enthusiasts during the early 1970s for having relied on three 2.5-kilowatt Jacobs wind plants for forty years to supply his household electricity. Courtesy Bob McBroom, Holton, KS.

wind-electric power. John Lorenzen, a Woodward, Iowa, farmer, began using wind machines to produce his own electricity in the mid-1920s. In 1940, he installed three 2.5-kilowatt, 32-volt DC Jacobs wind-electric plants on his property, and that was all he ever needed to operate his household appliances and workshop equipment for the next thirty-five years. He used the excess electricity in his batteries to hydrolyze water into hydrogen gas, which he stored in a tank. Whenever there were periods of no wind, Lorenzen used the hydrogen gas to power an engine that recharged his batteries in about ten minutes.[59] Lorenzen not only had years of experience operating wind-electric plants, but gradually he integrated the appropriate DC appliances into his household over a period of decades. For newcomers, the process of switching over a household to wind-electric power was often daunting. In 1976, after reading *The Mother Earth News Handbook of Homemade Power*, William Randolph of Xenia, Ohio, decided to operate his home on electricity generated by the wind. Since he could not afford a newly manufactured wind turbine, he crisscrossed the western Ohio countryside in search of a vintage machine. Through numerous conversations, he found a farmer who was willing to sell him a 1937-built, 32-volt DC, 650-watt Wincharger "Little Giant." The machine had not operated in many years and required total rebuild. In addition, Randolph rewired his home with heavy No. 10 copper wire in order to use the 32-volt DC power produced by the wind plant for lighting. He used power inverters to provide 110 volts AC for those household appliances that required it. The wind turbine was placed into service on November 3, 1976, and Randolph happily wrote in an article for *Wind Power Digest* the following spring that "Life is good and we are enjoying our measure of independence from the local utility."[60]

Jim and Jan Erdman, who were in their early twenties, spent $13,000 in 1977 for 120 acres just south of Tomah in northeast Wisconsin to set up a home for themselves and their three young sons. The closest electric power was a utility pole about a half-mile away. When the Erdmans inquired with the Dairyland Power Cooperative about running the power to their property, they were quoted a

Jan Erdman and her two young sons enjoy a summer day in 1979. The 2.5-kilowatt Jacobs wind-electric generator supplied 32-volt DC power to the family's off-grid home in Tomah, Wisconsin. Courtesy Jim and Jan Erdman, Menomonie, WI.

price of $3000. Another turnoff for the Erdmans was the fact that the cooperative received its power from a nuclear power plant.[61] However, the family understood that it still needed some form of electric power for the household. Jim Erdman came across an advertisement in a beekeeping magazine about a 32-volt DC, 2500-watt Winpower for sale nearby. He also purchased a 60-foot tower on which to erect the machine at a local estate auction. The next challenge for the Erdmans to overcome was how to store and use the 32-volt DC effectively. Jim Erdman learned that the city of Tomah was discarding its former telephone system batteries, and he was able to purchase a sufficient number of them at scrap value. He also wired the house for 32-volt DC and then began searching for appliances that would operate on this type of power.[62] He discovered that a number of the General Electric distributors in the mid-1970s still held inventories of 32-volt light bulbs. A 50-watt GE light bulb at the time could be purchased for 75 cents each as part of a minimum order of $10, or for 68 cents apiece if bought in a case of 120 bulbs.[63] The Erdmans also came in contact with Richard Coulson of Coulson Wind Electric, from whom they purchased an array of 32-volt DC compatible appliances, such as a vacuum cleaner, kitchen mixers, toaster, sewing machine and fan, as well as various workshop tools. For keeping food cold, they used a gas refrigerator, and water was pumped from a well by an electric pump jack and stored in a large water tank.[64] However, within a year of operation, the Winpower had developed a short in the armature of the generator, and the Erdmans purchased another rebuilt wind-electric plant—this time a 32-volt DC, 2500-watt Jacobs.[65] With the exception of heating water for bathing and washing clothes on a combination wood and gas stove, the young family became used to living with wind power. "We didn't have any problems. The kids turned the lights off when not in use. If there was no wind for a week, there were no waffles. When the kids heard the wind and

the batteries were bubbling, they knew they could have waffles," Jim Erdman said. (The 32-volt DC waffle iron was a Christmas gift to the family from Richard Coulson, but it had to be used sparingly since it required significant current to operate efficiently.)[66] Jan Erdman recalled that "our kids thought it was normal," and "when you walked into the house it appeared typical," but she noted that it took several years for the family to learn how to efficiently use the wind-electric plant.[67] When Jim Erdman changed jobs in 1989, the family relocated to Menomonie, Wisconsin, but continued to use a Jacobs wind-electric plant for their household electricity.[68]

In 1972, Elliott Bayly of Oak Creek, Colorado, used a 2.5-kilowatt, 32-volt Jacobs wind-electric plant from the 1930s to power the equipment of his radio station, KFMU 103.9 FM. He decided to use wind power after determining that a traditional utility transmission line to the remote site would cost him $15,000 compared to about $2000 for a Jacobs to charge batteries. Bayly, who earned an electrical engineering degree from the Massachusetts Institute of Technology, a master's degree from Stanford University, and his doctorate from the University of Minnesota by the late 1960s, converted his 110-volt AC radio equipment to use 32-volt DC power. The station broadcast music from 7 a.m. to 1 a.m., and was called "The Sound of the Wind."[69] However, Bayly had aspirations of manufacturing wind turbines of his own design and in the late 1970s became the founder of Denver, Colorado-based Whirlwind Power Company and later Wind Power Technologies in Duluth, Minnesota.[70] In 1978, Paul Shulins, a student at Plymouth State College in Plymouth, New Hampshire, unwittingly gained national media attention for erecting a 200-watt, 12-volt Wincharger on the roof of a campus building as part of a science project to provide supplemental battery power to the school's radio station, WPCR. The college station used the jingle, "We are on wind power. This hour is 100 percent wind propelled."[71] Plymouth State College stated at the time, "During the New Hampshire winter, WPCR has been operating about 3–8 hours per day on wind power. It is the only station in Plymouth capable of operating during

Jim Erdman acquired these used Exide batteries from the local telephone company in 1979 to store electricity generated by the family's Jacobs wind-electric generator. Courtesy Jim and Jan Erdman, Menomonie, WI.

Members of the Energy Task Force (left to right: Ted Finch, Travis Price, David Norris, and an unidentified visiting student) make plans to erect a wind-electric generator on top of an old tenement building at 519 East 11th Street in New York City during 1975. Courtesy Travis L. Price III, Washington, DC.

a general power failure."[72] The Intercollegiate Broadcasting System, Inc. of Vales Gate, New York, praised the young station manager for his clever use of wind-electric power and the attention that it brought to beleaguered collegiate radio stations. "With all the negative images being thrust upon 10-watt noncommercial educational stations by those calling them 'electronic sandboxes,' mention of a positive project of this kind helps call attention to the innovation and creativity that exists at many of these stations," the association wrote in a January 12, 1979, letter to Shulins, who graduated from Plymouth State College in the spring of 1979 and spent his career in the broadcasting industry.[73]

While purchasing utility-produced electric power was expensive in rural locations during the 1970s, it was an even greater financial burden in urban settings, especially for the inner-city poor. In New York City, for example, the cost of electricity had risen to 10 cents per kilowatt hour by 1975, which few individuals in poor neighborhoods could afford. Landlords reputedly responded by

shutting off electric power and other utilities to drive off tenants, and then they burned the buildings to collect the insurance money. In early 1975, a young solar architect, Travis Price, arrived in New York City from New Mexico on a Federal Energy Administration grant to study solar power as a means of improving living conditions for the inner-city's poor. He met Roberto "Rabbit" Nazario, a community activist from Puerto Rico, during a "Sweat Equity" meeting in New York's Lower East Side. The Sweat Equity program was designed to give tenants an opportunity to take over the city's abandoned buildings if they worked together free of charge to improve them, a process also referred to as "urban homesteading." Price and Nazario quickly formed the Energy Task Force in February 1975 to focus on rehabilitating a five-story building at 519 East 11th Street. Other architectural and engineering students and graduates joined the effort, among them Henry Deerborn and David Norris of Yale University, Chip Tabor from MIT, along with Michael Freedberg of the Eleventh Street Move-

This Jacobs wind-electric plant installed on the rooftop at 519 East 11th Street in New York City in 1975 led to congressional passage of the 1978 Public Utility Regulatory Policies Act (PURPA). Courtesy Stephen James Govier, Suffolk, England.

ment, and the building tenants. Price advised that the task force's first objective was to insulate the building, which included improving windows and doorways with caulking and weather stripping. In the summer of 1976, the task force took the process a step further by designing a rooftop wind-electric and water-heating solar system for the building.[74] For the wind-electric portion of the project, the task force observed that the Lower East Side was a plain of rooftops of mostly five- and six-story buildings, across which the wind blew relatively unobstructed. In addition, a two-month analysis found an average annual wind speed in the Lower

East Side of close to 9 mph. With these calculations in hand, the task force then applied for and received a $14,122 government grant to purchase, erect, and operate a wind-electric generator. It was decided to purchase a refurbished 2-kilowatt Jacobs wind-electric plant from the Midwest to be placed on top of a new 37-foot Rohn tower. Instead of charging batteries, the task force took the step of connecting directly to the Consolidated Edison utility by using a synchronous inverter that converted the wind-electric plant's DC power into AC power compatible with the grid. The benefit of this tie-in with the utility was that when the wind-electric plant produced more power than the building consumed, it pushed it into the grid. When the wind died down, however, the inverter allowed the utility power to flow back into the building.[75]

The wind-electric plant at 519 East 11th Street was erected in November 1976 onto concrete anchor blocks by task force members and building tenants without the assistance of a mechanical lift or crane. Freedberg said it was accomplished "with a lot of luck, sweat and muscle power."[76] The event immediately attracted widespread media attention and was widely praised by city, as well as national, politicians and celebrities.[77] Con Ed filed a lawsuit to try to stop the wind-electric plant's grid connection, citing concerns over compatibility between the two forms of electric power generation. With free legal representation from former US Attorney General Ramsey Clark, the New York State Public Service Commission allowed the connection to continue, forcing Con Ed in January 1977 to draft new guidelines allowing individuals to use wind energy and interconnect with the utility's lines. This action led to the congressional passage of the Public Utility Regulatory Policies Act (PURPA) in 1978, which granted wind-electric plant owners across the country the right to interconnect their systems directly with the utilities.[78] The task force estimated that through the wind-electric plant, solar water heaters, and insulation, the building cut its utility bill by about $4000 in the first year. The Jacobs machine alone supplied electricity for about one-third of the building's lighting for the hallways and common areas, saving the tenants about $200 a year.[79] However, the spread of wind-electric power across New York City was ultimately stifled through the late 1970s by a tariff permitting Con Ed to assess a system's owner a monthly bill based on the machine's installed capacity—in this case $6.28 per kilowatt—rather than on the customer's peak-connected demand load. "The questionable rationale here is that the utility must maintain standby capacity for the windmill customer, while the windmill owner is given no credit for his/her potential standby capacity for the utility," wrote Mary Christianson of the Energy Task Force in 1979.[80] In addition, the tenants at 519 East 11th Street eventually lost interest in maintaining the wind-electric plant, which had suffered weather damage over the years, and was taken down in the mid-1980s.[81]

While wind-electric plants continued to attract public interest during the 1970s, it was still a difficult technology to integrate into an American lifestyle that was deeply dependent on utility-generated power. "There is this sort of glamorized image of wind power that people carry which is good in a way to promote the thing, but there's often too much glitter attached to the image that doesn't have much to do with the practical aspects," Jim Sencenbaugh told the *Wind Power Digest* in 1976, adding that "It's no plug-in trip or Sears and Roebuck idea where you go down to the store and buy one out of a cardboard box and this thing erects itself, disconnects [from the electric utility,] and runs your whole house. There's a lot of careful planning involved[,] so the decisions are rational."[82] When he explained to prospective customers about what a small wind-electric plant could realistically power in the average household—"usually the lights, small appliances[,] maybe a microwave, a tv or stereo"—about 95 percent of them decided not to pursue this alternative energy source.[83] David Simms, whose family successfully used a 32-volt DC Jacobs machine to electrify a log home in rural Quebec during the mid-1970s, believed that going forward it was more important to design a house around the wind-electric machine rather than the other way around, which was the common practice of the day. "You have to be conservation-minded to get by in a low wind area like this, but you can do it," he said to Canadian publication *Harrowsmith*. "When there is no wind for a day or two, you start being really careful. You don't waste electricity. Wind power puts you more in touch with the house and what it takes to run it."[84]

4 Collective Energy

At the start of the 1970s, a small group of enterprising businessmen and engineers—many of whom were freshly graduated from universities—began exploring the development of clean wind energy conversion systems as a means of countering the nation's energy crisis. These individuals had the goal of establishing commercial production of wind turbines that they could sell to homeowners, farmers, and small business owners to reduce their dependence on increasingly expensive utility-produced electricity. It was not long before these wind-electric pioneers were creating informal relationships; sharing ideas and experiences with each other through written correspondence, telephone calls, and face-to-face meetings; and even collaborating on the development of wind turbines. Michigan wind energy enthusiast and businessman, Allan O'Shea, proposed a meeting in early 1974 of the fledgling industry's members to exchange ideas and share their experiences. In April that year, fifteen industry representatives converged on the Florida International University campus in Miami. "After exchanges, they agreed to meet again and to work together [in the form of a trade association tentatively known as the American Wind Energy Association] to promote and further use of the wind," wrote Craig Toepfer, another wind energy proponent, at the time.[1] The next meeting was hosted by O'Shea's Environmental Energies, Inc. in Detroit on September 20–21, 1974, two days before the weeklong World Energy Conference being held in the city.[2] Environmental Energies also issued a public statement regarding America's increasing interest in small wind-electric power by installing a Dunlite wind plant on top of a billboard sign in

downtown Detroit that would welcome the thousands of attendees expected to attend the World Energy Conference.[3] In the meeting invitation, O'Shea outlined the varying reasons for establishing an association of wind energy. "The parties involved need a 'central office' for channeling and funneling information and for receiving feedback on other companies and their projects. Secondly, we need to

In 1974 Allan O'Shea, who owned and operated Environmental Energies, Inc., in downtown Detroit, hosted a gathering of wind energy enthusiasts that led to the founding of the American Wind Energy Association. Courtesy Stephen James Govier, Suffolk, England.

stabilize and show our credibility to the public. An effort backed by an association always appears more credible. Third, the public needs an organization it can turn to for advice and knowledge on aerogenerating systems," he said.[4] An unexpected crowd of about fifty wind energy representatives crammed into a small basement meeting room of a Detroit police station across from Environmental Energies' store at 6 Mile Road and Grand River Avenue.[5]

During the two-day wind power meeting in Detroit, the attendees formally agreed to establish the American Wind Energy Association (AWEA), a nonprofit Michigan corporation, to "promote and further the use of the wind."[6] AWEA's goals included sharing technical and marketing information with members and the public, interacting with government agencies and legislative bodies involved in wind energy, conducting and participating in meetings and conferences, and establishing industry standards.[7] The newly formed organization elected O'Shea as its president and Solomon Kagin of Real Gas and Electric Company as vice president, while Susan Lee Gerhard of West Wind was chosen secretary and Nancy Horning of Environmental Energies was named treasurer. The AWEA board of directors included O'Shea, Bill Delp of Independent Power Developers, Chuck Pipher of Delatron Systems Corporation, and Charles Leverich of Creative Electronics. Other founding members of AWEA included John Sayler of American Energy Alternatives; Mike Mooney of Dyna Technology; Tim Horning of Environmental Energies; Jerry Cook of J&R Nursery; Derrick Partridge of Conservation Tools and Technology; Bob Edgerton and Kevin Moran of Edmund Scientific Company; Walter Schoenball of the German Wind Energy Association; Ray Twyman of Industrial Components; Marcellus Jacobs of the Jacobs Wind Electric Company; the law firm of Karr, Robert & Duane; Don Mayer of North Wind; Charles Wright of Rohn Tower Company; Geoffrey Gerhard of West Wind; Jim Kelly of Wind Machine & Pump; Chuck Tellas of Milan Screw Products; R. T. Mijanovich of Soleq; Cy Minella of 21st Century Living; William Ashe of the Wind Energy Society of America; Mike Evans of *Wind Power Digest*; Bill Barnes of Windlite Alaska; Afred Wurdelman; Roy Brewer of Win-

Environmental Energies, Inc., made a public statement in support of wind energy during the 1974 World Energy Conference in Detroit by installing a Dunlite wind turbine on top of a billboard that welcomed conference attendees. Courtesy Stephen James Govier, Suffolk, England.

power; and Bill Gillette and Al Lishness of Zephyr Wind Dynamo Company.[8]

Interest in AWEA spread rapidly, and when it held its second meeting at the Winco Division of Dyna Technology in Sioux City, Iowa, on April 11–12, 1975, as many as seventy-five people were in attendance and paid membership reached more than thirty-five. One of those attendees was Woodward, Iowa, farmer John Lorenzen, who delighted the audience with his story about how he had used three 2500-watt, 32-volt Jacobs Wind Electric Company plants as his only source of power for nearly fifty years.[9] It was also at this meeting that the association's bylaws were drafted by the executive committee and ratified by the voting members present. In addition to various presentations from government and industry representatives, the association agreed to hold meetings twice a year—spring and fall—in different locations throughout the country, often where a company member served as the host. In addition, various committees were established for legislative affairs, product ratings, product standards, and membership.[10] AWEA had already started publishing a quarterly newsletter to keep its members informed of association matters and upcoming conferences and events.[11] To join the organization, it cost $100 per year for those members with voting rights, to as little as $15 for those individuals who just wanted to receive the AWEA newsletters.[12] "The American Wind Energy Association is now a small but viable organization with momentum," O'Shea declared after the Iowa meeting. "Our most difficult task during the next year is to keep the momentum. This next conference will determine whether or not we can accomplish this. Money, membership, accomplishments are all distinct problems and have to be dealt with. Only

one entity will make or break the American Wind Energy Association and that entity is people."[13]

AWEA, as an association dedicated to the development of wind energy, was not alone in the field when it started. Five months before its first meeting in Detroit, another group with a focus on wind power had already formed—the Wind Energy Society of America (WESA). WESA was set up as a nonprofit technical society, based in Pasadena, California. The society's founding members hailed mostly from academia, the scientific community, and large corporations, and their goal for facilitating wind energy's widespread application was to focus on developments in power generation, utilization, and storage; aerodynamics, mechanical, and structural engineering; and meteorology and wind damage prevention.[14] WESA also aimed to develop a comprehensive technical library of documentation and video and audio recordings to be shared with universities, as well as to publish a wind energy engineering textbook for use in colleges and universities.[15] The society started its membership drive with a small exhibit booth at the Alternate Energy Sources Conference held at the University of California in San Diego in May 1974. This effort was followed by another exhibit booth at the San Diego Ecology Conference at the Mission Valley Shopping Center in April 1975, and a month later WESA presented a technical session on wind energy systems at the Solar Energy for Earth Conference at the University of Southern California. In June 1975 WESA garnered media attention at the Inventors Workshop International in Los Angeles, which further propelled interest in the group. WESA's membership stood at 125 by August 1975, many of whom were also members of AWEA.[16] By 1976, WESA counted 300 members, with 84 holding doctoral degrees. The association, however, failed to meet its goals due to lack of funds and was soon overtaken by the increasing stature of AWEA.[17]

Within the first full year of its existence, AWEA continued to face operational challenges and increased expectations from its growing volunteer membership. The association's third meeting was hosted by Real Gas and Electric in Santa Rosa, California, in October 1975, with more than 150

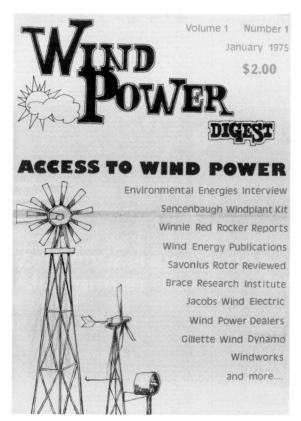

The *Wind Power Digest*, which started publication in 1974, became the voice for the emerging wind-electric energy industry in the United States. From the author's collection.

Attendees at the July 1976 American Wind Energy Conference at Madison, Wisconsin. Front row left to right: Hans Meyer (with infant son), Windworks; Richard Katzenberg, Natural Power, Inc.; Glenn Gazley, North Wind Power; Mike Evans, Wind Power Digest; William Drake, Enertech. Back row left to right: Pam Meyer, Windworks; Herman Drees, Pinson Energy; Don Mayer, North Wind Power; David Sellers, North Wind Power; Ben Wolff, American Wind Energy Association; Steve Blake, Sunflower Power; Forrest "Woody" Stoddard, University of Massachusetts; Clint "Jito" Coleman, Windworks; Jack Park, Kedco; Bob McBroom, Kansas Wind Power; and Earle Rich, Natural Power. Courtesy Earle Rich, Amherst, NH.

people in attendance. A new slate of officers was elected during the meeting. John Sayler of American Energy Alternatives became president, while Richard Katzenberg of the Natural Power Company was named vice president and Judith Woelke was picked to serve as secretary and treasurer. The central focus of this meeting was to coordinate local and state legislative activities involving wind-electric plant installations and to improve interassociation communications, as well as boost AWEA funds through increased membership. Sayler also made it his goal to begin an AWEA data center at Boulder, Colorado, with the purpose of creating a mailing list of individuals with an interest in the organization.[18]

The July 1976 AWEA meeting was hosted by tower manufacturer Unarco Rohn at Madison, Wisconsin, and brought the association into its first significant contact with officials from the year-old US Energy Research and Development Administra-

tion's Federal Wind Energy Program. This program was initially focused on the research and development of large-scale wind turbines but now wanted to encourage the proliferation of wind-electric plants of less than 100 kilowatts for farms and rural applications. The Energy Research and Development Administration (ERDA) also announced that it would begin testing commercially available small wind plants at a newly formed test site at the Rocky Flats nuclear weapons assembly plant near Boulder, Colorado, and compile and share the test results with the industry. "The overall objective of our program for the smaller systems is to stimulate manufacturing of these wind machines by the private sector and utilization of these systems by the public," George P. Tennyson, ERDA's program manager in charge of wind energy research, told the AWEA meeting attendees.[19] As Don Mayer of North Wind Power described the meeting's outcome for AWEA, "It was at this meeting in Madison that the Association began to develop a professional outlook toward the development of small wind energy conversion systems."[20]

North Wind Power hosted the October 1976 AWEA meeting in Warren, Vermont. The Wind Energy Exposition had by then generated widespread interest within the rapidly emerging industry. As many as 300 people attended the meeting. Companies that set up wind energy systems for the first time at an AWEA meeting included American Energy Alternatives, Helion Inc., Sencenbaugh Wind Electric Company, and Pinson Energy Corporation, in addition to North Wind Power. Newly minted wind industry and academic leaders, such as Hans Meyer of Windworks, Jack Park of Helion, Herman Drees of Pinson, Forrest S. "Woody" Stoddard of the University of Massachusetts, and Steve Blake of Sunflower Power, spoke on leading edge topics like wind plant aerodynamics and blade design, as well as proper siting of these machines for maximum efficiency in the wind. A number of government officials also provided updates on the emerging federal wind energy program.[21] Frank R. Eldridge, of the Mitre Corporation, discussed

a study which proposed that the federal government could develop a "wind seeding" program to stimulate wider commercialization and deployment of wind-electric systems to make them competitive with conventional electrical power generation systems.[22] The association also announced the launch of the *Wind Technology Journal*, a quarterly technical publication initially edited by Herman Drees and focused on wind energy conversion and utilization advancements.[23] The AWEA meetings, though operating on meager budgets at the time, within three years had proved to be "focal points for information dissemination and business deals" that helped many new companies get their start in the wind energy industry.[24]

It was also during the October 1976 meeting that AWEA made some difficult decisions about where it was headed as the primary association representing the up-and-coming wind-electric industry. The relationship between AWEA and the Federal Wind Energy Program strengthened with a $21,850 contract issued earlier that year to the association from the Battelle Pacific Northwest Laboratories in Richland, Washington, to participate in an eleven-

American Wind Energy Association members make presentations at the Rocky Flats Standards Workshop in Boulder, Colorado, in May 1977. From left to right: Ben Wolff, American Wind Energy Association; Mike Evans, Wind Power Digest; Richard Katzenberg, Natural Power, Inc.; Steve Blake, Sunflower Power Company; Don Mayer, North Wind Power; and Herman Drees, Pinson Energy. Courtesy Bob McBroom, Holton, KS.

month investigation into current wind-plant siting practices. Another $9730 contract was awarded to the association by Rockwell International, ERDA's contractor in charge of monitoring the performance and durability of commercially available, small wind-electric plants at the Rocky Flats test site in Colorado. Its purpose was to develop an index of manufacturers, distributors, and researchers involved in wind energy conversion systems. New leadership for AWEA was elected during the October 1976 meeting. Richard Katzenberg, formerly the association's vice president, was elevated to president, while Herman Drees, Don Mayer, and Ben Wolff of Windworks were named vice presidents. The AWEA board included these new officers, as well as John Sayler, the outgoing president.[25]

Most importantly, due to the Federal Wind Energy Program contracts and increasing enthusiasm for wind's potential within the national energy supply, AWEA needed to step up its interaction with the Carter administration to further bolster national wind research programs and more effectively lobby Capitol Hill lawmakers directly to implement legislation that fostered expansion of the wind industry. Thus, Ben Wolff immediately resigned as AWEA vice president to serve as the association's first executive director, and Mike Evans, editor and publisher of the *Wind Power Digest*, was named executive secretary to oversee membership promotion and publication of the association's newsletter, resulting in the relocation of AWEA's national office from Detroit to Bristol, Indiana.[26] Still, it was not an easy task to sharpen the direction for AWEA, considering its membership diversity, which by now included an array of wind turbine manufacturers and distributors, consumers, and researchers. Katzenberg was optimistic that AWEA could have an effective voice in Washington and even relocated his family to the city to help raise the association's profile, all the while maintaining his own company, Natural Power, in New Boston, New Hampshire.[27] "I hope this Association can take an active role in helping to enact legislation through the issuing of position papers that reflect the industry's needs and through the conducting of seminars to help those who vote on legislative packages," Katzenberg wrote in early 1977. "There is a place in this industry for

Government support and involvement. We need to help draw that fine line between (1) Governmental assistance and encouragement and (2) Governmental interference and over-involvement."[28]

A primary goal for both AWEA and the Federal Wind Energy Program was to develop common terminology, testing standards, and product standards for use by individual consumers of small wind-electric machines. As more small wind plants—many newly minted by their developers and lacking years of product testing—entered the market, there was significant concern that consumers could be in the vulnerable position of "buyer beware" and have no recourse if those machines failed to live up to their promised performance. The lack of industry-wide terminology and product standards that were backed by the federal government also deterred some banks from providing homeowners with loans to purchase small wind machines. Numerous municipalities also refused to issue permits to erect wind turbine towers in residential areas due to perceived safety concerns. "Right now, with our conflicting specifications and our nonstandard terminology we might be thought of as having installed our wind turbine on a tower of Babel, instead of on steel," George Tennyson, the federal government's head of wind energy research, told small wind turbine manufacturers attending the Rocky Flats Standard Workshop in Boulder, Colorado, on May 10, 1977.[29] Tennyson highlighted how a number of commercial wind plants recently introduced to the market experienced significant damage during high wind periods at the Rocky Flats test site. He also emphasized the importance of developing standard terms for defining aspects of wind turbine technology.[30] Tennyson explained that the Rocky Flats test site would allow for the collection of "accurate data" from individual wind turbines "using recognized techniques and instrumentation" that could then be used to compare the performance of these machines.[31] Tennyson believed that this goal to establish wind turbine standards could be achieved via the government's partnership with AWEA. "I want the American Wind Energy Association and its members, together with Farm and Rural Wind systems people in government and our prime contractor here,

Rockwell International, to be the vanguard that lays down the standards that will guide the whole industry, because I believe this group can and will provide a standard which will lead us to successful commercialization," he said.[32]

In 1977, AWEA entered contracts with Rockwell International to develop a "Consumer's Guide to Commercially Available Wind Turbine Generators" and a report outlining "the State of Standards for Wind Energy Conversion Systems" for $4760 and $5000, respectively.[33] At the start of 1978, ERDA (now renamed the Department of Energy) issued individual AWEA members contracts to design and build various small wind-electric machines for testing at Rocky Flats. The Department of Energy selected Windworks, Grumman Energy Systems,

Jay Carter Jr. of Burkburnett, Texas-based Jay Carter Enterprises, Inc., began producing a 25-kilowatt wind turbine during the late 1970s. He preferred a wind turbine manufacturing industry that operated without US government grant programs. Courtesy Bob McBroom, Holton, KS.

Alcoa, and United Technologies Research Centers to develop 8-kilowatt wind generators for farms and rural homes, while Enertech Corporation, North Wind, and Aerospace Systems Inc., in collaboration with Pinson Energy Corporation, were tasked with developing "high-reliability," low-maintenance, 1-kilowatt machines for testing. Since its inception in early 1977 through early 1978, the Rocky Flats site was already testing the performance of as many as eight different machines, ranging in output from 1 kilowatt to 15 kilowatts (see Chapter 5).[34] Many upstart wind energy companies, often operating out of small workshops on shoestring budgets, welcomed the government funds and related purchases of their machines for testing, with the belief that they would rapidly become successful commercial wind-plant suppliers. Many of these companies quickly realized, however, that accepting government money also came with a price. "We totally changed character when we had to suddenly comply with cumbersome government reporting standards and delivery timeframes, which wasn't easy for small entrepreneurial companies," Herman Drees declared.[35] While the AWEA leadership generally supported access to federal government assistance through design and research grants as good for the fledgling industry, some members disagreed and believed that the industry's success or failure should be based on free market principles. Jay Carter Jr. of Burkburnett, Texas-based Jay Carter Enterprises Inc., which developed a 25-kilowatt wind turbine, called the federal government's grant program "unfair and anti-free enterprise" and warned that "it will actually hinder the growth and development of good, practical wind and solar generators."[36]

At the March 1-4, 1978, AWEA meeting in Amarillo, Texas, a new group of officers was elected, including Herman Drees as president; Don Mayer, William Batesole of Kaman Aerospace Corporation, and Paul Vosburgh of Alcoa Allied Products as vice presidents; Steve Blake of Sunflower Power as treasurer; and Vaughn Nelson of West Texas State University as secretary. Past presidents John Sayler and Richard Katzenberg remained on the board.[37] The new board reflected the continuing change of AWEA's character from a group of small entrepre-

Wind energy industry exhibits, like this one of Pinson Energy's display table at the Vermont Energy Show in 1978, were generally simple affairs during the 1970s. Courtesy Herman Drees, Thousand Oaks, CA.

neurs to a more diversified membership, which increasingly included larger firms from the aerospace and heavy power industries. It was also during this meeting that the AWEA board approved the allocation of $6000 a month to cover the expenses of operating a permanent office in Washington, DC. Ben Wolff, the association's chief administrative officer, was put in charge of administering AWEA's activities in the nation's capital.[38] This included interfacing with the executive and legislative branches, as well as working with relevant federal agencies with an interest in wind energy development. He intended to concentrate AWEA's policy analysis and research at this office. In May 1978, Wolff found office space for the association at 1000 Connecticut Avenue, NW, in the Farragut Square neighborhood near the White House. Mike Evans of *Wind Power Digest* was also relieved from the responsibility of publishing AWEA's newsletters, as this activity was also centralized in the new Washington office. Wolff urged the members to use the AWEA headquarters' resources, as well as to visit the office during their trips to Washington.[39] Concern remained, however, as to whether AWEA had the financial wherewithal to stand up the programs that it planned to offer its members. "It's going to be hard for us to maintain the level of professionalism that we've tried to attain during this year of relatively low cash availability.

It's going to take a lot of effort by a lot of people to help us pull through this," Richard Katzenberg warned.[40]

In 1978, the wind energy industry experienced some success on the legislative front. Congress passed the Public Utility Regulatory Policies Act (PURPA), which required utilities to purchase electric power at reasonable market rates from small power producers with capacities of 100 kilowatts or less, using renewable energy, including wind turbines. Before PURPA, a utility could deny a homeowner the right to connect to the grid or could threaten to disconnect the homeowner's regular electric service. After PURPA, homeowners with small wind turbines were guaranteed the right to connect to their utility without discrimination. If the wind turbine produced more power than what was being used at the moment, the homeowner was allowed to sell the excess to the utility and could actually watch the power meter disk spin backward. While the electric power was being pushed back into the grid, the utility was not obligated before PURPA to purchase that excess power from the wind-plant owner. PURPA also spared small turbine operators from compliance with the Federal Power Act, which governed the operations of public utilities.[41] Paul Gipe, who became an early authority on small wind energy systems and a proponent of

Tom Sadler (left) of Pinson Energy and Rick Katzenberg (right), an American Wind Energy Association founder, shown in front of a Pinson Energy Cycloturbine on the grounds in front of the US Capitol Building during Earth Day in 1981. Courtesy Herman Drees, Thousand Oaks, CA.

beneficial government regulations for wind power, called PURPA "downright revolutionary" because it encouraged "decentralized energy investment" and altered their "view of energy conservation from one of conserving to save money to conserving to make money."[42] He also believed PURPA would financially motivate homeowners to invest in wind turbines with larger outputs than required to meet their household electric consumption in order to take advantage of power buybacks from the power companies.[43] The Federal Energy Regulatory Commission, which was responsible for developing the federal regulations for PURPA's implementation, forecast that the integration of small wind-electric plants into the public utility system had the potential to save 200,000 barrels of oil per day by 1995.[44] The state utility commissions oversaw the

implementation of PURPA by individual utilities. However, it took four years of overcoming legal wrangling and resistance from the utilities to bring PURPA into full implementation, so that small wind-plant owners across the country could benefit from the legislation.[45] Another piece of legislation important in promoting small wind-plant ownership was the 1978 Energy Tax Act. This legislation, which was one of five component pieces of the 1978 National Energy Act, amended the Internal Revenue Code to provide a 10 percent investment tax credit for wind, solar, and other renewables. The new credit was intended to be combined with an existing 10 percent business investment credit. Two years later, the renewable energy credit was extended to the end of 1985 and expanded to 15 percent through the Crude Oil Windfall Profits Act of 1980. For small wind turbine proponents at the time, the $10,000 limit would prove too little to stimulate the growth of the residential wind turbine market.[46]

While generally supportive of federal government wind energy development programs, AWEA leaders remained concerned by the splintered nature of the research undertaken by the different agencies, as well as frustrated by the insufficient political support to facilitate the near-term and long-term large-scale payoff for wind power. Richard Katzenberg, former association president and owner of Natural Power Inc., in the summer of 1979, wrote:

> My frustration is that the wind industry seems to be "lost in a maze" at the Department of Energy. It is not clear to me what criteria are used to evaluate various energy technologies. If we understood the criteria perhaps we could better understand the policy decisions being made. When I view the level of commitment to other energy technologies, and consider the very real technical and environmental hurdles that must be overcome, I'm completely bewildered.[47]

The association also believed that the Department of Energy should pursue an aggressive program that promised a contribution of seven quads of wind-generated power to the nation's total energy output by the year 2000, resulting in a savings of $230 billion in conventional power plant equipment and fuel expense.[48]

By 1980, AWEA's Washington office consisted of Executive Director Ben Wolff and two administrative staff members. A fourth employee, Thomas O. Gray, joined AWEA that year and was immediately tasked with overseeing the association's wind turbine standards program, which consisted of several subcommittees. Prior to joining the association, Gray, who hailed from Michigan, worked as a legislative assistant on Capitol Hill from 1974 to 1980, during which he was introduced to various renewable energy legislative acts.[49] As the AWEA's standards development manager, Gray became acquainted with the association's myriad wind turbine manufacturers. The Performance Subcommittee (formerly the Rating, Testing, Data Reduction, and Reporting Formats Subcommittee), which Michael Bergey of Bergey Windpower Company headed, worked with the Department of Energy at Rocky Flats to develop a set of voluntary industry standards for rating, safety, and reliability of small commercial wind machines in order to secure consumer, regulator, financial, and insurer confidence in the application of this technology.[50] Other AWEA subcommittees focused on small wind industry terminology standards and guidelines for small wind turbine–utility interconnections.[51] Within a year of joining AWEA, however, Gray's responsibilities compounded when Wolff resigned and he was asked to take over the financially troubled operation as executive director. Gray estimated that while the membership dues brought in about $25,000 a year, AWEA was about $100,000 in debt. Even the Internal Revenue Service threatened to close the operation due to unpaid taxes. In addition, AWEA's two employees left when their salaries could not be guaranteed for at least a month or possibly longer. Gray managed AWEA singlehandedly for at least two months without a salary and brought in his wife, Linda, to work for free as the association's bookkeeper.[52]

These 50-kilowatt US Windpower turbines made up part of California's first wind farm at Altamont Pass, east of Livermore, in 1981. Courtesy Herman Drees, Thousand Oaks, CA.

In addition to Gray's budget-tightening initiatives, AWEA's return to financial health was largely attributed to the changing direction within the wind energy industry at the start of the 1980s, specifically the rapid development of wind farms, or concentrations of multiple wind turbines in a single location that collectively fed electric power into the utility grid. These arrangements were made possible, as well as economically feasible, with the help of the available federal tax credits, state incentives, and PURPA. In general, wind farm developers at the time were groups of investors who arranged the finances to purchase the wind turbines from the manufacturers and secured the land leases on which to erect them. The electric power was then sold to the utility at a specific per-kilowatt rate over a fixed contract period.[53] "At present, this arrangement meets everyone's needs," Gray explained at an energy conference in June 1982. "The manufacturer obtains a market for his machines. The utility gets a potential power source without undertaking the risk of large capital investment in a new technology. And the windfarm developer, with some careful math work, can put together the tax credits on the equipment, depreciation allowances, and revenue derived from selling the electricity in a package which provides a competitive return on his investment."[54] While AWEA's membership had generally consisted of about forty small firms and a handful of large corporations, such as Boeing, General Electric, and Grumman Energy Systems, at the start of the 1980s, the wind farm operators coming on line, particularly in California and Hawaii, further diversified the association's membership with more financially affluent companies. These companies, which included the likes of US Windpower, Fayette Manufacturing Corporation, and WindMaster Corporation, looked to the AWEA to represent their interests on legislative and policy matters in Washington, DC. In addition to cutting operational expenses, AWEA's new membership helped pay off its debts by the end of 1983.[55]

The change in the membership composition became most apparent at AWEA's 1982 Wind Energy Expo in Amarillo, Texas. First, the conference hotel's available rooms and exposition space sold out weeks before the event. At 275 attendees, it

was the highest attended Wind Energy Expo, so far. Amarillo was picked as the event site for a second time because of its proximity to the US Department of Agriculture's wind energy test center at nearby Bushland. While representatives from the small wind turbine manufacturing sector were present, an even larger attendance was witnessed from the large power generation firms and public utilities that purchased electric power from the nation's first wind farms. Representatives from the Danish wind turbine manufacturers, which started exporting their machines to California's wind farm developers, and Oscar Holst Jensen of the Denmark's Risø National Laboratory were also present. The growing influence of larger companies in AWEA was also expressed through the association's board of directors, which by then included representatives from General Electric, US Windpower, Boeing Engineering, and Wisconsin Power and Light. Ted Anderson of Westinghouse Electric was reelected AWEA president, a position that he would hold from 1981 to 1984. Most of the producers of small wind turbines and related equipment had also switched their emphasis from standalone to grid-tied applications, and some were building 20 kilowatt and larger machines with the goal to participate in wind farm development.[56] Mike Evans, editor of *Wind Power Digest*, lamented after AWEA's Amarillo conference that "[i]t was obvious to this writer that the main thrust has moved towards utility interface machines and associated equipment. The standalone 'Survivalist' was mentioned little. Battery people were conspicuous by limited mention throughout all of the seminars and no representation at the exhibits. It appeared DC equipment was most definitely not the thrust of the conference."[57]

Despite the enthusiasm and financial prospects surrounding wind farms, political and legal threats to the long-term development of wind energy loomed at the start of the decade. First, newly elected President Ronald Reagan had already voiced disdain for renewable energy programs during his candidacy and vowed, if elected, to cut funding to them as early as possible in his administration. In 1982, the Reagan administration took aim at slashing the Department of Energy's wind energy research budget. The state of Mississippi in

These 65-kilowatt Nordtank Energy Group wind turbines from Denmark were part of a wind farm erected at Tehachapi, California, in the early 1980s. Courtesy Warren Bollmeier, Kaneohe, HI.

1982 challenged the constitutionality of PURPA by insisting that it infringed on states' rights in utility regulation. After PURPA's enactment in 1978, many state utility commissions hesitated to implement load management requirements and new rate structures to accommodate small wind-electric plants interconnected with the grid. In the early 1980s, solar was far too expensive for homeowners, so almost all of the first customer-owned generators were small wind turbines. Additionally, the Federal Energy Regulatory Commission (FERC) exempted small wind-electric producers from state utility regulation and required utilities to purchase the excess power put into the grid by the wind-electric turbine at a "just and reasonable" price. The legal challenge by Mississippi landed in the Supreme Court where it was rejected in a narrow 5–4 decision from the justices. The Supreme Court ruled in favor of PURPA once again in a second case brought before it in 1983 by the American Electric Power Service Corporation, which challenged the constitutionality of FERC's cogeneration rules.[58] The roadblocks to PURPA from utilities across the country, however, remained an ongoing hurdle for many small wind turbine owners. "It

took 15 years for the telephone industry to work out rights and standards for customer owned equipment. If we don't move fast at this early stage, it could take as long for the rights of cogenerators to be similarly implemented," warned wind industry journalist Donald Marier in 1983.[59]

An even bigger threat to the wind industry's momentum was the December 31, 1985, expiration of the federal investment tax credits, which had been established by the Energy Tax Act of 1978 and extended by the Crude Oil Windfall Profit Tax of 1980. Both large and small wind turbine manufacturers relied heavily on the tax credits, as well as government research support, to stimulate commercialization of their machines. However, even with these benefits, many smaller turbine producers still lacked sufficient orders to build and sell their machines cost effectively against conventional electric power provided by the utilities to consumers. Two years before the expiration of the tax credits, many of these companies were already worried about continuing in a market without the availability of the tax credits for their customers. Even companies that considered entering the wind energy industry at that time hesitated with those investments.[60] "Facing significant investments in product development and production tooling, these companies are concerned that the market for WECS (wind energy conversion systems) will be insufficient should the various federal, state, and local incentives be repealed or allowed to expire. These worries are exacerbated by uncertainties in future interest rates, fuel costs, and the ability to reduce WECS costs through mass production," wrote Michael Lotker, vice president of Renewable Energy Venture, a wind power development firm, in 1983. "In short, many potential manufacturers of WECS, after analyzing the market, come away unconvinced that the federal role is permanent enough to provide a solid basis for their own business investments."[61] Yet, there remained a glimmer of hope that Congress and the White House would enact legislation to extend the tax credits for wind energy before the expiration and allow the industry to continue to gain its footing. President Reagan's Energy secretary, Donald Hodel, in the spring of 1983 stated, "We are not going to be able to invest in every good idea, but the United States can't afford to slam the door on any technology and then find that the rest of the world has passed us by."[62]

In early 1984, legislative attempts to persuade lawmakers to extend the investment tax credits for wind energy had failed to gain traction, and panic was settling into the industry. The Solar Energy Industries Association, in an attempt to save its own manufacturers, split from the other renewable energy groups by proposing a tax credit extension solely for solar technologies. Although angered by this move, AWEA and the other renewable energy industries, such as small hydro, biomass, and energy conservation, rallied and began lobbying the House and Senate to secure legislative support for extending their respective tax credits. In October 1984, AWEA established a subcommittee, led by Robert Sherwin of Enertech, to raise $240,000 for the lobbying effort.[63] Several economic factors, however, were working against the political impetus on Capitol Hill to continue the investment tax credits for renewable energies, including sinking oil prices, a massive federal deficit, and calls for tax reform. Without the federal investment tax credits, state governments warned that their renewable

energy incentive programs could not make up the financial shortfall. For example, California's 10 percent renewable energy tax credit in 1985 would have needed to increase to 50 percent to make up the difference, which amounted to an estimated loss of about $150 million for the state.[64] By late spring of 1985, AWEA realized that a full-on extension of the investment tax credits for wind energy appeared unlikely. So, it was decided to propose a three-year extension of the tax credits with a stepped down rate of 10 percent, 10 percent, and 5 percent for each year, while the residential credits over the same proposed extension period would be stepped down at 35 percent, 30 percent, and 25 percent, with credit to be taken for equipment expenses of up to $20,000 instead of $10,000.[65] AWEA hoped that the Reagan Administration's continuation of subsidies for the oil and gas industries within its tax reform initiative would make it difficult for congressional lawmakers to ignore the call to extend

Thomas Gray, executive director of the American Wind Energy Association, speaks to a news reporter during the Wind Power '85 conference in San Francisco in August 1985. Courtesy Gray family, Norwich, VT.

the tax credits for renewable energies like wind. "The figures I have seen indicate that the conventional energy industries get enough tax breaks every six months to pay for the entire cost of the phasing out the renewable energy and conservation credits over five years," commented Gray of AWEA at the time.[66] In the late summer of 1985, AWEA stepped up its public relations through the media to explain the wind industry's successes. "Basically, we are doing everything we can," Gray said. "We have been trying to approach members of [C]ongress through companies in their districts."[67]

The majority of the installed wind power in the United States at the time was concentrated in three sites across California: Altamont Pass east of Livermore, San Gorgonio Pass near Palm Springs, and Tehachapi Pass 40 miles southeast of Bakersfield. In addition to California's generous tax incentives and mandate for the state's power companies to purchase the wind-generated power, the land availability and favorable winds of these

three sites made them fertile ground for rapid wind farm development.[68] As one rancher, who benefited from the rent money paid to him by the wind farm operator who erected turbines on his Altamont ranch land, said in an interview with *Smithsonian* magazine in late 1982: "Nobody ever wanted to live out here in the wind. It's hard on the cattle. And it's hard on your nerves. I've been cussing this wind for over 20 years. But now, it bothers me when it don't blow!"[69] Between 1981 and 1985 thousands of wind turbines—mostly horizontal axis designs—were installed at these three locations by a few different companies, from both the United States and Europe.[70] However, opponents of extending the tax credits pointed to instances of non- or underperforming turbines being erected merely to take advantage of the available incentives. Most wind farm developers in California sold turbines to lawyers, doctors, and other investors. The investment tax credits allowed under the law at the time helped individuals offset their personal income

This wind farm in California's Altamont Pass, which was constructed in the fall of 1983, included the Darrieus-style FloWind 17 turbines, each with a rated output of 142 kilowatts. Courtesy Herman Drees, Thousand Oaks, CA.

tax.[71] "Suddenly, wind-energy development became trendy in investment circles. Wealthy investors, with $50,000 or $100,000 to disburse, call their brokers, captivated by the idea of aiding the environment while padding their pocketbooks," wrote Robert W. Righter about the 1983 rush to build California's wind farms in his book, *Wind Energy in America*[:] *A History*.[72] The business magazine, *Forbes*, in early 1984, called California's wind farm industry, "The great windmill tax dodge," adding that "the windmill industry would disappear into desert sands without the credits, barring some dramatic decrease in the cost of wind-generated electricity, which doesn't seem likely very soon." At the time, wind turbines, including capital costs, generated power at a cost between 12 cents and 15 cents per kilowatt-hour, compared to about 4 cents to 4.5 cents per kilowatt-hour from coal, oil, and nuclear power plants.[73] In addition, politicians from across California, including Republican Governor George Deukmejian, Democratic Congressman Fortney "Pete" Stark, who served on the US House Ways and Means Committee, and Palm Springs Mayor Frank Bogert, opposed extending the tax credits for wind energy.[74]

Despite the lobbying efforts by AWEA and its allies, the tax credits for wind expired on December 31, 1985, and the industry entered a period of disarray. Gray forecasted that without the federal tax credits and the subsequent shrinking of California's incentives, wind farm investments in 1986 would plummet.[75] Well-capitalized firms with established wind turbines for the wind farm business, such as US Windpower (renamed Kenetech in 1988), Fayette Manufacturing, and FloWind Corporation, however, were able to hang on without the federal tax credits and even completed some substantive turbine installations, but at a slower pace. According to AWEA, in 1985 the industry installed 4950 turbines at a capital investment of $868 million. A year later, the industry erected only 1825 turbines at a capital investment of $254 million, and by 1987, 1392 turbines were raised at a capital investment of $158 million, further demonstrating the wind turbine manufacturers' heavy dependence on the former federal and reduced state tax incentives.[76] Smaller firms

offering residential turbines fared much worse. and many closed their operations after the tax credits expired, leaving the owners of these machines with worthless service plans and warranties.[77]

Michael Bergey, president of 10-kilowatt wind turbine manufacturer Bergey Windpower, was elected AWEA board president for 1985–1986. He watched much of his company's domestic business dry up during the second half of the 1980s but witnessed an uptick in exports primarily to developing countries for meeting remote village electric power and water-pumping needs.[78] Robert Sherwin, who became AWEA board president for 1986–1987, revived Enertech by forming Norwich, Vermont-based Atlantic Orient Corporation, which developed a new 50-kilowatt wind turbine for use in wind farms and other applications, such as supplementing diesel-powered generators commonly used in rural communities.[79] Northern Power Systems of Moretown, Vermont, expanded beyond its small wind turbine roots in the late 1980s by building larger machines with rated outputs of 125 kilowatts for the wind farm and community power businesses.[80] The market downturn also did not dissuade two ambitious entrepreneurs, David Calley and Andy Kruse of Flagstaff, Arizona, from entering the small wind turbine business in 1987 with the sale of a 300-watt machine, known as the WindSeeker, which initially sold for $300 apiece.[81]

Meanwhile AWEA continued to hold its annual conferences. The attendance at the 1986 Wind Energy Expo and National Conference at Cambridge, Massachusetts, dropped to below half the recordbreaking attendance of about 800 set in 1985 at the conference in San Francisco.[82] Under Gray's leadership, AWEA emphasized the role that wind energy could play in the national reduction of carbon emissions, which were linked to global warming by the international scientific community in the 1980s. AWEA estimated that a single 100-kilowatt turbine in a wind farm kept about 5 tons of carbon a month from being emitted into the atmosphere. "The 16,900 wind turbines in California windfarms last year produced about 1.8 billion kilowatt-hours of electricity. If coal had been burned instead to produce that electricity, more than half a pound of carbon would have been

A view of the densely packed wind farms at San Gorgonio Pass near Palm Springs, California, from the top of a larger wind turbine in 1995. Courtesy Herman Drees, Thousand Oaks, CA.

emitted into the atmosphere for each kilowatt-hour produced—or nearly 1.2 billion pounds (600,000 tons) altogether," Gray said in a July 7, 1988, AWEA press release.[83] AWEA used these statistics to encourage Congress and the White House to increase the amount of wind energy research funds for the Department of Energy, which by this point was less than $9 million, or one-tenth the amount when President Reagan took office in 1981. "The Administration's record in this area is shameful, and that of the Congress is not much better," Gray lamented at the time.[84] AWEA was also on the forefront of the nation's move to deregulate the electric power utilities during the late 1980s and lobbied for a portion of all new electric power generated to come from wind farms.[85] By the end of 1988, AWEA announced plans to increase its lobbying staff to strengthen its voice on these issues, not only in Washington with incoming President George H. W. Bush, but among the states, like California. "Wind energy has enormous potential in the US. Its cost is relatively low, especially when the environmental costs of other fuel sources are considered, and the

public strongly favors it as a new source of energy. We have to do more to communicate those facts in Washington and Sacramento," Gray said.[86]

By 1989, Gray had run AWEA for eight years. He saved the association from demise and even steered it through one of the lowest points in its nearly fifteen years in operation. In an interview nearly twenty years after his resignation as executive director, Gray said: "I had basically run out of bright ideas for things to kind of break the association and the industry out of the place that it was at."[87] During his tenure at AWEA, Gray had relocated to Vermont, and despite his enormous effort to make that situation work, he and the AWEA board of directors decided it was not a feasible way, long term, to run a Washington-based trade association.[88] After his departure as AWEA executive director, Gray accepted the position as executive director at the Washington, DC-based Council for Renewable Energy Education. However, the newly formed organization had difficulty attracting funding and dissolved within a year. Gray then became a consultant to the US Export Council for Renewable

Randall S. Swisher, who served as executive director of the American Wind Energy Association from September 1989 to January 2009, fostered development of the country's wind farm industry. Courtesy Randall S. Swisher, Silver Spring, MD.

Energy and marketing director for wind measurement technology manufacturer, Second Wind.[89] An early advocate for internet-based networks and communications, Gray also worked part-time for the Institute for Global Communications, which operated networks such as EcoNet and PeaceNet. All the while, AWEA retained Gray to continue writing and preparing the association's weekly newsletter, which at that time was distributed via EcoNet. The newsletter, which was distributed not only to AWEA members but to Capitol Hill lawmakers and the media, provided updates on the federal government's wind energy research programs and supporting legislation, as well as the latest activities within the industry. "I would say one of the major benefits [of the newsletter] . . . is that it allows us to give our viewpoint. It's kind of house-oriented. We can talk about what's going on, what trends there are, and what we think they mean, in the industry. . . . We could never do it if we didn't have our own trade organization," Gray explained.[90]

In September 1989, AWEA appointed Randall S. Swisher as its new executive director. Swisher, who was alerted to the job opportunity by Gray, was already a seasoned legislative advocate for renewable energy development and its emerging use by the electric utility industry. He joined AWEA after serving from 1984 to 1989 as a legislative representative for the American Public Power

Association, which represented municipally owned electric utilities. Swisher arrived in Washington, DC, from his home state of Iowa in 1970 as a university graduate student. With an interest in pursuing his dissertation on a consumer advocacy topic, Swisher joined Ralph Nader's DC Public Interest Research Groups (DC PIRG). "I joined the DC PIRG initially as a participant-observer for six months, with the intention of to write my dissertation based on that experience. The six months turned into six years and set the course for my career," he said.[91] In 1975, Swisher participated in DC PIRG's legal intervention in the licensing of the Potomac Electric Power Company's proposed Douglas Point nuclear power plant in Charles County, Maryland, approximately 30 miles south of the District of Columbia. After a two-year campaign, the proposed nuclear plant was canceled.[92] In 1979, he worked on Capitol Hill as a legislative assistant to Democratic Congressman M. Robert Carr of Michigan, before joining the staff of the National Association of Counties in 1980, where his role for the next four years was to work with county governments to foster renewable energy and energy efficiency programs. Swisher switched to the American Public Power Association, where he believed his interest in renewable energy development would be a better fit, but here too, he found this effort marginalized.[93] Swisher recalled that when he joined AWEA, the association sought an individual who "had familiarity with the legislative process and was a *comfortable* communicator."[94] For him, becoming AWEA executive director was more about professional fulfillment. "AWEA was first and foremost the opportunity to be a legislative advocate and to use AWEA as a vehicle to communicate the potential that this technology had to change the world in fundamental ways," he said.[95] However, AWEA in late 1989 was still a small association representing a struggling industry. In addition to Swisher's role as executive director, AWEA had a staff of four. The association's operating budget at the time was about $775,000, based mostly on funds for federal research, export promotion, and wind energy promotional materials, while annual membership dues generated $47,000.[96]

To replace what was lost when the wind energy tax credits expired at the end of 1985, AWEA in early 1989 began devising a "legislative push" to introduce production incentives for wind farm operators. To do this, the association recommended that "Congress allocate a certain percentage of an oil import fee or gasoline tax for a 10-year production incentive," which would be linked to electricity generated by wind farms.[97] That year, the association attempted to convince Capitol Hill lawmakers to add production incentives for renewable energy to proposed acid rain-reduction legislation. "There is every reason why such an approach makes good public policy. Clean energy sources such as wind, which do not pose the risk of major environmental damage, should be a larger part of America's energy mix," the association wrote in letters to then Senate Majority Leader George J. Mitchell (D-ME) and Senator Max Baucus (D-MT), chairman of the Senate Subcommittee on Environmental Protection, in early April 1989, adding that year, "U. S. wind electric turbines will produce some 1.8 billion kilowatt-hours of electricity, or more than enough for the residential needs of Washington, D.C., with no air pollution, contribution to global warming, or nuclear waste."[98] It took another three years of lobbying and refinement by AWEA before a production incentive to benefit the wind industry would come to fruition. In 1992, Congress passed the Energy Policy Act, which included a production tax credit (PTC) for electricity generated by renewable sources, such as wind, solar, biomass, and geother-

Nancy A. Rader, who became the longtime executive director of the California Wind Energy Association, is credited with crafting the concept of the Renewable Portfolio Standard in the early 1990s, which many US states have used to promote wind energy. Courtesy Nancy A. Rader, Berkeley, CA.

mal. The PTC was essentially a federal rebate on the taxes paid by the companies that owned wind turbines and provided a 1.5 cents per-kilowatt-hour tax credit on the electricity generated by these machines over a ten-year period.[99] Compounded by rising fuel costs and the scientific community's heightened concern about coal- and oil-burning power plants contributing to global warming, the PTC helped excite renewed public interest in wind energy in the early 1990s, which further influenced acceptance of this renewable energy source by electric utilities. However, there was no extension of the tax credits for small wind turbines, and the PTC did not apply to turbines intended for residential use. Also, the PTC was too weak for small turbines with much smaller annual energy production contrasted with the large wind farm turbines.[100]

While the electric utilities were undergoing deregulation in the early and mid-1990s, AWEA began lobbying for the requirement that a certain percentage of their power be generated by renewable energy sources. Inspiration for this requirement traced its roots to 1983 when Iowa Governor Terry Branstad signed a law requiring the state's seven investor-owned utilities to purchase 105 megawatts of their annual power from renewable energy sources, but the law was not enforced at the time.[101] Ten years later, AWEA hired energy consultant Nancy A. Rader (who became executive director of the California Wind Energy Association in 2000) to develop a renewable energy policy intended to provide a more effective incentive than PURPA that would fit with a deregulated market. Rader developed what was essentially a tradable set-aside and called it the Renewable Portfolio Standard (RPS). As AWEA's policy advisor and West Coast representative, she further refined the concept over the next two years with assistance from PhD candidate Brent Haddad (now professor at the University of California at Santa Cruz). AWEA then advocated the RPS as part of its "Criteria for Restructuring the Electric Industry."[102] In 1995, the California Public Utilities Commission adopted AWEA's RPS proposal, also championed by the Union of Concerned Scientists, which would require each seller of power in California to "build a diverse portfolio of generation resources comprised of at least 12 percent renewables."[103] Rader,

at the time, called RPS "a tradable, market-based mechanism that recognizes the inherent values of both renewable and conventional technologies."[104] California, however, failed to include the RPS in its utility deregulation legislation. Republican Congressman Daniel Schaefer of Colorado in 1996, who was then chairman of the US House of Representatives' Energy and Power Subcommittee, proposed legislation to develop a national RPS. Although the bill failed to gain widespread support, it helped to galvanize the concept.[105] AWEA, along with support from the Union of Concerned Scientists, began a campaign to promote RPS on a state-by-state basis. "[B]ecause we have limited resources, what we tended to do was to identify in two or four states a year where we felt we had an opportunity and we would hire a lobbyist in that state and try to then organize an RPS campaign," Swisher explained.[106] By the late 1990s, a handful of states began to implement renewable portfolio standards with varying power percentages and timelines. Iowa started enforcing its RPS with the state's utilities in 1997.[107] In 1999, due to the strong advocacy by Public Citizen, the Environmental Defense Fund, and others, George W. Bush, then Texas governor, signed RPS legislation as part of the

These 500-kilowatt NedWind wind turbines from Denmark were commissioned at San Gorgonio Pass near Palm Springs, California, by Dutch Energy Corporation in 1994. Courtesy Herman Drees, Thousand Oaks, CA.

state's deregulation plan. The simplicity of the plan rapidly propelled Texas to the forefront of US wind energy production and made it a model for other states to follow.[108]

The production tax credit and renewables portfolio standards initiatives did little to promote small wind turbine production for residential use. The primary beneficiaries of these incentives were suppliers of wind turbines with outputs of more than 100 kilowatts per machine for wind farm developers and operators. However, by the 1990s, the ramp-up of the PTC and RPS programs was also too late to save the financially plagued US wind turbine manufacturers and wind farm developers. Once formidable US companies ten years earlier, like FloWind, Fayette, and Kenetech (formerly US Windpower), closed their doors.[109] Kenetech, by far the largest developer of wind farms in the US, which introduced a successful variable-speed wind turbine to the market in 1993, was in financial peril as far back as 1995. A company spokesman that year said Kenetech faced "death of a thousand cuts syndrome" due to numerous wind farm development backlogs both in the United States and abroad. Kenetech was financially burdened by outstanding construction loans and did not get paid for its work until the wind farms were certified to be operational. It also experienced technical problems with its 33-KVS turbine. By 1996, the company filed for bankruptcy and ceased production.[110] AWEA faulted the US government's insufficient support of renewable energy programs for Kenetech's downfall. "Despite Kenetech's missteps, the fact remains that our government's inconsistent policies and its over-whelming emphasis on short-term fixes to problems at the expense of long-term policies has made managing a growing company in the renewable energy business far more difficult than it should be," the association said.[111] European wind turbine manufacturers, particularly those in Denmark, ben-efited from government incentives that encouraged the industry's technological acceleration during the 1980s and 1990s. European companies, such as Bonus Energy, Vestas, Nordtank, Micon, and NedWind, became dominant players and exported their large turbines to the US market.[112] By the 1990s, these companies started manufacturing

some components of their machines in the United States to shorten their wind farm delivery times. In late 1995, Micon installed a US-made 600-kilowatt turbine—the largest at the time—in Sibley, Iowa, for IES Utilities.[113]

By the mid-1990s and early 2000s, the bulk of AWEA's resources were dedicated to facilitating utility-scale wind energy development in the United States, which included protecting the PTC and lobbying for state-level RPS programs, preserving PURPA, securing funds for Department of Energy advanced wind turbine research programs, and participating in wildlife protection initiatives.[114] The PTC was already under attack by September 1995, when House Ways and Means Committee Chairman William R. Archer Jr. of Texas intro-duced legislation that included a provision to repeal the new wind tax incentive. The PTC was ultimately preserved, but the threats to its existence contin-ued.[115] After 1999, the PTC's preservation became a constant battle for AWEA as Congress generally reauthorized the legislation for a year or two at a time through the early 2000s. The result was "on again, off again" production cycles in the construc-tion of wind farms. Thus, when the congressionally mandated expiration date for the PTC neared, wind farm developers rushed to get their turbines erected. A wind project could not take advantage of the PTC benefits unless it was brought into service before the legislative expiration, and wind farm developers generally required six months of lead time to make arrangements to buy equipment, obtain permits, and secure financing before con-struction could begin.[116] In March 2002, Congress authorized a two-year PTC extension, which led to one of the biggest wind farm construction booms in the United States, from less than 2000 megawatts of installations in 1998 to more than 6000 megawatts of turbines erected across thirty states by the end of 2003, or enough to power 1.6 million average US households.[117] When the PTC expired on December 31, 2003, Congress made no immediate plans for its renewal, and so wind projects across the nation ground to a halt. AWEA's executive director, Randall S. Swisher, highlighted the frustrations of AWEA members during the spring of 2004 by stating, "Today, a wide range of

US companies are interested in the wind industry, but many are staying on the sidelines because of the on-again, off-again nature of the market produced by frequent expirations of the PTC."[118] AWEA estimated that more than 2000 megawatts of wind projects valued at $2 billion were idled by the failed PTC renewal during the first half of 2004.[119] Congress restored PTC in September 2004, and the large wind industry regained its momentum, installing 2500 megawatts worth of turbines valued at more than $3 billion in 2005.[120] The association's goal during the first decade of the 2000s was to replace the annual and bi-annual PTC renewals with five-year extensions.[121]

AWEA's small wind turbine members—those with machines of less than 100 kilowatts output for residential and small business use—that remained in business during the 1990s struggled to grow in the US market, with ever-larger turbines dominating wind farm development. The lack of an investment tax credit to stimulate residential sales of small wind machines did not help their cause. Meanwhile, surviving small wind turbine manufacturers took advantage of various federal government business and research grants available to them to further technological developments and find new markets for their machines. Northern Power Systems (NPS) in 1995, for example, received a $578,000 grant from the National Aeronautics and Space Administration (NASA) to design, build, and test a 100-kilowatt turbine for use at the Amundsen-Scott Station at the South Pole. The company planned to use lessons from this project to facilitate its development of turbines for remote station and village power. By the mid-1990s, NPS already operated four of its 3-kilowatt turbines along the Antarctic coast to provide power to a telecommunications station.[122] Another Vermont

The Kotzebue Electric Association erected its first three Atlantic Orient Corporation 50-kilowatt wind turbines in 1997 to reduce the Alaskan community's reliance on a diesel-powered generator. Fifteen AOC 15/50 turbines were installed at Kotzebue by 2005. Courtesy Kotzebue Electric Association, Kotzebue, AK.

turbine manufacturer, Atlantic Orient Corporation, in 1995 entered a contract with the Kotzebue Electric Association of Alaska, located about 30 miles north of the Arctic Circle, to supply three 50-kilowatt machines for a wind demonstration project to reduce a remote village's reliance on its diesel-powered electric generator.[123] With research and development assistance from the US Department of Agriculture in the early 1990s, Bergey Windpower of Oklahoma manufactured a line of 1.5 kilowatt water-pumping wind-electric turbines, which it exported to developing countries throughout the world as part of remote village water programs funded by the US Agency for International Development, World Bank, and other foreign government aid agencies.[124] Small wind turbine manufacturers also maintained a voice, albeit a smaller one, within AWEA during the 1990s, participating in various committees and on the association's board. Bergey Windpower's president Michael Bergey was asked to serve a second term as AWEA president in 1993–1994.

The end of the 1990s and the early 2000s witnessed the start of an upsurge in US sales of small wind turbines. There were individuals who purchased these machines out of concern that the computer systems operating the nation's electric utilities and other infrastructure might shut down at midnight on December 31, 1999. Increased sales of small wind turbines were also attributed to public concerns over rising fuel and electricity costs, political threats to the supply of oil imports, and increasing environmental awareness. Few individuals, however, could afford to purchase a residential small wind turbine—the price tag of a 10-kilowatt Bergey Windpower machine at the time was between $35,000 and $50,000—without available state government incentives to help offset the cost. States that offered small wind incentives at the start of the 2000s included California, Massachusetts, New York, New Jersey, North Carolina, Pennsylvania, Ohio, Vermont, and Wisconsin. In 2003, New York initiated a program that provided purchasers up to 50 percent cash back on the cost of a residential wind system in addition to low-interest loans. California offered small wind turbine buyers rebates based on the size of their systems; thus,

a $50,000, 10-kilowatt machine was eligible for a $22,500 tax rebate.[125] Small wind turbine manufacturers in the United States received an additional boost when Congress passed the Energy Improvement and Extension Act of 2008, which included an investment tax credit equal to 30 percent, or up to $4000, on the cost of a 100-kilowatt or smaller wind energy system. A year later, while the country was immersed in one of the deepest economic recessions since the Great Depression, congressional lawmakers passed the American Recovery and Reinvestment Act, which not only retained the investment tax credit for small wind but removed the $4000 maximum credit limit for purchases of these systems.[126]

With these newfound state and federal incentives for small wind turbine installations came dread among established companies and industry analysts that opportunists would take advantage of the reenergized market by offering customers unproved machines or misrepresenting their capabilities altogether. This activity had occurred during the late 1970s and early 1980s with the available investment tax incentives and tarnished the reputation of the small wind industry. Some schemes that took advantage of the tax incentives involved the erection of wind machines that were essentially nonelectric-generating whirligigs.[127] Paul Gipe, who began studying and writing about the wind energy industry during the mid-1970s, referred to the new US government tax incentives as "capital subsidies," preferring the European system of utilizing feed-in tariffs to stimulate and sustain small wind installations. "Capital subsidies, because they encourage sales of hardware and not the generation of electricity, are simply bad public policy. . . . Most countries that have used capital subsidies learned their lessons and abandoned the approach—years ago," Gipe wrote in 2008.[128] As predicted, the combined state and federal tax incentives sparked a renewed interest among the public for residential wind turbines as a way to reduce electric utility bills or, in some cases, attempt to disconnect from the utility altogether. For some individuals, the erection of a small wind turbine was a visual demonstration of environmental stewardship.[129] Pennsylvania Governor Edward G. Rendell, in an attempt to stimulate public interest in small wind turbines, authorized a $193,000 grant to Southwest Wind-power to erect fifteen of its machines on 35-foot towers across the state.[130]

An AWEA study in 2007 identified twelve established US manufacturers of small wind turbines and another eight firms preparing to enter the market.[131] By the following year, the association counted sixty-six firms in the United States either manufacturing or intending to build small wind turbines, with reported sales of 10,500 units.[132] In its 2009 *Small Wind Turbine Global Market Study*, AWEA reported: "The industry projects a 30-fold growth within as little as five years, despite a global recession, for a cumulative US installed capacity of 1,700 MW (megawatts) by the end of 2013," largely spurred by the federal investment tax credit.[133] However, some startups enticed customers by promoting the visual aesthetic of their machines, reintroducing inexpensive vertical-axis designs, and offering rooftop mounted turbines—all of which were considered inadvisable by longtime industry experts due to previously known inefficiencies.[134]

With the growing concern to ensure that only reliable and safe small wind turbines would be allowed access to state and federal investment tax credits, as well as other financing and permits for installation, industry called for a new machine certification program. AWEA, under the leadership of Michael Bergey, proposed the first set of wind turbine performance standards in 1980, which included the machine's startup wind speed, cut-in wind speed, maximum power, overspeed control, power availability, power curve, maximum tower loads, rotor speed, and mean power output.[135] The association updated these performance standards in 1988 and 2000. AWEA was designated an accredited standards-setting body in 1995 by the American National Standards Institute.[136] However, the measures remained largely ineffective because they allowed wind turbine manufacturers to voluntarily "self-certify" the performance of their machines. With growing frustration, Gipe wrote in May 2000: "Manufacturers have simply ignored the standard. And AWEA has ignored the fact that manufacturers have ignored the standard. Most small turbine manufacturers have no 'test reports' to release even

if they chose to do so. That is, their turbines have been tested to any accepted standard."[137]

In 2006, AWEA set up the Small Wind Turbine Subcommittee, which was chaired by Michael Bergey, and tasked it with developing test standards for a new turbine certification program. The subcommittee's work involved more than sixty participants, including turbine manufacturers and technical consultants, state government and national laboratory officials, and representatives from the Canadian Wind Energy Association and British Wind Energy Association. "Hundreds of hours of detailed discussion, debate, compromise, revision, and formal response" were involved before their completion in 2009.[138] The association derived most of new wind turbine testing standards from previous work performed by the International Electrotechnical Commission. AWEA's small wind turbine standards applied to machines with either horizontal or vertical axis rotors, as well as to related turbine controllers, inverters, wiring and disconnects, and installation and operation manuals. The standards also covered an array of turbine tests, including performance, acoustics, strength and safety, and duration.[139] To maintain the integrity of the testing and certification outcomes, AWEA forbade self-certification by wind turbine manufacturers and stated that this work must be performed by an independent certifying agency, such as the newly formed Small Wind Certification Council or a nationally recognized testing laboratory.[140] Depending on the wind conditions of the approved test site, testing and reporting took six months to a year to complete.[141] In 2009, Ron Stimmel, AWEA's small wind advocate, called the new certification program "a real game-changer" for the small wind industry. In addition to the AWEA's new small-wind-permitting guide for state and local policymakers, Stimmel said, "It will establish a level of transparency in the industry that consumers and regulators have never before had and will provide a great foundation for the industry's predicted growth."[142] In accordance with the AWEA 9.1–2009 standards, small wind turbine certification programs were offered by the Solar Rating and Certification Corporation and Intertek.[143]

Michael Bergey of Bergey Windpower, who was named president of the newly formed Distributed Wind Energy Association, speaks at the 2011 Small Wind Conference in Stevens Point, Wisconsin. Courtesy Michael Bergey, Norman, OK.

While small wind turbine manufacturers were once the foundation of AWEA and the successes with implementing the investment tax credit and small wind certification program, their voice within the association had become smaller as the 2000s got underway. By then, Michael Bergey and David Blittersdorf of NRG Systems (now with All Earth Renewables) were the last of the small wind industry advocates left on the association's board. "The situation at AWEA for small wind had been deteriorating for several years as the board makeup involved more lawyers and financial firms that were newer to the industry and solely focused on large-scale projects," Michael Bergey said. "In 2007 after 26 years on the board, the AWEA nominating committee did not nominate me for another term. I chose not to run, because I just wasn't being effective. We maintained our membership for a few years, but dropped it in 2009 because there were no real benefits."[144] By 2010 AWEA still maintained some small wind firms within its membership, particularly foreign-based firms that hoped the association would provide them a path into the US market, but it was a tiny part of AWEA's income and their numbers continued to decline.[145] Bergey, Blittersdorf, and others from the industry met at the Small Wind Conference at Stevens Point, Wiscon-

sin, in June 2010 and unveiled the launch of a new national trade association to promote the legislative, regulatory, and business interests of small wind manufacturers, distributors, project developers, dealers, installers, vendors, and advocates.[146] In 2004, the Midwest Renewable Energy Association began hosting the Small Wind Conference as part of its annual fair to "support honest, open dialogue between installers and manufacturers of home-scale wind-electric systems."[147] Many participants of the Small Wind Conferences embraced the proposed national association for the small wind industry.

The name Distributed Wind Energy Association (DWEA) was picked for the new, not-for-profit association. "Distributed wind," a term coined by the Department of Energy and wind industry in the early 1990s, involves the deployment of "typically small wind turbines at homes, farms, businesses, and public facilities to off-set all or a portion of on-site energy consumption," whereas wind farm operators sell their power to the utilities, which in turn, retail it to their customers.[148] All Earth Renewables and Bergey Windpower made the largest contributions toward launching the new association. Jennifer Jenkins, formerly of Southwest Windpower, was named executive director, with the election of Michael Bergey as president; Kevin

Schulte of Sustainable Energy Developments as vice president; Lisa DiFrancisco of North Coast Energy Systems as secretary; and David Blittersdorf as treasurer. The association appointed Lloyd Ritter of Green Capitol to be its Washington lobbyist. Committees were also activated to cover federal and state policy, permitting and zoning, communications, and standards to encourage the proliferation of distributed wind power. In addition, the DWEA enacted a strict code of ethics in which "unrealistic performance claims," "dishonest sales hype," and "questionable practices: short towers, rooftop, urban" were grounds for membership rejection or termination.[149]

AWEA's leadership initially opposed the formation of DWEA. During the 1990s and early 2000s, AWEA received some funding from the Department of Energy and the US Agency for International Development to promote the small wind industry, and it even maintained a small wind staff of one to four people, depending on the level of contract funding and to support various ongoing marketing and technical support initiatives.[150] Randall Swisher retired from the association in 2008, and AWEA's new chief executive officer, Denise Bode, took over in 2009, with a focus on continuing AWEA's primary objective of facilitating the interests of large, utility-scale wind power. The association barely noticed the loss of the small wind turbine manufacturers with a membership of nearly 2500 and its 2009 Windpower conference in Chicago drawing a record 23,000 attendees.[151] AWEA quickly accepted the new association for small wind power and agreed to cooperate on common issues. For example, DWEA members, like Bergey Windpower, continue to participate in AWEA's standards committee. The small wind certification standard, AWEA 9.1–2000, was updated in 2015 through the AWEA standards committee and adopted as the ANSI standard.[152]

DWEA generated its own excitement as a new association, surpassing more than 110 members in 2012.[153] That year the association's income, which was primarily supported by membership dues, reached $418,725, but its operating expenses were $395,703.[154] One of the largest segments of membership included the small wind turbine installers. These mostly small operations were particularly attracted to DWEA's installers committee, which provided webinars on tower climbing safety, the National Electrical Code Article 694 for wind-electric systems, and group insurance.[155] In

The Distributed Wind Energy Association begins to hold its annual conferences in Washington, DC, as shown here in 2013. Courtesy Michael Bergey, Norman, OK.

addition, DWEA scored myriad federal, state, and local policy victories in 2012. They were instrumental in extending the investment tax credits for wind turbines over 100 kilowatts. The group succeeded in having the US Fish and Wildlife Service recognize in its *Land-based Wind Energy Guidelines* that small wind turbines pose low avian risks. They published the DWEA Small Wind Model Ordinance with the National Association of Counties (NACo) and redrafted the NACo wind energy zoning guide. DWEA promoted increased funding and expanded coverage of the New York State Research and Development Authority's on-site wind turbine rebate program. The association participated in the launch of the Department of Energy–National Renewable Laboratory Small Wind Competitiveness Improvement Project, which is a cost-shared research and development program aimed at lowering the costs of small wind turbines. The group developed precertification criteria for midsized wind turbines for use in state incentive programs. Members of DWEA also met with senior officials at the State and Defense

departments about including small wind power at their US and overseas installations.[156]

In March 2015, DWEA released its *Distributed Wind Vision—2015–2030*, in which the association outlined the prospects for the distributed wind industry to achieve 30 gigawatts of "behind-the-meter" wind-electric power generation nationwide in the next fifteen years. The association determined that an estimated 23.7 million homes, farms and ranches, businesses, communities and towns, and remote locations scattered throughout the country could benefit from one or more small wind turbines. According to the report, small wind turbines of up to 50 kilowatts and manufactured for at least twenty years of service were suitable for fostering energy independence for most American households, farms, and small businesses.[157] However, DWEA's report warned that the distributed wind industry still faced "cost and consumer acceptance barriers that most emerging technologies face, but significant policy barriers also pose unique challenges for distributed wind."[158] Small wind systems providers continued to struggle against

Distributed Wind Energy Association members (left to right) Michael Bergey of Bergey Windpower; Jennifer Jenkins of the Distributed Wind Energy Association; Diego Tebaldi of Northern Power Systems; Shaun Bourdreau of Endurance Wind Power; and Kevin Shulte of Sustainable Energy Development, shown here in front of the White House spend a day in early 2015 meeting with US lawmakers and federal regulators to promote the small wind energy industry. Courtesy Jennifer Jenkins, Flagstaff, AZ.

cheaper photovoltaic solar systems, as module prices plummeted with the flood of imports from China and some US states' feed-in-tariffs programs favoring solar over wind. In addition, local building permits necessary to erect wind turbines often place excessive constraints on tower heights; to operate effectively, most of these machines require 80- to 180-feet-tall towers. "It can take more man-hours to obtain a permit to install than it does to manufacture, deliver and install a small wind turbine," commented the association, adding that there were more than 25,000 separate zoning jurisdictions in the United States.[159] Among the biggest policy threats looming for the distributed wind industry was the failure by Congress to extend the 30 percent investment tax credit, as well as to expand it to cover wind turbines with outputs of larger than 100 kilowatts, before the December 31, 2016, expiration. "The Distributed Wind Industry has only had a federal incentive since 2009 and it would be very unfortunate and a failure of smart energy, environmental, and industrial policy to let the ITC expire in 2016 and not address the unfair limitation to 100 kW," DWEA said in its report.[160]

With the assistance of a $488,634 Advanced Manufacturing Technology Consortia Program grant from the Department of Commerce's National Institute of Standards and Technology in May 2014, DWEA established the Sustainable Manufacturing, Advanced Research and Technology (SMART) Wind Consortium, which set out to explore ways to improve production processes, reduce component costs, increase output, and lower the overall price of small wind turbines to US consumers.[161] While US small wind turbine manufacturers have for years dominated both domestic and foreign markets with their machines, by 2010 there was increasing concern that cheaper-built Chinese copies threatened to displace them.[162] The consortium, which held both face-to-face meetings and conference calls between October 2014 and February 2016, consisted of more than 250 individuals who participated in various subgroups focused on wind turbine design, mechanics, and operations. Representatives came from twenty-five manufacturers and more than fifty component suppliers, four federal laboratories, and thirty academic institutions, as well as approximately twenty nonprofit organizations, government agencies, and other stakeholders from across the country.[163] Consideration was given to the success of the manufacturers of large wind turbines. "Due [to] its dispersed nature, distributed wind's techni-

Patrick Gilman, manager of the US Department of Energy's Wind Energy Technologies Office manager, addresses the Distributed Wind Energy Association conference in Washington, DC, on February 28, 2019. Courtesy Michael Bergey, Norman, OK.

cal challenges are substantial and distinct from those of utility wind," the consortium observed. "Utility-scale wind turbines have increased in size over the past decades and become more cost-effective in part due to increased investments in R&D. Many of those cost-effective concepts may be relevant and applicable for distributed wind technology."[164] Whereas, with the exception of Bergey Windpower and Northern Power Systems, most of the US small wind turbine manufacturers that participated in the consortium were relative newcomers to the industry, getting their start in the business during the early 2000s. They were also far from the levels of manufacturing scale of the large turbine and solar panel manufacturers.[165] The consortium's final report, which was published in May 2016, focused on a handful of areas where the competitiveness of the distributed wind industry could be improved. These included optimizing wind turbine design to achieve increased efficiency and lower costs; improving manufacturing parts, materials, and processes by encouraging collaboration with other industries and component standardization; streamlining processes to install and maintain wind turbines; and expanding partnerships with other industries, such as electric vehicle fleet operators for new small wind application opportunities and the surfboard industry to develop new blade composites. In addition, the consortium called for increased recruitment and training efforts to ensure a sustainable small wind industry workforce.[166]

The looming expiration of the investment tax credit for small wind at the end of 2016 threatened to derail the progress of DWEA and the small wind industry's work at large. While the association's members pressed congressional lawmakers to preserve the tax credit for small wind, there was no legislation to facilitate an extension. The expiration was immediately felt by the small wind industry. Some wind turbine manufacturers and component suppliers either scaled back or ceased production altogether, while many of DWEA's installer members shifted to solar panel installations.[167] DWEA continued to meet with national lawmakers about the importance of restoring the investment tax credit for small wind. On February

9, 2018, Congress passed the Bipartisan Budget Act of 2018, which included a five-year extension of the 30 percent investment tax credit for small wind turbines of 100 kilowatts and less, retroactive to January 1, 2017. While the legislation placed small wind on a parity with solar energy's tax extension passed into law at the end of 2015, it also included a phase-out period for the investment tax incentive. Small wind turbines installed after December 31, 2019, and before January 1, 2021, would receive a 26 percent tax credit, and those machines erected after December 31, 2020, and before January 1, 2022, would be entitled to a 22 percent tax credit.[168]

Meanwhile, the collective advocacy in the wind energy industry, though at times hard fought to hold together and often in the face of political and industry adversaries over the past forty-five years, has ultimately proved to be an effective model for the industry's preservation and progression. AWEA has watched its members become significant contributors to the nation's electricity production. By the end of 2017, an estimated 89,077 megawatts of wind power from more than 57,000 turbines had been installed across forty-three states, or enough to power twenty-six million American homes, with another 30,000 megawatts of wind farms in development at the start of 2018.[169] DWEA was undergoing struggles similar to those that the AWEA experienced during its earliest years of existence; primarily maintaining a sustainable membership. (AWEA merged with the larger American Clean Power Association in January 2021.) The fledgling association's members continued to face intense competition within the renewable energy industry, particularly from residential photovoltaic solar energy. Interest in small wind turbines as an electric power source for remote applications and communities has remained strong. The DWEA board of directors also believed that federal policy remained critical to the future of the US market and business prospects for its industry members. In addition, with the renewed investment tax credits and continued efforts by small wind turbine manufacturers to lower their manufacturing costs, Michael Bergey of Bergey Windpower said, "I'm fairly optimistic that we can grow this business substantially."[170]

5 US Government Programs

In May 1976, the rugged, windswept landscape known as Rocky Flats was designated to become the epicenter of US government research into the validation and promotion of small wind-electric systems. The 400-acre site was once part of the Lindsay family's cattle ranch about 15 miles northwest of Denver, Colorado, between Golden and Boulder, and had become a buffer zone to a nearby sprawling nuclear weapons manufacturing facility, also known as Rocky Flats, that was operated on behalf of the Atomic Energy Commission by the Dow Chemical Company from 1950 to 1974. Rockwell International took over operation of the Rocky

Flats facility in 1975.[1] When the newly formed US Energy Research and Development Administration (ERDA) began its search for sites to launch the small wind-electric systems test center, Dr. Donald A. Wiederecht of Rockwell International, who pondered ways to harness Rocky Flats' powerful winds, answered the solicitation. By year's end, the Rocky Flats adjacent to the much larger nuclear facility site won out against a number of other government contractor facilities throughout the country. What interested ERDA most about the location, besides its wind, was the immediate access to Rockwell International's technicians and engineers who were

This former ranch property shown in June 1973 became the site of the Rocky Flats wind energy test facility. US Department of Energy photograph courtesy Warren Bollmeier, Kaneohe, HI.

highly experienced with project management to oversee the newfound program called Technical Management Support for the Development of Wind Systems for Farm and Rural Use.[2]

ERDA was in a hurry to get the small wind-electric systems test site up and running, particularly as a burgeoning industry had already started to emerge in the United States and offer wind turbines for sale. There were many questions surrounding the technical credibility of small wind plants—mostly ranging from 1 to 10 kilowatts generally, but less than 100 kilowatts—being sold to consumers. In fact, there were no minimum design and operational standards to help consumers select and use these machines most effectively and to overcome potential obstacles for erecting them, such as complying with zoning ordinances and obtaining bank financing and insurance, as well as integrating these machines with the utility power grid. ERDA, with backing from the Ford Administration and Congress, took it upon itself to bring order to the fledgling small wind energy industry and to ensure that the technology offered to the market met certain standards and would provide durable and predictable performance. The concern was that without this governmental assistance, the industry might flounder before it could prove itself economically viable. In 1975, available small commercial wind plants generated electricity at 25 cents per kilowatt, or six times that of the average utility-produced rate.[3] "Right now the capital cost per kW of these units is too high to compete with most other forms of energy, except for special applications where there is high wind," Louis Divone, acting chief of ERDA's Wind Energy Conversion Branch, told a wind energy conversion workshop in Washington, DC, in June 1975. "What we face is the problem of how we increase the performance and decrease the cost to allow them to compete on a large scale."[4] The ERDA Federal Wind Program pegged the competitive cost goal for small wind machines at 1 to 2 cents per kilowatt-hour.[5] There was also the concern within the federal government that many small wind machines offered for sale had not been on the market long enough to demonstrate their reliability.[6]

The US government's interest in the wind as a contributor to the country's future energy sources started shortly after the 1973 OPEC oil embargo, during which time gas prices spiked and alerted the country to its increasing dependence on imported energy sources. In December 1973, ERDA's predecessor, the National Science Foundation, initiated a five-year wind energy research and development program for utility-scale wind turbine technology. The foundation, in turn, handed this program over to the National Aeronautics and Space Administration (NASA) Lewis Research Center in Cleveland, Ohio.[7] NASA's engineers already had years of experience in aeronautics, propulsion, and space power systems, as well as large project management, which the foundation believed could lead to a successful national wind energy program.[8] US wind energy research was further strengthened in 1974, when President Gerald Ford signed into law the Solar Energy Research, Development and Demonstration Act, which called for "an intensive research, development, and demonstration program" for solar energy technologies, including wind-electric systems. The goal of the legislation was not only to reduce the country's dependence on foreign fossil fuel imports, but to "expedite the long-term development of renewable and nonpolluting energy resources, such as solar energy."[9] Initial funding in the amount of $24.5 million to launch wind technology research came from the National Science Foundation's Research Applied to National Needs (RANN) program.[10] NASA's research started out to include small (less than 100 kilowatt) and large (100–250 kilowatt) wind energy conversion systems. In 1974, the space agency's Langley Research Center at Hampton, Virginia, erected on the roof of a two-story building a 15-foot tall, vertical-axis wind turbine, with two bowed balsa wood blades coated in fiberglass that resembled an "eggbeater." The design, which was expected to generate 1 kilowatt in a 15-mph wind, was based on a similar machine developed by the Canadian National Research Council in 1970–1972.[11] To encourage the commercial development of the small wind-electric industry, NASA in early 1975 proposed subjecting available machines to wind tunnel and field tests. The information collected from these tests could then be used by the industry to improve the technology.[12]

By the time ERDA was created in 1975, however, it became apparent that NASA's focus was on the development of large-output wind turbines, which the space agency believed "will have to be designed to provide electricity into existing power grids at costs competitive with alternative power sources."[13] Before starting their work, NASA engineers reviewed the large turbine experiments conducted in Europe during the late 1950s and 1960s. One of the most successful and inspiring designs was Denmark's Gedser wind turbine. The three-bladed machine, which was erected in 1957 and operated

until 1968, was capable of producing 200 kilowatts of power in a 33-mph wind. It was connected to a Danish utility and produced about 400,000 kilowatt-hours annually. The French also constructed two large three-bladed wind turbines in the 1950s at Nogent Le Roi, one producing 130 kilowatts and the other 300 kilowatts. However, it was the two-bladed test wind turbines developed and tested by Germany's Dr. Ulrich Hütter, which operated from September 1957 to August 1968, that most attracted NASA's engineers. Unlike the other European wind machines, Hütter's high-speed, variable blade-pitch rotor design allowed his turbines to produce 100 kilowatts of power in an 18-mph wind.[14] The installation cost per kilowatt, however, remained

In 1975, NASA completed construction of a 100-kilowatt, two-bladed rotor at its Plum Brook Station near Sandusky, Ohio. Courtesy NASA Glenn Research Center, Cleveland, OH.

exuberantly high compared to conventional electric power generation. For example, the installation cost for the Gedser wind turbine at the time was about $205 per kilowatt.[15]

At the start of 1974, NASA began work on the design and construction of a 100-kilowatt, downwind test machine, called the Mod-0, at its Plum Brook Station near Sandusky, Ohio. It was completed in August 1975. The Mod-0 consisted of a variable-pitched, two-bladed rotor with a 125-foot diameter, which in an 18-mph wind turned at forty rotations per minute. The rotating rotor shaft drove a gearbox that turned the shaft of an alternator at 1800 rotations per minute to generate the 100 kilowatts of power. The assembly sat on top of a 100-foot-tall, four-post, steel-trussed tower.[16] NASA continued its work on large turbines into the mid-1980s, further developing the technology and increasing the output into the megawatt range, in partnership with large corporate contractors

such as General Electric, Kaman Aerospace, Westinghouse, Grumman Aerospace, Boeing, and Lockheed. These two-bladed turbines were erected throughout the country, including at the MOD-0A (200 kilowatt) sites in Rhode Island, New Mexico, Puerto Rico, and Hawaii.[17]

ERDA's approach to small wind-electric power was wholly different than the work at NASA, which applied to machines that were built from the ground up through contracts with large aerospace companies. The Rocky Flats test site was a location where small commercially available and prototype wind-electric turbines, many of which were manufactured by upstart businesses, could be evaluated in the field. ERDA considered a wind machine to be "commercially available" if the manufacturer built at least ten units and placed at least one of them in operation, while a "commercial prototype" required the manufacturer to have reached a stage in which at least one machine was

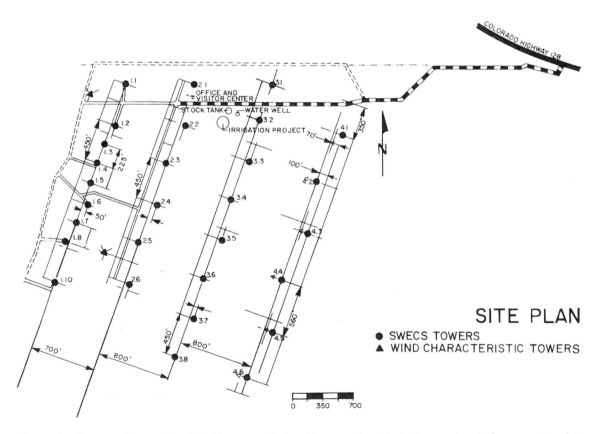

SITE PLAN

● SWECS TOWERS
▲ WIND CHARACTERISTIC TOWERS

The Rocky Flats test site consisted of thirty-two wind turbine test sites. Each site consisted of a support pad, tower, anemometer, and a utility shed with microprocessors. The test rows were oriented to align with the predominantly northwest wind. From the *Wind Energy Innovative Systems Conference Proceedings* (1979).

built and operating but not in production.[18] Within months of securing authorization to start the small wind program, Rockwell International assembled a staff of five engineers who were tasked with laying out the new Rocky Flats test center.[19] The site was divided into five parallel rows, which were perpendicular to the prevailing westerly wind. The first row consisted of six pads, spaced at 450-foot intervals, for towers of 40-foot heights. The second row, which was 700 feet from the first row, offered room for six 50- to 55-foot towers spaced 450 feet apart. The third row provided room for five towers with heights of 65 to 70 feet and spaced about 900 feet apart. The fourth row was set up for four taller towers, and the fifth offered space for two towers. The layout placed smaller machines of less than 6 kilowatts in the first and second rows, with turbines of increasingly larger rotor sizes and output placed in rows three to five.[20] The test facility purchased the test towers, measuring instruments, and electrical controls. Two 130-foot-tall meteorological towers to measure the site's general wind speed and direction, as well as air temperature, were installed by December 1976. Varying types of commercial towers—seventeen in total—were erected along first and second rows between 1976 and 1977. These towers included a four-post, 40-foot Aermotor windmill tower and another of similar structure manufactured by Australia's Dunlite Electrical Products Company. Rockwell International also purchased four 40-foot Unarco-Rohn self-supporting and guyed towers, as well as a combination of other 55-foot towers, including one that folded and another with an "Octahedral" shape. Each tower was equipped with an R. M. Young Propvane anemometer to measure wind speed and direction at each wind turbine. At the base of each tower was a small shed that contained turbine performance measurement instrumentation. A drilled well and stock tank were added to the Rocky Flats site to conduct wind-electric "mini-irrigation" tests. Two enclosed trailers for data collection and personnel were set up on the north end of the property.[21]

The Rocky Flats engineers began installing commercial wind turbines on the towers for testing in December 1976. The first machine to be installed was a French-built Aerowatt. The 4.1-kilowatt rated machine, which was placed on Rocky Flats' tilt tower, had two aluminum blades with a rotor diameter of 30.7 feet. The tail vane for the upwind machine was angled 50 degrees upward from the back of the generator.[22] In January 1977, a 2-kilowatt, three-bladed, upwind Dunlite machine of Australian manufacture was installed on the company-made tower.[23] Four more wind-electric plants were set up during the first two weeks of February 1977, including a 20-kilowatt, three-bladed, downwind Grumman Energy Systems Windstream-25 machine with a 25-foot rotor diameter; a 1.2-kilowatt, three-bladed, downwind Kedco machine with a 12-foot rotor diameter; a 6-kilowatt, three-bladed, upwind Swiss-built Elektro with a 16.5-foot rotor diameter; and a pre-1950s era, 3-kilowatt, three-bladed, upwind Jacobs Wind Electric Company machine with a 13.6-foot rotor diameter, which had been rebuilt by the North Wind Power Company.[24] In June 1977, a small, 7-watt machine with a tetrahelix-shaped rotor with an aluminum frame covered in nylon made by the Zephyr Wind Dynamo Company, was placed on top of one of Rocky Flats' meteorological towers to charge the data logger batteries.[25] A 15-kilowatt, three-bladed, downwind Zephyr with a 20-foot rotor diameter was installed in August 1977 and was followed with the installation of a 1-kilowatt, upwind Sencenbaugh Wind Electric machine with a 12-foot rotor diameter in September 1977.[26] By the end of 1977, Rocky Flats received a 1.5-kilowatt, upwind machine Amerenalt (renamed Altos Corporation), which included an 8-foot-diameter wind wheel made with twenty-four formed Clark Y airfoils attached to stainless steel spokes supported by two concentric aluminum rims, as well as a 2-kilowatt, upwind machine from the American Wind Turbine Company, which included a 15.3-foot aluminum wind-wheel with forty-eight narrow blades held in place by an inner and outer rim—resembling a large bicycle wheel.[27] Additionally, a 2-kilowatt Pinson Energy Corporation vertical-axis machine with three aluminum blades and a rotational diameter of 12 feet was scheduled for delivery in early 1978.[28]

This 2-kilowatt, vertical-axis wind turbine was delivered to Rocky Flats for testing by Aerospace Systems Inc. of Boston and Pinson Energy Corporation in early 1978. US Department of Energy photograph courtesy Warren Bollmeier, Kaneohe, HI.

Simultaneously to the erection of these wind machines, Rocky Flats personnel began acquiring and setting up data collection technology. Microprocessor data loggers were installed in the sheds next to each of the turbine towers to gather information on wind speed and direction, voltage and current output, yaw angle, and rotor speed of each machine. This data was to be transmitted regularly to a central onsite Data General Nova computer, which in turn compiled the data onto magnetic tapes for future analysis. Some of the wind turbine measurements sought from the data collection included power versus wind velocity, which were processed to arrive at coefficients of power versus tip speed ratios of the turbines, and the variance in the data. The goal was to use this data to extract deeper operational details about the machines, such

as the effect of wind passing through and around tower structures on blade vibration and stress, pitch control response of the rotor blades to gusts, and resonant frequencies produced by the wind turbines and their towers during operation. The meteorological towers had their own data loggers for collecting wind speed from the anemometers and direction from wind vanes. This data, too, was transmitted to the Nova computer and stored on magnetic tapes. However, it would take nearly a year before these systems began properly communicating data with each other.[29]

Within the first half year of starting the tests, most of the machines experienced some form of damage. The winds that swept across the flats were often amplified by a funneling effect caused by a gap in the Rocky Mountains, putting the test machines to their maximum operating speeds. Winds blowing across the flats had an average of 10 mph, but at least several times a year, especially during the winter months, could be in the range of 100 mph.[30] On February 22, 1977, the $34,000 Aerowatt machine was blown down in a 90-mph wind during the night. The incident was embarrassing for the newly launched Rocky Flats test site, as a television news crew arrived at 6:30 a.m. the following morning to film the operation of the first wind machines against the sunrise background, only to witness the Aerowatt laying on the ground. In fact, it was not uncommon for the test engineers to find parts of wind turbines twisted, bent, or broken on their towers or scattered around the site after such wind bursts.[31]

Some industry observers questioned the need for a federal small wind research program, since this electric-generating technology proved successful between the 1930s and mid-1950s, prior to the national completion of rural electrification.[32] Some small wind proponents were also skeptical of what they believed to be the Department of Energy's "nuclear and conventional energy bias."[33] Rocky Flats was also criticized for not serving as an ideal site for testing small wind turbines because of its destructive gale-force winds, which "made it look like the contractors didn't build good stuff," as one

This 2-kilowatt Dunlite wind turbine from Australia, which was erected at Rocky Flats in October 1977, met its power rating in 25 mph winds. But like many of the test machines at the site, the Dunlite was torn apart by a high wind burst in November 1977. US Department of Energy photograph courtesy Warren Bollmeier, Kaneohe, HI.

test engineer lamented.[34] But the damage to the machines witnessed by the Rocky Flats engineers at the start of the program told them otherwise. "Therefore, test results to date are in the form of qualitative observations of machine behavior and detection of many 'bugs' inherent in these new systems. The significance of these 'bugs' cannot be ignored, either in the time and attention required of Rocky Flats personnel in curing them or in the level of frustration and disappointment they would present to a consumer," the Wind Systems Program staff at Rocky Flats wrote in an October 1977 paper.[35] Rocky Flats personnel reported operational problems with the test site machines to the respective manufacturers. In some cases, faulty or dam-

aged parts were repaired on site; in the worst cases, the machines were taken off their towers, returned to the manufacturer for repair, and brought back to Rocky Flats for continued testing.[36]

Another unflattering result from the Rocky Flats tests for most of the early commercial wind machines was their failure to achieve their manufacturer's rated outputs. In many cases, the on-site engineers and technicians found that inadequate machine parts led to the inefficiency. In the case of the multibladed American Wind Turbine machine, Rocky Flats contributed its low-power output to "a resistive electrical load instead of a submersible pump motor which the machine was designed to power" and "the high start-up torque of the per-

Wind turbines undergoing testing in 1981 in the third row at the Rocky Flat test site, including (left to right) an 8-kilowatt United Technologies Research Center machine, an 8-kilowatt Windworker, a 40-kilowatt Storm Master turbine, an 8-kilowatt Grumman Energy Systems unit, and a 25-kilowatt Carter Enterprises turbine. US Department of Energy photograph courtesy Warren Bollmeier, Kaneohe, HI.

manent magnet generator."[37] One of the machines that performed most optimally at Rocky Flats was the three-bladed, upwind Dunlite Model 82, which the Australian manufacturer began producing some forty years earlier. "Data taken from the machine show good agreement with the manufacturer's data," Rocky Flats wrote in its report. "The machine achieved its rated output of 2000 watts at just over 25 mph. Its maximum output of 3000 watts was reached at just under the rated wind speed of 32 mph. While [there were] failures during the test period, all occurred in winds well above the manufacturer suggested survival wind of 80 mph."[38] Like most of the machines at the test site, the Dunlite suffered catastrophic damage in November 1977.[39] The Federal Wind Program was careful not to present these test results as a ranking system for commercially available wind machines; rather, they aimed to encourage the manufacturers to initiate improvements and help consumers make better informed decisions when purchasing a wind machine. "We don't want to be in a position to say that a particular machine is bad,

but we hope to provide information that will show which machines are best for certain applications. It is not our intent to attempt to show any machine in a bad light," said Rockwell International's Lou Seaverson in a 1976 presentation.[40]

In early 1977, Rocky Flats was tasked with encouraging the commercial development of higher electrical output wind machines of 8 and 40 kilowatts, as well as 1- to 2-kilowatt wind-electric machines for operating in rural and remote locations. Rocky Flats announced the three programs to the wind-electric industry on March 2, 1977, through the American Wind Energy Association members and solicited their applications for the individual contracts. Each prototype program was structured as a two-phase effort over a nearly two-year period. During the first phase, the manufacturer finalized a design of the prototype, including providing cost estimates to produce one thousand units per year. The second phase involved the fabrication of the prototype machine at the manufacturer's location and then shipping it to

Engineers at Rocky Flats began testing 40-kilowatt turbines at the start of the 1980s, like the McDonnell Douglas "Giromill" and Kaman Aerospace prototypes shown here. US Department of Energy photograph courtesy Warren Bollmeier, Kaneohe, HI.

Rocky Flats for field testing. Most of the machines for the programs were expected to be ready for testing by late 1979 and early 1980.[41]

ERDA and Rocky Flats engineers determined that wind turbines with a rated output of 8 kilowatts in a 20-mph wind were suitable for most residential and farm building applications. In addition to advancing the industry's technical knowledge and capability to develop this machine size, Rocky Flats' cost goal for the system was $750 per kilowatt installed, excluding batteries, inverter, and other secondary components. Four companies were awarded contracts to develop the 8-kilowatt prototypes. Grumman Energy Systems received a $310,000 contract to design a downwind machine with a three-bladed rotor diameter of 33.25 feet. Although considered part of the 8-kilowatt class, Grumman's design was rated to generate 10.5 kilowatts of power in a 20-mph wind and up to 20 kilowatts at higher wind speeds. Rocky Flats awarded a $295,000 contract to Windworks for its 8-kilowatt prototype downwind machine with a three-bladed, 33-foot diameter rotor. Windworks' machine included a direct-drive permanent magnet alternator, hydraulic blade-pitch control system, and "free-flapping" blades made of aluminum and fiberglass. Another contract, valued at $410,000, was awarded to United Technologies Research Center for its downwind machine with 31-foot rotor diameter. The machine used a "composite bearingless rotor" design used in helicopter applications, combined with a "centrifugal force mechanical actuation system" for controlling blade pitch. The Aluminum Company of America received the fourth contract, valued at $303,000, to design and build a vertical-axis machine of a Darrieus-style, three-bladed rotor with a 33-by-33.7-foot diameter.[42]

Through its own analysis and communications with the US Department of Agriculture's Agricultural Research Service, the Rocky Flats engineers determined that wind turbines with a rated output of 40 kilowatts in a 20-mph wind were well suited for large-scale water-pumping applications on farms and ranches across the Southwest and Great Plains. They based their analysis on the fact that wind-generated electricity could be used in a supplemental capacity to power deep-well, heavy-duty turbine pumps, which were predominantly powered by natural gas-fired engines at the time. By the mid-1970s, irrigation-dependent crop farmers in the Texas High Plains encountered higher prices for natural gas—in the range of $2 per 1000 cubic feet—and were expected to use larger volumes of this fuel in the coming years as water tables dropped between 1 and 4 feet a year.[43] According to Rocky Flats engineers, "Both this region of Texas and other areas using a high level of irrigation lie in a portion of the United States which is comparatively 'rich' in wind energy resources. Therefore the possibility exists for utilizing the wind as an alternative energy source for water pumping applications."[44] Another potential application considered for 40-kilowatt-size wind turbines by the Federal Wind Energy Program was providing electric power to small remote villages and businesses in mountainous regions and on coastal islands.[45] Two firms received contracts to develop 40-kilowatt prototype turbines. Kaman Aerospace received a $370,000 contract to design and build a two-bladed downwind machine with a rotor diameter of 64 feet and rated to produce 40 kilowatts in a 20-mph wind.[46] A contract for a wholly different wind turbine design was awarded to McDonnell-Douglas Aircraft Corporation. The aircraft manufacturer, along with technical assistance from Valley Industries' Aermotor Division, proposed to the government a wind machine called the Giromill (short for cyclogiro windmill), which consisted of three 42-foot-long vertical blades rotating around a central tower at a 58-foot diameter. The angles of the blades to the wind were individually modulated to receive the wind most efficiently. The Giromill was also designed to rotate independent of wind direction.[47]

Rocky Flats engineers also forecast the need for 1- to 2-kilowatt wind machines to provide electric power for remote applications, such as cabins, communications repeater stations, galvanic protection to buried pipelines, remote seismic monitoring stations, water pumping, and offshore navigational aids. "A system with an electrical output of about 1 kW would meet the requirements of these applications, provided the system could operate reliably over a wide range of geographic locations," they said.[48] Emphasis was placed on the durability of these machines, since the expense of traveling to a remote site to conduct maintenance work might exceed the system's actual cost. This meant the machines had to successfully operate in some of the harshest weather conditions on earth, including Arctic winter winds and icing conditions in the case of Alaska and the Rocky Mountain range, and coastal hurricanes where winds can reach upwards of 150 mph and coat materials with corrosive saltwater.[49] In January 1978, Rocky Flats awarded three contracts for what was called the "high-reliability" wind turbine program. They included a $280,000 contract to Aerospace Systems, in collaboration with Pinson Energy, to develop and construct a 1-kilowatt, vertical-axis, three-bladed Cycloturbine with a 15-foot rotor diameter, capable

of self-starting, maintaining rotor control during operation, and shutting down in high winds. A $150,000 contract was awarded to Enertech for a 2-kilowatt, two-bladed, downwind machine, with a hub design that caused the rotor blades to stall in excessive winds. Lastly, North Wind Power received a $260,000 contract to develop a three-bladed, upwind machine with a 16.9-foot rotor diameter equipped with an automatic tilt-back mechanism that caused the rotor assembly to shut off in a horizontal position for high-wind protection.[50] In addition to operational reliability and durability, Rocky Flats engineers set the cost goal for these machines at $1500 per kilowatt, excluding tower and storage equipment.[51] All three prototype machines were to be performance tested at the Rocky Flats test site by May 1980.[52]

Small wind turbine manufacturers shown testing their machines at Rocky Flats in 1981, including (left to right) Kedco Inc., North Wind Power Company, ASI/Pinson Energy Corporation, Bergey Windpower Company, and Enertech. US Department of Energy photograph courtesy Warren Bollmeier, Kaneohe, HI.

Even though these prototype wind turbines were built by engineers with knowledge of aerodynamics and access to the latest design technology of the day, it did not take long for technical weaknesses within these machines to become evident at the Rocky Flats test site. The engineers' method of "free stream" wind testing, as opposed to controlled wind tunnel and vehicle-mounted testing programs, meant that turbines could be operating in modest 10- to 20-mph winds for a period and then suddenly face gale-force winds of more than 80 mph that took these machines to their technical limits and inflicted damage.[53] At first, the McDonnell-Douglas Aircraft-inspired Giromill showed promising performance at Rocky Flats but began to experience routine breakdowns in its blade actuator. On December 24, 1980, the Giromill's lightning rod snapped off and damaged a blade and support arm, and then the day after Christmas the rotor on the stub tower blew over. Although the approximate $1.5 million test machine was repaired and returned to service in April 1981, the ongoing technical problems led to it being taken down in the summer of 1982. Both McDonnell-Douglas

and Valley Industries' Aermotor Division were disappointed to have to abandon the program, but neither company could justify the cost of continuing.[54] Other companies credited their participation in the Rocky Flat prototype tests with providing the necessary technical foundation to produce their later commercial wind turbines. Windworks in mid-1980 began selling a 10-kilowatt machine based on the design of its 8-kilowatt prototype that was tested at Rocky Flats.[55] Enertech used its Rocky Flat test participation to refine the commercial production of a three-bladed, downwind machine with a rated output of 1.5 kilowatts, as well as larger turbines shortly thereafter.[56] In 1980, North Wind Power also began offering a commercial variant of its high-reliability, three-bladed, upwind 2-kilowatt machine with the tilt-back shutoff mechanism for remote applications.[57]

Prototype wind machine test contracts continued to be offered by Rocky Flats to the small wind industry during the late 1970s and early 1980s. Tumac Industries and North Wind Power received contracts to design and fabricate prototypes with 4-kilowatt outputs, while Enertech and United

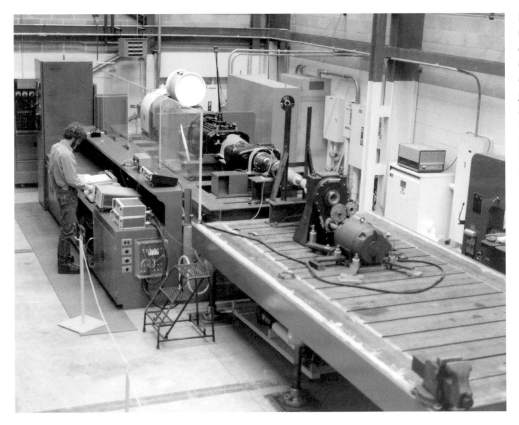

Rocky Flats engineers used this dynamometer during the early 1980s to define the performance characteristics of gearbox and generator assemblies. US Department of Energy photograph courtesy Warren Bollmeier, Kaneohe, HI.

Technologies Research Center were contracted to build 15-kilowatt test machines. Rocky Flats was also considering how to expand the scope of its tests and studies to include hybrid systems (combining wind power with other renewable energy sources, such as photovoltaic solar panels), rotor performance and control, and component fatigue and reliability, to name a few.[58] In 1980, Rocky Flats established a dynamometer test facility to test the performance of generators with outputs of up to 40 kilowatts (and shortly thereafter to 100-kilowatt outputs). A fixed direct current motor was used to turn a variable speed transmitter and torque transducer, which was attached to the rotor shaft of the wind-electric generator. A computer was used to measure and record the input torque, rotational speed, gearbox temperatures, vibration, and power-out characteristics during the tests.[59] The test center's engineers also sought to understand better the power connections between small wind-electric machines and utilities. A special "mini-grid" or "electrical island" was established at Rocky Flats to perform and evaluate interconnectivity tests in a controlled environment.[60]

Rocky Flats expanded wind machine testing beyond the confines of its Colorado site by introducing a Field Evaluation Program in 1980. The purpose of the Department of Energy program was to help newfound wind-electric businesses spread the installation of their machines within the community. These companies faced myriad challenges, including utility hesitation to grid interconnections,

Rocky Flats engineers subject wind turbines to controlled velocity testing at the American Association of Railroads test facility in Pueblo, Colorado. This test involved mounting the wind turbine system to a flat railcar, which was pushed by a locomotive down a track at predetermined speeds. US Department of Energy photograph courtesy Warren Bollmeier, Kaneohe, HI.

prohibitive local and state installation codes, difficulty securing bank loans and insurance coverage for small wind systems, and a general lack of public knowledge about how wind generators operate. Rocky Flats was authorized to purchase as many as 120 commercially available small wind machines and award them to people across the United States and territories. The first of these units were erected in New England during the autumn of 1980. Rocky Flats designated seven field engineers to carry out this work.[61] A Rocky Flats official observed at the time that "[t]hese engineers are envied by some wind systems people for their exotic travel schedules. . . . But the engineer who inspects potential installation sites in the balmy Virgin Islands one week could find himself spending the next week shivering above the arctic circle. One field engineer spent several memorable November days in the crawl space under a house in Kotzebue, Alaska."[62]

One of the more interesting, offsite tests for Rocky Flats, which commenced in 1980, involved mounting stub towers to the beds of two flat railcars on which wind generators were attached. The railcars were located at the Department of Transportation's 52-acre rail test facility outside of Pueblo, Colorado. The facility allowed the engineers to conduct controlled velocity tests of commercial wind machines, while the railcars were pushed along a 2.5-mile precision track at 20 mph. The tests were similar to those presented in wind tunnels or road vehicle tests, a practice first utilized by wind-electric plant manufacturers during the 1920s and early 1930s to test propeller designs and carried on by a number of the new manufacturers in the mid- to late 1970s. Rocky Flats engineers, however, believed the rail-powered controlled velocity tests were superior to wind tunnel and vehicle tests by offering finer control over low-speed tests, eliminating ground-plane turbulence, and the ability to obtain other test data such as blade pitch and yaw tracking. The railcar tests were made available first to wind machines with rotor diameters of 5 meters (16 feet) or less, with the goal to increase the rotor sizes up to 10 meters (32 feet). Five machines were tested on these railcars between 1980 and 1981.[63]

Rocky Flats also supported other small wind energy research programs outside its direct oversight.

These included wind machine test monitoring and support at the University of Massachusetts and Oklahoma State University. The National Science Foundation provided grants to both university programs in 1975 and transferred oversight responsibility to Rockwell International at Rocky Flats in 1976. Under the direction of Professor William E. Heronemus and a cadre of faculty and students, the University of Massachusetts' Department of Engineering at Amherst in the early 1970s set out to design and build a 1500-square-foot house that would offer 80 percent heating self-sufficiency through the combination of a wind turbine and flat glass plate solar collector. Dubbed the "Wind Furnace" project, by 1976 the university developed a three-bladed downwind machine with a 32.2-foot diameter and rated output of 24 kilowatts. The pitch of the machine's fiberglass blades was controlled by sensing the blade tip speed to wind speed ratio. The unit, which was placed on top of a 60-foot guyed, monopole steel tower, stood next to the test house. The electricity generated by the wind turbine was directed to a load controller, which in turn energized heating elements submersed in a 1500-gallon basement water tank. The tank furnished both the heat and hot water for the house. The Wind Furnace concept, as well as the university's wind turbine design, proved successful but did not take hold within commercial home construction industry as Heronemus and his team had hoped.[64] Meanwhile, in 1974, Dr. William Hughes of Oklahoma State University began performance testing of a 15.3-foot-diameter, multibladed machine designed a year earlier by Thomas O. Chalk of St. Cloud, Florida. The wind wheel, which resembled a bicycle rim, operated as part of the generator in that the field poles were embedded in the rim, eliminating the requirement for gears between the rotor and generator. Hughes's goal was to further the commercialization of Chalk's machine by including automatic systems for rotor directional control and regulating voltage output, and then double the size of the wind wheel to 30 feet for larger applications. At the time of these tests, a group of investors using Chalk's design established the American Wind Turbine Company in Stillwater, Oklahoma.[65]

The Rocky Flats test site accepted wind turbines for testing with a variety of different rotor types, like this multibladed, bicycle-shaped one by the American Wind Turbine Company, in 1978. US Department of Energy photograph courtesy Warren Bollmeier, Kaneohe, HI.

The Rockwell International staff at Rocky Flats, in coordination with subcontractors and the American Wind Energy Association, developed an array of reports related to the emerging small wind energy industry. These reports educated private individuals who were interested in acquiring small wind-electric systems for personal use, as well as informed municipalities, utilities, banks, and insurance companies about the operations of these machines and how they could be safely integrated into the community. One of the simple but most useful reports for consumers during the mid-1970s was the *Commercially Available Small Wind Systems and Equipment* booklet, which listed about forty-five US companies in the process of manufacturing

and selling wind-electric machines, as well as about a half dozen mechanical water-pumping windmill manufacturers. The booklet also listed numerous wind machine dealers and distributors, as well as suppliers of wind anemometers and recorders, towers, batteries, inverters, and do-it-yourself plans.[66] Some reports produced by the Federal Wind Energy Program offered a snapshot of the small wind-electric industry's well-being during the late 1970s. A 1979 report produced under subcontract to consulting firm JDB & Company of Washington, DC, declared: "The harsh reality is that, for several reasons, most of the existing SWECS (small wind energy conversion systems) will not succeed in the market place," adding that despite the federal

A view of the Rocky Flats wind systems test center administrative building, with a 100-kilowatt Alcoa vertical-axis wind turbine nearby. US Department of Energy photograph courtesy Warren Bollmeier, Kaneohe, HI.

government's support for developing a commercial market for small wind machines, most of the companies were still plagued by capital, management, staffing, and cash flow problems.[67] From a technological perspective, Rocky Flats also reported at the end of the decade that "current SWECS, while nearing the threshold of competitiveness with conventional energy sources, are inhibited from reaching their lowest cost potential by the use of off-the-shelf components, less than optimum rotor designs, and (in some cases) overly complicated control systems."[68] The test engineers believed the Department of Energy's wind machine prototypes and advanced concepts demonstrated "significant energy cost improvements over commercially available systems."[69] Whether the test results were good or bad, Rocky Flats believed it was important to make the information publicly available. In fiscal year 1980–1981 alone, the test center released twenty-three technical publications.[70]

By end of the 1970s, Rocky Flats had emerged from a barren patch of no-man's land into an internationally recognized small wind systems test facility, with a staff of sixty-eight engineers, technicians, and administrators.[71] Further demonstrating its commitment to Rocky Flats, the Department of Energy in 1980 began construction of a 22,700-square-foot building to house its offices, laboratories, machine shop, and equipment assembly area.[72] The Federal Wind Energy Program had a budget of $67 million in fiscal year 1980, $14 million of which was allocated to Rocky Flats. "Where does this money go?" asked Darrell Dodge, an information specialist at the Wind Systems Office, at the time. "Most goes to subcontractors, as directed by DOE. Another large portion goes to purchase small wind systems for testing or evaluation under several programs. The rest is spent on the Rocky Flats Plant for salaries, test equipment and facilities. Some of these funds pay for plant overhead expenses but all of these funds are ear-marked by DOE for the Wind Systems Program. They do not drain on plant weapons funds."[73] Despite the test site's proximity and contractual connection to nuclear weapons component manufacturer Rockwell International, the Department of Energy "maintained a consistent flow of information to the news media" about the small wind test activities at Rocky Flats and granted occasional access to the site for coverage by local and national television crews.[74] Rocky Flats also gave the public a glimpse into its activities through displays, which were exhibited at numerous energy

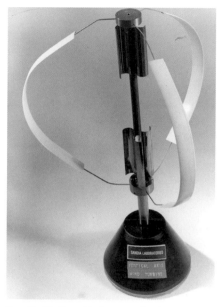

Sandia National Laboratories in Albuquerque, New Mexico, began to experiment with Darrieus wind turbines in early 1970s, starting with this 12-inch desktop model to these larger rooftop prototypes. Courtesy Herman Drees, Thousand Oaks, CA.

expositions and fairs throughout the country. An estimated 3.5 million people viewed these displays in 1979 alone.[75]

Rocky Flats and NASA weren't the only federal government agencies involved in wind energy research during the 1970s. In 1974, engineers at Sandia National Laboratories in Albuquerque, New Mexico, initiated a study of the Darrieus vertical-axis wind turbine, which included evaluating the aerodynamics and structural performance of this "egg-beater"-style machine, with the objective of demonstrating that the design could be operated practically and easily integrated with the power grid. Another goal for the Sandia engineers was to develop a vertical-axis machine that could be manufactured more simply and less expensively than the majority of horizontal-axis machines being produced for the commercial market. Some earlier efforts were made to develop Darrieus-style machines, such as the work undertaken by Canada's National Aeronautics Establishment during the 1960s.[76] Sandia's first vertical-axis machine stood at only 12 inches tall and sat on an engineer's desk. This was scaled up to a 6.5-foot (2-meter) diameter wind tunnel test model. In the spring of 1974, Sandia engineers erected a 16-foot (5-meter) tall, three-bladed test unit on the roof of the agency's administration building. Since the vertical axis around which the blades rotated served as the tower, the generator was mounted on the ground rather than on top of the tower in the case of horizontal-axis machines. The blade cores for the rooftop model consisted of thin steel coated with a layer of foam and fiberglass. While the test unit showed promise, the blades had to be hand-turned two to four times faster than the wind before the machine turned them on its own. There was also a problem with runaway blades, which the engineers rectified by introducing a disc brake system similar to that used in automobiles.[77]

After two years of rooftop testing, Sandia in 1978 designed a two-bladed machine with a vertical axis 82.5-feet (25.15-meter) tall and a diameter of 56 feet (17 meters) capable of generating up to 100 kilowatts in a 31-mph wind. The design and scale of the unit offered an alternative to the large-scale, horizontal-axis machines under development by NASA and its contractors at the time.[78] In 1979, the Department of Energy awarded Alcoa a contract to build four of these large machines for testing. Construction began in January 1980 and lasted until March 1981. Only three test machines, however, were ultimately built by Alcoa due to budget constraints. The machines were erected at different sites for specific tests: Rocky Flats for structural performance; Bushland, Texas, for agricultural applications; and Martha's Vineyard, Massachusetts, to feed a power grid.[79] From 1974 to 1980, the laboratory published more than thirty-five technical reports from its research of the vertical-axis wind turbines.[80] Sandia's vertical-axis design work inspired FloWind Corporation to take the lead on constructing up to 500 commercial units of this type from the mid-1980s to early 1990s for California's wind farms at Altamont and Tehachapi passes.[81]

One of those federal programs, which was more closely aligned to Rocky Flats but involved practical application tests of commercially available small wind technology, was managed by the US Department of Agriculture's Agricultural Research Service at Bushland, Texas, starting in 1976. The remoteness, as well as high-energy use, of many Midwestern American farms made them attractive candidates for standalone wind-electric systems. An Agricultural Research Service study commissioned with the Manhattan, Kansas-based Development, Planning and Research Association determined that 996,700 farms across twenty-one states in the mid-1970s were potential candidates for wind-electric systems and that two-thirds of those sites were concentrated in Texas, Missouri, Iowa, Oklahoma, Kansas, and Nebraska.[82] The biggest, most costly use of energy on these farms at the time, particularly across the Southern Great Plains, was for operating high-output irrigation pumps. According to the Agricultural Research Service, traditional mechanical windmills generally pumped less than 10 gallons of water per minute, while several hundred gallons of water per minute were required for practical irrigation. Many farmers by the early 1970s used natural gas-powered, internal combustion engines installed at well heads to pump this water.[83] For its irrigation tests, the agricultural agency was especially interested in the prospects of integrating vertical-axis wind machines with irrigation systems due to their simplicity and reasonable efficiency over horizontal-axis machines that were just entering the commercial market at the time. In partnership with physicists from West Texas State University in nearby Canyon, Texas, a two-bladed vertical-axis wind machine of the Darrieus design manufactured by Dominion Aluminum Fabricating (DAF) Limited of Ontario, Canada, was erected at the Bushland site over a preexisting 320-foot-deep well in early 1977. The wind machine was 55-feet tall with a rotor diameter of 37 feet and was placed upon a four-post steel tower 30-feet tall. It had a rated output of 65 kilowatts in a 30-mph wind. The system was designed to take over from utility-line power feeding a continuously operating pump when the winds exceeded 13 mph. In late 1977, Bushland acquired a second, smaller DAF vertical-axis machine with a rated output of 4 kilowatts for trials.[84]

The Agricultural Research Service also conducted a series of wind power application tests of universities throughout the country. An irrigation contract was entered into with Kansas State University at Manhattan using a 30-foot-tall, 20-foot-diameter DAF machine to perform shallow well-pumping tests in tailwater pits that collect the runoff from field irrigation. The wind-electric-powered water pump was used to recirculate the tailwater back across the fields. Another test, which took place at Iowa State University in Ames, focused on how to use wind energy to reduce heating costs for farm households. This three-year test incorporated a 15-kilowatt, horizontal-axis Grumman Windstream 25 machine to heat water electrically. At Virginia Polytechnic Institute in Blacksburg, a 10-kilowatt, horizontal-axis Swiss-built Elektro wind machine was used to run the refrigeration equipment and air-circulating fans for an apple storage cooling system. Another Kaman subsidiary, Kaman Science Corporation, collaborated with Colorado State University at Fort Collins to integrate a 20-foot diameter DAF machine with the school's campus dairy farm milk cooling system. In addition, the Agricultural Research Service and Cornell Univer-

The US government began promoting the use of wind-electric power for agricultural applications during the late 1970s and early 1980s. Shown here in this Department of Energy photograph is a 15-kilowatt Grumman Windstream 25 turbine on an updated New York farm in 1978. From the author's collection.

sity at Ithaca, New York, installed a vertical-axis Pinson Cycloturbine and coupled it with a stirring motor inside a water tank. The purpose was to test the ability to heat water through the frictional force of agitation.[85] From these tests, the Agricultural Research Service engineers at Bushland, as well as at the universities, generally found that the wind machines performed "below predicted levels" and encountered numerous malfunctions.[86] Researchers at Bushland, as well as Kansas State University and Colorado State University, were disappointed with the initial performance of the DAF machines tied to their experiments, pointing to "non-streamlined parts" and suboptimal electrical outputs. They often resorted to making their own repairs or improvements to the machines without technical support from the Canadian manufacturer.[87]

By the start of the 1980s, the Agricultural Research Service's Conservation and Production Research Laboratory and West Texas State University's Alternative Energy Institute had expanded their field testing of wind-electric systems performance and reliability analysis at both the Bushland and Canyon test beds, as well as on nearby ranches and business locations.[88] These tests broadened the earlier studies on deep-well irrigation pumping, building and water heating, and load demands, as well as integrating wind-electric machines with rural electric cooperatives.[89] The researchers also considered the prospects of using wind-generated electricity as supplemental power for low-temperature crop-drying equipment and saltwater conversion to fresh water via reverse-osmosis.[90] Other wind machines erected for testing at the time included three two-bladed, 25-kilowatt Carter 25s at the Canyon well field; a three-bladed, 15-kilowatt United Technologies Research Center 108-foot-diameter machine to supplement power at a Phillips Petroleum stripper oil well near Borger, Texas; a three-bladed, 25-kilowatt Enertech 44; a three-bladed, 25-kilowatt Wingen 25 used to power an irrigation pump at the Stewart Kirkpatrick farm near Tulia, Texas; and a three-bladed, 10-kilowatt Windworker.[91] In the mid-1980s, the Department of Energy asked Sandia National Laboratories to find ways to further reduce the cost and improve the performance of vertical-axis wind turbines. Sandia's engineers determined that a two-bladed machine with 112-foot diameter on a 165-foot vertical tower was optimal. The machine was solely for test purposes and erected at the Agricultural Research Service's Bushland site. With a variable speed—between 28 and 38 rotations per minute —the turbine, which was set into motion by an electric motor, was rated at 500-kilowatt output in 25- to 40-mph winds. The blade ends attached to the end of a rotating center axis tube served as a main shaft, which, in turn, drove a speed-increasing transmission and then the generator. Sandia engineers used a modular design for the tower tube and blades sections and equipped the machine with more than 150 sensors to measure its performance. (The test machine remained at Bushland until its removal in the spring of 1998.)[92]

The Department of Energy's Federal Wind Energy Program encompassed two other important programs. Congress established the Solar Energy Research Institute (SERI) in 1974 at Golden, Colorado, to serve as a national center to conduct research and promote the development of solar energy.[93] One of the myriad tasks assigned to SERI included the Department of Energy's Technical Information Dissemination Program for wind energy. SERI provided the public with numerous wind energy research and analysis reports performed by the federal government.[94] SERI researchers also conducted systems and operations analysis for small wind energy systems, including the use of computer modeling to determine economic values for wind energy system installations, evaluating noise and predicting television interference from wind turbines, identifying a passive blade cyclic pitch concept for testing at Rocky Flats, and completing a range of legal and institutional reports covering requirements such as utility interconnections, safety analysis, financing, state and local regulation and incentives, product liability, and legal and regulatory issues.[95] In 1974, the Department of Energy contracted with Battelle Memorial Institute's Pacific Northwest Laboratory in Richland, Washington, to conduct extensive research, along with numerous university laboratories, into wind characteristics. The analysis generated by the laboratory and its contractors assisted companies with designing wind machines and operating them at the most optimal sites throughout the United States and its territories. In 1980, the Pacific Northwest Laboratory published *A Siting Handbook for Small Wind Energy Conversion Systems*, which small wind turbine installers used for years.[96]

National concern about overdependence on foreign oil raised its head again in 1979 when OPEC member Iran suddenly decreased its oil output due to internal strife and its displeasure with Western countries for their support of the deposed shah. Oil prices rose to $39.50 per barrel, and fuel supplies contracted over a twelve-month period, resulting in long lines of cars at US gas stations—a visual reminder of the 1973 oil crisis.[97] The nuclear power industry also took a hit in late March 1979 when the reactor at Three Mile Island, 11 miles south of the Pennsylvania state capital of Harrisburg, experienced a partial meltdown and heightened concerns about the public safety of this energy source.[98] In light of these energy crises, as well as political pressures resulting from a beleaguered US economy, President Jimmy Carter addressed the nation on July 15, 1979, in his "Energy and the National Goals" speech, also known as "A Crisis of Confidence." He stated, "It's clear that the true problems of our nation are much deeper—deeper than gasoline lines or energy shortages, deeper even than inflation or recession."[99] To enhance the country's "energy security," Carter asked "for the most massive peacetime commitment of funds and resources in our Nation's history to develop America's own alternative sources of fuel," which included renewable energies.[100]

In 1980, Congress advanced the nation's wind energy program by passing the Wind Energy Systems Act. The legislation specifically called for a further reduction in the cost of electricity produced by wind energy systems to a level competitive with conventional energy sources; installation of at least 800 megawatts of wind-electric power by 1988, of which 100 megawatts would be attributed to small wind systems; and increasing the growth of the commercial wind-electric industry.[101] For fiscal year 1980, the Department of Energy's Federal Wind Energy Program was appropriated its largest level of funding at $63.4 million, up from $59.6 million in fiscal year 1979 and $35.5 million in fiscal year 1978.[102] But the latest oil crisis was short-lived due to a rising global surge in oil production and wind energy's champion, President Carter, was replaced by a new president, Ronald Reagan, in 1981, who was less interested in promoting renewable energy development. In fact, Reagan quipped during an election debate with Carter that "[t]he Department of Energy has a multibillion-dollar budget, in excess of $10 billion. . . . It hasn't produced a quart of oil or a lump of coal or anything else in the line of energy."[103] Despite this political shift and shrinking budget, allocations for wind energy research—$34.4 million in fiscal year 1982—the Federal Wind Energy Program (renamed the Department of Energy's Wind Energy Technology Division) appeared confident in its ability to con-

The NASA Mod-5B wind turbine, constructed by Boeing, shown here in 1987 on Oahu, Hawaii, had a rated power output of 7.2 megawatts in wind speeds of 30.6 mph. Courtesy NASA Glenn Research Center, Cleveland, OH.

tinue its work. Rocky Flats, for example, announced plans for five small wind energy research projects, which included aerodynamics, structural statics and dynamics, power subsystems, systems integration, and industry-funded testing and analysis at the test site.[104] In 1984, the Department of Energy began assessing the fragmented nature and efficiency of its overall wind energy program due to budget cuts and staff reductions, and by October of that year it merged SERI with Rocky Flats.[105]

At the start of 1985, the Department of Energy released its five-year wind energy research plan for the years 1985–1990, which included the work of the SERI Wind Energy Research Center at Rocky Flats, Sandia Laboratories, and Pacific Northwest Laboratory. The program's overall fiscal year 1985 budget was $29.1 million, up from $26.5 million in fiscal year 1984. The plan acknowledged NASA's preparation to exit wind energy research activities after completing its Mod-5 program, which aimed to build a large wind turbine with a rated output of 3 megawatts.[106] In July 1987, NASA and its contractor Boeing completed construction of a massive two-bladed, 3-megawatt machine at Oahu, Hawaii.[107] However, the space agency already started turning over its wind energy research to SERI in late 1985.[108] Rocky Flats also determined that new research should focus on expanded funds for the wind industry and university programs. This resulted in the development of the Cooperative Field Test Program, which involved cost-sharing arrangements for field testing already available commercial wind machines in wind farms and other applications. The goal for the program was to provide Rocky Flats researchers with data and to help wind machine developers resolve specific technical problems, particularly in the areas of siting, aerodynamics and structural loads, and fatigue. There was also the initiation of a new university research program, which called for academic institutions to be more closely

US Department of Energy wind systems engineer Warren Bollmeier inspects a 2-kilowatt Enertech turbine at the Rocky Flats test site in 1983. Courtesy Warren Bollmeier, Kaneohe, HI.

involved with Rocky Flats' activities. Lastly, Rocky Flats started a wind-diesel hybrid system project, which would allow its researchers to determine appropriate configurations, hardware, and controls for electric-generating systems that used a combination of wind turbines and diesel generators.[109] Over the next five years, the Department of Energy focused its research efforts on further driving down the cost of wind-electric power from 10 to 15 cents per kilowatt-hour (it was $1.50 per kilowatt-hour in the late 1970s) to 4 cents per kilowatt-hour through decreased capital costs, enhanced performance, and longer operational life of wind turbines. The department worried that without this per-kilowatt cost reduction, the wind industry would be unable to compete more widely against conventional power sources, such as coal, oil, and natural gas, in the event that the wind tax incentive program ended.[110]

The US wind industry's momentum, however, was severely curtailed when the Reagan Administration and the Congress allowed the wind investment tax credits to expire in late 1985. For small wind-electric systems owners, the tax credits covered 40 percent of the cost of a wind turbine, or a maximum of $4000, which helped to stimulate commercial sales of these machines. With that incentive suddenly removed, three-quarters of the nearly forty US companies manufacturing small wind turbines either closed their doors immediately or dropped out of the market shortly thereafter due to loss of sales.[111] The budget for the Department of Energy's wind energy program by fiscal year 1988 was also slashed to about $8 million, causing research efforts to shrink and stall.[112] Many industry observers at the time believed this lack of support for wind energy research during President Reagan's second four-year term ceded this research to the Europeans, namely, the Danes and Germans; even Japan stepped up its entrée into the wind energy industry.[113] In a symbolic demonstration of the Reagan Administration's disregard for renewable energies in favor of boosting fossil fuels, the thirty-two solar water-heating panels installed on the roof of the White House by the Carter Administration in June 1979 were quietly removed by contractors during resurfacing work in 1986.[114]

When President George H. W. Bush entered the White House at the start of 1989, the Department of Energy's wind energy research program remained a paltry $8 million for the fiscal year and was increased to $11.2 million by fiscal year 1991.[115] During the Persian Gulf War (August 2, 1990–February 28, 1991), the American public again became aware of the country's uneasy dependence on oil from the Middle East. Iraqi military forces under Saddam Hussein temporarily seized the oil assets of Kuwait and threatened to take other neighboring countries' oil fields, until they were driven back into Iraq by the US-led coalition forces. The Bush Administration became convinced that, in the interest of national security purposes, the United States needed to refocus its efforts on renewable energy technology development. To lead this effort, the White House on September 16, 1991, designated the contract-managed Solar Energy Research Institute, which included the Rocky Flats wind test site, to become the National Renewable Energy Laboratory (NREL), covering all forms of renewable energies. "Cost-effective renewable energy technologies can contribute in their way to a strong and growing economy domestically, by spurring competition and innovation in US markets, and in our balance of trade, by displacing more expensive imported energy and providing new services and products for export," Bush said during the announcement.[116] As the sole federal renewable energy laboratory, this meant that wind technology research had to compete for its share of the annual budget with other renewables, such as photovoltaic solar, fuel cells, and biomass. NREL's Wind Technology Division received $21.4 million in fiscal year 1992 and $24 million in fiscal year 1993 at the end of the George H. W. Bush presidency.[117]

Rocky Flats, now designated as the National Wind Technology Center, became the epicenter for advanced wind turbine research, but with an increasing focus on larger, utility-scale machines. Starting in the mid-1990s, the facility underwent an upgrade to its existing buildings, including the construction of a new 10,000-square-foot industrial user facility to support collaborative research and development with the commercial wind industry. The new building was designed to perform durability tests on turbine blades more than 90-feet long. (It was later expanded to test blades more than 164-feet long.) In addition to the site's administrative building, a third structure housed dynamometers that served as a de facto wind turbine simulator by allowing researchers to test electrical, mechanical, and control systems of turbine gearbox equipment.

President George H. W. Bush on September 16, 1991 announced the designation of the Solar Energy Research Institute as the US Department of Energy's National Renewable Energy Laboratory. Standing next to President Bush is Deputy Energy Secretary W. Henson Moore. Courtesy National Renewable Energy Laboratory, Boulder, CO.

The site maintained its series of turbine test pads—over twenty—arranged across four evenly spaced rows. The first row consisted of smaller turbines with rated outputs of 10 to 50 kilowatts, in addition to a pad to test village power systems. The second row contained the so-called Advanced Research Turbine (ART) Facility, consisting of two large turbines—a three-bladed upwind and a two-bladed downwind machine—placed on top of 82-foot-tall standalone tubular towers. The gearboxes were designed to allow researchers to test different drive train components. The third row contained wind turbines of 100 kilowatts and larger. The fourth row was reserved for megawatt turbines. A goal for the National Wind Technology Center was to develop certification test capabilities that would meet both domestic and international installation and operation requirements to satisfy standards programs.

At the start of 1993, newly elected President William J. Clinton and Vice President Albert Gore Jr., both were outspoken proponents of reducing greenhouse gas emissions by supplanting fossil fuel-power generation with renewable energies. Clinton's Climate Change Action Plan called for more than fifty new and expanded incentives to reduce the country's emission levels.[118] The administration, however, dedicated only modest funds toward research at the National Renewable Energy Laboratory during the mid-1990s. In 1995, the laboratory was forced to reduce its overall workforce of 900 employees by 10 percent to meet the anticipated fiscal year 1996 budget shortfall of $30 million to $57 million from a fiscal year 1995 operating budget of $237 million.[119] On October 22, 1997, President Clinton announced the Climate Change Technology Initiative, which called for $3.6 billion in new and existing tax credits for acquiring renewable energy technologies and $2.7 billion in renewable energy research funding, but the plan failed to gain traction with the White House plagued by scandal and an uncooperative Congress.[120] The National Wind Technology Center's research continued, despite the budget constraints, even announcing a $4.5 million small wind turbine project in November 1996 which aimed at analyzing the long-term performance and reliability of machines with rated outputs of 5 to 40 kilowatts.

The four companies picked to design, build, and test prototype small wind turbines for the project included Bergey Windpower Company of Norman, Oklahoma; Cannon-Wind Eagle Corporation of Tehachapi, California; Windlite Company of Mountain View, California; and Wind Power Technologies of Duluth, Minnesota.[121] Promotional activities related to wind energy became part of the Department of Energy's recently formed Office of Energy Efficiency and Renewable Energy. Wind energy in the United States, in general, during the Clinton Administration received a boost with the 2000 announcement of the Wind Powering America Initiative, which sought to generate 5 percent of the nation's electricity through wind power by 2020. While the majority of this wind energy would be generated by large turbines, the initiative promoted small wind projects for communities, individual homes, and small businesses.[122]

President George W. Bush, like his father, viewed the promotion of renewable energy sources, such as wind, as a key component of the country's national security, especially in the aftermath of the September 11, 2001, terrorist attacks, and the nation's reduced reliance on foreign energy imports. While previously serving as the governor of Texas, Bush had promoted the development of massive wind farms across the western part of the state. Texas quickly became the largest producer of wind energy by the early 2000s, unseating California as the traditional leader in this space. The Energy Policy Act of 2005 called for enhanced wind energy research by the federal government, and in January 2006, during his second term in office, President Bush announced the Advanced Energy Initiative, which further boosted research initiatives at the National Renewable Energy Laboratory in Colorado.[123] At the time, the laboratory's Wind Energy Program declared that its mission was to "support the President's National Energy Policy and Departmental priorities for increasing the viability and deployment of renewable energy; lead the Nation's efforts to improve wind energy technology through public/private partnerships that enhance domestic economic benefit from wind power development; and coordinate with stakeholders on activities that address barriers to use of wind energy."[124]

President George W. Bush (fourth from left) led an Energy Panel discussion at the National Renewable Energy Laboratory (NREL) on February 21, 2006. In addition to the president, the panel included seven energy experts: Dan Arvizu, NREL director; Dale Gardner, NREL's associate director for system integration; Lori Vaclavik, executive director of Habitat for Humanity of Metro Denver; Bill Frey, business director of biobased materials for DuPont; Larry Burns, vice president of research and development for General Motors; Patty Stulp, president of Ethanol Management Company; and Patricia Vincent, chief executive officer of the Public Service Company of Colorado. Photograph by Jack Dempsey/NREL. Courtesy National Renewable Energy Laboratory, Boulder, CO.

While most of this focus was placed on promoting the proliferation of higher efficiency, megawatt wind turbines, the US government's Wind Energy Program did not lose sight of continued research on small wind energy. It was now referring to small wind energy as "distributed wind systems," meaning that the electric power generated by these machines was for on-site consumption. "Outside of the use of large, bulk power generation facilities, distributed wind technologies have been a focus of the [wind energy] program since its inception [in the 1970s] and show great potential for engaging local populations in addressing America's energy future," the laboratory said in 2006.[125]

Meanwhile, the US Department of Agriculture's Conservation and Production Research Laboratory continued to conduct wind energy research, with particular emphasis on utilizing wind turbines for water-pumping applications, on its designated 15-acre test site at Bushland, Texas, through the mid-1990s and early 2000s. Led by Dr. R. Nolan Clark and two primary engineers, Brian D. Vick and Byron A. Neal, the research was focused primarily on coupling wind turbines with electric water-pumping systems and evaluating their performance against mechanical water-pumping windmills, as well as against each other. Two decades of tests at Bushland and selected offsite locations in West Texas with various-sized wind turbines, including 1.5-kilowatt and 10-kilowatt Bergey Windpower, 1-kilowatt Southwest Windpower Whisper 100, and other commercial machines, showed that wind-electric-powered water-pumping systems, with correctly matching pumps, technical adjustments to the turbines, and higher hub heights to capture faster wind speeds, were capable of outperforming their mechanical counterparts in terms of pumping output and economic value.[126] For example, the 1.5-kilowatt Bergey 1500 wind turbine was connected to a 1.1-kilowatt electric motor and powered a 740-watt submersible pump. More than 700 hours of tests were conducted, and the Bergey, on an average daily water volume, exceeded the mechanical windmill by about 1000 gallons, or 45 percent more water, almost year-round.[127] Other benefits of wind-electric pumps over mechanical windmills included reduced maintenance and the ability to erect the turbine towers away from wells since power could be transmitted through electric cables to well pumps at distances of up to a half mile.[128]

The USDA researchers in the 1990s, however, found that mechanical windmills still outperformed similar rotor-sized wind-electric, water-pumping systems during periods of low winds—usually during the midsummer months when water for livestock is most needed. This was because the high "solidity," or the ratio of blade area to the blade sweep area, of mechanical windmill wheels allowed them to start pumping water faster at lower wind speeds. This performance was confirmed during tests of the 1.5-kilowatt Bergey, with its three-bladed, 10-foot rotor diameter on a 60-foot tower, against an Aermotor windmill, with a 10-foot diameter wind-wheel of eighteen sheet metal blades on a 33-foot tower. The researchers discovered that the centrifugal pump that was matched with the Bergey was not as efficient as the positive displacement piston pump of the Aermotor windmill.[129] The USDA mostly alleviated this operational disparity during low-wind periods with the same Aermotor windmill when it tested a three-bladed, 2-kilowatt Havatex 2000—with advanced 5-foot-long airfoils and a permanent magnet alternator with rare earth magnets—that was connected to a helical pump between February 1999 and April 2000.[130]

In general, convincing American farmers and ranchers to give up their long-held mechanical windmills for wind-electric-powered water-pumping systems proved difficult, despite the fact the USDA researchers demonstrated that a properly set-up wind-electric system could be 25 percent

The US Department of Agriculture's Bushland, Texas, wind turbine test site in July 2004, showing an Enertech 44, AOC 50, 10-kilowatt Bergey, and two water-pumping Aermotor windmills. Courtesy of Brian D. Vick, Amarillo, TX.

cheaper than a new mechanical windmill, even when factoring in the cost of the taller wind-electric towers.[131] For Bergey Windpower, its work with the USDA allowed it to offer a new line of wind-electric water-pumping systems, which it sold to remote villages in developing countries throughout the world during the 1990s. By the end of the decade, an estimated several hundred wind-electric pump systems, ranging from 1 to 10 kilowatts in size, had been installed in more than twenty mostly developing countries.[132]

The USDA researchers broadened the scope of their wind energy tests in the 2000s. One of their tests combined a 900-watt wind turbine and solar photovoltaic array to create an off-grid "hybrid" water-pumping system. Despite the technical challenges of integrating the wind turbine and solar photovoltaic array, as well as the seasonal differences in wind and sun availability in the Southern

Great Plains, the researchers found that the combination of power sources offered an increase in daily water pumped by 1500 liters or 28 percent in August, when demand for water for livestock is greatest, than if the wind turbine and solar panels powered the electric pump individually. During the winter, when there is less demand for water, the researchers suggested that the wind turbine's electric output could be used for other activities, such as heating the water tank, battery charging, or producing hydrogen for fuel cell-powered farm equipment.[133]

By 2005, the USDA researchers, along with their counterparts at the West Texas A&M University's Alternative Energy Institute at Canyon, had expanded into other areas of wind energy research, including conducting a study into improving electric power generation of US wind farms by focusing on the average wind turbine output density as opposed to rated megawatt capacity of the wind farm.[134] Other tests focused on wind turbine blade performance. In 2006, the USDA and university researchers modified the design of a prototype blade with improved airfoil shapes, but it initially experienced a failure at the "root," or where the blade attaches to the turbine's rotor hub. Once this problem was resolved, they built a set of three 5-foot-long small wind turbine blades using the redesigned root.[135] The blades were attached to the rotor hub of the Bushland site's 1.5-kilowatt Bergey machine, which was connected to a submersible pump for pumping water into a livestock tank. The test blades improved the machine's peak efficiency by 37 percent and reduced its cut-in wind speed from 13 mph to 11 mph.[136] Power curves and other data were collected on several wind turbines for grid-tie electricity and various off-grid loads using resistors, capacitors, and incandescent light bulbs. A 300-watt Southwest Windpower Air Module machine was used to develop and test a controller for an off-grid ultraviolet light water purification system that was used by Baylor University after some modifications to purify water in Africa. The researchers coupled Bushland's Enertech 44 and Atlantic Orient Corporation 15/50 machines with a large battery bank, diesel generators, and a Northern Windpower controller to simulate electrical

loads of remote Alaskan and Canadian villages.[137] In 2006–2007, Sandia Laboratories used one of Bushland's three 115-kilowatt Micon 108 wind turbines to performance test a set of three 30-foot carbon fiber blades. Until that time, most wind turbine blade structures were covered with either fiberglass or epoxy to ensure both lighter weight and durability. Carbon fiber promised both lighter blade weights and increased durability.[138]

By 2009, however, USDA in Washington was losing interest in continuing its wind energy research, preferring to divert these resources to developing biofuels. Dr. R. Nolan Clark obtained funds for another five years of wind energy research at Bushland before retiring in 2009. Engineers Byron Neal and Brian Vick continued with their research at the wind laboratory through June 2012 and November 2013, respectively. During those final years, small wind turbine noise research was performed by studying the effects of blade flutter and electrical loading on both 1- and 10-kilowatt turbines, as well as exploring ways to increase the efficiency and output of wind farms.[139] The wind energy research program at Bushland officially closed in 2014, and the wind turbines and other equipment were either transferred to other research facilities or sold as government surplus.[140] Meanwhile, USDA has continued to provide financial support to agricultural producers and rural small businesses with wind energy projects through its Rural Energy for America Program (REAP) loan guarantees and grants. Loan guarantees covered up to 75 percent of a renewable energy project's cost, or up to $25 million, while grants were issued up to 25 percent of the project's cost or a maximum of $500,000. A combination of loans and grants could cover up to 75 percent of an eligible project's cost. Since starting the program in 2003 through 2016, the total number of REAP grants issued for wind projects surpassed $71 million, with the largest concentration of these funds being issued for projects in Iowa and Minnesota.[141]

Although US government studies into vertical-axis wind turbines ended by the early 1990s, Sandia National Laboratories—the former lead research contractor for this wind machine technology—shifted to horizontal-axis turbine rotor and blade analysis. In particular, Sandia researchers during the early 2000s launched a structural mechanical adaptive rotor technology (SMART) program. Most large wind turbines in operation by 2010 used passive rotors. "Sandia's SMART rotors have active surfaces similar to airplane wings, with actuators that change their shape, allowing greater control and flexibility," explained Jonathan White, a lead wind energy researcher at the laboratories, in 2011.[142] In July 2011, Sandia announced the relocation of its wind energy test facility at Bushland to the Texas Tech University Wind Science and Engineering Center (renamed the National Wind Institute in December 2012) at Lubbock. Sandia was attracted to the 67-acre West Texas site for its consistent, year-round wind, as well as existing facilities, including a 200-meter-tall meteorological tower and a 9000-square-foot assembly building.[143] In concert with the Department of Energy, Sandia, university,

and other contract researchers continued to evaluate innovative rotor technologies, as well as turbine-to-turbine interactions, which wind farm operators could use to increase their turbine productivity.[144] Sandia's wind energy research lead at the time, Jonathan White, estimated as much as 15 percent of wind energy production and revenue was diminished due to "complex wind plant interaction."[145] In July 2013, the Sandia Scaled Wind Farm Technology (SWiFT) facility was commissioned at Texas Tech University to perform turbine-to-turbine wind interactions. The site includes three 225-kilowatt Vestas 27 wind turbines (two of which are owned by Sandia and the Department of Energy, and one by Vestas Wind Systems), with the possibility of increasing that number to nine or more in the years ahead.[146] For part of the year, Vestas used its wind turbine at the SWiFT facility for its own advanced development work, while the remainder of the time the turbine was available for collaborative projects with the other two turbines.[147] Vestas leveraged the controller developed by Sandia for its four-rotor, twelve-bladed turbine project.[148] Beyond turbine interaction, there were opportunities for SWiFT to demonstrate the next generation of intelligent rotor blade technology and grid controllers.[149]

The Obama Administration remained a proponent of wind power, although its priority among renewable energy research and development was largely pointed toward solar photovoltaic technologies. On February 4, 2011, the Department of Energy launched the SunShot Initiative to lower the cost of solar energy by 75 percent, with the goal of making this form of renewable energy cost competitive on a per-kilowatt-hour basis against other forms of utility-generated energy by 2020.[150] Consequently, small wind energy companies—those producing machines at less than 100 kilowatts of output—have continued to lose market share to increasingly cheaper solar PV installations for homes and businesses since the start of the decade and several firms have closed their doors. To help enhance the competitiveness of small wind-electric systems providers against other renewable energy technologies and to meet national certification standards, the Department of Energy's Wind Energy Technologies Office and the National

By utilizing the US Department of Energy's Distributed Wind Competitiveness Improvement Project grants, Pika Energy of Westbrook, Maine, was able to prepare its 1.5-kilowatt wind turbine for commercial production in 2014. Courtesy National Renewable Energy Laboratory, Golden, CO.

Renewable Energy Laboratory in 2012 initiated the Distributed Wind Competitiveness Improvement Project (CIP). The program provides eligible small and midsized wind turbine systems manufacturers with funding to help facilitate component improvements and systems optimization, manufacturing process improvements, prototype testing, certification testing, and type certification.[151] CIP participants are small businesses, so depending on the scope of the project, the federal funds help to cover up to 80 percent of the mostly 18- to 24-month contracts. Requiring the participating companies to share in the cost of their CIP projects ensures the Department of Energy that they are serious about achieving the project goals.[152]

By 2017, CIP contracts, valued at more than $5 million, had been awarded to twelve different manufacturers and component suppliers. The program has been deemed a success by the Department of Energy's Wind Energy Technologies Office, with several small wind turbine manufacturers receiving multiple CIP awards between 2013 and 2017.[153] Northern Power Systems in Barre,

Vermont, used one of its CIP grants to achieve a 15 percent increase in energy output for its NPS-100 100-kilowatt turbine by increasing the blade length and improving blade aerodynamics.[154] Multigrant recipient Pika Energy of Westbrook, Maine, reduced its blade manufacturing cost 90 percent by developing a tooling and cooling system to produce blades using injection-molded plastic, as opposed to its previous method of hand-shaping its blades. The company was able to use the CIP awards to achieve performance certification for its T701, three-bladed, 1.5-kilowatt turbine and bring the new machine into commercial production in 2014. Due to the optimized manufacturing processes, Pika Energy was able to reduce the end-user cost of its wind turbine by more than $3000.[155] Bergey Windpower Company of Norman, Oklahoma, used one of its CIP awards to develop the new Excel 15 wind turbine, which is 40 percent more efficient than its existing 10-kilowatt machine, produces 85 percent more energy, and still sells at a similar price.[156] In 2017, Bergey received another CIP award to develop a standardized, 100-foot-tall,

The US Department of Energy's National Wind Technology Center at Golden, Colorado in 2013. Courtesy Dennis Schroeder, National Renewable Energy Laboratory, Boulder, CO.

self-supporting lattice tower.[157] The successful implementation of these new production and technology innovations will depend on the resiliency of the wind-electric companies, along with ongoing federal wind research programs and successive congressional reauthorizations of the Residential Renewable Energy Tax Credit.[158]

For more than forty years, the Department of Energy has offered access to test facilities throughout the country and has provided hundreds of millions of dollars in technology development funds to the small wind-electric industry. While just a small portion of the department's overall mission and budget, over the years these wind energy research programs allowed numerous small, entrepreneurial companies to advance their wind turbine technologies from concept to commercial availability, something that many of these firms would have never been able to accomplish sufficiently on their own. Since the early 1970s, small wind turbine manufacturers have faced steep economic, regulatory, and political hurdles not experienced by their pre-1950 counterparts that once flourished across the American heartland. Despite periods of installation increases over the past four decades, the industry's goal for long-term, increased proliferation of small wind-electric systems has remained unmet against other renewable technologies, such as large megawatt wind turbines and solar PV systems, which compete with traditional utility-produced electric power. Although the number of small wind turbine manufacturers in the United States has decreased since 2010, the Department of Energy's National Renewable Energy Laboratory and Wind Energy Technologies Office, as well as partner research institutions, remain committed to the promotion and sustainability of this segment of the renewable energies industry.

6 Jacobs's Return

By the mid-1950s, the United States' once robust wind-electric power industry was grinding to a halt. The efforts the Rural Electrification Administration had made since the 1930s to provide utility-produced alternating current to rural communities and farms chipped away at the wind-electric industry. An abundance of cheaply produced electricity from large, centralized utilities using oil, coal, and nuclear energy sources reached most American homes, farms, and businesses by power lines in the years immediately following World War II. Two of the wind-electric industry's most prominent leaders—brothers Marcellus L. and Joseph H. Jacobs, founders of the then thirty-year-old Jacobs Wind

Electric Company—knew that continued manufacturing of their prized direct current wind plants would come to an end. They had already outlasted nearly all their competitors. "Alternating current is all over the place . . . often at artificially low prices. That's a tough combination to beat and I quit trying to fight it in the 50's. I could see the handwriting on the wall back around '52, '53, '54 . . . and we closed the factory in 1956," Marcellus Jacobs said in a 1973 interview, reflecting on the changes that had been made to the wind-electric industry by the 1950s.[1]

Prior to closing the company's Minneapolis factory, however, the Jacobs brothers had considered converting their wind-electric generators

The Jacobs Wind Electric Company factory of Minneapolis in 1950, six years before it closed. Courtesy Paul R. Jacobs, Corcoran, MN.

from DC to AC output. In 1951, Marcellus Jacobs even proposed to Congress that the electric power utilities should install 5-kilowatt, AC wind-electric turbines on top of the thousands of large steel towers crisscrossing the country in order to feed supplementary power into the grid. This would be an alternative to building large 800-kilowatt wind turbines like the one proposed by government scientist Percy H. Thomas. "The idea we have been working on for some time is to take a line, for instance, from Minneapolis to Great Falls, Mont., and install a series of these five-kilowatt plants at several mile intervals, which could be directly connected into existing power lines, as boosters, and secure a maximum monthly kilowatt-hour output per dollar of building and installation cost. No additional special transmission lines or other extra cost would have to be added," Marcellus Jacobs told the US House Committee on Interior and Insular Affairs in a written statement for a hearing regarding legislation proposing that the Interior Department develop large-scale wind turbines. Although power companies by that time were already feeling the strain of peak demand, especially during winter, they weren't interested in Marcellus Jacobs's idea, and instead installed numerous backup diesel generators around the country to handle the overloads.[2]

The same year the Jacobs Wind Electric Company plant shuttered, the Jacobs brothers and their families moved to Fort Myers, Florida. A significant reason for the move was to find relief for Joseph Jacobs's persistent sinus problems in Florida's warmer climate.[3] Although they had exited wind-plant manufacturing, the brothers maintained the original Montana corporate entity of the Jacobs Wind Electric Company.[4] Joseph Jacobs settled into retirement in Florida but died suddenly in 1963 in an automobile accident. Still in his early sixties, Marcellus Jacobs became involved in real estate. Florida was then experiencing an increase in population, as more people, ready to retire, moved there from colder, northern states. However, Florida's swampy conditions presented a challenge to builders, causing canals to be dug to channel water away from newly constructed communities and to generate fill in low-lying coastal areas. The canals

also provided access to waterfront lots. The problem with this method was that water in long finger canals often became stagnated, choked with vegetation, and smelly with dead fish. The ever-inventive Marcellus Jacobs set out to solve this problem. In the early 1960s, he developed a seasonal flow system to circulate fresh water through waterways in the rainy seasons.[5] Then, starting in the late 1960s, he discovered a year-round tidal-powered flushing design for the Island Park subdivision, which he started in the late 1950s 10 miles south of Fort Myers where he resided.[6] He patented his discovery in the early 1970s.

Wind-electric power, however, was never far from Marcellus Jacobs's mind. He kept up with the latest news and predictions on world energy consumption throughout the 1960s. The starting point for the Jacobs family's return to the wind-electric industry came in the summer of 1969 during the family's eight-week road trip. Marcellus and Edna Jacobs's son, Paul, who had just graduated from the University of Florida with a degree in geography, drove the family in his 1962 Oldsmobile 98. They departed Fort Myers in July, driving across the south to Arizona where they would first visit Marcellus's then ninety-nine-year-old mother, Ida Jacobs, who was living in Phoenix at the time. From there, they planned to drive through California, where they would visit relatives, and onto Alaska along the Alcan Highway. However, it was in eastern New Mexico, about an hour's drive from the Texas border on Interstate 40, which also paralleled old Route 66, where the Jacobs stopped to look over an early 1950s 3-kilowatt Jacobs on a 50-foot tower still standing next to the site of a demolished Route 66 store. "What we were seeing was about eighteen years old then and looked ready to turn on and run some more. That fact got my attention," Paul Jacobs recalled, adding that moment "was the start of my postgrad education into the family small wind-plant design business that had started about 1922 at the Jacobs ranch east of Vida, [Montana], when they first rigged a water pumper head on a short tower to produce shaft power for their small shop."[7] Although Marcellus Jacobs already viewed his son as a "junior business partner" on matters of land development and canal construction in the years immediately following his brother's death, Paul Jacobs noted that this was also the start of a detailed dialogue between father and son for the rest of the trip:

That stop also got my father started relating many of the stories that he had accumulated in building up the [Jacobs Wind Electric Company] in the 1930s and 40s with their hundreds of dealers serving sites like this one which was out where grid power was not economic then. . . . Besides telling me about his firm's market before [the] REA (Rural Electrification Administration) and his dealers, on this nearly two-month trip, M. L. related details as to how those machines worked as part of a self-contained system (today called "islanding") that brought 20th century power to rather remote, distributed sites like this, where there was no powerline yet when we were there in 1969. The long-gone crossroads store (probably removed in the mid-1960s for I-40) must have used the power for its coolers for food, for pumps (water and gas), and for the electric power for lights and radio that linked them into the new, post-kerosene power era. From the evidence seen in 1969 and M. L.'s experience he related helping his dealer in Albuquerque get established in the 1940s (after World War II), this area was so remote it only went into a 20th century lifestyle with electric power after 1950 and was by modern standards a third world technology zone until after World War II.[8]

Marcellus Jacobs also explained to his son in great detail how the company developed its second-generation workhorse wind plant in Minneapolis in 1932. This reminiscence triggered new thoughts on how to modernize the Jacobs machine, so that it could meet the new ways in which electricity was being consumed.

For the next five years, father and son conducted "thousands of hours of discussions, sketches, and analysis" on what the elder Jacobs knew from the

This later second-generation, 3-kilowatt Jacobs wind plant was relocated from El Paso, Texas, to the Fort Myers, Florida home of Marcellus Jacobs in 1973. Courtesy Paul R. Jacobs, Corcoran, MN.

earlier markets for distributed DC wind-electric systems, and "we started looking at how to bring the basic functions of our tested design features forward to utilize newer technology that was there by 1969," Paul Jacobs said. "This included the diode, early power electronics and offset hypoid gearboxes, like the one in the back of my '62 Olds that allowed a lower shaft tunnel from the engine to the rear wheels."[9] The bevel gear system was actually part of the Jacobs Wind Electric Company's first-generation plants, which the Jacobs brothers did away with by 1932. The reason was that without the offset torque to balance the pinion torque on the ring gear, they discovered that a bevel gearbox at that time was an inefficient way to harness wind in stronger wind resource areas.[10] However, Marcellus and Paul Jacobs believed the gearbox now had operational merit in a new Jacobs machine design:

> M. L. and I by the early 1970s were looking at use of the first generation [Jacobs wind plant] design "with a gearbox," only now one that would not crank itself out of the wind on better wind days, to solve the major cost and size/weight problems that the old, custom direct-drive generator design had run into as the relative price of copper was now much higher and the increased power demand for the 1970s lifestyle at remote sites made the 3 kW (kilowatt) sized units much too small for the larger loads. Other incentives to go back to the gearbox was the opportunity to get rid of the major headaches and cost of using a brush system to convey the power generated from the free yawing generator, on the top of the tower, to permanent wires down the tower and the opportunity to buy inexpensive, quality VAC generators (with field controls) that were being introduced to the farm market for PTO (power takeoff) generators for quick emergency hookup to a tractor when the REC's (rural electric cooperative) grid died in a storm.[11]

Paul Jacobs obtained a master's degree in business administration in June 1971 from the University of Florida, with a focus on real estate and urban land development, and spent the remainder of that summer setting up the family's tidal flushing business, called Environmental Canal Systems, Inc. Still living at home at the time, he and his father continued to discuss their future wind-electric business plans. Later that summer, Marcellus Jacobs purchased a Winnebago motorhome and with his wife and Paul took a month-long trip north to Minnesota, west to Seattle, south to California, and then back to Florida, during which they saw more old wind plants and continued their discussions for the new Jacobs wind-plant design. Upon their return from this late summer trip, Paul Jacobs went to work as a real estate appraiser, while during the evening hours he assisted with marketing Environmental Canal Systems' tidal flushing systems. "We could see the tidal power business would not take off easily as new regulations more or less stopped the digging of more finger canals in Florida," Paul Jacobs said.[12] But the timing was right for the change in corporate pursuits. Before the end of 1972, Paul, along with his parents, took a second cross-country trip in their Winnebago that included Minnesota, Montana, and Utah and back through Colorado, New Mexico, and Texas. During this trip, Marcellus Jacobs located a 1953 3-kilowatt Jacobs in West Texas near Fort Stockton, which he had shipped back to Fort Myers. The following year the machine was installed at Marcellus Jacobs's home and included with it a 110-volt DC battery bank. "This was the start of our return to a more full-time focus on wind again," Paul Jacobs said.[13] Marcellus Jacobs had already exited the real estate business, while his son set up his own real estate appraisal and consulting firm in February 1974.[14]

By late 1973, Marcellus Jacobs had reviewed his records from the mid-1950s, which included file cards of all the company's newer wind plants sold throughout the United States, including their sites and owners. Paul Jacobs explained that his father used this information to send letters and make telephone calls to the owners of these wind plants in West Texas and New Mexico in order to try to purchase them. If the original owners had since died or moved away, Marcellus Jacobs called the local sheriff's office to track down the current property owners.[15] With purchase leads in hand, father and son set out on a road trip to West Texas

Wind plants manufactured by the Jacobs Wind Electric Company prior to 1955 were still found on towers throughout the western United States, like this second-generation, 3-kilowatt Jacobs in New Mexico from 1952 to 1975. Courtesy Paul R. Jacobs, Corcoran, MN.

and New Mexico in March 1974. West of Del Rio, Texas, they picked up their first Jacobs machine of the trip from rancher Bob Allen. In total, they accumulated six Jacobs generators and related parts, which they transported back to Fort Myers on a trailer. The towers, however, had to be taken apart and shipped back separately to Florida by motor carrier. "All the time we were on that several week trip, we further discussed what we could learn of the changing market for wind power, the condition of various parts of the machines that were purchased—checked them for wear and how well [they] survived what was a basic period of little or no service," Paul Jacobs said.[16] Through subsequent trips to the Southwest, the Jacobs would bring back to Florida another two dozen wind plants.

It was also in 1973 that the seventy-year-old Marcellus Jacobs became a celebrity of sorts among a new generation of wind-electric enthusiasts and developers. He had spoken up about the proved track record of the wind power industry, which he had helped to formalize more than forty years earlier during a June 2–3, 1973, Wind Energy Conversion Workshop in Washington, DC. Then the *Mother Earth News* in late 1973 published a landmark article, based on a folksy interview with Marcellus Jacobs, about how the Jacobs Wind Electric Company developed its highly successful wind plants four decades earlier.[17] The article included a photograph of Marcellus Jacobs in his Fort Myers office, pointing out the technical features of the Jacobs wind plant using a scaled model.[18] Yet, he was careful not to divulge the work that he and his son were already undertaking to develop a new wind plant. "You're not actively engaged in wind plant work of any kind at this time?" the

Mother Earth News writer asked near the end of the interview. Marcellus Jacobs replied, "No, I have other interests now." The writer asked one more time, "You mean you don't think about wind-driven generators at all?" He then offered, "Well . . . I did buy one of my old plants out in New Mexico this summer . . . and I've still got quite an assortment of DC equipment and appliances packed away. I'm doing it mostly for my son, you know . . . but I imagine I'm going to have a little fun setting that windplant up and running it this winter."[19]

In 1974, Marcellus and Paul Jacobs set up a new Jacobs Wind Electric Company as a Florida corporation and quietly embarked on formalizing their new Jacobs wind-plant design. Paul Jacobs, whose real estate business was also thriving, secured the legal documents, and Marcellus Jacobs from his sale of Island Park land put up the money

for the new company. "I tried to function not only as M. L.'s business partner, but also as his de-facto chief of staff," he said.[20] At the same time, Marcellus Jacobs set up in his shop the tools and circuits for the 110 volts-DC battery bank needed to motorize and test the now several dozen post-1948 Jacobs generators that he and Paul had picked up during their trips to Texas and New Mexico. He also refurbished the blade-actuated governors with his small remaining stock of original parts from the Minneapolis plant. He then sold most of these reconditioned machines to customers in the Great Lakes and New England areas. "The goal was to bootstrap funds for new parts upgrades for a new generation of wind plant, nominally I would call it 'third generation.' ([The] 1920s Montana machines were the first; 1930s to 1950s were second.) . . . Overall, I would say he rebuilt the firm in the 1970s

Marcellus Jacobs, shown with his son Paul, demonstrating the proper method for erecting a Jacobs wind plant tower. From *How to Build—From the Ground Up—The Tower Supplied with the Jacobs Wind Electric Plant*, Fort Myers, Florida (1975). Courtesy Paul R. Jacobs, Corcoran, MN.

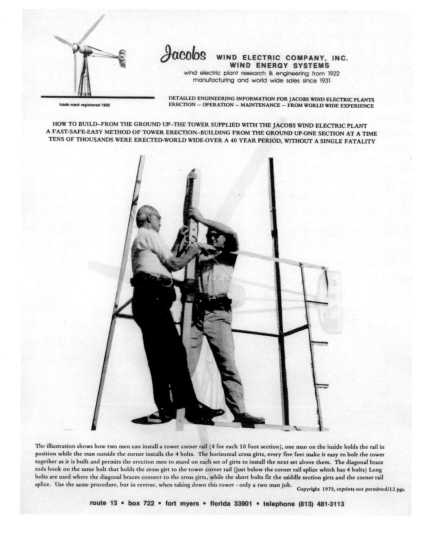

using the experience with how the 1950s era parts worked in the rough Southwest environment. We figured it was the best way to compete with the rebuilders, by making something they could not do on their own. M. L.'s experience was his competitive advantage," Paul Jacobs said.[21] In October 1974, Marcellus Jacobs met with Senator Lee W. Metcalf (1911–1978) of Montana, who served in the US Senate from 1961 to 1978, and revealed the company's plans to produce a new, larger wind plant.[22] Two months earlier Senator Metcalf had entered the *Mother Earth News* article about Marcellus Jacobs into the *Congressional Record*.[23]

By this time, young people were scouring the countryside of the United States and Canada in search of early wind-electric machines to put back in service, with the Jacobs wind plants being most coveted. Many of these machines could still be found on their towers in operable condition, even after being idle for more than twenty-five years in most cases. *Public Power* magazine in the 1970s wrote about a retired midwestern farmer named John Lorenzen who purchased three 2.5-kilowatt, 32-volt DC Jacobs machines around 1940 and continued to use them for his source of electric power despite having access to utility power lines nearby.[24] Letters from around the country flooded Marcellus Jacobs's mailbox from people who obtained Jacobs wind plants and sought his advice about how to return them to operation. Paul Jacobs recalled how his father often spent several hours each night at his typewriter replying to this correspondence.[25] Marcellus Jacobs was also known for his corrective, as well as instructive letters to editors of publications whenever he came across misrepresentations of Jacobs wind plants. In one of these letters, published in the December 1977 *Alternative Sources of Energy* magazine, he provided his prescription, including diagrams, about how properly to erect a windmill tower from the ground up versus the method of building the structure laying on the ground and then lifting it into place by tractors and guy wires.[26] Marcellus Jacobs was also a voice of reason and correction at meetings of the American Wind Energy Association, of which the Jacobs Wind Electric Company was a charter member.[27]

Robert W. Righter, a wind energy historian, stated that Marcellus Jacobs developed his wind plants in the "craftsman tradition." He meant that instead of a university-trained engineer's approach to building products, Marcellus used the "common sense of applied mechanics" to develop his wind turbines. The craftsman approach builds on years of experience through field testing mechanical concepts, during which one learns what works and what doesn't, and continues this process over and over until an optimal product is developed. The Danish wind turbine designers later used this same method to great success, supplanting their American and German counterparts, according to historian Righter and German scholar Matthias Heymann.[28] Paul Jacobs agreed with this assessment. "As craftsmen then, M. L. and I had assembled an actual new third generation working model, done by hand the way all previous Jacobs models were first made," he said.[29] Consequently, the Jacobs approach to explaining wind plant technology, as well as how they designed them, did not sit well with many classically trained American engineers. Paul Jacobs noted how the engineers who staffed the NASA and Department of Energy wind energy programs during the 1970s were "highly dismissive" of Marcellus Jacobs, with his lack of formal engineering education. Yet they were fearful of him.[30] Paul recalled how Joseph M. Savino, an engineer with NASA's Lewis Research Center, during a visit to their Fort Myers office in 1974 to look over some reconditioned second-generation Jacobs wind plants, remarked to Marcellus Jacobs that it was his own 1950 letter to Congress that helped terminate funding to the government wind energy program and they did not want that to happen again.[31] In reality, it was most likely the Korean War that diverted the government's attention from wind energy development in the early 1950s.

Unlike other startup wind turbine manufacturers in the mid-1970s, the Jacobs prided themselves on not taking a penny of government funds to develop their new line of wind plants. However, it was admittedly difficult for them to self-fund the development of their new wind plant, to the tune of more than $240,000 by early 1978. They also did not directly participate with their machines at the Department of Energy's small wind power test center at Rocky Flats, Colorado, nor did the Jacobs Wind Electric Company obtain a federal grant to assist in the development of its new 8-kilowatt machine, like Windworks, United Technology Research Center, Alcoa, and Grumman Aircraft Company at the time. However, both Marcellus and Paul Jacobs visited Rocky Flats on several occasions in 1977 and 1978 to meet with the staff. They used these opportunities to point out in person, as well as via follow-up written correspondence, the inadequacies in the governor, blades, and other related parts of a 3-kilowatt, 110-volt DC North Wind Power Company rebuilt Jacobs test wind plant. That's why the Jacobs were shocked to read in a 1979 semiannual report from the Rocky Flats Wind System Test Center disparaging references to the Jacobs Wind Electric Company related to the North Wind Power machine. Marcellus Jacobs promptly wrote to Terry J. Healy, manager of the Rocky Flats Wind Systems Program, on July 12, 1979, calling the report "patently false, biased, deceptive, and a fraud perpetrated on the American Public at Taxpayer's expense" and demanded an immediate retraction of all references to the Jacobs Wind Electric Company in the report. The report noted that the North Wind Power wind plant experienced blade, hub, and field winding failures during tests in the summer of 1977. Marcellus Jacobs took the report apart paragraph by paragraph, citing overall that the North Wind Power machine used an assemblage of poorly made "imitation" parts and that the company might have disturbed the field windings of the Jacobs generator when it took it apart for rebuilding. "Sure our plants could be destroyed, blades broken, etc. when a tornado wind directly struck the plant and tower," Marcellus Jacobs wrote in a 1978 letter after one of his trips to Rocky Flats to view the troubled North Wind Power wind plant. "Such winds are sometimes mixed in with a high wind front, but the thousands of rank and file plants 'took it' year after year." Feeling they were being stonewalled by the Rocky Flats staff to correct the report, the Jacobs on July 13, 1979, sent a letter to Robert Anderson, chairman of Rockwell International, the Department of Energy's contracted operator of the Rocky Flats site, with their demands. By mid-August 1979, Rockwell

Marcellus Jacobs (center) in 1979 at his Florida research and development site with a second-generation, 3-kilowatt Jacobs wind plant in the background and a third-generation, 8-kilowatt prototype machine in the foreground. Standing to his right is Fernando Lee, a former Jacobs dealer in Brazil during the 1940s and 1950s. Courtesy Paul R. Jacobs, Corcoran, MN.

International's vice president and general manager, R. O. Williams Jr., assured the Jacobs in writing that the final report expunged any references to the Jacobs Wind Electric Company.[32] The cleansing of the Department of Energy's report came just in time as the Jacobs prepared to formally announce the launch of their new 8-kilowatt, 110-volt AC wind plant to the market in the fall of 1979.[33]

For the previous five years, the Jacobs spent countless hours designing and experimenting with their new 8-kilowatt wind plant at Marcellus Jacobs's home.[34] While keeping certain foundational design elements of the second-generation Jacobs machines, such as the three-blade propeller system, post-1948 blade actuated governor and tail vane, the third-generation machine included many new electromechanical characteristics. Based on their earlier discussions about introducing a gearbox to the new machine's design, they came up with a hypoid gear drive encased in a welded gearbox assembly at the top of the tower. The horizontal blade shaft, which was tilted at 10 degrees to balance gear torque against the 23-foot diameter propeller's back thrust pressure, turned the gears that rotated a vertical shaft connected to a brushless alternator mounted inside a 6-foot stub tower. A spring snubber control on the folding tail vane automatically folded the powerhead and gearbox around to the side of the tower in winds of more than 40 mph.[35] The test machines, of which four were built, were rated to generate 6–8 kilowatts of 110 volts AC in a 25-mph wind. With a synchronous inverter, the Jacobs could offer a wind plant that meshed with a typical American home's grid-tied electrical system, avoiding the need of a battery bank. The new wind plants were also designed to minimize maintenance and withstand severe storm conditions.[36] The Jacobs subsequently filed seven patent applications with the US Patent and Trade Office between 1975 and 1979 to protect the new wind-plant design.[37]

Since the Jacobs did not possess the equipment to manufacture parts for their new 8-kilowatt prototypes, they relied on numerous third-party suppliers. Marcellus Jacobs, who considered himself a "product engineer," enjoyed cultivating relationships with skilled machinists and electrical engineers. Sometimes this meant repeated discussions and experimentation to refine parts until they met the Jacobs's specific requirements. For the production of the prototypes, the Jacobs picked Fairfield Manufacturing Company of Lafayette, Indiana, to manufacture the gearbox assemblies; Fidelity Electric Company of Lancaster, Pennsylvania, for the alternator assemblies; Tuthill Corporation of New Haven, Indiana, for the governor ball sockets; American Spring and Wire Specialty Company of Chicago for governor springs; Real Gas & Electric Company of Guerneville, California, for the propeller blades and synchronous inventers; Tol-O-Matic of Minneapolis, Minnesota, for the caliper disc brakes; and Unarco-Rohn of Peoria, Illinois, for the six-foot stub towers. Other small parts, such as bearings, seals, bolts, vane swivel material, and

the lower end of the mast pipe support, could be purchased from general wholesalers.[38] The Jacobs prototypes cost nearly $40,000 each, not counting special engineering, patent filings, and other related expenses.[39] By autumn of 1978, the Jacobs had to figure out how to finance the final leg of bringing their new wind plants to market. During a visit in September 1978 with their governor machining vendor, Jim Tyrrell of Braun Engineering in Bloomington, Minnesota, the Jacobs struck up a conversation with Tom Braun, the company's designer and engineer, about how Minneapolis-based Control Data Corporation (CDC) had expressed interest in entering the wind energy business. CDC was already a globally recognized manufacturer of multimillion-dollar mainframe computers and electronic control systems, and could prove a reliable source for synchronous inverters, a struggle for the new Jacobs Wind Electric Company to find. Tom Braun's firm had performed various custom machining and engineering work for CDC and had connections with senior managers there. "[Braun] on his own did some calls and found a contact at CDC who started the discussions with M. L.," Paul Jacobs said.[40]

The Jacobs and CDC management reached a verbal agreement on January 22, 1979, for the two companies to build ten preproduction wind plants rated at 10 kilowatts. Marcellus Jacobs spent the next two months traveling the country, meeting with component suppliers to make design adjustments for the larger wind plants and obtain price quotes for their parts. He estimated that each prototype should not exceed $6000 apiece to make, and that each machine should have a base value between $8000 and $10,000 and a collective market value of $80,000 to $100,000.[41] Satisfied with the figures, CDC through its Control Data Capital Corporation (CDCC) subsidiary reached a "statement of understanding" with the Jacobs Wind Electric Company on April 25, 1979, to establish a one-year trial period during which together they would build the ten preproduction units. Based on the understanding, the Jacobs Wind Electric Company would purchase the wind-plant parts from the vendors under its own name, while CDCC provided the funds. The Jacobs would have all their travel and other out-of-pocket expenses covered by CDCC during the one-year trial period. CDCC also agreed to pay the Jacobs a one-time front-end fee of $25,000 to use their patents (three issued and four pending at the time). The companies estimated the value of the Jacobs patents to be $350,000. To assemble the wind plants, CDC prescribed a 10,000-square-foot workspace in the Minneapolis area. The two parties also agreed that the first five units would be kept for internal testing and evaluation, while the others would be sold to select outside customers. In addition, the statement of understanding proposed the next steps if the trial period was deemed successful, including establishing a new company to be called the Jacobs Wind Electric Company, in which CDCC would have a minority interest and the Jacobs would hold the rest. Marcellus and Paul Jacobs would also be responsible for the new company's manufacturing and sales. However, if the trial period proved unsuccessful, CDCC agreed to retain the ten prototype units, while the Jacobs maintained their patents and the related research generated under the agreement.[42]

For the next year, a small CDC staff at Bloomington, Minnesota, led by Robert W. Darr for Donald W. Pederson, CDCC's president, studied the Jacobs' 8-kilowatt prototype machines and had drawings made with engineering specs for the eventual manufacturing of the wind plants. CDC and the Jacobs Wind Electric Company in late 1979 attracted the public's interest in their new machines by issuing a press release to the media. Marcellus and Paul Jacobs, for their part, conducted the initial marketing of the new machines from their office in Florida, and by early 1980, had secured in writing about 1000 requests for the new wind plants, along with 200 dealer applications. But CDC struggled to comprehend the new wind-plant designs. Therefore, in the spring of 1980 the company asked the seventy-seven-year-old Marcellus Jacobs, along with his son, to come to Minnesota to directly oversee the research and development staff that had been assigned to the effort. In the fall of 1979, one of the three 8-kilowatt machines built in Florida that CDCC purchased was installed at inverter manufacturer Autocon's location in Plymouth, a suburb of Minneapolis, where it was used to send power into the company's inverter test shop. (Autocon was a CDC-owned division.) A second 8-kilowatt demonstration unit was erected around the same time at another CDC facility in Aberdeen, South Dakota.[43]

In April 1980, CDC established a third Jacobs Wind Electric Company as a Delaware corporation. Marcellus and Paul Jacobs, who kept the Florida-based Jacobs Wind Electric Company that they formed in 1974, were made the majority shareholder of the new company. CDCC, which became the minority shareholder, supplied the financing to operate the new Jacobs Wind Electric Company, while the Jacobs provided the intellectual capital under a detailed sales agreement with CDCC. Marcellus and Paul Jacobs were named president and executive vice president, respectively, of the Delaware corporate entity. Both men also held two of the three board seats, with Marcellus Jacobs as board chairman. CDCC Chairman Edward Strickland held the third seat. To oversee the operation more directly, Paul Jacobs in 1982 bought a place in nearby Corcoran, and Marcellus Jacobs, who maintained his residence in Fort Myers, rented an apartment in the Minneapolis area.[44]

In April 1980, the new Jacobs Wind Electric Company rented a single dock bay of a block of three in a building on Fernbrook Lane in Plymouth. The operation quickly increased to encompass all three bays. Each bay, which included a large door and loading dock, had office space in the front and an open area for warehousing and assembly work. In addition to Marcellus and Paul Jacobs, the new company started with two other employees—a draftsman and a CDC mechanical engineer who had worked on the project for a year at CDC's headquarters. The engineer secured another three or four assemblers that summer to help set up the initial assembly line and motor test stand. The shop was arranged to receive components, assemble the wind plants, and motor-test them on the stub tower assemblies, running the power through each inverter before they were shipped. Within a year, the company added up to twenty employees at the site to work in the office, order parts, perform assembly and testing, and ship the wind plants. The company also made space at the location to bring its blade

Jacobs Wind Electric Company's third-generation, 10-kilowatt turbine, which was introduced in the fall of 1980, maintained the three-blade propeller system from 1948 but introduced a gearbox mounted inside the stub tower. Courtesy Paul R. Jacobs, Corcoran, MN.

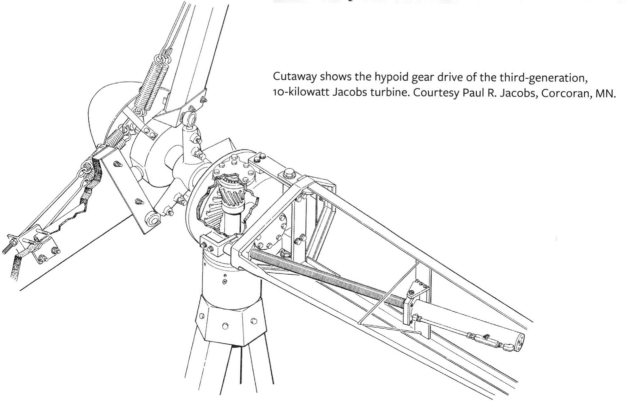

Cutaway shows the hypoid gear drive of the third-generation, 10-kilowatt Jacobs turbine. Courtesy Paul R. Jacobs, Corcoran, MN.

preparation in-house. This included receiving the manufactured wooden blades from a supplier in California, painting them using a Minneapolis paint shop, and balancing and finishing them with the necessary hardware at the Fernbrook Lane site. Paul Jacobs estimated that the Jacobs Wind Electric Company worked with as many as 200 different suppliers, but he noted that half of those were for components used for building the company's Mastermind Synchronous Frequency Chargers. "To make all of the highly specialized parts under one roof would have been heavily capital intensive and very expensive, since we tried to buy small quantities of other firms' large parts production runs," he said. "On our own we could not get that advantage of scale. Further, by keeping our design flexible, we could keep upgrading the design and strength of components so we could react quickly to field and dealer reports on problem components. That is how we were able to take the 8 kW (1979) prototypes and by fall of 1980 ship our first 10 kWs." The company also did not have the room to maintain a large onsite inventory of parts and subassemblies, so most components were ordered to arrive just in time for assembly. "M. L. was careful to use good vendors like Onan (for alternators) and Rohn (stub towers) when he could find them, plus quality small parts vendors for bearings, toggles, inverter components, etc.," Paul Jacobs said. "We were not inclined to spend $10 to make a custom part we could buy in quantity for $3."[45]

The first one hundred 10-kilowatt machines were sold via a dealer network personally approved by the Jacobs. Each dealer sent at least one representative to the company's Plymouth location for instruction on how properly to install and service the machines. The wind plants retailed to dealers for about $20,000 apiece. However, like any new technology, the Jacobs knew the true test of their first 10-kilowatt wind plants would be demonstrated through their operation under actual field conditions. They also anticipated a certain amount of component failure and purposely avoided any exports of the first 10-kilowatt machines to dealers outside the United States so that they could address these problems quickly. The majority of repairs, which were mostly made in the field by the dealers,

This third-generation, 10-kilowatt Jacobs unit was installed in Rhode Island in 1980. Courtesy Paul R. Jacobs, Corcoran, MN.

A third-generation, 10-kilowatt Jacobs wind turbine was erected on this Wisconsin farm in 1981. Courtesy Paul R. Jacobs, Corcoran, MN.

involved replacing the synchronous inverters. Like the new machines, the inverter electronics were still in their infancy and susceptible to misfires and lockups. It was not uncommon for some dealers to replace several inverters at a single wind-plant site. It is estimated that CDC spent about $5 million on new product launch repairs between the summer of 1980 and 1982. Paul Jacobs explained that the new Jacobs Wind Electric Company used the tried-and-true, but simple, process of "field and concomitant modification" in deploying the first 10-kilowatt machines to the market. "[T]his iterative approach required ever more field testing and replacement of any [and] everything that broke with better, upgraded replacement parts, until in a normal product run of several thousand units one can get all the obvious bugs out, as we had done with the [Montana] firm's products by about 1940. . . . What M. L. explained to me was that in the early [and] mid 1930s there had been problems with certain components on the early VDC machines that had required 'concomitant modification' that was expensive for their [Montana company] to fund. That is how our [Montana] firm 'bought' its good reputation [and] good will of our original customers [and] dealers," he said. Within two years, the new Jacobs Wind Electric Company successfully installed more than three hundred 10-kilowatt wind plants through a network of more than seventy-five dealers.[46]

The "concomitant modification" approach was also applied to the Jacobs Wind Electric Company's suppliers, and it often resulted in vast improvements to components. Particular attention was paid to correcting problems related to the inverters. The Jacobs Wind Electric Company started in 1980 using a "two-box" system, which included an Autocon custom field controller that fed raw, variable frequency AC power from the wind plant to an Achieval Wind Energy inverter that used a DC power conditioning choke with silicon-controlled rectifier to convert the variable frequency power back to a 60 hertz AC compatible with the grid. The Jacobs worked with Achieval Wind Energy in a Boston suburb on many upgrades to its inverter technology, and by mid-1981 they introduced the one-box Mastermind Synchronous Frequency

Charger specifically made for the Jacobs 10-kilowatt wind plants. By 1983, the Jacobs Wind Electric Company had its own design of Mastermind inverters assembled by Minnesota suppliers and tested at its Fernbrook Lane facility.[47] Upgrades were also made to the wood curing and paint techniques for the blades, and the governors were improved with larger blade shaft diameters and lower friction bearings.[48] There were instances, however, when the Jacobs could not reach a suitable resolution with a supplier over a component problem and had to look elsewhere. This was the case with Unarco-Rohn, its longtime supplier of towers throughout the 1970s. Marcellus Jacobs had tried to encourage the manufacturer between 1980 and 1982 to build a stiffer tower for the new 10-kilowatt Jacobs wind plants and use the Jacobs historic tower anchor system, which was learned from the water-pumping windmills, and cease using anchors that consisted of "J" bolts attached to the concrete tower footers. There was also occasional breakage of the cold welds in the tower legs. The Jacobs Wind Electric Company ultimately switched to another tower supplier, Advanced Industries of Iowa.[49]

At the start of 1982, Marcellus and Paul Jacobs believed the new Jacobs Wind Electric Company was headed in the right direction, as it continued to perfect its 10-kilowatt wind plants and place increasing numbers of them into service. A reporter for *The Minneapolis Star*, in a January 7, 1982, article, depicted a confident and energetic Marcellus Jacobs, noting:

> Sitting in the [*sic*] his office last week, the 78-year-old Jacobs looked like the successful business executive that he is. He cut a smart figure in a three-piece suit, as he spoke breezily of investing $1 million in the company. . . . As he speaks, Jacobs' fingers twirl around like the propellers on his wind plant, and his mind spins restlessly, still searching for the perfect design. After all these years, he still wakes up at 2 in the morning and jots down ideas to improve the wind plant.[50]

Newspaper reporters enjoyed Marcellus Jacobs' unscripted remarks about how the United States squandered its available wind energy, telling a journalist of *The Dispatch* in Minnesota, "We're OPEC agents, that's what we are, wasting our resources," and berating the Department of Energy's wind energy program as "outrageously wasteful." Marcellus Jacobs firmly believed that for the next five years energy costs would rise, making wind power an even more sought-after alternative for homeowners. *The Dispatch* stated, "Who knows? Perhaps a Jacobs plant may someday provide electricity to the White House."[51] However, the way in which the new Jacobs Wind Electric Company operated in the CDC organization, as well as Marcellus and Paul Jacobs's role in it, was about to change by the second half of 1982. Unaware of the coming corporate changes at the Jacobs Wind Electric Company, author Jon Naar in his 1982 book, *The New Wind Power*, intuitively wrote: "Being a legend in one's own lifetime is a rare distinction, yet not without its burdens, as Marcellus Jacobs has learned during

a career that spans virtually the entire history of modern wind-electric power."[52]

In 1982, the company started production on a larger 17.5-kilowatt wind plant. Several of these machines were tested in the field by the end of the year, including one at the Plymouth factory site and another at Paul Jacobs's home in Corcoran which was designated as a test site by the company. The goal was to ramp up to full production of the 17.5-kilowatt model in 1983.[53] Before this could occur, however, the Jacobs needed additional financing to expand the operation, and so they decided to sell their majority share in the Jacobs Wind Electric Company Delaware Corporation to CDC in late 1982.[54] This move effectively took the day-to-day operations and planning out of the Jacobs's hands and placed it under a new group of CDC managers, which included Robert D. Schmidt as board chairman, president, and chief executive officer;

Marcellus Jacobs shown with a third-generation, 10-kilowatt Jacobs unit at the company's Plymouth, Minnesota, factory. Courtesy Paul R. Jacobs, Corcoran, MN.

Robert Hennessy as president; Emil Ebner as vice president of sales and dealer relations; and a host of new CDC engineers and manufacturing personnel. CDC even brought back one of its inverter specialists, Jack Hauser, from retirement to try to resolve the company's persistent inverter problems.[55] CDC immediately fired the Jacobs's two outside factory representatives who over the course of two years visited more than 100 new wind-plant sites. Their work helped to raise the quality of dealer installations and reduce the rate of warranty expenses for the company. The Jacobs strongly warned CDC against this move. "This cut off our ability for the next couple years to effectively troubleshoot new problems by giving us few trained persons to send out in 'fireman mode,'" Paul Jacobs said.[56]

CDC also had a different vision for the direction of the Jacobs Wind Electric Company. They wanted to be suppliers of machines to the emerging, but lucrative, wind farm business. In 1983, the company had shipped eighteen 17.5-kilowatt wind plants to its Hawaii dealer, Wind Power Pacific, to erect at the Kahau Ranch in North Kohala on the Big Island. This would become the first multiple installation of Jacobs wind plants.[57] To further solidify this Hawaii effort, CDC, at the request of the Hawaii dealer in December 1982, entered a joint venture with Renewable Energy Ventures (REV) to "develop, own, and operate wind parks in the United States and selected foreign nations using Jacobs Wind Energy Conversion Systems."[58] The move fueled rumors among Jacobs's dealers that individual wind plant sales would be supplanted by wind farm business. Many were already complaining about the recent price increase per Jacobs wind plant to about $35,000. CDC's Ebner tried to avert the dealers' concerns by stating in January 1983: "There are two routes to the marketplace. . . . One is the dealer route which we are going to strengthen considerably; the other is the wind farm investor route."[59] Marcellus Jacobs also attempted to allay dealer worries at the February 10, 1983, meeting of the California Wind Energy Association by announcing the company planned to increase the size of its dealer network.[60] Yet, it became increasingly clear throughout 1983 that the wind farm business was CDC's top priority. CDC had the Jacobs Wind

These third-generation 17.5-kilowatt Jacobs units were installed at a wind farm at the Kahau Ranch in North Kohala on Hawaii's Big Island. From Earth Energy Systems Inc., *Bringing Energy to the World* (1985).

Electric Company sign a master purchase order agreement whereby for a fixed price per wind plant REV agreed to buy three years of Jacobs-produced machines to guarantee a steady market and cash flow to help develop larger models.[61] "The overall object of the joint venture is to install 2,250 WTGs (wind turbine generators) by 1985 in order to exploit federal tax credits available until then. These machines would be located principally in Hawaii, California, and Alaska," the agreement said.[62] As Paul Jacobs further explained, "in essence, since CDC financed and controlled the whole process, it was to be a 'right pocket to left pocket' set of transactions. M. L. and I as minority directors on the [Delaware] firm's board[,] we were told this method would speed up the learning curve at REV controlled sites and potentially reduce the cost of the warranty upgrades by having a limited number of sites, but many machines at a site."[63]

At the start of the 1980s, California had become a hotbed for wind farm development, and every major wind turbine producer in both the United States and Europe wanted a piece of the action.

CDC, with its control of the Jacobs Wind Electric Company and joint venture partner, REV, sought to take part. Federal and state tax credits covered 40 to 50 percent of the capital investment for wind plants used in the state's first wind farms. The 1978 Public Utility Regulatory Policies Act (PURPA) also required utilities to purchase the power at the "avoided cost" rate, which at the time was the equivalent the utility paid to generate power from fossil and nuclear fuels. Pacific Gas and Electric Company, for example, paid between 3.5 and 7 cents per kilowatt-hour for wind-generated electricity.[64] There were three primary areas in California for wind farm development: Altamont Pass east of Livermore, Tehachapi outside Bakersfield, and San Gorgonio Pass near Palm Springs. While continu-

ing to develop projects in Hawaii, REV, which CDC fully took over in 1983 and placed under the management of its Jacobs Wind Electric Company, set its main sights on the San Gorgonio Pass wind farms. By 1984, CDC used REV to line up a group of investors for its biggest wind farm project to date. The REV Wind Power Partners 1984–1, as the $32 million project was called, was located in North Palm Springs and consisted of 376 wind plants—a combination of Jacobs 17.5-kilowatt and Energy Sciences, Inc.'s 80-kilowatt machines—to produce an estimated 17 megawatts. The power was to be sold for the next thirty years to Southern California Edison. REV was also involved in another smaller $8 million wind farm that year in North Palm Springs, which it said consisted of 198 wind plants for a collective output of 3.7 megawatts, and a third similar size wind farm development in the area in 1985.[65]

The Jacobs Wind Electric Company also explored the development of so-called hybrid power systems, which included a combination of wind plants, photovoltaic solar arrays, engine generators, and battery backups, for continuous electric power for remote, standalone operations. To ramp up this business quickly, CDC in late 1984 acquired another longtime wind-plant manufacturer, Dyna Technology, and its 250,000-square-foot manufacturing plant in LeCenter, Minnesota, for $3.5 million. Dyna Technology, then a subsidiary of Keystone Products Corporation in Clifton, New Jersey, was the maker of Winco (formerly Wincharger) wind turbines ranging in size from 200 watts DC to 2.5 kilowatts AC and engine generators. CDC put Winco together with its Jacobs Wind Electric Company and renamed the business Earth Energy Systems Inc. The company designed most of these hybrid systems using the 10-kilowatt Jacobs; Winco engine generators; solar panels from providers such as ARCO Solar, Solarex Corporation and Kyocera Solar Systems; batteries from producers like Exide, GNB, Allied C&D Power Systems, and VARTA Batterie; and Winco inverters and others from manufacturers such as Powermark, Soleq Corporation, and Heart Interface. The system was linked together through a CDC microprocessor-based controller, which allowed the system to switch from

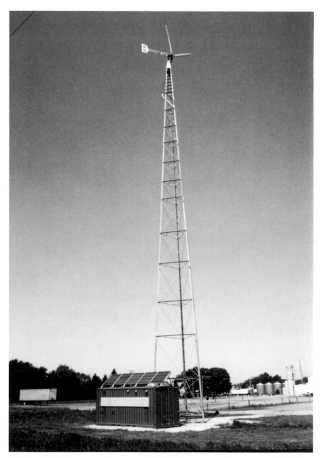

This Earth Energy Systems hybrid power plant consisted of a 10-kilowatt Jacobs wind turbine and an 800-watt photovoltaic solar array, in addition to a 10-kilowatt generator and a battery housed inside a 20-foot marine container, to provide continuous electrical power to off-grid applications. From Earth Energy Systems Inc., *Hybrid Power Plant*[:] *Class A up to 10 KW Continuous* (ca. 1985).

Earth Energy Systems began manufacturing a 20-kilowatt Jacobs wind turbine for the wind farm market. Courtesy Paul R. Jacobs, Corcoran, MN.

one energy source to another—wind, solar, or diesel power—depending on their availability. A smaller hybrid system designed by Earth Energy used a retooled 1.2-kilowatt, four-bladed Wincharger on towers ranging from 39 feet to 120 feet in height. How many of these hybrid systems Earth Energy actually sold was not determined. By 1985, however, CDC had moved Earth Energy's manufacturing and engineering activities to Dyna Technology's LeCenter facility, while maintaining research and development at the facility on Fernbrook Lane in Minneapolis. The company claimed at this time to have about 1600 wind plant dealers and distributors worldwide.[66]

The Jacobs, meanwhile, continued to lead research and development of the company's large wind plants, often personally conducting performance tests of components on two wind plants at Paul Jacobs's Corcoran property.[67] This work even included the development of the 20-kilowatt wind plant, which came online in 1985. However, it was the pervasive hardware and electronics problems that Marcellus and Paul Jacobs discovered through their tests that troubled them greatly, and were manifesting themselves frequently in the wind plants being installed at the Hawaii and California wind farms. For example, during side-by-side comparison tests of inverters, they discovered a routine 20 to 30 percent variation in kilowatt hours over a twenty-four hour period, and in some cases the kilowatt-hour production fell by as much as 60 percent, compared to the published projections used to sell the wind plants to the wind farms.[68] Marcellus Jacobs personally drafted more than thirty consulting reports between February 14, 1984, and June 13, 1985, covering a range of issues from propeller blade flutter, governor ball socket breakage, tower manufacturer mistakes resulting in metal fatigue, control springs for dual-fold tail vanes, and governor hub lubrication. The mostly two-page reports offered clear and concise recommendations to make the necessary fixes and were sent to as many as twenty senior managers and engineers, including the board, of Earth Energy.[69] In his February 12, 1985, consulting report, for instance, Marcellus Jacobs pointed out that the governor hub nuts and lock washers were "incorrectly designed and must be changed at once" for all the 20-kilowatt wind plant governors installed at the Jacoby-Kerr wind farm project at Palm Springs. "I do not know how this improper design ever got past the men in the assembly department, let alone be shipped. . . . When the men at the wind farm put the required torque pressure on the nut to tighten it against the

Marcellus Jacobs remained devoted to the Jacobs Wind Electric Company brand and perfecting its wind turbines until his death in July 1985 at the age of eighty-two. Courtesy Paul R. Jacobs, Corcoran, MN.

Paul Jacobs departed Earth Energy Systems shortly after his father's death, returning to the real estate consulting and brokerage business in both Florida and Minnesota, but he never lost his passion for the wind energy industry. Courtesy Paul R. Jacobs, Corcoran, MN.

lockwasher, the tapered end acted as a center punch point and simply spread the lockwasher open and pushed on through it, striking the governor hub casting. . . . Every governor with this nut design can sooner or later work the nut into the washer, spread it, and loosen the hub," he wrote.[70] However, the Jacobs realized that their consultative reports for making these necessary repairs and improvements to the wind-plant components were mostly ignored by the CDC management in charge of Earth Energy. They were not interested in "the norm of following an iterative product development approach that relies on real world field testing at stressful sites to evolve a design," which was the practice followed by Marcellus and Joseph Jacobs decades ago and was now being used successfully by Danish wind-plant manufacturers, Paul Jacobs said.[71]

At eighty-two years old in 1985, Marcellus Jacobs could have walked away from the business, but that was not his nature. He remained resolute in fixing the wind plant component problems, especially since these machines were branded with the Jacobs name. This all-consuming determination, Paul Jacobs said, contributed to his father's automobile accident in Plymouth on June 28, 1985, when he unwittingly pulled out in front of another car and was broadsided. Marcellus Jacobs suffered multiple internal injuries from the accident, which resulted in his death on July 15, 1985.[72] Throughout his life, Marcellus Jacobs remained an iconic figure to those in the wind power industry. As physicist Claude Eggleton noted in a 1985 *Wind Power Digest* article memorializing his achievements:

> Marcellus Jacobs was a true pioneer in the wind energy effort, devoting his entire adult life to it. He helped initiate the first great effort in the 20's and 30's, through his innovative designs and business development, establishing a record for quality engineering. He lived to see and participate actively in the rebirth of the wind industry today. A true expert, his valuable advice was sought by many individuals and companies, however but by few government officials. He was honored by his peers, receiving, for example, the Wind Energy Society of America's Four Winds Award for Excellence, at a University of Southern California meeting,

in 1975. His best memorial will probably be his imaginative well engineered generators and the industry he helped founded.[73]

After his father's death, Paul Jacobs found himself pushed aside at Earth Energy. However, the situation at both CDC and Earth Energy had already been going from bad to worse by mid-1985. Shearson Lehman Hutton, Inc., which was contracted by the REV partnership to sell the wind farm machines, abruptly canceled its work on the planned $45 million REV Wind Power Partners 1985–1 wind farm in North Palm Springs, which was supposed to consist of 500 Jacobs turbines for a collective output of 25 megawatts. At the same time, the Reagan Administration was making good on its promise to end federal tax breaks for wind energy, causing many American wind turbine manufacturers to exit the industry. It was reported that Earth

Energy in 1985 was losing about $1 million a month, while CDC's mainline business was also hit by a $567.5 million loss that year as personal computers began to challenge its longtime mainframe business. CDC dismissed Earth Energy's president, Robert Schmidt, at the end of the year, along with most of the company's staff. It is estimated that, including the canceled orders for the 500 wind plants for the failed REV Wind Power Partners 1985–1 wind farm, there had been enough wind-plant parts and subassemblies processed through Earth Energy's warehouse for the Delaware firm by the end of 1985 to assemble as many as 2000 machines with outputs of 10, 17.5, and 20 kilowatts. CDC, in late 1986, wrote off more than $90 million to dispose of the Earth Energy businesses.[74] The assets of the former Earth Energy were parsed out over the next three years, with Minneapolis-based Wind Turbine Industries Corporation (WTIC) obtaining the bulk

A third-generation, 17.5-kilowatt Jacobs wind turbine installed at the home of Paul Jacobs in Corcoran, Minnesota, in 1983. The unit's alternator was rebuilt in 2012 and is still operating today. Courtesy Paul R. Jacobs, Corcoran, MN.

A REPCO employee's view looking down from the top of a Minnesota co-op member's 120-foot tower while providing service to a 20-kilowatt Jacobs unit in 2012. Photograph by Stephen Jacobs. Courtesy Paul R. Jacobs, Corcoran, MN.

of the wind-plant manufacturing parts and tooling inventory, including the surplus parts and subassemblies from the canceled California projects, while the Florida corporation of the Jacobs Wind Electric Company, in the hands of Paul Jacobs and Marcellus Jacobs's widow, received only a cash settlement of $500,000 and ambiguous control of some of the same intellectual property (trademarks) conveyed to WTIC. This arrangement left outside observers confused as to who represented the "Jacobs" brand and who could upgrade the designs. The patents that he and his father had originated before and during their operation with CDC and Earth Energy were licensed to WTIC and to the purchasers of the various REV wind farms, but they were never divested by Earth Energy.[75]

Paul Jacobs remained active in the wind energy industry after 1988. He was asked to serve as an expert witness in 1991 by the investors of the failed REV 1984–1 wind farms in Palm Springs against CDC, Earth Energy, REV, and Shearson. He was also asked to develop a technical report of the problems related to the 17.5-kilowatt machines, which led to an out-of-court settlement of $12 million in 1992.[76] He also rekindled his real estate consulting

and brokerage business in both Florida and Minnesota. In 1993, he took on the Florida Department of Transportation for patent infringement in the use of the Jacobs Wind Electric Company's patents during the 1982 construction of a tidal gate and culvert flushing system in the upper Tampa Bay area. Although the Jacobs Wind Electric Company was able to prove the patent violation, the case was not finalized until 2010 because of a later malpractice lawsuit lodged against the law firm that represented the case.[77] In 1986, the Minnesota Department of Energy and Economic Development recognized Paul Jacobs for his significant contributions to the Wind Resource Assessment Project Advisory Committee, and he participated as a research fellow to the University of Minnesota's Department of Landscape Architecture in the College of Design, where he became involved in rural technology and cooperative development research.[78]

Like his father, however, Paul Jacobs remained passionate about wind-electric power and continuing the legacy of the Jacobs Wind Electric Company. Based on his own research while consulting to Earth Energy in 1984 and 1985 and additional findings in California in 1988–1989, he secured a patent in 1990 for a motion-damping apparatus for wind-driven propellers.[79] But it was a conversation that he had in the early 1990s with John Noland, a longtime attorney for the Jacobs Wind Electric Company's Florida corporation, that created a new chapter for the company. While the Jacobs Wind Electric Company had exited wind-plant manufacturing in the mid-1980s, numerous former customers of these machines sought help with technical matters and utility integration. This need was further compounded as the more than six hundred 17.5- and 29-kilowatt Jacobs wind plants at the former California wind farm sites were being taken down and resold as rebuilt units to individual customers throughout the country. "[I]t was John's idea that it could be advantageous to organize a separate new entity as a co-op," Jacobs said.[80] But it would be no ordinary rural electric cooperative, which sells central station electric power to farmers and other rural customers. This new, first-of-its-kind entity would instead assist rural producers of electric power using their own distributed wind, solar, biomass, and gas sources to sell their excess power to the rural electric co-ops and public utilities. "Basically, as was the experience my father ran into beginning in the 1940s, the REC co-ops did not want to buy power from their members, only sell power to them," he said.[81] In late 1993, Paul Jacobs, along with M. G. Edds, a former research engineer who worked for the Jacobs Wind Electric Company in 1984–1985, and several other wind energy proponents, founded REPCO (Rural Energy Producers Electric Power Cooperative), a nonprofit North Dakota rural electric power co-op, also operating in Minnesota. REPCO has two types of members: those who buy a membership and obtain a small amount of technical help, and those who request hands-on services, such as preventive maintenance and wind-plant repairs, which are provided through a network of former Jacobs Wind Electric Company employees. While the co-op is open to owners of any type of small wind turbine, it has been dominated by those who own and operate Jacobs machines.[82] "So far, in the past twenty years, we have made over a dozen major rebuilds of the 17.5- or twenty-kilowatt wind plants on the towers for members," Paul Jacobs said. Because of the small number of skilled technicians in REPCO, the hands-on work generally does not extend beyond a several-hundred-mile radius of the Minneapolis area. REPCO has also done some Internet-based consultations with Jacobs wind-plant owners in Canada and as far away as Ireland and Italy. However, REPCO's membership remains modest. "I suspect REPCO will attract more members and maybe some more with solar, but it is a hard sell initially to folks who are still focused on getting their meters just to operate in reverse," Paul Jacobs said. He added: "[T]he public is still interested in the idea at least of distributed renewable energy, even if the current political forces are now trying to roll back forty years of progress since the PURPA laws were passed in Congress to deregulate the grid and open it to competition."[83]

7 Bergey's Rise

Since the early 1970s, numerous small wind turbine manufacturers in the United States have come and gone. This is testament to the fact that it is difficult to design, build, and deliver a successful and affordable wind energy conversion system for the commercial market. Regulatory and financial obstacles also often proved too difficult for many companies to remain in the industry. Bergey Windpower of Norman, Oklahoma, has been the exception. For more than forty years, the company weathered the industry's ups and downs to become a recognized leader in residential-size wind turbines. Bergey Windpower attained this stature by spending the past four decades continuously engineering improvements to their machines. The company's long-term success, however, has more to do with the acumen of its father and son founders. Without their educational backgrounds, fiscal responsibility, belief in making the world a better place, and collective desire to build an efficient wind turbine, Bergey Windpower might have disappeared years ago.

Karl Halteman Bergey Jr. was always more interested in how things worked in the air rather than on the ground. Born in 1922 to Karl H. and Mary S. Bergey in Lewistown, Pennsylvania, a small town about fifty miles northwest of the state capital, Harrisburg, he would listen from his second-floor bedroom for the buzz of airplanes taking off from one of two nearby airstrips and then race outside to catch a glimpse of them flying overhead.[1] For young Karl Bergey and many others like him during the late 1920s, there was an intense excitement in the prospect of aviation and travel through the skies. Young aviators, some of whom trained to fly aircraft

during World War I, as well as celebrity fliers like James H. Doolittle, Noel Wien, Charles A. Lindbergh, and Amelia Earhart, attracted attention in the press and the imagination of many American people with their spectacular achievements from inside the cockpits of early aircraft. These aviators, some of whom became known as "barnstormers" for their daredevil antics in small, itinerant "air shows," needed only a flat, grassy patch to drop into small towns and entertain their inhabitants. Some townspeople were fortunate enough to take short flights with these pilots.[2] Commercial air passenger services also emerged during the late 1920s, including one that Karl Bergey recalled operated regularly between Pittsburgh and New York, but this form of travel was out of financial reach to most Americans who were still feeling the negative effects of the

A teenage Karl Bergey, shown participating in a model aircraft club in Pennsylvania, was fascinated by aerodynamics and flight. Courtesy Michael Bergey, Norman, OK.

Great Depression. Bergey, who was too young to partake in actual flight during the late 1920s, read everything he could about aviation from newspapers, magazines, and books, and became deeply involved in the construction of flying model aircraft.[3]

In the late 1920s and early 1930s, airplanes became a passion for many young men. If you couldn't pilot an actual aircraft, why not build a scaled-down version of one that you could launch into the air yourself? And that was what many young aviation enthusiasts like Karl Bergey did. The earliest model aircraft were constructed of lightweight materials, such as balsa wood, paper, and silk, and they were powered by twisting a large rubber band stretching through the fuselage that, once taunt and released, would rapidly spin a wooden propeller. The plane would then be immediately released into the air by hand. A properly designed airplane of this type could quickly climb into the air before the rubber band lost its tension, hang on a breeze for a while

Karl Bergey entered the undergraduate aeronautical engineering program at Pennsylvania State College in 1941, where he studied under renowned aerodynamics expert, Dr. David J. Peery, who headed the college's Department of Aeronautical Engineering. Courtesy Michael Bergey, Norman, OK.

and then glide or crash back to the earth. The earliest of these planes were known as "free flight" models, or those operated without assistance of wireless remote control. Model aircraft clubs sprang up across the country, and their members often competed in local, state, and regional competitions for prizes and bragging rights.[4] Bergey enjoyed free flight models, but quickly became interested in the emerging use of palm-sized, gas-powered model aircraft engines that came onto the market in the early 1930s. His favorite was the popular Brown Junior, a small single-cylinder engine of about one-fifth horsepower that ran on a mixture of gasoline and oil and could lift model planes with wingspans of 2.5 to 15 feet at much greater speeds and heights than their free-flight predecessors. The Brown Junior was the brainchild of model aviation pioneer William L. Brown IV of Philadelphia, Pennsylvania, who together with his father, William L. Brown III, decided to set up a small factory in Philadelphia in about 1935 to manufacture the engines for sale to hobbyists. One of those engines was purchased by Bergey, who not only idolized the engine's design characteristics and operation, but also its designer.[5]

In the autumn of 1941, Bergey was accepted into the undergraduate aeronautical engineering program at Pennsylvania State College (now Pennsylvania State University). By then, working in the university's machine shop was his childhood idol, William L. Brown IV, with whom he enjoyed discussing the mechanical aspects of aviation. However, a larger influence on Bergey's education was Dr. David J. Peery, professor and head of the college's Department of Aeronautical Engineering, who in 1949 would write and publish the groundbreaking textbook in the field, *Aircraft Structures*. Peery's book, which influenced a generation of aeronautical engineers including young Karl Bergey, emphasized "basic structural theory which will not change as new materials and new construction methods are developed." The book focused on the basic mechanics of airplane structures, in addition to load distribution on aircraft parts and aerodynamics.[6] Bergey's studies were interrupted in 1943 when he enlisted in the US Navy and trained to become an aerial navigator, but he was then assigned as an engineering officer to the Patuxent Naval Air

Aeronautical Engineer Karl Bergey (second from right), with Fred Weick holding the Federal Aviation Administration type certification for the first Piper Cherokee in 1960, in front of the Piper Aircraft Development Center in Vero Beach, Florida. Courtesy Michael Bergey, Norman, OK.

Station in Maryland. The war came to an end by the time he completed his training, and he returned to Pennsylvania State College in the autumn of 1945 to continue his formal education.[7] After earning an undergraduate degree from Pennsylvania State College in 1948, Bergey worked at Grumman Aircraft Engineering Corporation at Bethpage on Long Island for a year and a half before enrolling in a master's degree program for aeronautical engineering at the Massachusetts Institute of Technology. While studying at M.I.T., he conducted design and analysis work for Professor Otto Koppen, designer of the Heloplane, a form of vertical-lift, fixed-wing aircraft. He also kept in touch with his mentor and friend, Peery, and after completing his master's degree in 1951 Bergey joined North American Aviation in Southern California (which is now part of the Boeing Company). There he was involved in structural analysis, preliminary design, and project engineering for a hypersonic bomber.[8] It was also during this time that he met his wife, Patricia.[9]

Bergey moved to Vero Beach, Florida, in September 1957, where he took a job as assistant chief engineer at the Piper Development Center. The center was founded and headed by aviation pioneer Fred E. Weick, whom Bergey had admired from his days as an engineering student. Weick and his team were tasked with designing the Piper PA-28 Cherokee aircraft. Karl Bergey described the experience in an interview as both "exciting" and "nerve-racking." "I would go home for lunch, throw up and worry if everything worked," he recalled.[10] Despite the anxiety, the aircraft design worked,

and the Piper Cherokee would become one of the most appreciated single-engine aircraft ever built. In fact, it is still in production today. He also served as design project engineer for all Cherokee models through the Arrow, Cherokee Six, and the first of several multiengine versions, including the PA-35. In addition to designing aircraft, Bergey became a certified pilot, owning and operating a Piper Vagabond PA-15, which he acquired in 1952. "In the early days at Vero Beach, a few of us had access to several Lock Haven-built aircraft, both single and twin [engine]. Later on we had essentially unlimited access to Cherokees."[11]

In 1968, Bergey decided to leave Piper, or as he put it, he was "sweet talked into" becoming a vice president for research and engineering of the Aero Commander Division at North American Rockwell in Oklahoma. The work did not sit well with him, however, and after two years he left the company to join the teaching faculty of the University of Oklahoma Department of Aerospace, Mechanical and Nuclear Engineering in Norman.[12] Bergey quickly established himself as a charismatic and gifted instructor whose classes were highly sought after by the university's aerospace engineering students. He also challenged his students to become involved in various projects. One of those projects that received an abundance of attention on campus was the University of Oklahoma's Urban Car. The goal of the 1971 senior class design project led by Bergey was to build a nonpolluting, hybrid-electric vehicle. Involved in the project was Bergey's son, Michael, who at the time was a student at Norman

High School.[13] The two-seat car included an electric motor and lead–acid batteries for energy storage. To extend the range of the vehicle, as well as provide heating and air conditioning, a small internal combustion engine was added.[14] The Urban Car had a top speed of 55 mph and could travel up to 20 miles on the charged batteries. With the hybrid motor in operation, the operating range was extended to 40 miles. Bergey's students entered the car in the Urban Vehicle Design Contest at the General Motors Proving Grounds at Milford, Michigan, in August 1972, where despite some motor control problems, it placed in the middle third of the sixty-five cars entered by designers from throughout the United States and Canada.[15]

Bergey's interest in wind power was born out of a conversation with a university colleague in 1971. "I remember it very clearly," he said in an interview. "The national concern about energy consumption started as the impact of the oil crisis got underway. Tom Love, head of the [aerospace engineering] department, and I were chatting one day about all the wind in Oklahoma and, he said, 'why don't we do something about it?' That's when the lightbulb went off."[16] He believed that the university could be a center for wind generator design and development; in a paper in June 1971, he wrote: "The application of current aerospace engineering and fabrication techniques to wind power generators would appear to offer worthwhile improvements in efficiencies and specific power costs."[17] Bergey had no difficulty putting together a group of engineering students in 1972 to design and build a wind turbine that could be operated at night to charge the batteries of the Urban Car.[18] Under Bergey's guidance, his students conducted technical and economic feasibility studies and created a program to build a wind generator. The result was a machine with blades 12 feet in diameter. V-belts were used to couple the rotor to an automobile alternator. With average winds, the wind generator could charge the Urban Car's batteries in eight hours.[19]

In 1973, Bergey and a half-dozen engineering students developed a computer modeling program to predict the performance of wind generator systems using actual wind data obtained from the National Oceanic and Atmospheric Administra-

tion's National Severe Storms Laboratory on the university campus. A television tower in Norman was also outfitted by the laboratory to measure wind characteristics at six levels, ranging from 146 feet to 1458 feet. Bergey and his students found the data to be particularly useful for analyzing the siting and performance of wind generators.[20] They also determined that for the small wind-electric power industry to prosper and proliferate, the development of "low-cost blade and tower structures, reliable control systems and efficient storage methods" was necessary. Bergey added: "I believe that we must avoid the trend toward technological overkill."[21] Based on the technology available in the early 1970s, Bergey was aware that the "maximum theoretical energy recovery of any wind turbine is

University of Oklahoma Professor Karl Bergey influenced engineering student Mark Worstell to develop this two-blade, downwind pitching wind turbine for the Student Competition on Relevant Engineering (SCORE) at Sandia Laboratories in New Mexico during the summer of 1976. Courtesy Herman Drees, Thousand Oaks, CA.

59[.3] percent. Blade inefficiencies and mechanical losses reduce theoretical recovery to a maximum of about 40 percent. The overall efficiency of a complete wind rotor generating system, from zephyr to powerline, is not likely to be more than 35 percent, and may be less."[22]

Bergey also believed that the small wind-electric power industry in the early 1970s was still too much in its infancy to stand on its own as a commercial industry and needed financial support. "The critical element is the willingness of government and industry to make the necessary commitment to policies that will encourage the development and use of wind power . . . and could take the form of development grants and subsidies," he wrote in the summer of 1973. "Both are consistent with past government efforts to promote the use of emerging technologies."[23] Karl Bergey made these recommendations to the US government's National Science Foundation, which undertook the first discussions of developing a federal wind energy program prior to the formation of the Energy Research and Development Administration (later the Department of Energy).[24]

Bergey's eldest son, Michael L. S. Bergey, would play a pivotal role in the next phase of the family's evolution as a small wind-electric power pioneer. While Michael Bergey enjoyed participating in the University of Oklahoma's Urban Car project during high school, he wasn't immediately set on becoming an engineer like his father when he entered the university's undergraduate program in 1972. "I started as a zoology major, then thought I'd be an architect and then a lawyer. I resisted engineering for about two years. When I started taking engineering classes, I actually found them interesting," he said.[25] In 1975, he joined a university project led by engineering student Mark Worstell to build a two-blade, downwind pitching wind turbine with a 14-foot diameter.[26] The machine, called POWERS I, was entered in the Student Competition on Relevant Engineering (SCORE) at the Sandia National Laboratories in New Mexico during the summer of 1976. SCORE was set up in 1971, a year after the 1970 Clear Air Car Race, as a way for university engineering students to compete against each other on design and hardware fabrication projects. The

Karl Bergey's eldest son Michael, who attended the University of Oklahoma, led a student design team to develop this 11.2-kilowtt "Vertical Axis, Articulated Blade, Wind Turbine" for the SCORE competition in June 1977. Courtesy Michael Bergey, Norman, OK.

competitions of 1976 and 1977 placed a focus on alternative energy projects, namely, wind turbines and solar collectors. Michael Bergey recalled that there were about a dozen different wind turbine designs shown at the SCORE 1976 event, but what caught his attention was a vertical-axis wind turbine, called the Cycloturbine, presented by a Massachusetts Institute of Technology team led by graduate engineering student Herman Drees. "That got me interested in vertical-axis turbines," Bergey said.[27] He then set out to develop a vertical-axis wind turbine of his own.

Michael Bergey spent his senior year of undergraduate engineering studies obsessed with developing an optimal performing vertical-axis wind turbine, with the goal of presenting it at the following summer's SCORE Energy Resources Alternatives competition. He used computer modeling to analyze the efficiencies, namely, the lift and drag in the wind, of both the traditional bow-shaped Darrieus blades and a system of straight but articulating blades attached parallel around a vertical axis, known as the "Giromill" configuration.[28] The term *Giromill* (short for cyclogiro windmill) was coined by McDonnell-Douglas Company of St. Louis and the Aermotor Division of Valley Industries of Conway, Arkansas, which together set out in the mid-1970s to develop a 40-kilowatt prototype machine of this design.[29] Bergey, along with fellow engineering students Steven Van Swearingen and Jim Frazier, carried out numerous hours of research and design work at the University of Oklahoma Wind Energy Research Center, which was a single room at the university's North Campus engineering laboratory. It was a race against the clock, with his graduation looming and the SCORE competition scheduled to take place shortly thereafter in June 1977. However, Bergey and his team produced a machine that they called the "Vertical Axis, Articulated Blade, Wind Turbine," or VAABWT, in time for the SCORE competition. It was also referred to as the POWERS II. The self-starting machine consisted of three 9.75-foot blades with a rotor diameter of 22 feet. The machine's novel features included its blades covered with a super-thin aircraft-grade aluminum and its electromechanical blade pitch control system. Immediately below the blades were the pitch control motors, thrust and flanged bearing, disc brake and calipers, gearbox, slip-ring assembly, timing belt, and generator rated at 11.2 kilowatts. The turbine was designed to operate on top of a steel, four-post, modular tower nearly 30 feet in height.[30] Michael Bergey admitted that there was little time to test the machine sufficiently at the university before it had to be crated and shipped to the competition site in Richland, Washington. "Once it was set up, we got a strong wind which gave us excellent points for performance," he said.[31] The University of Oklahoma wind turbine placed first in the wind division of the competition, and second overall, losing the top spot to a biofuel project.[32]

The VAABWT was not without its problems. Bergey noted the turbine's pitch control was faulty and needed to be replaced by a new system that included a proportional controller and added wind direction sensors to be placed in front of each blade. "This system, if it works, will be the last word in versatility and should be able to yield some very interesting performance data," he declared.[33] In addition to about a dozen other minor fixes, Bergey said the VAABWT's blades and support arms had to be modified to reduce drag related to the "arm-blade intersection," and the blades themselves were

Michael Bergey used the University of Oklahoma's Wind Energy Test facility's data acquisition system in late 1977 to test modifications to his vertical-axis wind turbine. Courtesy Michael Bergey, Norman, OK.

lengthened to 15 feet. The VAABWT modifications were developed by Jim Frazier, who turned the project into his master's thesis. He used the University of Oklahoma's Wind Energy Test facility's data acquisition system to test the modifications to the wind turbine.[34]

It was not long after graduating from the University of Oklahoma with a degree in mechanical engineering that Bergey also began to receive job offers. Some oil companies at the time were offering newly minted engineers salaries ranging from $1200 to $2000 a month. While the money was tempting, Bergey did not accept any of the positions. Instead, he and his father, who continued to teach aerospace engineering students at the University of Oklahoma, began discussing the prospects of starting their own small wind-electric company. "My father had industry experience and knew how to put a company together. He also knew how to raise money," Michael Bergey said. "He asked me how little I could live on and I came up with a quick answer—$400 a month."[35] They formed Bergey Windpower Company with $40,000 in late 1977.[36]

But starting a company from scratch proved challenging. The wind energy market in the late 1970s was already full of small companies inspired by the energy crisis from earlier in the decade, many of which were struggling to stay in business due to expensive, lackluster-performing machines. "We tried to do some work with one of our original horizontal axis machines. We put a Jacobs [generator] on it to create a direct-drive 3.2-kilowatt machine, but the shaft bent and we had to rebuild it. Then dad started designing a machine from scratch," Bergey said.[37] This horizontal-axis machine, called the POWERS III, aimed to simplify and reduce the cost of the wind-generating system as well diminish maintenance. The design consisted of a 12-foot, downwind "dual rotor system," combining a multiblade inner rotor with three airfoil blades in an outer rotor. The turbine had "a low start-up wind speed, a wide operating range, moderate aerodynamic efficiencies and a tendency to be self-limiting at high wind speed conditions."[38] Michael Bergey acknowledged that the POWERS III concept "didn't do very well."[39]

The fledgling company didn't give up, however, and tested a number of new wind power technolo-gies at the time, such as special airfoils, passive blade pitching, rotor speed controls and low-speed permanent magnet alternators. In its startup days, Bergey Windpower enjoyed unfettered access to the University of Oklahoma's Wind Energy Center located on the North Campus next to the Norman Municipal Airport. The facility included a workshop, computer systems, and a "high-bay hanger for indoor assembly and testing."[40] Despite this access to the latest testing and production equipment, both father and son long maintained the philosophy that "simple is better." They sought to remove as much wear-and-tear as possible on moving parts in their turbine designs and increase efficiencies wherever they could in the technol-ogy.[41] "We found these stators and put together a direct-drive, permanent magnet alternator. We were continually trying to get a gallon out of a pint-size bottle," Michael Bergey said.[42]

By late 1978, Bergey Windpower was close to completing what it believed to be a successful horizontal-axis wind turbine design that could be delivered to the commercial market. They equipped the prototypes with three 5-foot-long aluminum blades, which vibrated and fluttered in the wind. To correct the problem, Bergey installed angled pitch weights in front of the blades that eliminated

The fledgling Bergey Windpower Company leased the University of Oklahoma's World War II-era naval air station building at 2001 Priestly Avenue in Norman to begin manufacture of the BWC 1000 wind turbines. Pictured from left to right: Jim Frazier, Larry Campbell, Alan Busche, Michael Bergey, Beverly Binns, Farrel Droke, and Pat Smith. Courtesy Michael Bergey, Norman, OK.

the vibration and caused them to passively pitch as the rotor speed increased. The rotor design also removed the need for shafts, gears, and other coupling mechanisms between the blades and the electric generator. Karl and Michael Bergey patented their new blade-pitching technology and branded it Powerflex.[43] For directing the rotor into the face of the wind, as well as taking it out of service when the winds reached in excess of 32 mph, the Bergeys developed a furling vane tail that resembled those of earlier all-steel mechanical water-pumping windmills. Working with a Boston company, Achieval Wind Electronics, they also sold solid-state electronic inverters for electric grid-tied and battery-charging capabilities. Their turbines could be installed on either tilting or stationary pole towers, as well as on top of fixed trussed towers. In addition, the Bergeys determined that their first machine should be rated at 1 kilowatt, which they deemed suitable to supplement electric power for the average American home.[44] Yet testing would last nearly two more years before the company felt ready to offer its turbine to market.[45]

Meanwhile, through Karl Bergey's connections at the university, the company secured office space in a campus building at 2001 Priestley Avenue in Norman, Oklahoma, once occupied by the US Navy. "It was cheap rent and included the electricity. We held

onto that space for thirty years before they asked us to move, so they could tear down the building," Mike Bergey remembered.[46] Bergey Windpower had officially incorporated in 1979 and secured a Small Business Administration loan that helped facilitate the expansion of its operations and support for research and development.[47] In addition to Karl and Michael Bergey, other key employees included fellow University of Oklahoma graduates Jim Frazier, Farrel Droke, Allen Bushe, and Pierre Veragen.[48] The company also subsequently hired University of Massachusetts wind program graduate and ten-year wind power industry veteran L. Ward Slager to serve as its marketing director.

By 1980, the Bergeys felt confident enough to introduce their first wind turbine to market, called the BWC 1000. These machines were easily recognizable with their orange-painted rotor hubs and distinctive tail vanes. "BWC 1000" was painted in black along the length of the white generator housing. The logo on the tail vane consisted of three concentric ovals with the last one trailing off with an arrow pointing downwind. They were marketed for their "high reliability, low maintenance, and automatic operations in all weather conditions."[49] The machine was sold in two versions: the BWC 1000 with a battery-charging direct current output of 12 volts, 24 volts, 36 volts, 48 volts, and 120 volts,

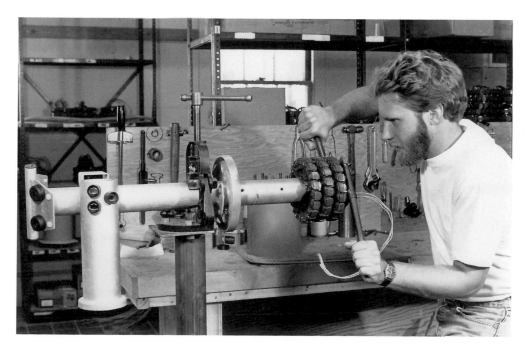

Bergey Windpower employee Alan Busche shown assembling a direct-drive permanent magnet alternator of a 1-kilowatt BWC 1000 in 1980. Courtesy Michael Bergey, Norman, OK.

which could be used for a variety tasks in remote areas; and the BWC 1000-S for direct connection into residential and industrial buildings which used the company's Powersync inverter to provide 115 volts, 60-cycle alternating current in "synchronization" with utility-generated power.[50] The 24-volt DC BWC 1000, without the tower or installation, when introduced to the market cost $2895, with the 12-volt DC variant at $2995; and the BWC 1000-S for grid-intertie cost $3195. The prices included the wind-electric generator and rotor, in addition to the related electronics to charge batteries or to connect to the power grid. Towers, batteries, and installation added another $2000 to $5000 to the cost of the machines.[51] With the available federal tax credit for small wind at the time, which covered 40 percent of the installed cost of the wind power unit up to $10,000, as well as additional state tax credit programs, the net cost of the actual Bergey Windpower machine could be below the company's list price.[52]

The earliest run of BWC 1000 units, however, was not without problems. "We produced the first eighty to one hundred turbines and started getting blade failures. The single-surface (sheet metal) aluminum blades were all retrofitted with new fiberglass blades" at the company's expense, Bergey said.[53] The new fiberglass blades were made in a continuous process called "pultrusion" and featured a custom airfoil developed by Karl Bergey. The company also learned that when its machines were placed on 40-foot towers, their performance was significantly reduced by wind turbulence. Bergey Windpower began recommending that their customers use 80- to 120-foot-tall towers for optimal results with its machines.[54]

The timing of Bergey Windpower's entry into the small wind-electric power market also was not the best. There were already established players with slightly larger output machines, such as Wind Works, Enertech Corporation, North Wind Power Company, and Sencenbaugh Wind Electric, which had already been selling their machines on the market since the late 1970s. "We were a poor company with limited advertising in trade magazines. Our marketing was aimed at recruiting dealers, which we did mostly at the trade shows," Michael Bergey

said.[55] The company soon built a pool of dealers, who came to the company's Norman, Oklahoma, offices for training workshops, covering "siting, installation techniques, cost and performance estimation, trouble-shooting, and, of course, sales techniques."[56] "About half of the seminar may be devoted to classroom work and discussions, but there is also a large amount of 'hands-on' instruction, including machine assembly/disassembly and actual system installations," Michael Bergey further explained at a Rural Electric Wind Energy Workshop in June 1982. "In essence, what we [are] trying to do is produce a cadre of wind energy professionals."[57] The trained dealers then purchased the Bergey Windpower equipment from the company at a 20 percent discount and made their profit on the difference between the wholesale and the list prices and on the installation work.[58]

The BWC 1000 ended up putting the Bergey Windpower Company on the map of the small wind-electric power industry. A marketing coup for the fledgling company was to see its machine prominently displayed on the cover of the July 1982 issue of *Popular Science* related to the article "33 Windmills You Can Buy Now!" article.[59] Neil Holbrook of Power Towers, Inc., one of California's largest small wind-electric systems installers in the early 1980s, considered Bergey Windpower's 1-kilowatt machine to be among the most "commercially reliable."[60] He counted the BWC 1000 among the four machines he sold at the time, including the 25-kilowatt machine from Carter Wind Systems, the 10- to 17.5-kilowatt Jacobs Wind Electric Company turbine, and the 4-kilowatt Entertech Corporation machine.[61] The company estimated that it sold more than 600 BWC 1000 units between 1980 and 1990 across the United States and to customers in more than fifty countries.[62] The majority of sales were for the grid-tied BWC 1000-S machines, which by generating 100 to 200 kilowatt-hours a month offered an average savings of 10 to 15 percent on a residential customer's electric bill.[63] "I don't think, however, that this is a reflection of an inherent superiority over non-interconnected systems," Michael Bergey said at the time. "The dominance results from the fact that interconnected systems are the 'path of least resistance' for both the

Bergey Windpower's new 10-kilowatt wind turbine, BWC Excel-S debuted in 1983 and included the company's patented PowerFlex passive blade system. Courtesy Michael Bergey, Norman, OK.

customer and the wind machine manufacturer. The public wants systems that are easily retrofittable and that will directly reduce their utility bills."[64]

It was not long before Bergey Windpower dealers began pressuring the company to build a larger machine. In 1982, the Bergeys began design work for a 10-kilowatt turbine. The development of the turbine, which was supported by a Small Business Administration guaranteed loan obtained by Karl Bergey, lasted about eighteen months.[65] The new machine, called the BWC Excel, was essentially a scaled-up version of its successful 1-kilowatt machine. For example, the BWC Excel's blade diameter was 23 feet, slightly more than double the BWC 1000. The blade assembly used the same Powerflex technology, as well as a more powerful version of the company's Powersync inverter, supplied by Achieval Wind Electronics, for power-grid connections.[66] When the company was ready to officially release the 10-kilowatt machine in June 1983, it already had more than 100 orders.[67] The

BWC Excel had an initial list price of about $10,000 for just the grid-tied generator and $11,000 for the battery bank-charging unit; add on a 100-foot tower and the Powersync inverter, and the cost was about $16,000.[68] In addition to households, the machines were marketed to farms and small businesses, and came with a three-year warranty, the longest in the industry at the time.[69] The Bergeys prided themselves on manufacturing the machine completely in-house, while most of their competitors assembled components supplied by outside vendors with little in-house basic manufacturing. Only the fiberglass protrusions for Bergey Windpower's blades were ordered from a supplier in Bristol, Virginia, but they were finished by the company's employees in Norman.[70]

Bergey Windpower also picked up interest in its machines from overseas markets. By 1983, the company estimated that 20 percent of its BWC 1000 units were exported to other countries. Some of its largest overseas clients were located in the Middle East, Far East, and Caribbean. It was also approached by a Japanese firm, as well as the Chinese government, to license and produce the BWC 1000 in those countries to shorten the order-to-market cycle.[71] Michael Bergey told *Alternative Sources of Energy* magazine at the time that "[t]he U. S. wind turbines are technically ahead of anybody else. . . . When people start thinking about windpower, they look at the French, the English and the Australian products and then they look at U. S. systems and very quickly come to the conclusion that the U. S. is where you buy windmills."[72] However, instead of standalone wind machines, many overseas customers sought a combination of renewable energy wrapped into one, known as "hybrid" systems, which included not only wind-electric technology but added photovoltaic panels and fossil-fuel-powered generators to provide more well-rounded power solutions for remote communication sites and village electricity.[73] These hybrid systems also had the advantage of being stowed safely in standard 20-foot and 40-foot-long steel marine containers, which once lifted off the ship and placed on the dock could be transported by trucks to their setup locations. Some hybrid systems even incorporated use of the container.[74]

Since the company's inception, Michael Bergey had been a proponent of setting performance standards for testing and rating wind-electric systems. He noted in the late 1970s that lack of performance standards diminished consumer understanding and confidence in wind energy technology. "Present industry practices do not allow for direct product comparisons and can often be misleading or woefully inaccurate," he said, adding that "Stimulation of the market will be brought about through fostering consumer confidence in wind energy products by ensuring that useful specifications will be presented in a consistent manner and that manufacturers have properly addressed important design issues such as safety and reliability."[75] Even though he was a relative newcomer to the American Wind Energy Association, Bergey was named chairman of the organization's Performance Rating Subcommittee in 1979. At that time, the most common way to compare wind-electric systems was via their "rated power," an unreliable indicator since these ratings were at different wind speeds, chosen by the manufacturer and therefore could not be meaningfully compared. The AWEA's Performance Rating Subcommittee believed that the standards for comparing small wind-electric machines should include the mean power output/annual energy output, maximum lateral force, noise level during operation, and power availability.[76] For Michael Bergey, standards gave the emerging industry a

Bergey Windpower tested engineering concepts of its 10-kilowatt BWC Excel wind turbine at the US Department of Energy wind energy test center at Rocky Flats, Colorado, in the early 1980s. US Department of Energy photograph courtesy Warren Bollmeier, Kaneohe, HI.

"most effective tool for exposing and discouraging the wind energy shysters that have begun appearing now that the tax incentives have grown so generous."[77] Annual Energy Output, which Michael Bergey modeled after the US Environmental Protection Agency's Estimated Gas Mileage rating for cars, assumed standardized wind conditions and for the first time provided a way to compare small wind turbine performance on an "apples-to-apples" basis. The work of the AWEA Performance Rating Subcommittee, as well as European organizations, became the basis of the IEC 61400–12 standard that is the worldwide method for rating the performance of wind turbines of all sizes today. In 1982, AWEA recognized Bergey for "Leadership in the Development of a National Performance Rating Standard for Small Wind Systems."[78]

In the early 1980s, a new industry had emerged in the form of erecting groups of wind machines in high-wind areas to produce utility-scale electric power known as wind farms. Karl Bergey believed manufacturers of wind turbines with minimum rated output of 10 kilowatts could become significant suppliers to wind farm developers. At that time, machines of less than 100 kilowatts were being deployed to test the wind farm concept, with an aim of constructing groups of megawatt turbines, like those developed by the Boeing and NASA partnership, for even greater output. The first wind farm in the United States was placed on the ridge of Crotched Mountain near Greenfield, New Hampshire, by the US Windpower Company of Burlington, Massachusetts, in late 1980 and consisted of twenty of the company's 30-kilowatt, three-bladed, downwind machines on top of 60-foot towers.[79] In 1983, Karl Bergey conducted a study to demonstrate the economic and technical benefits of using 10-kilowatt machines versus those with outputs of 1 to 2.5 megawatts, such as the model developed by Boeing under US Department of Energy sponsorship. Through mathematical modeling, he concluded that 10-kilowatt machines arranged at staggered heights across a 1200-acre wind farm would increase power density by as much as 60 percent, contrasted to the megawatt machines on fixed height towers. He also noted that the cost per installed kilowatt for the small

machines would be 20 percent less than those of the large turbines. Karl Bergey even suggested that small wind turbines could work together with an array of large megawatt machines by serving as "an economically attractive 'fill-in' and 'fill-out' capability for windfarms made up of relatively larger units."[80] Most importantly, if small wind turbine manufacturers like Bergey Windpower could supply their machines to the burgeoning wind farm industry, they could become multimillion-dollar enterprises. "The dramatic reduction in cost for individual EXCEL-sized units would encourage homeowners, ranchers, and small businessmen to purchase home-sized units with or without the benefit of tax credits," Karl Bergey wrote. "Energy costs could be less than 4.0 cents per kilowatt-hour, a great bargain today, and [an] even greater bargain in the years ahead."[81] The promise of tens of thousands of small wind turbines for wind farms, however, quickly faded as developers started with less expensive 50- to 100-kilowatt turbines and then sought ever larger machines for these projects as the technology matured.

Starting in 1984, the Bergeys knew the small wind-electric power industry was in trouble. While the federal tax credits, as well as those offered by various state governments, for installing these machines were still in place, sales for these machines had started to wane. Interest rates had also increased, making it less affordable for individuals to purchase wind turbines. As one journalist put it in the spring of 1983, "For the time being, circumstances are conspiring to keep a good technology down. Stable oil prices and the lingering recession are two of the chief villains. And SWECS (small wind-electric systems) remain very expensive."[82] Meanwhile, Michael Bergey was named president of the American Wind Energy Association in 1985. At the group's annual conference in San Francisco in August that year, he oversaw a ceremony at the nearby FloWind wind farm at Altamont Pass, marking the first million barrels of oil saved by the country's wind farms. An observer of the conference said the industry mood was "surprisingly upbeat given the uncertainty on the extension of the tax credits which the industry faces."[83] However, the worst for Bergey Windpower

A Bergey Windpower 10-kilowatt wind turbine installed at Ain Tolba in northeast Morocco in 1988, eliminating the community's reliance on diesel engines to pump water. Courtesy Michael Bergey, Norman, OK.

and its competitors became a reality in the autumn of 1985, when Congress allowed the federal tax credit for small wind to expire and the Reagan Administration showed little interest in continuing these programs, and oil prices also dropped below $10 per barrel. Interest in small wind-electric power disappeared almost overnight. Many small turbine manufacturers and their dealers closed up and left their customers holding empty service plans and warranties.[84] In 1986, the Bergeys saw their revenue fall by 80 percent based on a sudden 90 percent drop in sales.[85] "We shrunk to almost nothing," Michael Bergey said.[86] However, unlike many of their counterparts within the small wind-electric power industry, the Bergeys weren't about to close their doors. For the next two and half years, neither Karl nor Michael Bergey drew a salary from Bergey Windpower. Karl was able to live on his university salary, while Michael did some outside consulting work. The company maintained a staff of five but watched its University of Oklahoma co-founders—Farrel Droke, Allen Bushe, Jim Frazier, and Pierre Veragen—leave to seek employment elsewhere.[87]

To try to stimulate business for Bergey Windpower during those difficult days in the industry, Michael Bergey sought to participate in various federal government-sponsored energy programs for developing countries. One of those programs in which the company became most active involved the US Department of Agriculture's Agricultural Research Service in Bushland, Texas, where tests were underway to maximize water-pumping output with wind-electric machines. This technology was first tested by the West Texas facility in the late 1970s as an alternative to expensive natural gas and diesel-powered irrigation pumps utilized in the American West. A wind-electric pumping system at the time generally consisted of a three-phase motor on a submersible pump, which was driven at variable speed directly from a wind turbine's three-phase AC alternator. The machines proved unreliable, and interest in this form of rural water pumping diminished by the early 1980s. During the late 1980s, engineers of the USDA test site and Bergey Windpower worked together in support of a US Agency for International Development-funded village water supply project in Morocco. The Bergey wind turbines were simpler and found to be more reliable than the previous turbines, which allowed the researchers to focus on the pumping application. Bergey Windpower and the USDA refined the technology, developed electronic controls, and field tested several different pumping systems with simulated depths of 100 to 400 feet. In 1988, Bergey Windpower installed the first commercial wind-electric water-pumping system using its

10-kilowatt turbine in Ain Tolba, Morocco, to replace an inoperable diesel-powered pump for the village water supply. A second Bergey 10-kilowatt water-pumping system was installed at Dar El Hamra under the same USAID project and the two turbines supplied clean drinking water to five villages and about 4000 people. For Michael Bergey the most rewarding part of this first rural water supply project was the reduced number of young girls fetching water because water was now available in their villages. The complete units cost about $20,000 each and had the ability to pump as much as 600 gallons per minute.[88]

In 1990, the company replaced its BWC 1000 with a 1.5-kilowatt unit, called the BWC 1500.[89] The USDA's Bushland facility soon after acquired a BWC 1500 and a 10-kilowatt BWC Excel for $8000 and $26,000, respectively, along with a traditional mechanical Aermotor windmill for $6000, to conduct further water-pumping tests. More than

700 hours of tests were conducted, with the systems operating at seven different pumping heads ranging from 55 feet to 190 feet. At its test site, the USDA found the wind-electric system's average daily water volume exceeded the mechanical windmill by about 1000 gallons, or 45 percent, nearly year-round. The agency concluded that wind-electric pumps operated better than mechanical windmills when the average wind speed exceeded 11 miles per hour but were comparable to each other in average wind speeds of 8 to 11 miles per hour. The wind-electric water pumps also required less maintenance than their mechanical counterparts. Another benefit of wind-electric systems was the ability to erect the turbine towers away from the wells. This ability proved beneficial in hilly terrain where turbines needed to be placed at the highest elevations to catch the wind optimally, while wells were located in the valleys. Electric cables could stretch from the turbines to the well pumps of distances up to a half

A Bergey Windpower BWC Excel-PD (Pump Drive) machine in 1993 powering the Magic Circle Energy Corporation's Jones No. 2 stripper well in Jet, Oklahoma. The turbine pumped oil from a depth of 7500 feet at a variable stroke rate that depended on the wind speed. Courtesy Michael Bergey, Norman, OK.

mile. By the late 1990s, more than 100 wind-electric pumping systems, from 1 to 10 kilowatts in size, had been installed as part of this program in more than twenty, mostly developing, countries.[90]

Bergey Windpower participated in other rural electrification projects throughout the developing world during the 1990s using its 1.5-kilowatt and 10-kilowatt turbines.[91] In 1992, the company worked with Integrated Power Corporation, a subsidiary of Westinghouse, and the Indonesian government to develop hybrid power systems to provide twenty-four-hour electric power service to two remote villages, Tanglad and Julingan. The systems consisted of two Bergey 10-kilowatt turbines, along with photovoltaic solar panels and a diesel generator. The Bergey turbines were installed on tilt-up towers, which allowed them to be easily lowered to the ground for maintenance and typhoons.[92] In 1996, Bergey Windpower worked with the US Department of Energy in southern Mexico to replace a beach resort's diesel generator with a wind/battery/diesel hybrid power system. A modified Bergey EXCEL with a 48-volt winding was used in the system and included various protections for the turbine's fasteners and bolts against saltwater corrosion.[93] One of the more interesting test programs for Bergey Windpower in the mid-1990s was its participation with the Department of Energy's National Renewable Energy Laboratory and the University of Colorado to develop wind-

electric-powered ice-making machines for fishing villages in developing countries. Without electricity, these villages relied on traditional techniques, such as smoking, air drying, and salting, to preserve fish for later consumption. The government and university researchers determined that "Wind-electric ice making holds great promise in meeting an important need in remote third world villages."[94]

The Bergeys had hoped the developing country market for their wind-electric machines would boost production at their Norman factory back to 200 units per year, a mark that they achieved in the early 1980s. Not only would this help the company boost revenues, but it would also allow it to reduce overall cost per machine.[95] Reaching those levels of production based solely on exports during the early 1990s proved elusive. Yet Bergey Windpower remained adaptive in its product offerings and operations. In 1994, it introduced an 850-watt turbine for smaller off-grid applications, such as remote homes. This machine lasted on the market until 1998 when the company discontinued it for an upgraded 1-kilowatt machine, called the XL.1 (now Excel 1). Bergey Windpower also ended production of its 1.5-kilowatt turbine, the BWC 1500, five years later.[96] The company continued to look for ways to reduce manufacturing and shipping costs, particularly overseas. Since the late 1980s, Bergey Windpower licensed production of its turbines in certain countries, such as India,

In 1994, Bergey Windpower introduced an 850-watt turbine for smaller off-grid applications, such as remote homes. Courtesy Michael Bergey, Norman, OK.

The Bergey Windpower staff in 1993. Back row (left to right): Ken Parker, Pieter Huebner, Kay Furries, and Norman Bortz. Front row (left to right): Michael Bergey, Gary Shotts, Doug Sellers, and Karl Bergey. Courtesy Michael Bergey, Norman, OK.

Australia, and China. The company intended to reduce costs through its license with the Indian government by only adding 40 to 50 percent of the machine's content and the rest coming from a local manufacturer.[97] In 1998, the Bergeys established the Xiangtan Bergey Windpower Company as a joint venture with a 15,000-employee, state-owned enterprise, Xiangtan Electrical Machinery Group Company (XEMGC), in the Hunan Province of China to handle manufacturing of their turbines for sale in that country and to provide certain parts for assembly of units in the United States, while Bergey Windpower maintained product research and development at its Norman headquarters. The joint venture marked the first Chinese government-approved foreign cooperation in wind energy. In 2002, the joint venture was dissolved with XEMGC, and the Bergeys themselves established a wholly owned subsidiary, Beijing Bergey Windpower Company, with an office in Beijing and a factory about 40 kilometers outside the city.[98]

In 1999, domestic interest in small wind-electric power was rekindled as household electric bills continued to rise throughout the United States. People were also increasingly concerned about the country's continued reliance on fossil fuels to generate utility-level electric power and its contribution toward global warming. Some people simply wanted independence from the power grid through the ability to generate their own electricity.[99] According to a November 27, 2001, Gallup poll, 91 percent of Americans supported "investments in sources of energy such as solar, wind, and fuel cells."[100] In 1992, Congress established a robust incentive for wind farms, but it did not include small wind turbines for homes and farms. While the federal tax credits for small wind power remained nonexistent through the 1990s and at the start of the twenty-first century, several states, starting with California, established incentive programs of their own that helped reduce the financial outlay for solar and small wind turbines. In 2002, the American Wind Energy Association Small Wind Turbine Committee, which Michael Bergey chaired, reported that 2001 sales for US small wind turbines reached 13,400 units valued at an estimated $20 million, or a 40 percent increase over the previous year.[101] The machines ranged in size from small 400-watt units to charge sailboat batteries, to 3- to 15-kilowatt units for household use, and up to 100 kilowatts for small commercial operations.[102] Most of the turbines sold were the very small units for

In the early 2000s, Bergey Windpower experimented with larger size wind turbines, like this 50-kilowatt prototype unit on top of a 120-foot monopole tower in front of a Walmart store at McKinney, Texas, in June 2005. Courtesy Bergey Windpower, Norman, OK.

smaller, more efficient homes that use less electricity than the average-sized home."[106] By the start of 2010, however, the small wind power industry had already started losing the cost advantage it once held over photovoltaic solar technology—$4 per watt versus $8 per watt, respectively, in 2001—by the start of 2010. The Chinese government poured tens of billions of dollars into the country's solar panel factories, effectively flooding the global market with inexpensive panels. Solar panels also had the advantage of a much lower profile since they could be attached to roofs or ground mounts with little fanfare and no moving parts. In addition to the higher upfront purchase cost of a wind turbine, erecting them on a 60- to 100-foot-tall tower in residential areas usually encountered stiff local government permitting challenges, ultimately deterring many of those individuals and businesses initially interested in small wind-electric power.

battery charging on boats and for cabins. Bergey Windpower experienced an increase in US sales for its wind turbines, particularly the 10-kilowatt machines. By 2004, the company reached sales of $4.5 million.[103] A larger, three-bladed, 50-kilowatt machine, the BWC XL.50, which was introduced to the market by Bergey Windpower in the spring of 2001, was less successful and soon discontinued by the company.[104]

Bergey Windpower anticipated a further expansion in the sales of its turbines in the United States when Congress in 2009 passed a 30 percent investment tax credit on small wind turbines of up to 100 kilowatts.[105] In 2011, the company introduced its Excel 6, a 6-kilowatt machine with a slightly smaller rotor diameter than the 23-foot-diameter rotor of its Excel 10 unit. With a retail cost of $21,995, including an inverter, Bergey Windpower said the Excel 6 could "provide most of the electricity for

Michael Bergey had the rare distinction to serve as president of the American Wind Energy Association twice—1985 and 1994—and was named the association's "Wind Industry Man of the Year" in 1994. Courtesy Michael Bergey, Norman, OK.

Michael Bergey remained active in AWEA and served a second term as president in 1994, earning him the association's "Wind Industry Man of the Year" award. He had hoped that AWEA could knock down the barriers the small wind turbine manufacturers faced over the years. He served on the association's board for twenty-seven years, including two stints as president. While he praised the AWEA for the work that it had done for the small wind industry over the preceding forty years; "we were always lost in the shadow of big wind interests," he said.[107] So he left the AWEA by 2010 and helped form a new trade association that was better suited to promote the small wind industry, the Distributed Wind Energy Association (DWEA). He has served as board president, with the exception of a two-year period, since its founding. Distributed wind energy markets are mostly located in rural and commercial areas with an acre or larger properties, and as explained by DWEA, "Distributed wind offsets local energy consumption near the point of use (avoiding transmission system expansion), promotes more energy choices for Americans, and has substantial potential to increase private sector energy investment."[108] The association maintains five committees which focus on federal policy, market access, permitting and zoning, state policy, and standards for the small wind industry. Of the roughly seventy members in the association, about a dozen are small wind turbine manufacturers.[109]

By 2012, Bergey Windpower watched its exports exceed US sales. That year Michael Bergey noted that the company's overseas sales of turbines were strongest in the United Kingdom due the country's generous feed-in-tariff and significant wind resources. In Asia, the company's sales increased to both South Korea and Japan due to robust feed-in tariffs for small wind turbines. Bergey Windpower, however, gave up chasing the Chinese market due to already intense domestic competition, untenable accounts receivable problems, and the government's ability, twice exercised, to cancel property rental agreements to benefit its own development projects. The company closed its manufacturing subsidiary in China in 2012 and transferred production of its 1-kilowatt turbines back to its Oklahoma factory.

Karl (left) and Michael Bergey, shown here at the offices of Bergey Windpower in Norman, Oklahoma, in October 2015. Photograph by author.

In addition, the company continued to experience increased sales of its wind turbines used for micro-grid projects in developing countries. These systems utilized a combination of wind, solar, and diesel generator power to supply electricity to rural villages and telecommunication sites. Bergey turbines have often been widely used in "green" retrofits to remote cell phone tower sites in an effort to save diesel fuel and operating costs.[110]

For more than forty years, Bergey Windpower has repeatedly beat the odds, while many of its competitors have disappeared during the severe downturns in the small wind-electric power industry. Both Karl and Michael Bergey admit that it has been a brutal business, but neither has entertained the thought of selling or closing operations. Instead, they have continued to persevere, adjusting production and continuing to improve their wind turbines and related technology. The Bergeys believe that one of their biggest challenges is to bring down the cost of wind turbines against solar electric power in low-wind areas by 30 to 50 percent over the next three to five years.[111] Another way the Bergeys think the small wind-electric industry can better compete against solar is to copy that industry's successful third-party leasing programs. In other words, there are people who are interested in small wind turbines but cannot pay the upfront costs associated with the machines. In 2015, United

Wind started a "WindLease" program in New York that offers turnkey financing, operation, and maintenance of small wind turbine installations and forecasts that it could become a more than $20 billion industry nationwide in the next five to ten years. Michael Bergey called the program "one of the brightest market opportunities," and on January 20, 2017, he announced that United Wind would purchase 100 of its 10-kilowatt machines, the largest order ever for the company and as measured in units for small wind turbines in US history.[112] "It's a major victory for U.S. manufacturing and the small wind turbine industry," Michael Bergey said.[113]

Bergey Windpower has also continued to reap the benefit from continued engagement with federal government-sponsored wind research programs. From 2013 to 2017, the company participated in the US Department of Energy's Distributed Wind Competitiveness Improvement Project (CIP). The program annually awards cost-shared subcontracts to individual wind turbine manufacturers, with a focus on component improvements and systems optimization, manufacturing process improvements, prototype testing, certification testing, and type certification. The duration of the subcontracts would range from eighteen to twenty-four months.[114] Bergey Windpower's research and development of a new, higher-output wind turbine benefited significantly from four successive CIP subcontracts received between 2013 and 2017. The company used the research funds and interactions with the Department of Energy's Wind Energy Technologies Office and National Renewable Energy Laboratory staff to develop a new three-bladed, 15-kilowatt turbine, called the Bergey Excel 15, and tower system to replace its twenty-year-old Excel 10 machine. Bergey Windpower was able to demonstrate a levelized cost of energy of 13 cents per kilowatt-hour, or nearly half that of the 10-kilowatt Excel 10.[115] The company erected its first Excel 15 turbine for a customer in North Dakota on February 26, 2019, followed by a second installation in Florida on March 4, 2019.

Karl Bergey, Bergey Windpower's founder and chairman, died on May 27, 2019, at the age of 96. Although he retired from teaching aeronautical engineering classes at the University of Oklahoma in 2017, he continued to come to work each day at Bergey Windpower. Karl Bergey leaves behind a successful family-owned company, with son Michael Bergey as president. The company carries on Karl Bergey's wind turbine engineering philosophy, laid out forty years ago, which values mechanical simplicity but also demonstrates a capability to capture useful energy from an often unpredictable and turbulent source, the wind. According to Bergey family members, former University of Oklahoma students, and company employees, Karl Bergey liked to quote French author Antoine de Saint-Exupéry: "Perfection is achieved not when there is nothing more to add, but when there is nothing more to take away." "A wind turbine is a challenging bit of engineering with the variability of environmental conditions. An 80-mph wind will sift out the wheat from the chaff," said Michael Bergey about the successful introduction of the Excel 15 to the market. "We learned our lessons over the years."[116]

At the start of 2019, Bergey Windpower introduced its newest wind turbine, the 15-kilowatt Bergey Excel 15. The photograph shows one of the first installations on Long Island, New York. Courtesy Michael Bergey, Norman, OK.

8 Reinvention and Innovation

While interest in residential wind-electric power during the early 1970s was rekindled by the increase in energy costs and emerging environmental awareness, the capacity to manufacture these machines in the United States at that time was all but nonexistent. The once vibrant US small wind turbine industry disappeared following World War II, along with the knowledge and equipment to produce these machines. With the exception of rebuilders, Dyna Technology of Sioux City, Iowa, which had taken over the once prolific business of Wincharger Corporation, also known as Winco, during the early 1960s was the only US company that still sold the traditional 200-watt, 12-volt, direct current (DC), two-bladed wind generators. These machines, however, were limited to small, remote power applications utilizing battery-powered appliances.[1] At the start of the energy crisis in 1973, wind turbines with larger, more useful electrical outputs for American homes had to be imported from abroad. These imports mostly included the 2-kilowatt Dunlites from Davey-Dunlite Company of Adelaide, Australia; the 6-kilowatt Elektros from Elektro GmbH of Winterthur, Switzerland; and the 4-kilowatt Aerowatts from Aerowatt of Paris, France.[2] These imported machines had existing operational track records in their respective origin markets; Dunlite's roots, for example, traced back to the 1930s to meet Australia's rural electric needs, Elektro introduced wind plants after World War II to provide power to remote outposts and ski chalets in the Swiss Alps, and beginning in the mid-1960s the Aerowatt machines found both commercial and marine applications across France. Because of their costs, they represented considerable investments for

most American homeowners in the early 1970s.[3] In addition to the list price, these machines had to be carefully crated and shipped in marine containers for the trip across the ocean to the United States, which added to the per-unit cost. There was also the inherent risk of damage to components during shipping. In 1973, the 4-kilowatt Aerowatt was priced at more than $10,000 and this did not include the cost of the electronic controls, tower, and installation.[4] The 6-kilowatt Elektro at the time had a price tag of more than $6000; a Canadian

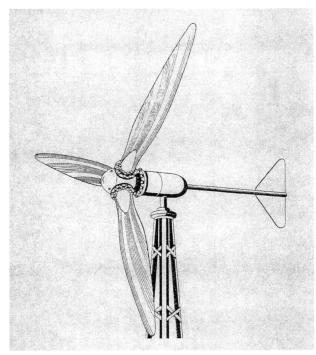

At the start of the 1970s, newly manufactured wind-electric turbines for residential use in the United States were imported, like the Swiss-built 6-kilowatt Elktro. From Elktro GmbH, *37 Years Experience with Elektro Windmills*, Winterthur, Switzerland (ca. 1970).

homeowner who imported and installed one of these units lamented that it cost him as much as two Volkswagen Beetle automobiles.[5] An imported 2-kilowatt Dunlite could be purchased from a US distributor for about $3000.[6] Additional equipment, such as towers, batteries for electric power storage, or inverters for direct utility grid connection, as well as the installation work, could easily add several thousand dollars to the cost of these wind turbines.

The biggest concern with imported wind turbines was what to do when problems occurred during installation and operation. Canadian Ed Harper, who purchased his 6-kilowatt Elektro in February 1975, spent more than a year repairing the 450-pound generator, which was found to contain numerous steel chips from the Swiss factory that damaged the internal bearings. He documented the problems and corresponded with the company, which stated that the deficiencies were the result of "a period of severe labour shortage and at the same time increased production." Due to the sudden interest in wind power resulting from the energy crisis, Elektro's production jumped from about 25 machines a year to more than 200 by 1974.[7] Henry Clews of Holden, Maine-based Solar Wind, which became Elektro's primary importer of wind plants into the United States during the early 1970s, also experienced fracturing steel hub plates, which caused the wood blades to fall to the ground. By the end of 1974, Elektro introduced new propeller hubs with stronger hub plates.[8] Other American customers of Elektro complained about the unclear installation and operation instructions, which were only available in German, not to mention the weeks it took to send and receive written correspondence from the Swiss company by airmail.[9]

By 1975, the burgeoning small wind-electric consumer market in the United States had become ripe for a newfound domestic wind turbine manufacturing industry to emerge. The earliest developers hailed from a variety of backgrounds—engineering, electrical and machine shops, and business. A common thread among these individuals was the desire to design and build a modern wind generator that would become an immediate commercial success. This optimism was further fueled during

the decade by new federal government programs that sponsored the research and advancement of small wind energy, a national law that allowed small wind turbines to connect to the public utility grid, a congressionally authorized investment tax credit for wind, and the formation of the American Wind Energy Association to represent the collective industry. It appeared that hundreds of thousands of farms and ranches across the country, as well as numerous suburban homeowners, were suitable candidates for small wind turbines to supply useful electric power. Re-creating an American industry that reached its pinnacle of manufacturing, sales,

US wind-electric turbine manufacturers, like Bergey Windpower, Carter Wind System, North Wind Power, Enertech, and Windworks, were optimistic about their prospects at the start of the 1980s. Pictured here is a 2-kilowatt Enertech machine at the federal government's Rocky Flats wind test site in Colorado. US Department of Energy photograph courtesy of Warren Bollmeier, Kaneohe, HI.

and installations of tens of thousands of small wind plants during the 1930s due to a lack of rural electrification faced different and much greater obstacles in the 1970s. By now, US consumers were mostly tied to public utility systems and, although they were begrudgingly paying more in their monthly bills, they continued to rely on this constant flow of electricity to power their households. The utilities also made it difficult for households to install wind plants, despite the tax incentives and mandatory hookup regulations instituted by the late 1970s. By the start of the 1980s, homeowners spent a minimum of $8000 to $10,000 to install a small wind turbine, not to mention assuming the risk that a machine might fail to perform optimally or break down shortly after installation. Some individuals took matters into their own hands and built small wind turbines from scratch in their own workshops. Despite the negative factors associated with commercial-built turbines, wind energy expert Paul Gipe in 1981 warned against pursuing homemade machines. "Homebuilts are a thing of the past. Don't even attempt it unless you're skilled with your hands, know what you're doing, and have plenty of insurance. They end up costing more than anticipated. They're generally undependable. And they're also dangerous," Gipe wrote.[10]

The small wind industry optimistically expanded into the early 1980s and, according to the Department of Energy, about forty-five US companies were manufacturing small wind-electric machines ranging from less than 1 kilowatt to 100 kilowatts in output by 1981.[11] Unlike their counterparts from the earlier part of the century, this generation of wind turbine developers had access to new engineering and manufacturing techniques, wider availability to newer and lighter weight materials such as fiberglass and aluminum, and solid-state electronics and early-stage microprocessor technology. By the end of the 1970s, Dakota Wind & Sun of Aberdeen, South Dakota, and Independent Energy Systems of Fairview, Pennsylvania, completely shifted away from selling rebuilt pre-1950 Jacobs wind turbines due to the increasing lack of parts, and they began to manufacture their own variations of these machines.[12] Other companies, such as Bergey Windpower, Carter Wind Systems, North Wind Power, Enertech, and Windworks, continued improving their machines, while upstarts and dealers-turned-manufacturers joined the market with new product offerings.[13] Even Winpower of Newton, Iowa, a company that stopped manufacturing wind plants after World War II, reentered the market in 1984 with a new three-bladed,

Winpower of Newton, Iowa, which stopped manufacturing wind plants after World War II, reentered the market in 1984 with this 3.5-kilowatt machine, but ended the line a year later as the small wind market collapsed. From Wind Power Systems, Inc., *Introducing . . . The Winpower 3500*, Newton, IA (1984).

3.5-kilowatt machine and sold about twenty-five machines the first year. But within a year, like many manufacturers who experienced dismal returns on investments in their new wind-plant product lines, Winpower abruptly exited the wind energy business in 1985, selling off its designs, patent rights, and parts inventory.[14] Some small wind turbine manufacturers, which did not benefit from the rush to build wind farms in California, found receptive buyers overseas, particularly in the Caribbean and South Pacific islands, Africa, the Far East, and the Middle East, as part of government-funded rural electrification projects.[15] The small wind industry's mass exodus from the United States continued with the steep reduction in energy costs and the Reagan Administration's refusal to extend the investment tax credits for wind energy at the end of 1985, which reduced customer interest in this alternative energy.[16] Regardless of the US small wind industry's meteoric rise and fall between 1975 and 1985, many companies during this period nonetheless made important contributions to the advancement of turbine design and operation, which later influenced the engineering of large, utility-scale units that came to dominate the wind power industry. This chapter examines the histories and contributions of some of these firms.

The most distinguishing and eye-catching aspect of any wind turbine is the rotor and blade arrangement. It is also this component that controls how the wind turbine operates in the wind. German physicist Albert Betz in 1919 mathematically determined that an ideal wind turbine rotor theoretically could extract a maximum 59.3 percent of the power in the wind and convert it to mechanical power, which became known as the Betz Limit. Since the wind turbine rotor extracts kinetic energy from the wind, it will simultaneously reduce the speed of the wind. In reality, the Betz Limit could never be achieved due to frictional forces and aerodynamic losses encountered by the rotating blades.[17] The earliest wind turbine blade shapes and rotor designs were largely influenced by the aircraft propeller technology that emerged out of World War I. Unlike the aircraft propeller that was driven by an engine to provide lift and propel an aircraft forward through the air, a wind turbine's rotor system was

designed to begin rotating when the wind interacted with the blades and to increase rotational speed as the wind escalated. Rotors generally included various spring-controlled counterweights to limit rotor rpm to ensure the blades did not overspeed in the wind. Rotor efficiency of early horizontal-axis turbines were far from optimal mainly due to the simple geometrical characteristics of the blades, which were made from planed wood boards or sheet metal folded into an airfoil shape. From the early 1930s, the Jacobs Wind Electric Company became an industry leader with its ruggedly efficient three-bladed rotor design. (See chapter 2.)

While use of wind-electric power in rural locations diminished prior to World War II due to the

Display on the grounds of the University of Stuttgart (shown in 2015) commemorating Ulrich Hütter's research into wind turbine rotors with tapered, low-solidity blades. Photograph courtesy Julian Herzog, Stuttgart, Germany.

widespread distribution of utility-generated electricity, a handful of American and European scientists and engineers—mostly from the aerospace industry—continued to research wind turbine blade and rotor designs. To enhance rotor efficiency, experimentation was performed on blade twist and taper. Ulrich Hütter, an engineering professor at the University of Stuttgart in Germany in the early 1950s, used his extensive research of aircraft wings, as well as access to fiberglass and plastics, to fabricate optimally shaped wind turbine blades. In 1957, he constructed a downwind 100-kilowatt wind turbine, which consisted of a two-bladed, 110-foot diameter rotor. With this machine, Hütter demonstrated the benefits of rotors with finely tapered, low-solidity blades.[18] Rotor solidity is the ratio of the area that the blades occupy and the area that they sweep during rotation. Low-solidity rotors operate at a higher rotational speed and generate less torque, making them ideal for turning generators.[19] Hütter also realized the benefit of mounting the rotor hub on a hinge, so that it could operate like a teeter-totter tipped on its end at 90 degrees. This arrangement reduced blade root loads by allowing the blade tips to translate upwind and downwind relative to the rotor plane. Moreover, slender blades weigh less and thus had the potential of lower manufacturing cost.[20] Due to dwindling

institutional interest, the University of Stuttgart discontinued Hütter's wind turbine tests in 1964, but his rotor research returned to the forefront in the 1970s and significantly influenced the designs of future wind turbines.[21]

In 1966, Thomas E. Sweeney, director of the Advanced Flight Concepts Laboratory at Princeton University, embarked on the development of a lightweight framed and Dacron fabric-covered airfoil-shaped blade, dubbed the Sailwing, "to study the advance ratio characteristics of the blade at various blade pitch and blade twist angles."[22] From his research and tests, Sweeney constructed a two-blade Sailwing machine with a 10-foot diameter that successfully endured a year's worth of New England weather. In early 1972, he constructed a second two-blade Sailwing wind turbine on the university campus, which had a 25-foot diameter to test sail deformations and gyroscopic and other dynamic effects of the wind. Tip-mounted weights were attached to the blades for automatic pitch control to regulate their rotation in the wind. Sweeney also attached a small electric generator to the machine, resulting in the production of about 7 kilowatts in a 20-mph wind. The Princeton Sailwing's red and white painted strips, resembling a candy cane, only added to its visual appeal.[23] While Sweeney's Sailwing captured the attention of wind

Thomas Sweeney, director of the Advanced Flight Concepts Laboratory at Princeton University, developed the concept of a lightweight "Sailwing" for use on wind turbines. Photograph by Jan M. Drees (ca. 1973). Courtesy Herman Drees, Thousand Oaks, CA.

Based on a November 1972 *Popular Science* photograph, a group of young men who worked for Windworks raised this lightweight, do-it-yourself wind turbine into place at Richard Buckminster Fuller's island property off the coast of Maine. Courtesy Stephen James Govier, Suffolk, England.

power developers in the early 1970s, it was rapidly overtaken by further advances in horizontal-axis rotor and blade design. Sweeney, however, continued to research and develop his Sailwing wind turbines at Princeton University until the early 1980s, with units installed in California, Hawaii, New Jersey, and six countries.[24]

Another early 1970s wind turbine with a lightweight airfoil-shaped blade, albeit with a focus on doing it yourself and inexpensive materials, was produced by Windworks Inc. of Mukwanago, Wisconsin. The company started working on wind energy conversion systems in 1970 under the sponsorship and guidance of R. Buckminister Fuller, who promoted geodesic architectural designs. Windworks first offered plans for a simple wind turbine that used a rotor of cloth sails, reminiscent of the water-pumping windmills that were used on the Mediterranean island of Crete. By 1972, the company settled on an airfoil blade shape that incorporated commercially available honeycomb paper for its core, covered with a fine weave fiberglass cloth skin. Windworks said the blades could be made in lengths of up to 15 feet, employing "simple techniques and conventional power tools."[25] The company used three blades as part of a downwind turbine mounted on a 12-foot-

tall wooden tower, which Hans Meyer, head of Windworks, explained how to build in the November 1972 issue of *Popular Science*.[26] Windworks offered two ways to control its wind turbine's rotor speed, including a damping spring and a flyball governor that feathered the blades as they rotated in the wind.[27] While Meyer claimed the wind turbine could be constructed for as little as $200, he wrote in the *Popular Science* article that "[i]t won't supply 100-amp, 60-cycle juice for your home circuits, but it can be a fascinating trial of a non-polluting, independent power source fully capable of handling light-duty jobs," such as providing electricity for small power tools, lighting, and heating.[28] Windworks' do-it-yourself wind turbine designs captured the attention of home experimenters and tinkerers, but it's unclear how many of these machines were actually built.

Several entrants to the fledging US wind energy business in the early 1970s used multi-bladed wind turbine rotors reminiscent of the first commercial-built wind-electric machines introduced in the late teens to early 1920s. These early upwind machines coupled water-pumping windmill wheels with gears or belts to rotate the shafts of electric generators. Their heaviness in design, construction, and high solidity, however,

This "bicycle wheel" wind turbine, based on the 1973 rotor design of Thomas Chalk, was erected at the University of California at Berkeley in 1977. Drawing based on Battelle Northwest National Laboratories photograph. Courtesy Stephen James Govier, Suffolk, England.

This North Wind Power Company 2-kilowatt wind turbine, shown here at the federal government's Rocky Flats wind energy test site in Colorado in 1979, used a "modernized" version of Parris-Dunn Corporation's late 1930s tilt-back design for braking. US Department of Energy photograph courtesy Warren Bollmeier, Kaneohe, HI.

made them inefficient wind-electric generators, and so the manufacturing of these machines was discontinued by the mid-1920s. In 1973, Thomas O. Chalk of St. Cloud, Florida, designed and produced a wind turbine rotor resembling a giant bicycle wheel. An inner and outer ring, which contained twenty-four evenly spaced airfoil-shaped blades, were attached to the horizontal shaft by a system of wire spokes. A belt that encircled the outer rim turned the shaft of a generator placed immediately below the wheel. Chalk's wheels were generally 15 feet in diameter and, because of their aluminum construction, weighed only about 70 pounds. The wheel was turned in to and out of the wind by a spring-tensioned, self-governing tail vane.[29] Dr. William Hughes of Oklahoma State University, who tested one of Chalk's first

machines, called the design "near genius" in the July 1974 issue of *Popular Science*, which included a cover illustration of the wind turbine.[30] Chalk, who received no government funds to develop his wind turbine, commented at the time that "it doesn't necessarily require a multi-million dollar study to come up with an innovation. Occasionally, a guy working in his garage or basement will come up with a solution to a problem that has eluded experts for years."[31] Similarly, a battery-charging wind turbine of 12 or 24 volts output, which included an 8-foot multibladed rotor, was developed by American Energy Alternatives of Boulder, Colorado, in 1973.[32] Despite their aesthetic appeal and high efficiency claims, both companies' wind turbines struggled to compete against increasingly higher-output machines with two- and three-blade rotors. To keep his enterprise afloat, Chalk offered a mechanical water-pumping version of his wind-electric machine in the late 1970s.[33] Chalk attempted to reinvigorate his presence in the wind-electric market in 1980 by designing a 3.5-kilowatt machine that used his original bicycle wheel-style rotor, but his company failed to achieve full-scale manufacturing of the wind turbine.[34]

Most residential-size, horizontal-axis wind turbines—less than 10 kilowatts in output—that entered the market during the late 1970s and early 1980s embraced a three-blade rotor design, which employed a tail to guide the rotor into the wind. The rotor diameters generally ranged from 8 feet to 23 feet.[35] Wind turbines with larger rotor sizes can produce more power due to their greater swept area. By doubling the blade length, it was possible to increase the power captured by the rotor four times.[36] Common blade materials were wood, fiberglass, and aluminum, with laminated wood being the most popular due to its lower cost, ease to shape, and durability. Fiberglass offered similar cost, shaping, and durability characteristics as wood, while aluminum, which provided its own production benefits in terms of shaping, became the least used material for blade manufacture due to a proneness for premature fatigue.[37] Wind turbine manufacturers varied more so on how they mechanically braked their rotors in the high winds. Both Sencenbaugh Wind Electric and Kucharik Wind Electric used folding tails for their 500-watt DC wind turbines, a technique perfected a century earlier by the water-pumping windmill industry. Aeolian Energy, Dunlite, Future Energy R&D, and Natural Energy Systems Unlimited (a US distributor of Elektros) used rotor hubs with weights and springs that allowed their blades to feather as the wind speed increased.[38] Bergey Windpower, along with tail furling, used weighted flexible fiberglass blades for rotor control, while the Jacobs Wind Electric Company enlisted its proved spring-based, automatic variable pitch centrifugal governor, which it introduced in about 1948 as a replacement for an earlier flyball governor.[39] North Wind Power Company, as part of a Department of Energy design competition, set out in 1977 to develop a 2-kilowatt severe weather-resilient turbine with a three-blade,

upwind rotor that incorporated a "modernized" version of Parris-Dunn Corporation's late 1930s tilt-back design for braking. Parris-Dunn called this overspeed protection "slip the wind." In a similar fashion, North Wind's rotor and generator assembly started tilting back toward the tail vane whenever the winds exceeded 21 mph. Full shutdown of the North Wind machine occurred when the rotor was tilted a full 90 degrees to the tower. The company called this dampening mechanism the Variable Axis Rotational Control System, or VARCS.[40]

Wind turbine designs with outputs of more than 10 kilowatts tended to favor downwind rotors with two, three, and four blades, which integrated various mechanical and aerodynamic character-

During the early 1980s Windworks manufactured a 10-kilowatt, downwind turbine, which included a unique pitched and hinged rotor design. Courtesy Craig Toepfer, Chelsea, MI.

istics reflective of Hütter's earlier work. Hütter, who was known to guard his research closely, was routinely approached by young wind turbine engineers for his advice. The National Aeronautics and Space Administration and its aerospace and power generation industry partners in the 1970s settled on Hütter's two-blade, downwind rotor when starting their work on the experimental plus-100 kilowatt machines.[41] In 1974, a group of engineering students from the University of Massachusetts at Amherst, under the direction of Professor William E. Heronemus, who were familiar with Hütter's research, set out to design and test their own three-blade, downwind turbine with a 32.5-foot diameter rotor. With its optimally shaped, low-solidity fiberglass blades, which were pitch regulated and turned at variable speed by computer control, the WF-1 has often been characterized as the first "modern" wind turbine to be built in the United States. In 1976, its design became the foundation for Burlington, Massachusetts-based US Windpower's commercially built 50-kilowatt wind turbine, of which more than a thousand were manufactured and installed starting in the early 1980s for the country's first wind farms in New Hampshire and California.[42]

Nearly a dozen other commercial firms in the United States to emerge during the second half of the 1970s settled on downwind rotors for their wind turbines. In 1976, Jay Carter Enterprises of Burkburnett, Texas, pioneered a downwind turbine that consisted of two fixed-pitch fiberglass blades attached to a teetering hub. After numerous tests and refinements, the company started commercial production of its machines in 1979. The Carter machine's 32-foot diameter rotor was designed to produce 25 kilowatts in a 25 mph wind.[43] Zephyr Wind Dynamo Company of Brunswick, Maine, in 1975–1976 also began testing downwind turbines with three-bladed, fixed-pitch rotors that generated 7.5-kilowatts.[44] At the same time, Kedco Inc. of Inglewood, California, manufactured a downwind, three-blade machine, based on the research and designs of Jack Park, that was rated to generate 1.2 kilowatts in a 21-mph wind.[45] Based on research performed with the Department of Energy between 1977 and 1980, Windworks manufactured a down-

wind, three-bladed 10-kilowatt machine with a 33-foot blade diameter for the commercial market. The turbine included a unique pitched and hinged rotor design, which used a hydraulic cylinder that activated mechanical linkages connected to each of the blades to control blade pitch.[46] In 1982, Windworks became a subsidiary of Wisconsin Power & Light Company, further propelling the production and sales of its Windworker turbine, which included the highly visible installation of one of these machines near the west-side entrance to the Golden Gate Bridge in 1983.[47] Another wind-electric company that benefited from Department of Energy research funds during the late 1970s for the design and development of residential-size, downwind turbines was Enertech of Norwich, Vermont. After spending its first several years in operation as a distributor of other small wind turbines, the company began manufacturing three-bladed machines with 1.5-kilowatt and 1.8-kilowatt outputs, and in the early 1980s, it started production of a downwind, 20-kilowatt turbine with a 44-foot diameter rotor for heavier duty applications, such as farms and small businesses. With all three sizes of these machines, Enertech included specially designed blade-tip brakes that served as an emergency overspeed backup brake for its turbine rotors.[48] Fayette Manufacturing Corporation produced a three-bladed, downwind turbine with a 34-foot-diameter rotor and rated output of 30 kilowatts. The machine, which found a place in California's early wind farms, controlled rotor speed with blade tips that positioned at right angles to the direction of the blades' movement.[49] Another small downwind turbine developer, WhirlWind Power Company, in the late 1970s offered a 2-kilowatt machine with a two-blade rotor that was guided into the wind using a pilot rotor yaw drive, which consisted of a small four-leaf-clover-shaped rotor and small tail vane that sat just behind the large rotor and was oriented 90 degrees to it. The pilot rotor began to turn when the wind direction changed, driving a pinion on a bull gear at the top of the tower to move the main rotor into the wind again. WhirlWind, which went on to make higher output machines of this design in the early 1980s, said in its marketing literature that the pilot rotor "orients [the] machine

United Technologies Research Center spent the latter half of the 1970s developing a two-blade, downwind turbine that used a lightweight composite bearingless rotor pioneered by the helicopter industry. US Department of Energy photograph courtesy Warren Bollmeier, Kaneohe, HI.

In 1979, Terrance Mehrkam of the Energy Development Company installed this 250-kilowatt turbine to provide supplemental electric power to the Dorney Amusement Park near Allentown, Pennsylvania. Courtesy Bob McBroom, Holton, KS.

to [the] wind direction as if it had a tail, [and] turns [the] machine sideways in high winds as if it had a folding tail."[50]

Some wind turbine manufacturers in the mid-1970s sidestepped the residential market altogether by setting their sights on higher output machines purely for industrial or utility-integration purposes. Three of these firms—Kaman Aerospace Corporation, Grumman Energy Systems, and United Technologies Research Center—hailed from the aerospace industry and got their start in wind power by participating in research contracts with the Department of Energy. They developed downwind rotors of varying design. Kaman, which designed its prototype 40-kilowatt turbine's downwind, full-feathering rotor with cost in mind, used two blades because it was determined that a third blade would increase the cost of the turbine by $2000. The free-yawing downwind rotor design was selected since the company's engineers determined an upwind rotor would also "add complexity and cost to the system."[51] In 1976, Grumman began offering a 15-kilowatt, three-blade downwind turbine with a 25-foot diameter to government agencies, electric utilities, and universities for testing. This turbine also used full-span blade pitch to control rotor speed.[52] United Technologies spent the latter half of the 1970s working with the Department of Energy at Rocky Flats, Colorado, to develop a two-blade, downwind machine of eight kilowatts that utilized a lightweight composite bearingless rotor pioneered by the helicopter industry. United Technologies' wind turbine rotor was made in three parts: a flexible carbon composite beam at the center of the rotor and clamped to the rotor hub that could twist and bend, and two pultruded fiberglass blades attached to each end of the beam. The beam was twisted through centrifugal forces on pendulum-like weights that were located out of the rotor plane and connected to the beam with straps which pitched the blades into a stalled position for rotor speed and power output control.[53] In late 1980, a group of former Department of Energy employees from the Rocky Flats test site stepped out on their own to form Energy Sciences Inc. (ESI), with the goal of designing an optimum downwind turbine based on their years of experience with testing other commercial machines. ESI settled on a two-blade, fixed-pitch, downwind teetering rotor, a design that was influenced by teetering helicopter rotors. The turbine used spring-loaded blade tips for rotor runaway speed control. In 1981, the company began testing its 50-kilowatt turbines in harsh weather conditions at the Molokai Ranch in Hawaii. Satisfied with the results, ESI introduced its wind turbines to the US wind farm market. ESI sold about 700 50-kilowatt turbines by 1985 and was in the process of developing a machine with a peak output of 275 kilowatts before withdrawing from the market in the late 1980s.[54]

Two notable developers of vaneless, horizontal-axis wind turbines that got their start in the mid-1970s—Energy Development Company of Hamburg, Pennsylvania, and WTG Energy Systems in Buffalo, New York—built commercial machines of more than 200 kilowatts without the assistance of US government funds. Terrance Mehrkam, founder of the Energy Development Company, designed and erected his first wind turbine, a two-blade, downwind unit with 36-foot diameter rotor and 16-kilowatt output in 1976 to provide electricity to a farmhouse and a workshop used for making metal finishes. He estimated at the time that he reduced his annual utility electric bill by two-thirds, or to $1000. His wind turbine rapidly gained the public's attention, and orders soon began. Mehrkam settled on a four-blade, downwind rotor design, and his first commercial machines came in two sizes (10 kilowatts and 20 kilowatts) and incorporated commercially available, off-the-shelf components. However, Mehrkam became obsessed with building turbines with larger outputs. He manufactured 40-kilowatt machines and then a 140-kilowatt size. One of Mehrkam's high-profile 40-kilowatt wind turbines was erected in 1978 on a 60-foot tower at Hunts Point in New York City's Bronx borough to provide electricity to motors that drove aeration blowers at an experimental urban composting center that was operated by the Bronx Frontier Development Corporation.[55] In 1979, Mehrkam installed a 250-kilowatt turbine with a 75-foot downwind rotor that provided supplemental electric power to the Dorney Amusement Park near Allentown, Pennsylvania. At the time, the machine had the largest rated output in the US market. While his 45-kilowatt turbine was the most popular—ten were installed throughout the United States in 1979–1980—Mehrkam started to design a six-blade, downwind machine with a 160-diameter rotor with the capability to generate 2 megawatts of power. However, Mehrkam never got to finish this project. A significant flaw of his turbines was the lack of a system to stop the rotor in overspeed conditions. Merkham died suddenly in 1981 at the age of 34 by falling off one of his wind turbine towers at Boulevard, California, while attempting to manually turn the rotor out of the wind, a proce-

dure he was known to have performed a number of times before.[56]

After two years of research and development, WTG Energy Systems introduced its 200-kilowatt wind turbine on the Massachusetts island of Cuttyhunk in June 1977. The company's founder and president, Allen Spaulding Jr., who lived on the tiny island, installed the turbine to demonstrate that wind energy could reduce the reliance by small island communities on diesel-powered generators. Instead of the downwind rotor pursued by others, WTG settled on a vaneless, three-blade, upwind design based largely off Denmark's successful

WTG Energy Systems introduced this 200-kilowatt wind turbine on the Massachusetts island of Cuttyhunk in June 1977 to reduce the inhabitants' reliance on a diesel-powered electric generator. Illustration based on postcard photograph by Mitt Price, Islip, New York (ca. 1980). Courtesy Stephen James Govier, Suffolk, England.

This 3-megawatt turbine, which was constructed by Schachle & Sons, was erected for testing by the Southern California Edison Company at the San Gorgonio Pass wind farm site at Palm Springs in 1979. Courtesy Herman Drees, Thousand Oaks, CA.

Carter Enterprises spent three years refining the rotor and blades of its 25-kilowatt downwind turbine before selling its first machine in 1980. A Carter test unit shown at the federal government's Rocky Flats wind energy test site in Colorado in 1981. US Department of Energy photograph courtesy Warren Bollmeier, Kaneohe, HI.

Gedser wind turbine. The Gedser machine, which was erected in 1957 under the direction of Danish engineer Johannes Juul with financial support from America's post-World War II Marshall Plan, had a maximum output of 200 kilowatts. The turbine's rotor was "stall-regulated," and its blades included tip brakes. A motorized yaw mechanism kept the rotor in the upwind position during operation.[57] Juul's turbine operated successfully from 1957 to 1967 and was restarted in 1977–1978 with assistance from the US Department of Energy for conducting additional tests.[58] WTG's more simplistic rotor included 40-foot-long fixed-pitch blades with 4-foot hydraulically released tips to brake the rotor in excessive winds. Instead of traditional mechanical controls to manage the turbine's operation in the wind, WTG used a computer system.[59] The construction of WTG's wind turbine on Cuttyhunk Island was captured and preserved by documentary filmmaker David Vasser in the film, *Generation on the Wind*.[60] From 1979, the wind turbine at full operation often met up to 98 percent of the island population's electrical requirement.[61] By 1980, WTG sold turbines of similar size and output to the Nova Scotia Power Company in eastern Canada for a wind-hydro pump project and to the Pacific Power and Light Company in Oregon for a demonstration project. The company also designed a 20-kilowatt upwind turbine with a 28-foot-diameter three-blade rotor and had started design of 2.5-megawatt machine before exiting the market in the mid-1980s.[62] Another giant upwind turbine at the time was constructed in May 1977 by the Moses Lake, Washington-based company, Schachle & Sons. The turbine had a 72-foot diameter, three-bladed, upwind rotor and a 140-kilowatt output. Southern California Edison Company ordered a 3-megawatt turbine from the company for testing at the San Gorgonio Pass wind farm site in 1979.[63]

Horizontal-axis wind turbine rotors during the 1970s and 1980s experienced their share of malfunctions, despite abundant research and development efforts. Carter Enterprises, for example, spent three years refining the rotor and blades for its 25-kilowatt downwind turbine before selling its first machine in 1980. "It has taken a lot of research and development; a great many things have to

be learned to develop a reliable wind generator," Jay Carter Sr. said at an industry conference in June 1982. Even two and a half years after starting commercial wind turbine sales and with more than sixty units in the field, he acknowledged that "We are still learning, still making mistakes and still having failures, but we have also come a long way."[64] Many manufacturers, however, felt pressure to rush their machines to market without sufficient field testing in order to take advantage of the various government tax incentives for homeowners and wind farms. Wind Power Systems, which manufactured a 40-kilowatt downwind turbine, in late 1981 shipped fifteen of its machines to a Tehachapi Pass wind farm, where the three-blade rotors promptly failed. Although the problems were later fixed, Ed

Engineer Ed Salter developed ultra-low solidity, flexible rotor blades for the 40-kilowatt, downwind Wind Power Systems turbine, as shown here (ca. 1980) at the US Department of Energy's Rocky Flats test site in Colorado. Courtesy Herman Drees, Thousand Oaks, CA.

Salter, the designer of the wind turbines who was best known for his ultra-low solidity, flexible rotor blades, said the machines "needed at least two more years of intense development and testing in order to survive the swirling 60 mph plus storms that frequented the Tehachapi Pass."[65] Rotor weaknesses generally manifested themselves quickly in stormy weather conditions. With downwind, free yawing machines, there was the potential in turbulent winds for the rotor to swing into an upwind position where the blades might bend backwards and strike the tower. Rotor brakes were prone to failure in high winds, resulting in rotor runaway and destruction. Equipment failures often resulted in damaged reputations, excessive and costly field repairs, and even lawsuits. Downwind rotors were also noisy. This was due to the blades' passage behind the tower, or through the lower velocity flow in the wake of the tower, causing the blade to dynamically stall and emit a thumping sound.[66]

Parallel to the developments of horizontal-axis wind turbines during the 1970s and 1980s were efforts to build vertical-axis turbines. The basic configuration of vertical-axis wind turbines dates back more than 1000 years at the border area of northern Iran with Afghanistan, where these types of windmills were used to grind grain into flour. Their panemone rotor design worked similarly to water wheels stood on their ends, except there was a wall to shield the reed-covered paddles on one side, channeling the wind to push the exposed paddles in a uniform direction. Since this machine worked by directly pushing on the paddles, it provided force through aerodynamic drag.[67] Early on, the Chinese also developed drag-type machines for lifting water. These vertical-axis windmills consisted of a lightweight, wood-framed wheel with evenly spaced vertical cloth sails. The sails that went in the direction of the wind were angled edge-on to reduce drag and eliminated the requirement for a blocking wall.[68] Across the US Northeast and Great Plains during the late 1800s and early 1900s were examples of both "shop-made" and commercially manufactured vertical-axis windmills that pumped water.[69] In the late 1920s, Finnish engineer Sigurd J. Savonius experimented with vertical-axis rotors and settled on a structure that consisted of two partially displaced cylindrical halves which were mounted to a vertical shaft. This design, which operated on both drag and lift, allowed wind to flow through the middle of the rotor, substantially increasing the wind energy extraction efficiency compared other vertical-axis rotor configurations. In the early 1970s, many home-built, vertical-axis, drag-type wind turbines enlisted Savonius-style rotors, since they could be easily and cheaply made by cutting a 55-gallon oil drum in half lengthwise. While the rotor developed high torque, its rotational speed was slow and therefore found to be unsuitable for electric power generation.[70]

In the mid-1960s Canadian government engineers, Peter South and Raj Rangi, revisited the 1920s vertical-axis, eggbeater-shape rotor designs of French scientist Georges J. M. Darrieus with the goal to develop a modern, yet inexpensive commercial wind turbine. In 1970, they constructed a 30-inch diameter model of their Darrieus-style wind turbine, which consisted of three equally spaced, 1-inch chord aluminum blades curved to a troposkein shape around a vertical axis and subjected it to wind tunnel tests. The troposkein shape resembled a sideways jump rope when in motion and resulted in the centrifugally induced stresses in the blades being entirely tensile, which were far easier to design structurally than stresses resulting from bending. To expand on their results, South and Rangi built a larger test model with a 14-foot-diameter rotor and two blades made of 24 gauge cold-rolled sheet metal. The engineers concluded that the maximum power coefficient compared favorably with horizontal-axis wind turbine rotors.[71] They noted, however, "The wind turbine, although nearly self-starting, requires additional means of starting from a stopped position."[72] The researchers concluded with the simplicity and minimal use of materials that vertical-axis wind turbines of the Darrieus design could be manufactured at one-sixth the cost of a comparable size horizontal-axis machine, and "if produced in reasonable quantity, could supplement the conventional sources of power at comparable cost, provided the annual mean winds are 10 m.p.h. or higher."[73] During the early 1970s, the Sandia Laboratories in Albuquerque, New Mexico, was similarly tasked by the US

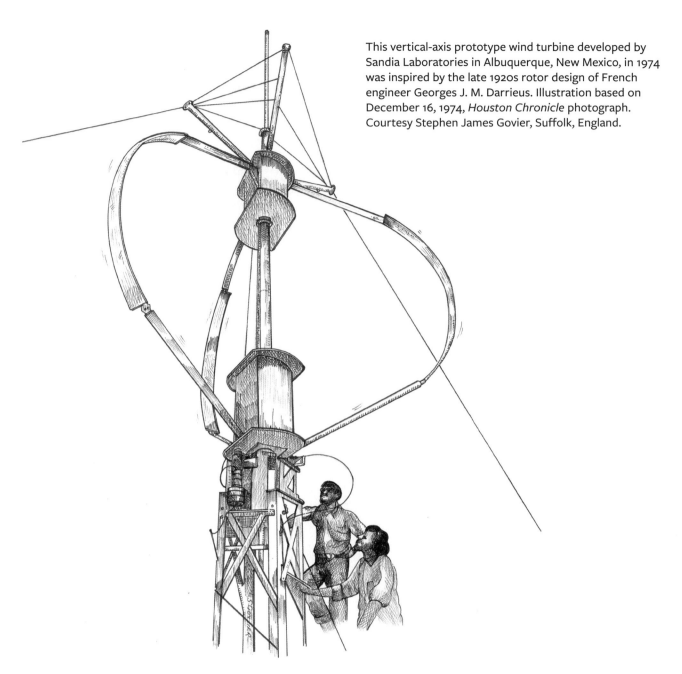

This vertical-axis prototype wind turbine developed by Sandia Laboratories in Albuquerque, New Mexico, in 1974 was inspired by the late 1920s rotor design of French engineer Georges J. M. Darrieus. Illustration based on December 16, 1974, *Houston Chronicle* photograph. Courtesy Stephen James Govier, Suffolk, England.

Department of Energy to develop an operating wind turbine prototype using the Darrieus rotor concept which could then be utilized by wind turbine manufacturers.[74]

Unlike the Savonius rotor, the Darrieus design operated entirely on the aerodynamic principle of lift rather than drag. Darrieus turbines were sometimes referred to as cross-flow machines, a general class of turbines in which the entire length of the blades moved transversely, or crossed, in the wind direction. When a Darrieus rotor turned in the wind, the blades acted analogously to the sail of a sailboat tacking around in a circle. In operation,

each blade had two passes at the wind for each revolution around the central axis. The middle section of the blades, where the swept area was widest, moved several times faster than the speed of the wind. When the blades traveled in the upwind or downwind sectors relative to the central shaft, they produced lift vectors that pointed substantially downwind but were angled slightly toward the direction the blades traveled, thereby producing torque around the central axis. In the sectors where the blades traveled into or away from the wind direction, they experienced minimal drag but not lift. The Darrieus turbine's high rotational speed made them suitable for driving electric generators. Unlike horizontal-axis wind turbine rotors, Darrieus rotors did not require a yaw system to orient them into the wind. Despite experimentation by Sandia Laboratories and others, however, the rotors failed to be self-starting and required an electric motor, or its induction generator operating as a motor, to start them turning in the wind. Sandia Laboratories experimented with using Savonius rotors attached to the central shaft to make the rotors self-starting. Another factor limiting the effectiveness of the Darrieus rotor was its near-ground-level placement, whereas horizontal-axis wind turbines could be placed high up on towers to catch the higher wind speeds aloft.[75]

Despite the Darrieus rotor's shortcomings, the design persuaded two large aluminum companies—US-based Aluminum Company of America (Alcoa) and Dominion Aluminum Fabricating Indal Ltd. (DAF) of Canada—to enter the wind turbine industry in the mid-1970s because of their ability to manufacture the extruded aluminum blades. At the time, the Darrieus rotor blades were among the largest aluminum extrusions made by the industry. Alcoa's longest blade measured 80 feet and had a width, or chord, of 29 inches. The blade profile for these vertical-axis wind turbines was symmetrical along the chord line, which allowed the blade to work equally efficiently when it was upwind or downwind from the central axis. The long extrusions were bent slightly about every foot into a near troposkein-shaped curve with a large hydraulic press. It was a slow, labor-intensive process. Prototypes of the Alcoa and DAF Darrieus machines were placed at various windy locations to gather data regarding performance, mechanical loads, and dynamic behaviors, as well as to develop mechanisms that optimally controlled the turbines. The companies were particularly interested in how the rotors behaved during stalled operations since the maximum power regulation depended on it. Unlike horizontal-axis turbines, the curved

These two Dominion Aluminum Fabricating Indal Ltd. Darrieus-style wind turbines—50 kilowatts (left) and 500 kilowatts—were erected at the Southern California Edison test site in San Gorgonio Pass in the early 1980s. Courtesy Herman Drees, Thousand Oaks, CA.

These stacked three-bladed, 5-kilowatt, Darrieus-style wind turbines manufactured by Dynergy Corporation were erected in 1980 at the *Clearwater Times* newspaper office at Clearwater, Florida. Courtesy Coy Harris, Lubbock, TX.

it difficult to purchase cost-effective aluminum extruded blades from Alcoa, which at the time was also a competitor with its smallest, 8-kilowatt Darrieus wind turbine.[77] When Alcoa exited the wind industry, Paul N. Vosburgh, who led the company's effort to build Darrieus wind turbines, launched Albuquerque, New Mexico-based Forecast Industries in 1982 to build 185-kilowatt Darrieus wind turbines of the Alcoa design, but the firm failed to gain a large foothold beyond installing about forty machines for a wind farm in California's San Gorgonio Pass.[78] FloWind, a

blades of the Darrieus rotor could not be pitched to spoil the aerodynamic lift. To stop the rotors from turning, mechanical braking systems were required. Operational failures with the prototypes dissuaded Alcoa and DAF from fully commercializing their Darrieus turbines, and so, in the early 1980s, they reverted to their core business of making aluminum.[76]

Attempts were made to manufacture residential-scale Darrieus wind turbines, such as Laconia, New Hampshire-based Dynergy Corporation's three-blade, 5-kilowatt machine and a three-blade, 12-kilowatt silo-mounted machine developed by Clarkson College under the direction of Professor John Rollins for dairy farms. Demonstration units were erected around the country, including a triple-stack of Dynergy's machines placed at the *Clearwater Times* newspaper office at Clearwater, Florida, and a single unit set up at the Public Service Company of New Hampshire's Seabrook nuclear power plant. But these companies found

The Cycloturbine developed by Pinson Energy Corporation in the late 1970s came with two or three straight blades that pitched cyclically as the rotor turned, giving it a self-starting capability and speed control. Courtesy Herman Drees, Thousand Oaks, CA.

subsidiary of Seattle-based Flow Industries, became a successful commercial manufacturer of Darrieus wind turbines for wind farms in the early 1980s. With technical support from Sandia Laboratories and use of Alcoa-extruded blades, FloWind manufactured hundreds of 142-kilowatt and 250-kilowatt machines for placement in California's wind farms at Altamont Pass and Tehachapi in the mid-1980s. FloWind's turbines operated for about ten years of their expected twenty-year lifecycle. The turbines developed fatigue cracks in their aluminum blades and were taken out of service.[79] Despite various research efforts, Darrieus wind turbines also failed to take off in Europe. Denmark's Vestas A/S, for example, designed two self-supporting (without guy wires), three-blade prototype Darrieus machines in 1979, but their disappointing performance led the company to abandon the initiative.[80]

One vertical-axis design that attempted to circumvent the disadvantages of the curved-bladed Darrieus turbines was introduced in the late 1970s and early 1980s by Pinson Energy Corporation. Herman M. Drees, Pinson Energy's founder, had designed and constructed a full-scale vertical-axis wind turbine while attending graduate school at the Massachusetts Institute of Technology. He made performance analyses and wind tunnel tests on a model for his master's degree thesis. Dubbed the Cycloturbine, this small wind turbine was manufactured at Pinson Energy's workshop at Cape Cod, Massachusetts, for the residential and light industry markets. The Cycloturbine could be mounted on

This 40-kilowatt Giromill, a joint project by McDonnell Aircraft Company and Valley Industries' Aermotor Division, was tested at the federal government's Rocky Flats wind energy test site in Colorado but never made it into commercial production. US Department of Energy photograph courtesy of Warren Bollmeier, Kaneohe, HI.

a tower. The machine was developed with two or three straight blades that pitched cyclically as the rotor turned, giving it a self-starting capability, and collectively through centrifugal action for rotor speed control. The blades were mounted on pivot bearings at the ends of a pair of struts and were pitched cyclically with push-pull rods connected to a bearing that was off-set, or eccentric to the axis of the vertical shaft. The orientation of the eccentric bearing was controlled with a wind vane so that the blades' pitching was timed properly with respect to the wind direction. More than 120 Cycloturbines were installed, mostly in New England, with about ten installed in several locations throughout the country as part of the Department of Energy's wind power demonstration program that was offered to small wind turbine manufacturers at the time.[81] Lacking operating capital, the company ceased operations in the mid-1980s. As Drees said, "the Cycloturbine was expensive to make and maintain because it had too many moving parts."[82]

In 1978, McDonnell Aircraft Company of St. Louis, with the technical and manufacturing support of Valley Industries, also of St. Louis, which owned the Aermotor Windmill Company, took part in a Department of Energy program to design, build, and test a vertical-axis wind turbine similar in rotor design to the Cycloturbine, but with a larger, 40-kilowatt rated output. Called the Giromill, its rotor consisted of three evenly spaced vertical blades each measuring 42-feet tall with a rotational diameter of 58 feet. The blades optimally pitched in the wind, using electrical actuators tied to a computational fluid dynamics model, instead of the simpler sinusoidal pattern used to pitch the Cycloturbine's blades. In July 1980, the Giromill was built for testing at the Department of Energy's Rocky Flats site. Problems with the machine quickly arose. For example, the blade actuators suffered routine breakdowns, affecting test quality. Before the end of the year, the rotor fell off the test tower. The Giromill was repaired and returned to testing on April 24, 1981. However, tests were discontinued at Rocky Flats by the summer of 1982, and work to further develop the Giromill was summarily abandoned by McDonnell Aircraft.[83] Robert V. Brulle, the company's engineer in charge of the Giromill

program, remarked more than twenty-five years later that "Although we showed an efficiency rating higher than conventional horizontal-axis windmills, it did not balance out the complexity between the Giromill and a conventional windmill."[84]

How a wind turbine produced electricity was determined by the type of generator it utilized. Pre-rural electrification wind turbines of the 1920s–1950s period relied on direct current (DC) generators that were drawn from a rotating armature through a set of carbon brushes. Through a simple control panel, this power was transferred from the generator to a set of storage batteries. DC generators came in a variety of sizes—6, 12, 32, and even 110 volts—depending on the user's requirements. A small 6-volt Wincharger charged batteries for powering radios or one to two light bulbs, while a 32-volt Wincharger provided electricity for a wider variety of household appliances and tools for the workshop and farm. The use of DC wind turbines mostly ended by the 1950s, when the federal government's rural electrification program extended utility-generated 120-/220-volt, alternating current (AC) to remote towns and farms across the country. In the early 1970s, DC wind turbines from this era were plucked out of barns and off towers and placed back in service by rebuilders. This measure helped fill an immediate demand from individuals retreating to rural settings where it was often too expensive to pay a local utility to extend power lines. However, these early machines required storage batteries and DC-powered appliances. Most Americans who sought a residential wind turbine during the early 1970s were not attracted to such an austere lifestyle and simply wanted a way to reduce their reliance on existing utility connections, as well as their monthly electric bills. This left small wind turbine suppliers scrambling to find ways to interconnect DC-generated power with the AC power of the utilities. One answer to this dilemma came in the form of synchronous inverters.

Synchronous inverter technology had been utilized in various industrial power applications since the early twentieth century. One of the most prominent longtime uses was found in elevators. As an elevator travels upward, its electric motor receives power from the utility, but when the elevator descends, its motor becomes a generator and through synchronous inversion returns a portion of the electricity it produces to the utility grid.[85] The challenge at the start of the 1970s was how to apply synchronous inversion between intermittent wind-generated DC power and the steady 60 Hertz, 120-/220-volt AC power from the utility grid. Alan W. Wilkerson, an electrical engineer from Wisconsin, took up the challenge to develop a synchronous inverter for tying wind turbines to the utility grid. Wilkerson used the latest semiconductor components, and the heart of its patented design was the thyristor bridge, which converted DC to AC voltage. He also had to consider safety features in his synchronous inverter, the most important of which was protection of the linemen working on utility equipment further down the line against back-fed power from the wind turbine or on preventing unregulated

Electrical engineer Alan Wilkerson developed the highly popular Gemini synchronous inverter in the mid-1970s for tying wind turbines to the utility grid. Courtesy Herman Drees, Thousand Oaks, CA.

DC from getting into the household wiring and damaging appliances. Wilkerson's safety measure was to introduce an AC-powered contactor, so that when a utility power outage occurred, the inverter would become isolated from both the wind turbine's DC power and the utility's AC line. Fuses were also added to both the DC and AC sides of the inverter to prevent overload from either source. Wilkerson licensed the sale of his Gemini synchronous inverters to Windworks. While other electronics firms offered synchronous inverters, Gemini was by far the most popularly used by residential wind turbine manufacturers in the late 1970s.[86]

Early synchronous inverters were prone to overloads, which damaged their electronics. This required the inverters to be disconnected and returned to the manufacturer for repair, a costly and time-consuming endeavor. Wind turbine manufacturers and installers in the late 1970s began collaborating with the inverter suppliers to resolve reliability issues in the field.[87] When synchronous inverters operated properly, however, wind turbine owners delighted in watching their electric meters run backwards as the wind blew. The first residential wind turbine to connect to a public utility grid in the United States occurred in 1974 when Windworks installed a reconditioned Jacobs and Gemini synchronous inverter at the home of Wisconsin Congressman Henry S. Reuss. Using a reconditioned 3-kilowatt, 110-volt DC Jacobs, Craig Toepfer, owner of Ann Arbor, Michigan-based Windependence Inc., a year later became the second residential wind turbine owner in the country and the first in Michigan to connect to a public utility grid using a Gemini synchronous inverter. In a 1980 interview, he recalled that "Detroit Edison was actually pretty good about the hook-up. . . . They gave us power meters and were out there with their own tape recorders monitoring the whole thing."[88] The Windependence system operated reliably and was later moved to Toepfer's home in Wisconsin when he joined Windworks. The system continued to operate reliably on a "net billing" program with Wisconsin Electric. In two years of reliable operation, the Jacobs/Gemini system produced more energy than was provided from the utility power lines, including surcharges, and resulted in a \$0.96 check from Wisconsin Electric for the excess electricity at the time the system was decommissioned.[89] Although prohibited under PURPA rules, utilities continued to resist synchronous inverter hookups through the late 1970s and early 1980s, mostly because of disagreement over how much they would pay per kilowatt-hour for the excess electric power put into the grid by the wind turbine owner.[90]

Other small wind turbine manufacturers during the early 1970s turned to alternators, like those found in automobiles, to generate their electric power. Alternators consist of a ring of either permanent magnets or field coil electromagnets that spin around a stationary armature to generate AC power. Unlike standard DC generators, alternators were simpler and required less maintenance due to their lack of brushes. However, since an alternator's rpm varied with the speed of the wind turbine rotor, it could not effectively produce the consistent 60 Hertz frequency AC, precluding the ability to directly feed electric power into the utility grid. The alternator's AC output instead had to be rectified

to DC first and then run through a synchronous inverter, which in turn converted it back to AC at a frequency level matching that of the utility grid.[91] Attempts were made to utilize pure AC synchronous generators in wind turbines, which in theory could bypass the requirement for power conditioning electronics to connect to the utility grid. Synchronous generators require precise rpm control from the driving device—rotor blades in the case of wind turbines—to generate a constant-frequency electrical output for feeding into a utility grid. The first wind turbine in the United States with a grid-connected synchronous generator was erected in 1941 by the S. Morgan Smith Company on the peak of Grandpa's Knob in Vermont and fed power directly to the Central Vermont Public Service Corporation, but the experimental machine accomplished this through a complex system of gears and hydraulics that controlled the pitch of the two 65-foot-long, stainless steel blades.[92] Small wind turbine manufacturers during the 1970s and early 1980s were constrained from directly connecting to the grid by the cost and technical complexity

Craig Toepfer's combination Jacobs wind turbine and Gemini synchronous inverter at his home in Wisconsin in 1981 often produced more energy than was used from the public utility Wisconsin Electric. Courtesy Craig Toepfer, Chelsea, Michigan.

required to successfully integrate synchronous generators with their systems.[93]

Some wind turbine manufacturers started integrating AC induction generators into their designs during the late 1970s. Induction generators are just common electric motors, like those used to run a refrigerator compressor or an oscillating room fan. When operating as a motor, AC passes through opposing magnets to create a rotating magnetic field in the motor's stator or casing, causing the rotor shaft to spin. When operating as a generator, the shaft is driven by a turbine, turning the coils in the rotor that are within the stator's magnetic field and delivers AC current directly to the utility grid. In other words, instead of using the torque from the motor shaft to provide power to an appliance, the motor will act as a generator by applying torque to the shaft via a wind turbine rotor, reversing the direction of the electrical power flow. In addition, brushless induction generators operate at a near-constant rpm that is determined by the frequency of the utility grid. In the case of the United States, the induction generator's speed is commonly 1800 rpm for a 60 Hertz frequency. Induction generators eliminate the need for an inverter or other power-conditioning devices to connect to the utility grid.

In the late 1970s, Enertech became the first US wind turbine manufacturer to use an induction generator in its 1.5-kilowatt machine and shortly thereafter its 1.8-kilowatt model and much larger 20-kilowatt unit. The company's three-blade rotor was connected to a low-profile, inline gearbox with the output shaft connected to an induction generator. In between the gearbox and induction generator was an electromagnetic disk brake. Enertech's induction generator was connected to the utility grid with a contactor that was controlled by a system of relays. An anemometer, which was attached to the turbine tower, registered wind speed and, when it was sufficient for the wind turbine to generate power, a contactor closed and the rotor was motored up to speed with the induction generator to spin its shaft at 1800 rpm. As the wind increased, the relative flow angle on the wind turbine's blades increased, and the power output became larger until the blades entered aerodynamic stall, passively regulating the maximum power output of the wind turbine. In strong winds, the power that the turbine generated decreased as the blades reached a deep stall. If the anemometer registered excessive wind, relays triggered the application of the turbine's rotor to brake and simultaneously open the contactor to the generator.[94] This induction generator layout and operational method was similarly used by Carter Wind Systems and ESI for their utility-interfaced wind turbines in the early 1980s.[95]

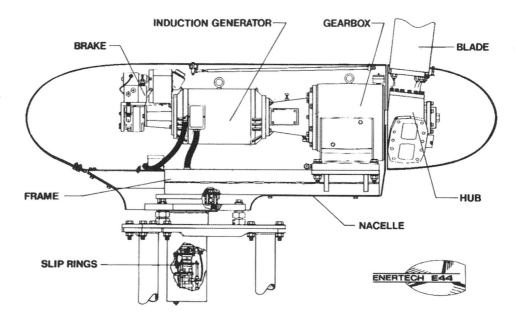

Enertech was the first US wind turbine manufacturer to use an induction generator. This cutaway showed the induction generator arrangement for the company's 20-kilowatt machine. From Enertech, Norwich, VT, *Enertech 44* (ca. 1983).

BRAKE

INDUCTION GENERATOR

GEARBOX

BLADE

FRAME

SLIP RINGS

NACELLE

HUB

ENERTECH E44

Induction generators, however, were not without their operational shortcomings. As wind turbines became larger and heavier, it was no longer practical to motor the rotors up to speed using the induction generator. Instead, the wind turbine rotors were sped up by the action of the wind and, at the right rpm, a contactor connected the spinning generator's electrical output to the grid through what was called a "soft starter." Thyristors were used to buffer the slight difference in electrical frequency and timing between the wind turbine's induction generator sinusoidal AC output and the utility grid. With the increasing size of wind turbine rotors, it became important to control the mechanical loads on the blades. Since wind turbines with induction generators operate at near constant rpm, they were considered a "stiff" mechanical system, which made it difficult to "bleed" off the internal blade stresses caused by gusty winds. The answer was to have the wind turbine's rotor operate at a variable but controlled speed, so that the excess energy causing the high blade loads was absorbed as an increase in rotor kinetic energy or rpm. Variable rpm wind turbines were made possible by controlling the

magnetic field in the induction generator rotor. Induction generators that have this capability are known as doubly fed. This arrangement required brushes to excite the generator independently from the generator stator. When incorporating a field-controlled generator rotor, the electrical frequency of the generator was no longer synchronous with that of the utility grid. To solve this problem, the induction generator's power was rectified to DC and run through an inverter containing insulated-gate bipolar transistors, which allowed for the rapid acceleration of an electrical frequency equal to the utility grid.[96]

Another important component of the wind turbine is the platform on which it operates: the tower. For centuries, windmill builders understood the importance of locating these machines where the sails will catch the most wind. In the late 1850s, the earliest American commercial windmill manufacturers began placing their units on top of four-post wooden lattice towers, and by the 1880s, they were shifting to all-steel lattice towers. Ladders were affixed to these towers to allow the windmill owner to climb to the top to perform maintenance or

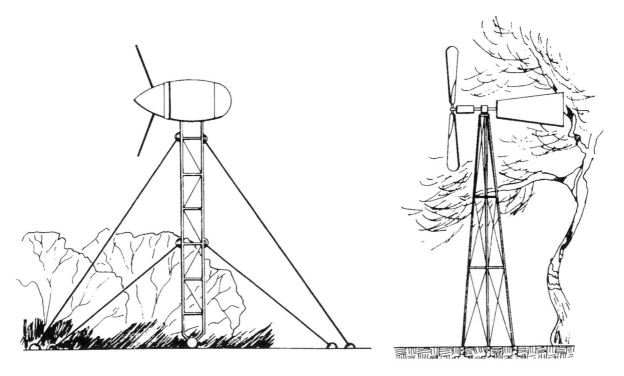

The two most common forms of wind turbine towers to emerge in the 1970s were guy-wire supported (left) and free standing (right). From US Department of Energy, *Home Wind Power* (1981).

repairs.[97] Several manufacturers sold tilting towers during the late nineteenth century, which allowed the windmill heads to be lowered to the ground for servicing, but the higher cost of these towers dissuaded their large-scale application.[98] While steel water-pumping windmill towers were sold in heights of up to 80 feet, their basic design remained relatively unchanged for decades. The towers became the primary platform upon which the earliest American wind-electric generator manufacturers during the early twentieth century erected their machines. To reduce the overall purchase price of their wind-electric generators, some farmers repurposed steel water-pumping windmill towers or assembled their own out of scrap timber and metal. For the small, radio-battery wind chargers, manufacturers offered cost-conscious customers 5- to 10-foot-tall lightweight metal towers that could be mounted to the roof of a barn or shed.[99] Attachments were also designed to affix small wind chargers to the side of a tower already occupied by a water-pumping windmill.[100] Few wind turbine manufacturers during the 1920s–1950s, however, produced their own towers, preferring to source them from water-pumping windmill factories. Wincharger Corporation of Sioux City, Iowa—the largest US manufacturer of wind turbines during the pre-rural electrification years—offered for sale its own extensive line of steel lattice towers with its machines, as well as for radio antenna applications.[101]

When interest in wind-electric power reemerged during the early 1970s due to the rapid rise of conventional energy prices, sellers of rebuilt wind turbines from the first half of the century continued to use the traditional water-pumping windmill or early Wincharger towers with their installations. This began to change as the decade progressed and newly minted commercial wind turbine manufacturers entered the market. Wind turbine engineers began to consider the tower as part of their machines' overall design and function. Two basic tower configurations dominated the industry: cantilevered or "free-standing" and narrow lattice towers held in place with guy wires secured to the ground. Although both types of towers could be climbed, some guyed and free-standing towers were manufactured with a hinge near the base for lowering the wind generator to ground level for

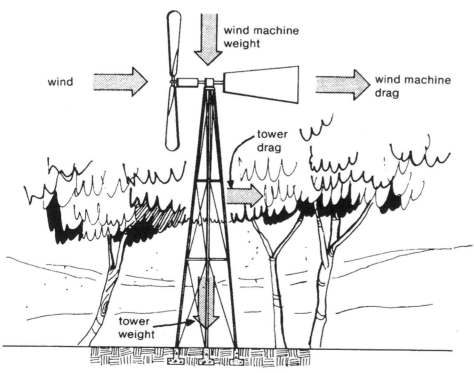

Various physical forces test the integrity of wind turbine tower designs. From US Department of Energy, *Home Wind Power* (1981).

maintenance and repair. Similar to purchasing wind-electric generators during the first half of the century, careful consideration had to be given to the tower height to ensure the machine obtained unobstructed access to the wind. The general rule is that the wind turbine's blades should be at least 30 feet above any obstacles such as trees and buildings. In many counties and municipalities, anyone who considered erecting a wind turbine during the 1970s had to first comply with local zoning ordinances. These rules often impeded the building of wind turbines within a community setting.[102]

Since the 1950s, Ulrich Hütter had been a recognized authority on the dynamic forces that impact wind turbine towers.[103] These forces included the downward pressure imposed on the structure by the weight of the wind turbine itself and the impact of constantly fluctuating winds pushing against it and the turbine rotor. Depending on the type of wind turbine, the rotor exerted different forces on the tower. Lift forces generated by the blades generally pointed downwind and could cause a tower to tip over if it was not securely fastened to its foundation. Other forces that had to be carried by the tower builder included periodic forces stemming from, for example, a rotor mass imbalance, wind shear, and in the case of downwind rotors, the blades passing through the tower's wake. That is why it was critical to design wind turbine towers using proper materials and material dimensions and to avoid resonant vibration frequencies between the rotor and tower over all operating conditions of

The Octahedron Module Tower, shown with Cycloturbine at the federal government's Rocky Flats wind energy test site in 1980, included a triangulated lattice structure that was influenced by the geodesic architecture of F. Buckminster Fuller. US Department of Energy photograph courtesy Warren Bollmeier, Kaneohe, HI.

the turbine system. Resonant vibration frequencies amplify structural loads and over time could lead to material fatigue because these would exploit weaknesses that cause material cracks, failure of bolts, or poorly welded joints in a tower structure.[104] "If forces were to act upon [the tower] in such a way as to get it to vibrate at its resonant frequency, you would actually be able to see it undulate (vibrate) at resonance. If this occurs, you are witnessing an action that will amount to total destruction in very short order!" wrote wind energy consultant Robert Kirschner in 1980.[105] Equally important to a wind turbine tower's structural integrity was how it was secured to the ground. Soil conditions determined the type of footing, which often involved concrete, necessary to anchor the tower and prevent uprooting. In the case of towers secured by guy wires, installers had to ensure the proper gauge of wire, tension fasteners, and anchors were used. It was not uncommon for guy structures to fail when a wire anchor came out of the ground or became detached from the tower.[106]

One of the unusual looking, but highly durable, tower designs during the 1970s was the Octahedron Module Tower manufactured by Natural Power Inc. of New Boston, New Hampshire, under the direction of design engineer Michael Duffy. The tower's triangulated lattice structure, which was influenced by the geodesic architecture of F. Buckminster Fuller, was first developed by Windworks Inc. of Mukwonago, Wisconsin.[107] The freestanding tower utilized Schedule 40 welded mild steel pipe, usually conduit, which was hot-dip galvanized to prevent corrosion. The tower could be ordered from Natural Power to include a vinyl-epoxy finish over the galvanized coating for additional protection in highly corrosive marine environments. The conduit also made it lighter than traditional towers constructed of angle iron. The ends of the conduit were flattened and folded into an angle, fastened together with threaded studs, and tightened with nuts on each side of the joint. The Octahedron tower was easily assembled from the ground up without the requirement of a crane or gin poles and sold in 25-, 34-, 43-, 52-, 70- and 82-foot heights. To access the wind turbine for maintenance or repair, the towers included a ladder of prefabricated

The most popular small wind turbine tower from the 1970s and 1980s was the triangulated lattice guyed and free-standing towers manufactured by Unarco-Rohn of Peoria, Illinois. A free-standing Rohn tower was used for erecting this Jacobs third-generation, 10-kilowatt turbine in Hawaii in 1981. Courtesy Paul R. Jacobs, Corcoran, MN.

steps that were U-bolted to the vertical tubes. Natural Power claimed the tower could withstand winds of 125 mph while supporting a turbine with a 20-foot-diameter rotor.[108] The towers experienced occasional catastrophic failures, however, due to preexisting fractures in the metal tubes where they had been flattened and folded into an angle. When a bend or connection failed, the towers tended to buckle, sending the wind turbine tumbling to the ground.[109] Despite its simple design and installation attributes, Natural Power discontinued the production of the Octahedron Module Tower before the end of the decade.[110]

Another tower manufacturer to enter the wind energy business in the early 1970s was Rohn Manufacturing (renamed Unarco-Rohn in the mid-1970s), a division of Unarco Industries Inc., of Peoria, Illinois. By the end of the decade, the company would become the nation's top supplier of towers to wind turbine manufacturers and installers. Dwight R. Rohn started Rohn Manufacturing in 1948 to produce towers for home television reception. The company's plant expanded in 1962 to accommodate its growing portfolio of steel towers, which it cut, welded, and galvanized onsite.[111] In the early 1970s, one of the company's veteran employees, Charles A. Wright, along with a young new hire, Philip W. Metcalfe, recognized the market potential that the burgeoning wind turbine industry could bring to the tower manufacturer. Wright was well acquainted with wind turbine installations, having erected numerous Wincharger machines between 1936 and 1944. He entered the radio tower installation business in 1945 and began working for Rohn Manufacturing in the late 1960s.[112] Wright surmised that Rohn Manufacturing's array of triangulated lattice guyed and free-standing antenna and communications towers—some as tall as 800 feet—could support the newest wind turbines. He and Metcalfe set out to introduce the company's towers to the fledgling US wind turbine manufacturing industry. By the spring of 1974, they had sold towers to Solar Wind, Windlite-Alaska, Sencenbaugh Wind Electric, and Real Gas & Electric Company.[113] The two men quickly expanded Rohn Manufacturing's customer base among other wind turbine manufacturers by participating in American Wind Energy Association conferences, even attending the founding meeting of the group in Detroit on September 20–21, 1974.[114] Three Rohn towers that became popular among wind turbine manufacturers during the 1970s included the 25G guyed tower, which could be raised to heights of 120 feet for machines with rotors of up to 12 feet in diameter; the 45G, a wider guyed tower that could be built to heights of 200 feet for machines with rotors of up to 16 feet in diameter; and the heavier three-leg cantilevered, self-supporting tower which was recommended for machines with rotor diameters of up to 24 feet.[115] By 1979, Unarco-Rohn recorded sales of $230,000 to wind turbine customers, which included 126 guyed and 82 self-supporting towers. The company expected its tower sales to jump to $500,000 in 1980 to manufacturers of machines ranging in size from

1 kilowatt to 10 kilowatts and to $1 million by 1981 with the commercial introduction of wind turbines with output from 15 kilowatts to 49 kilowatts.[116] "Most of these companies don't particularly want to get into the tower business and with the proper effort, we should be able to get most of this business," Metcalfe wrote to Unarco-Rohn's senior management at the time.[117] During the early 1980s, the company also became a significant provider of towers to wind farm developers in California.[118]

The renewed interest in small wind-electric power resulted in a requirement for various support technologies. Before investing in a wind turbine, customers were encouraged to conduct a wind site analysis, which required the use of wind measurement instruments to determine the average wind speed and direction over a period of several months, and to compare that information to recent annual wind records from local weather stations. Knowing the wind characteristics of a site determines whether it could sustain a wind turbine and is important to selecting the height of the tower and type and size of the wind turbine. In addition to recognizing potential wind obstructions, such as hills, trees, and buildings, a visual of longtime wind activity might show the direction in which a tree's branches have grown, known as flagging. In general, a site was considered suitable for a wind turbine if the average wind speed exceeded 10 mph with minimal turbulence.[119] The most common cause for unsatisfactory wind turbine operation is improper siting of the machine. "If a wind generator is sited improperly it will surely come back to haunt you," warned Robert B. Keller of Colorado's largest small wind turbine installer, Mountain Valley Energy Inc. of Boulder, in 1981. "Not only will the generator produce a minimum amount of power which will undoubtedly make for a dissatisfied customer, but it also has the possibility of giving the manufacturer and the wind industry in general a bad name because of possible adverse publicity on the machine's power output."[120] Scientists at the Pacific Northwest Laboratory in Richland, Washington, who conducted comprehensive wind studies in the 1970s and early 1980s, stated that anyone interested in a small wind turbine "should realize that a relatively small investment for locating the best available site

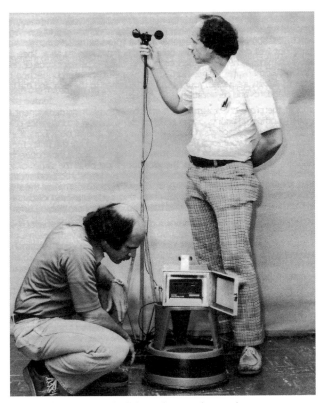

Bonneville Power Administration engineers inspect an anemometer for measuring wind speed. From Roy E. Reinhart and Angelina Quinata, Bonneville Power Administration, US Department of Energy, *Small Wind Energy Conversion Systems* (1981).

can easily yield savings of several thousand dollars over the lifetime of the system."[121]

Although mechanical anemometers used to measure wind speed had been around for many years, scientific instrumentation and electronics firms at the start of the 1970s used newly available semiconductor technologies to refine the accuracy and portability of these devices. Anemometers were available that showed the velocity of the wind using simple visual outputs in miles per hour or recorded on paper strip-charts, while others measured average wind speeds at a recording site over a select period.[122] Natural Power Inc. of New Boston, New Hampshire, which was started in 1973 by Richard Katzenberg, who served as president of the American Wind Energy Association during the mid-1970s, took the step of concentrating its electronics business on the fledgling wind and solar energy markets.[123] The company developed a popular anemometer-recorder system for the wind-electric

industry, which it said "proved to be exceptionally accurate in a wide range of wind velocity" during wind tunnel tests conducted at the Massachusetts Institute of Technology.[124] Sophisticated anemometers at the time were significant investments, costing upwards of $500 each, and were mostly purchased by wind turbine dealers and installers. Realizing this cost, some state governments with an interest in boosting the installation of small wind-electric systems offered anemometer loan programs to individuals seeking to conduct their own wind assessments.[125] Natural Power allowed customers to rent its anemometer equipment by the month.[126] Among the various electric controllers and inverters developed and sold by Natural Power in the 1970s, the company manufactured a unique digital tachometer that measured the rpm of a wind turbine's main shaft in the wind, which was useful information for blade designers.[127] By the early 1980s, Natural Power struggled to compete against larger wind instrumentation manufacturers and shut down its operation in late 1984.[128] Meanwhile, firms such as NRG Systems of Charlotte, Vermont, which was founded by David and Jan Blittersdorf, and Second Wind Inc. of Somerville, Massachusetts, overseen by Walter Sass and Ken Cohn, became leading players in advancing wind measurement tools for the wind energy business.[129]

Despite the numerous innovations developed by the small wind turbine manufacturers during the 1970s and 1980s, examples of these machines rapidly disappeared from the landscape following the industry's mid-1980s collapse. "Because these small wind turbines did not produce much electricity due to their size, they were discarded by owners as soon as they needed repair. When they did need work, the owners found out the manufacturer had gone out of business and there were no parts. So, these early small wind turbines are quite rare now," explained Coy Harris. Harris headed former Wind Engineering Corporation, which sold small Darrieus wind machines in the mid-1970s before manufacturing its own 25-kilowatt horizontal-axis downwind turbine later in the decade. With the exception of wind turbines manufactured in the early 1980s for the first US wind farms, most companies produced only limited numbers of machines for

This 25-kilowatt Wingen 25 turbine, which was manufactured in 1981 by Wind Engineering Corporation and the only one of twenty machines that is known to have survived, is on display at the American Windmill Museum in Lubbock, Texas. Courtesy Coy Harris, Lubbock, TX.

the commercial market.[130] Many of the early wind turbines used by the wind farms were also taken down and discarded once operational malfunctions occurred. In addition, the lightweight materials often used in the construction of the machines, such as fiberglass and wood, decayed during prolonged exposure to weather. Other materials, such as a generator's copper coils and aluminum blades, were often sold for scrap. While small wind turbines from the 1970s and early 1980s are still found on towers here and there throughout the United States, the 17.5-kilowatt Jacobs is most often the only machine manufactured during this era that remains in operation. Surviving firms, such as Bergey Windpower Co. and North Wind Power, saw their first-generation machines replaced by new models in the decades that followed.

Today, carefully preserved American water-pumping windmills and pre-rural electrification wind turbines dominate most public and private collections in the United States, whereas in Denmark and Germany there are museums with large collections of wind-electric machines spanning the entire twentieth century.[131] One of the largest public collections of windmills in the United States, which contains multiple examples of 1970s- and early 1980s-period small wind-electric turbines, is housed at the American Windmill Museum in Lubbock, Texas. The museum was started in 1993 by Texas Tech University professor and windmill historian, Billie Wolfe, and Coy Harris, who operated a mechanical engineering firm after closing the Wind Engineering Corporation in the 1980s.[132] Today, the museum grounds consist of 28 acres on which are constructed two large, steel-framed buildings with a combined capacity of 61,000 square feet to house the most significant and rarest American- and foreign-produced windmills. Another fifty working windmills and wind turbines can be found scattered throughout the grounds, including a 164-foot-tall, 660-kilowatt Vestas wind turbine, which supplies electric power to the museum. However, most of the American Windmill Museum's small wind turbines are displayed inside to ensure their long-term preservation. A center-piece wind turbine from the late 1970s and early 1980s in the museum's collection is the Windgen 25 machine manufactured by Wind Engineering Corporation. The three-bladed machine is equipped with an electric motor that allows visitors to push a button and observe its rotation. Also on display are two small Darrieus wind turbines—a 4-kilowatt DAF and a 5-kilowatt Dynergy. Both vertical-axis machines were sold by Wind Engineering Corporation during the mid-1970s. The museum occasionally receives donations of small wind turbines from the 1970s and 1980s, which are kept in an onsite storage yard for potential future restoration and display. "If you are in the history business, as we are, these machines should be kept for future study," Harris said.[133]

9 Spinning Forward

Despite the loss of federal tax credits and the drop in the price of oil to about $10 a barrel in the mid-1980s, California's state energy policies carried the fledgling US wind energy industry almost single-handedly through the rest of the decade with the continued construction of wind farms at Altamont Pass east of San Francisco, San Gorgonio Pass near Palm Springs, Tehachapi Pass east of Bakersfield, and the Boulevard area east of San Diego. By 1990, California's wind farms included more than 14,000 wind turbines with a collective generating capacity of about 1350 megawatts or enough residential electric power for 900,000 people.[1] National interest in this renewable energy source increased during the early 1990s for a confluence of political, economic, and environmental reasons. On August 2, 1990, Iraqi dictator Saddam Hussein ordered a military invasion of Kuwait, causing global oil prices to jump from $16 to $19 per barrel to more than $40 per barrel in the months leading to the January 16, 1991, drive by US-led coalition forces to remove the Iraqi army from the Persian Gulf country. Although the price increase was short-lived, the event heightened the US public's distrust and wariness of foreign oil suppliers, particularly those in politically unstable Middle East countries.[2] At the start of the decade, American homeowners also became increasingly concerned about the rising cost of their electric bills. According to the US Energy Information Administration, the nominal retail price for residential electric power increased from 5.4 cents per kilowatt-hour in 1980 to 7.83 cents per kilowatt-hour at the start of the new decade. By 1990, the average US household spent about $745 a year for utility-sourced electricity, and the cost of

those bills was expected to increase further as more electrical appliances were introduced to the home.[3] At the same time, the public clamored for effective policies to curb air pollution. Declaring that every American "deserves to breathe clean air," President George H. W. Bush on November 16, 1990, signed into law amendments to the Clean Air Act, which established phase-outs for production of ozone-depleting chemicals and mandated reductions in industrial pollutants, such as sulfur dioxide and nitrogen oxide, that contributed to acid rain.[4] Wind power offered an avenue through which the United States could reduce its dependence on fossil fuel and facilitate a cleaner environment for healthier living.

Wind energy engineer Warren Bollmeier, who attended the American Wind Energy Association's conference in Washington, DC, on behalf of the Honolulu, Hawaii-based Pacific International Center for High Technology Research in September 1990, noted: "The mood of the conference was very upbeat and business-like: this is in strong contrast from last year's conference in San Francisco."[5] The cost of energy among the nation's wind farms at the time was in the 7 to 9 cents per-kilowatt-hour range, but the industry aimed to reduce that cost to 5 cents per kilowatt-hour during the decade. "The industry appears to have completed its consolidation and may, in fact, be on the front end of a long[-]term growth phase; there seems to be a shift from the concern about foreign intrusion to a strong sense that the current industry make-up of developers, manufacturers, vendors and other interests is strong enough now to survive," Bollmeier wrote.[6] A new incentive for the US wind energy industry was put in place by Congress in 1992—the Production Tax Credit (PTC), which provided a 2 cents per-kilowatt-hour tax credit for electricity generated by wind turbines over the first ten years of a wind farm's operation. Another positive development for the US wind energy industry during the 1990s was the introduction of many state-based Renewable Portfolio Standards (RPS) programs, which set specific timelines and amounts of electricity that must be supplied by utilities through renewable energy sources, such as wind and solar.[7]

US Windpower introduced a three-bladed, variable-speed wind turbine in the early 1990s, as shown here in Copenhagen, New York. Courtesy Stephen James Govier, Suffolk, England.

The bulk of US government-sponsored and private-sector wind energy research and development at the start of the 1990s focused on the electrical, mechanical, and aerodynamic design improvements to increasingly larger wind turbines—with output ranging from 50 to 550 kilowatts—for use in wind farms. In 1992, wind turbine manufacturer US Windpower, the Electric Power Research Institute, and the Utility Wind Interest Group began field testing a three-bladed, variable-speed wind turbine. Since the speed of the prototype's blades were able to change with the wind speed, the institute declared that the machine could generate a 15 percent increase in energy output from the wind when contrasted with the same-size turbine operating at a constant speed.[8] The National Renewable Energy Laboratory in Golden, Colorado, in the early 1990s initiated the Advanced Wind Turbine program to help the US wind industry incorporate the latest technology into existing commercial, "utility-grade" machines. Three wind turbine manufacturers were part of the program, including Atlantic Orient Corporation, which took over production of the Enertech 44s; Northern Power Systems' North Wind 100s; and R. Lynette & Associates (or American Wind Turbines, Inc.), which continued the manufacture of the ESI-80s.[9] The three companies were enthusiastic about the prospects of improving their machines not only to compete with cheaper fossil fuel-generated utility power, but against the wave of larger wind turbines from Europe entering the US wind farm market. Clint (Jito) Coleman, president of Northern Power Systems, told attendees at the American Wind Energy Association's Windpower conference in 1991 that "the Advanced Turbine may be a significant breakthrough in both performance and cost of energy. The combination of low weight turbine and high performance, tailored specifically for an energetic Great Plains site, will make this configuration a very promising choice for utility planners in the near future."[10] While these wind turbines were soon erected for both wind farms and standalone applications throughout the world, the three turbine producers aimed to demonstrate the vitality of their machines in some of the coldest places on the planet. During the 1990s, the wind turbines were erected at various Alaskan villages

This 250-kilowatt proof-of-concept turbine being installed at the National Wind Technology Center in Boulder, Colorado, in April 2000 was designed and fabricated by the Wind Turbine Company of Belleview, Washington. Photograph by Warren Gretz. Courtesy National Renewable Energy Laboratory, Boulder, CO.

north of the Arctic Circle to provide supplemental electric power to longstanding diesel-powered generators.[11]

In the 1990s, there also emerged a race among Denmark's leading wind turbine manufacturers, such as Nordtank Energy Group (renamed NEG Micon after a merger with Danish competitor Micon in 1997), Vestas Wind System, and Bonus Energy, to build and export to the United States utility-size machines with rated outputs of 1 megawatt and larger.[12] Not to be left out, the US Department of Energy in 1994 announced a $40 million program to foster the development of US-built, 1-megawatt wind turbines. A competitive bid to design a 1-megawatt machine was announced, and

in 1996, the department selected two firms—Enron Wind Corporation subsidiary Zond Systems, Inc. and the Wind Turbine Company—to undertake this work. Zond's design was an upwind, three-bladed turbine, including a "smart" controller, airfoils based on a National Renewable Energy Laboratory design, and a variable-speed generator. The Wind Turbine Company set out to develop a "dual-speed," two-bladed turbine with multiple generators and "flapping" rotor that echoed design attributes of the experimental Smith-Putnam turbine erected in Vermont in 1943.[13] "The turbine's two blades will attach to the rotor hub with hinges, which allow the blades to flap in and out of the plane of rotation depending on the forces impacting the rotor during high winds," the US Department of Energy said.[14] The Zond and Wind Turbine Company turbines were expected to be delivered to the utility market at the start of the next decade.[15]

For those companies still engaged in the production of residential-size and off-grid wind turbines—5 to 40 kilowatts primarily—their focus at the start of the 1990s was on reducing manufacturing costs in order to continue selling competitively priced machines. Bergey Windpower and Southwest Windpower, for example, turned to importing electrical components from suppliers in China. Still, at the time, a 10-kilowatt wind turbine—considered the ideal size for the average American household—cost upwards of $35,000. In addition, these companies had the difficult task of restoring the public's confidence in small wind machines, since the exit of numerous manufacturers during the late 1980s had left many turbine owners holding worthless service plans and warranties. Idled or abandoned small wind turbines that remained on their towers were symbols of those failed enterprises. To ease the small wind industry's recovery, some states began offering tax incentive and rebate programs to help offset the purchase price of residential wind turbines.[16] Homeowners also needed to consider the wind resource, economic benefits, and local regulations and ordinances before erecting a wind turbine. First, a property of at least an acre in size required access to a reliable wind source, since most small turbines began generating only moderate amounts of electric power when winds reached 8 to 10 mph. Tapping unencumbered winds called for tall towers, but reaching suitable heights for a wind turbine—generally 35 feet above any nearby impediments such as trees or buildings—was often hindered by restrictive zoning ordinances, particularly for properties located within municipalities. While the 1978 Public Utility Regulatory Policies Act mandated that utilities buy back excess electricity produced by residential renewable energy sources of less than 80 kilowatts, some utilities remained hesitant to allow those types of connections with their power lines. If the homeowner was able to meet all the criteria and successfully install an efficient small wind turbine, it was possible for that system to lower utility bills by 50 to 90 percent.

A Bergey Windpower BWC 1500 wind-electric, water-pumping system in Oesao, West Timor, Indonesia on a Japanese International Cooperation Agency-funded irrigation project in 1992. Courtesy Michael Bergey, Norman, OK.

The American Wind Energy Association noted that an all-electric homeowner with a 10-kilowatt wind turbine could experience monthly utility bills of $8 to $15 for part of the year.[17]

Since overall commercial sales of residential wind turbines remained slow at the start of the 1990s, manufacturers often turned to the federal government for business opportunities. Bergey Windpower, for example, took advantage of available contracts through the US Department of Agriculture and US Agency for International Development to utilize its 10-kilowatt machines to power electric water-pumping systems for remote villages in developing countries.[18] In 1995, the US Department of Energy announced its Small Wind Turbine Project to help advance the technologies used in currently available commercial machines with rated outputs of 40 kilowatts. In describing the program, the US Department of Energy said, "The small turbines will use cutting-edge technology—towers, rotors, generators, and controls—developed for utility machines. New technology will make small turbines more reliable and cost effective as well as easier to transport and install than today's machines."[19] It was estimated that more than two billion people in mostly developing countries were still without access to electricity in the mid-1990s, and the US Department of Energy believed that small wind turbines could be the cost-effective means to providing electric power to these populations. In 1997, the Small Wind Turbine Project entered contracts with four companies to develop new machines: Bergey Windpower's 40-kilowatt, three-bladed, upwind, variable-speed turbine for village power; WindLite Corporation's 8-kilowatt, three-bladed, upwind, direct-drive, variable speed turbine for battery charging, communications, lighting, and water pumping in cold and wet environments; Wind Power Technologies' 16-kilowatt, three-bladed, upwind, variable-speed turbine for charging batteries for homes, farms, and small businesses; and Cannon Wind Eagle's 30-kilowatt, two-bladed, downwind machine for use in villages, ranches, and remote industrial sites. These machines were expected to be available commercially by the end of the decade.[20]

Southwest Windpower, an up-and-coming small wind turbine manufacturer, captured the market's attention in the mid-1990s with the introduction of its advanced small 12-volt wind turbine, known as the Air. The small company, founded by Flagstaff, Arizona-based entrepreneurs David Calley and Andrew Kruse, got its start in 1985 when the Reagan Administration withdrew federal tax incentives and most of the government research funding for wind energy. The company struggled to get a foothold in the wind energy market. Its first machine was a 300-watt, two-bladed, upwind battery charger called the Windseeker. The first ten units in 1987 sold to Inter Island Solar of Hawaii for $300 apiece, and as the company continued its sales, the price was raised to $595 per machine. In the early 1990s, Southwest's owners became entangled with a bad investor and nearly closed the company. With its survival on the line, in early 1994, Calley and Kruse wrote a new business plan, raised $20,000 from family and friends, and started designing a new small or "micro" wind turbine for the battery-powered market. The simple yet durably designed turbine, called the Air, became an immediate success in the marine and remote applications markets. The Air turbine consisted of a 46-inch-diameter, upwind rotor with three

flexible carbon fiber blades. The machine's brushless permanent magnet alternator, which could be set to produce 12-, 24-, and 48-volts DC, was housed inside a cast aluminum body. Weighing 13 pounds, the unit could be easily installed on top of a simple pole tower. Between 1995 and 1998, Southwest produced about 18,000 Air turbines. While the first machines weren't without their hiccups, including a noisy rotor, the company quickly remedied these problems through minor changes to materials and components. By 1999, Southwest produced about 15,000 Air turbines annually and through savvy marketing sold them in more than thirty countries, generating over $1 million in revenue a year and establishing itself as one of the most recognized small wind turbine manufacturers in the world by the end of the decade.[21]

Brent Summerville prepares to field test a Southwest Windpower Air turbine next to the company's Whisper 500 machine at Appalachian State University's Beech Mountain test site in 2006. Courtesy Brent Summerville, Boone, NC.

Most commercial residential wind turbines offered to the market at the start of the 1990s were "turnkey" products, meaning that they were sold to customers as a package, ready for assembly, and set up by a commercial installer or the homeowner, if they were skilled at erecting towers. These machines generally included limited warranties. However, there were still plenty of industrious individuals at the time who sought to build their own wind turbines from scratch. This interest in home-built machines was confirmed by a 1993 American Wind Energy Association membership survey.[22] Plans for constructing residential wind turbines have been available since the early 1970s. *Home Power* magazine, which got its start in 1987, became the leading periodical of record for residential renewable energy project "do-it-yourselfers."[23] Mick Sagrillo, a noted residential wind turbine expert, generally warned against pursuing home-built machines due to

This 20-kilowatt Jacobs wind-electric turbine was erected on the grounds of the Midwest Renewable Energy Association's annual Energy Fair at Custer, Wisconsin, during a workshop led by Mick Sagrillo in 2007. Photograph by author.

their high failure rate. Instead of "wasting time to re-invent the wheel," he recommended purchasing proven commercially built systems.[24] "This high failure rate is mostly due to the lack of understanding surrounding two principles that are fundamental to sound wind generator design. The first is known as 'cube law,' which states that the power available to a wind generator's rotor is a function of the cube of the wind speed. . . . The second principle that is overlooked is that the swept area of the rotor is a function of the blade length squared. . . . While it is tempting to take maximum advantage of the power in the wind that is blowing past your house, a lack of careful consideration of these two fundamentals can be catastrophic," Sagrillo wrote in 1993.[25] While examples of do-it-yourselfers-turned-commercial wind turbine enterprises occurred throughout the twentieth century, it generally took these individuals years of trial and error to perfect their machines.

An alternative to either a home-built or newly manufactured machine was to install a refurbished wind turbine. The Jacobs remained one of the most highly sought-after, second-hand wind turbines in the 1990s. The machines were so well built that it did not take much effort to reinstate one that had been left dormant on a tower or in a barn for years. The wind turbines manufactured by the Jacobs Wind Electric Company prior to the closure of its longtime Minneapolis factory in 1956, namely, the 2.5-kilowatt, 32-volt DC machines, were still desirable for off-grid, battery-powered applications. However, it was those Jacobs wind turbines manufactured by the company between 1980 and 1985, with rated outputs of 10 to 17.5 kilowatts and grid-tie capability, that became most attractive to homeowners. From the late 1980s, the machines became available through Prior Lake, Minnesota-based Wind Turbine Industries Corporation (WTIC).[26] However, these Jacobs wind turbines were still a considerable investment for most middle-class homeowners in the 1990s. Daniel Whitehead, who resided at Morrison, Illinois, said his 17.5-kilowatt Jacobs, which he purchased from WTIC, as well as a 120-foot tower and inverter, cost him $16,000 in 1993, and he performed the entire installation himself. "Add another $10,000 if you had a dealer do the install," he said.[27] Shortly

after erecting his own Jacobs, Whitehead started Illowa Windwork to install these Jacobs machines for customers across Illinois and Iowa through the remainder of the 1990s. Instead of sourcing Jacobs machines exclusively through WTIC, he learned that used 17.5-kilowatt machines were available from wind farms in San Gorgonio Pass in southern California. "I went to Palm Springs and purchased fifteen machines from a wind farm that had taken down 500 Jacobs units," Whitehead recalled. "[I] had to pick carefully because they just torched off the bottom legs [of the towers] and let them fall. They had hundreds of them stacked in piles. It was so sad . . . I would rebuild them and sell these much cheaper."[28]

By the end of the 1990s, the primary US small wind turbine manufacturers included Atlantic Orient, Bergey Windpower, Northern Power Systems, Southwest Windpower, WindLite, WTIC, WindTech International, and Wind Power Technologies. These companies not only delivered machines throughout the United States, but they also shipped them across the world. According to the American Wind Energy Association, US-built small wind turbines accounted for 30 percent of the global market at the end of the decade. The association concluded that "the potential for small wind turbines has barely been tapped" and "[a] vibrant U. S. market would make an enormous difference in building volume

that could lead to further improvements in the cost and competitiveness of small wind systems."[29] Even the association stated that by 2000 the small wind industry in the United States was in the position to grow. The electric utilities were restructuring and increasingly considered wind energy among their

A 10-kilowatt Bergey wind turbine was installed at this home in Great Falls, Montana, in 2003. Courtesy Michael Bergey, Norman, OK.

power-generation portfolios. Thirty states had already established net metering policies that encouraged residential wind turbine connections to the grid. There was also increasing access to state rebate and grant programs that reduced the capital costs of small wind turbines.[30] Without these financial incentives, machines with rated outputs of 5 to 15 kilowatts would cost between $17,000 and $24,000, minus the expenses for shipping, installation, and maintenance. "The wind system will usually recover its investment through utility savings within 6–15 years and after that the electricity it produced will be virtually free," the association said in 1999. "Over the long term, a wind turbine is a good investment because the wind system increases property value similar to any other home improvement. Many people buy wind systems in preparation for their retirement because they don't want to be subject to unpredictable increases in utility rates."[31]

In 2002, the American Wind Energy Association's Small Wind Turbine Committee released its twenty-year "roadmap" report for the industry. It hoped to benefit from increasing public interest in renewable energy due to steadily rising retail prices for electricity and growing concern about global warming due to the heavy dependence on fossil fuels. The committee's report noted that the market for small wind turbines was growing 40 percent per year, with sales of US-manufactured machines in 2001 reaching an estimated 13,400 units with a value of $20 million. These turbines ranged in size from 400 watts for small loads such as sailboats, cabins, and battery storage, to 3 to 15 kilowatts for homes, and up to 100 kilowatts for small commercial applications. The association forecasted that small wind turbines could contribute 3 percent or 50,000 megawatts to the country's electrical supply by 2020. With 60 percent of the country covered by sufficient winds, the association noted that approximately fifteen million American homes, as well as one million small businesses, could effectively utilize and benefit from electric power generated by small wind turbines. At the start of the 2000s, small wind turbines—in the 5- to 15-kilowatt range—were about half the cost of comparable-size photovoltaic solar systems: $4 per installed watt for a small wind turbine versus $8 per installed

watt for electric-generating solar panels. The American Wind Energy Association declared in its 2002 report that "[w]e do not foresee a time when solar electric systems will be less expensive than small wind systems." However, for the small wind industry to fulfill its 2020 goal of 50,000 megawatts, it needed to continue lowering the installed cost of these machines from about $3500 per installed kilowatt in 2002 to as low as $1200 per installed kilowatt by 2020. This would be accomplished mostly by decreasing the manufacturing costs of the machines, while continuing to make advancements in the technology.[32]

Since the 1992 implementation of the Production Tax Credit and expansion of state-based Renewable Portfolio Standards, which set specific timelines and amounts of electricity that must be supplied to utilities through renewable energy sources, wind farms have continued to proliferate throughout the United States. By the early 2000s, public interest in "community-scale" wind projects, in the several megawatt range, began to grow. This activity was further stimulated by the 2005 Federal Energy Policy Act's establishment of Clean Renewable Energy Bonds, which allowed entities such as rural electric cooperatives, Native American tribes, and nonprofit organizations that were ineligible for the PTC to apply for low-interest bonds to finance wind projects for local economic development. Since the large wind farms absorbed much of the megawatt-size wind turbines, however, community wind projects mostly relied on turbines in the range of 50–900 kilowatts. The National Renewable Energy Laboratory defined "small-scale" community wind projects as those that were "connected to 13.8 kV (kilovolts) or lower distribution lines, either behind the meter offsetting a portion or all of the electricity used on-site by a load in the community, or using a dedicated transformer with all energy sold to the interconnecting utility."[33] According to the laboratory, by 2006, an estimated 270 megawatts of community wind projects had been installed throughout the United States, and of that about 110 megawatts were defined as small scale or 1 megawatt and less.[34] One of the most successful demonstrations of a community wind project at the time was located in Hull, Massachu-

The Town of Hull, Massachusetts, in 2001 instituted a community wind program with the installation of this 660-kilowatt Vestas wind turbine. This was followed in 2006 with the erection of a second, 1.8-megawatt Vestas machine. Courtesy Kristopher Nixon, Hull, MA.

setts, where a group of citizens interested in wind energy's prospects successfully convinced the town to invest about $800,000 in the installation of a 660-kilowatt Vestas wind turbine in 2001, which was followed in 2006 with the erection of a second, 1.8-megawatt Vestas machine. It was estimated that the two wind turbines combined met 11 percent of Hull's ratepayer's electricity demand.[35]

Since the size and scope of community wind projects demanded fewer and larger machines, small wind turbine manufacturers continued their focus on residential and small business installations at the start of the 2000s. By then, more than half of all US states had instituted incentive programs, such as low-interest loans, cash rebates, and tax deductions, which in some cases covered as much as 60 percent of the $35,000–$50,000 cost of the average grid-tied, 10-kilowatt wind turbine. In 2003, New York offered a program that provided consumers up to 50 percent cash back on the cost of a residential wind system in addition to low-interest loans, while California offered small wind turbine purchasers rebates based on the size of the system. Thus, a $50,000, 10-kilowatt turbine might be eligible for a $22,500 rebate. From 2000 to 2005, more than 200 grid-tied small wind systems benefited from California's rebate program.[36] The only available federal incentive program for residential-size wind turbines at the time was offered to farmers and ranchers through the US Department of Agriculture's Farm Bill Section 9006 grants. Since the application process was rigorous, few farmers and ranchers were eligible for the program. In 2004, only twelve small wind systems, valued at $585,000, were funded by the USDA grants.[37]

Despite the costs, consumer interest in residential-size wind systems accelerated in the United States. The American Wind Energy Association estimated that about 6800 machines were sold in 2006, of which approximately 2500 were rated at 1 to 10 kilowatts for grid-tied use, and the remainder were less than 1 kilowatt for mostly off-grid applications. Of these US installations, 168 machines were imported from foreign manufacturers. Another 9170 US-manufactured small wind turbines were exported that year to overseas markets. In 2006, the association identified twelve established US small wind turbine manufacturers and another eight preparing to enter the market with machines.[38] The start of the Great Recession in December 2007 failed to dent the small wind turbine industry's momentum. In fact, the US small wind industry during 2008 grew 78 percent to $77 million, mostly due to an infusion of private equity investment that stimulated an increase in the number of turbines manufactured. Approximately 10,500 small wind turbines were sold in the United States that year. The American Wind Energy Association counted seventy-four companies in the United States that either manufactured or planned to manufacture small wind turbines, with the largest firms being Southwest Windpower, Proven Energy (of Scotland), Northern Power, Entegrity Wind Systems (of Canada), and Bergey Windpower.[39] An added boost to the small wind industry came in October 2008 when Congress passed the Emergency Economic Stabilization Act, which for the first time in twenty-three years included a federal investment tax credit for small wind system purchases. At first, the tax credit, which was good for eight years, covered 30 percent of the cost of a wind turbine, but it was capped at $4000. However, in February 2009, as part of a $787 billion economic stimulus package, known as the American Recovery and Reinvestment Act, which President Barack Obama signed into law, the cap was lifted on the 30 percent tax incentive for small wind systems.[40] With the cap removed on the federal tax credit, the American Wind Energy Association forecasted "a 30-fold growth [in small wind systems] within as little as five years, despite a global recession, for a cumulative US installed capacity of 1,700 [megawatts] by the end of 2013."[41]

Established small wind turbine manufacturers, such as Bergey Windpower and Southwest Windpower, focused on engineering improvements to their machine designs. With research assistance from the National Renewable Energy Laboratory, companies considered alternative power and load control designs to eliminate furling, which yaws the turbine out of the wind but contributes to the noise from turbine rotors increasing and decreasing rotationally when turning in and out of the wind. Alternative approaches to furling included the development of "soft-stall" rotor speed control, constant speed control, variable pitch and hinged blades, centrifugally actuated blade tips, and mechanical brakes. Reducing rotor noise emissions was also pursued through lower tip-speed ratios and lower peak-rotor speeds. Much attention was placed on advancing blade manufacturing. Instead of labor-intensive, hand-layered fiberglass blades, the industry embraced new, more efficient production techniques, such as injection, compression, pultrusion, and reaction injection molding. Small wind turbine manufacturers began to replace

When Southwest Windpower's Skystream 3.7 came onto the commercial market in 2006, it quickly became one of the most recognized residential wind turbines for its unique swept blades. Courtesy Andy Kruse, Boulder, CO.

traditional ferrite magnets in their generators with lighter weight, more compact rare earth magnets and reduced generator cogging to lessen cut-in wind speeds of their machines. Induction generators were introduced by the industry in the early 1980s for small wind turbines with the aim of eliminating power electronics that increased generator costs and complexity, as well as reduced reliability. Advances also continued with grid-tied inverters, including options for single-phase and three-phase configurations.[42]

One of the most recognized residential wind turbines from the decade was Southwest Windpower's Skystream 3.7. Together with engineers from the National Renewable Energy Laboratory's National Wind Technology Center in Boulder, Colorado, the company began design work on the turbine in 2001. The wind turbine's most distinctive feature was its three swept fiberglass blades that had airfoils, with a thin trailing edge to reduce drag and noise during rotation. The downwind rotor had a 12-foot diameter. The permanent magnet generator used a slotless stator to eliminate cogging for a smooth turning in low wind speeds. The generator included a built-in power inverter that delivered AC current to the home. The machine was first rated at 1.8 kilowatts, but shortly thereafter rated up to 2.4 kilowatts, which was enough to cover 40 percent of the electric power used by the average 2000-square-foot home. The wind turbine included software that allowed the homeowner to monitor its performance by computer. The first prototype was erected at the National Wind Technology Center for testing in 2003. Southwest Windpower introduced the Skystream 3.7 for commercial sale in June 2006. The company offered the turbine with two simple monopole towers: 33.5-foot or 70-foot tall. The cost to purchase and install the Skystream 3.7 ranged from $8500 to $11,000. Through savvy marketing and the machine's attractive appearance, the wind turbine was an instant success, despite several early technology flaws that the company quickly fixed.[43] The Skystream 3.7's popularity was further enhanced by the abundance of media coverage at the time of its introduction: *Popular Science* declared the wind turbine to be among its "Best of What's New" for 2006 and *Time* considered it to be

Potomac Wind Energy erected this Southwest Windpower Skystream 3.7 turbine in 2009 along the Potomac River in Charles County, Maryland, using a young, 2000-pound draft horse. Courtesy Carlos Fernandez-Bueno, Dickerson, MD.

among its choices of "Best Inventions of 2006."[44] Combined with the sales of its Air and Whisper machines, the Skystream 3.7 made Southwest Windpower by far the largest US manufacturer of small wind turbines by 2009.[45]

Many incentives and the public's fascination with small wind turbines encouraged a host of new companies to rush into the market with machines of ill-conceived design and questionable performance. Dismissed by the early 1980s as operationally inferior to the three-bladed, horizontal-axis wind turbines, small vertical-axis machines returned to the market in the early 2000s, emphasizing their aesthetic appeal for the burgeoning urban rooftop market. Developers used various rotor designs, including vertically aligned straight blades, Darrieus or eggbeater-shaped blades, Savonius or S-shaped blades, and helical-shaped blades. One of those upstarts was Fort Worth, Texas-based SkyDrill Power Systems, whose founder and chief executive officer, Barry Sterling, told the *Fort Worth Business Press* in 2009: "[W]e were trying to come up with an efficient system that could be used in urban areas, because when most people think about wind energy they think about the large horizontal-axis wind turbines out in West Texas. . . . We really were attempting to find something that would be both aesthetically pleasing and functional—something that you can look

Boston's Logan International Airport became the first commercial airport in the United States, in July 2008, to generate wind-electric power by installing a rooftop array of twenty 20-kilowatt AeroVironment wind turbines. Courtesy Massachusetts Port Authority, Boston.

A Mariah Windpower Windspire Giromill shown on the grounds of the US Botanical Gardens in front of the Capitol building in Washington, DC, in 2008. The machines were often ineffectively sited and failed to deliver on wind-electric generation promised by the manufacturer. Press release photograph in author's files.

out my window and see some of these things on tops of roofs."[46] To many established wind turbine manufacturers, however, vertical-axis machines were nothing more than whirligigs with marginally discernible electric-power generation capabilities. Despite technical and efficiency concerns associated with vertical-axis machines, the American Wind Energy Association reported in 2009 that approximately forty-five companies, of which twenty-one were US-based, either manufactured or planned to manufacture these types of small wind turbines.[47]

Municipal governments and corporations increasingly endorsed residential-size, grid-tied wind turbines not only to provide supplemental electric power to facilities, but also to utilize them as public demonstrations of their commitment to renewable energy. "They're trying to do something that we in the wind industry abandoned twenty, thirty years ago," longtime wind energy historian and consultant Paul Gipe told the publication *Orion* in December 2005. "We want wind energy to be accepted and rooftops is not the way to do it."[48] In August 2008, New York City Mayor Michael Bloomberg proposed erecting small wind turbines on roofs citywide. Similar rooftop promotional initiatives were proposed by San Francisco and Boston and also by Pennsylvania's state government.[49] While urban and residential areas might be located in windy areas, such as along the coasts or on the Great Plains, the wind lowest to the ground becomes dispersed and turbulent as it encounters buildings, trees, and other structures, which makes it inefficient and sometimes destructive for rooftop and utility pole-mounted turbines. If a tower could not be set at least 30 feet above and 500 feet away from the nearest structure, then the wind turbine's performance would be significantly minimized. Roof-mounted wind turbines also tended to induce unwelcome noises and vibrations within a building's structure.[50] Despite the Distributed Wind Energy Association's insistence on preventing poorly designed and ill-placed rooftop wind turbine installations, the US Green Building Council exacerbated their proliferation during the early 2000s through its Leadership in Energy and Environmental Design rating system, which rewarded architects and building managers for integrating energy efficiencies or "green" practices in both old and newly constructed buildings.[51]

One of the most notable vertical-axis turbines for urban landscapes at the time was developed by Reno, Nevada-based Mariah Power. Called the Windspire Giromill, the machine consisted of three narrow vertical blades of about 30 feet long, evenly spaced within a 2-foot radius spire-shaped rotor of aluminum construction. To prevent the rotor from turning more than 500 revolutions per minute, the company incorporated "electronic braking" software into its design. The machine consisted of a permanent magnet generator with a 120-volt AC output and was rated at 1.8 kilowatts in an 11-mph wind. In 2008, the Windspire retailed for $3995, with another $1000 for the installation.[52] In May 2008, a Windspire turbine was showcased at the US Botanical Garden next to the Capitol building as part of the "One Planet—Ours! Sustainability for the 22nd Century" exhibit. Buoyed by the publicity and continued outside investment, the company's founder and chief executive officer, Mike Hess, targeted urban markets for his machine, particularly commercial buildings. "I want to make the product ubiquitous, so we see it all over—on top of skyscrapers, right on a lamppost driving the lights on the street," he told National Public Radio in August 2009, adding that his company sold twenty machines for installation on the roof of Adobe's corporate headquarters in San Jose, California.[53] However, during a six-month test at the National Renewable Energy Laboratory's National Wind Technology Center in 2008, the Windspire experienced repeated hardware and inverter malfunctions. Mariah Power responded to the test results by stating that it had corrected the design flaws and "introduced other product upgrades since the first units were shipped."[54] The machines were often ineffectively sited and failed to deliver on the wind-electric generation promised by the company. Many Windspires were taken down within several years of their installation.[55] "In the annals of *greenwashing*, no wind turbine company has done more to develop the technique to a fine art. It's as if the company's entire business model was built on finding participants who wanted a greenwashing

In 2011 WindTronics, using the Honeywell brand, introduced a 1.5-kilowatt roof-mounted wind turbine with disastrous results. This example of WindTronics machine is preserved at the American Windmill Museum in Lubbock, Texas. Photograph by author.

project," Gipe wrote in June 2013, around the time Mariah Power closed its operation.[56]

Yet more overhyped small wind turbine designs were placed in the market prior to 2010, which quickly proved to be failures. Akron, Ohio-based Green Energy Technologies introduced a turbine for either rooftop or pole mounts in mid-2009 called the WindCube. The machine measured 22-feet square with a 7-foot depth. The rotor included five blades with a diameter of 15 feet. A 3-foot-deep shroud channeled the wind toward the blades. Green Energy Technologies, which rated the turbine at 60 kilowatts, claimed that the turbine began producing power in a 5-mph wind. The WindCube retailed for $279,000, but with available state grants and the federal tax credit, that price was expected to be reduced by two-thirds. The company's target market for the turbine was commercial buildings, and it installed its first unit on the roof of Crown Battery at Sandusky, Ohio, in 2009, adding that its partner, Roth Brothers, was prepared to manufacture 100 WindCubes a month by 2010.[57] After two years of research, WindTronics of Muskegon, Michigan, introduced its 1.5-kilowatt roof-mounted wind turbine in April 2011, which it branded with Honeywell. The 185-pound WT-6500, which was available for

sale through Ace Hardware for $4500, consisted of a 6-foot-diameter rotor with twenty slender blades enclosed inside a plastic ring. Around the inner ring were staged magnets and stators, which generated electricity as the gearless rotor turned. WindTronics claimed that the grid-tied machine could start generating power in a 2-mph wind. Despite the initial accolades in the press, the WT-6500 failed to deliver results, and, like the manufacturer of the WindCube (renamed Wind Sphere), WindTronics quit the market within several years.[58]

Small wind turbine technology experts attributed the increase of inferior machines on the market to the availability of generous incentives from numerous states and the federal government. Homeowners and businesses, as well as state incentive fund managers, however, generally lacked the ability to evaluate sufficiently the performance claims of those small wind turbine manufacturers. In the United States, the National Renewable Energy Laboratory's National Wind Technology Center and Det Norske Veritas (DNV) (formerly Global Energy Concepts) offered small wind turbine manufacturers the ability to field test and certify the performance of their machines, but the process could take a year or longer to complete, and few took advantage of the services. There was also no legal requirement for the companies to certify their machines. In 2006, the American Wind Energy Association, in conjunction with the Department of Energy, the National Renewable Energy Laboratory, and state agencies from Iowa, Illinois, Nevada, Oregon, and Wisconsin, set out to establish a new program that would encourage small turbine manufacturers to certify the performance of their machines and enhance consumer confidence. In December 2009, the association released the AWEA Small Wind Turbine Performance and Safety Standard (9.1—2009), which included not only the National Wind Technology Center, but more than two dozen organizations approved to provide field test services by the end of 2010. The tests covered the structural and operational integrity of the machines, as well as power performance, safety, and acoustic noise emissions as part of the certification process.[59] The first two turbines to complete their certification in 2011 included Bergey Windpower's 10-kilowatt Excel 10

A Bergey Windpower factory worker at Norman, Oklahoma, finishes blades for the company's wind turbines in 2012. Courtesy Michael Bergey, Norman, OK.

and Southwest Windpower's 2.1-kilowatt Skystream 3.7, with another twenty-eight machines undergoing field testing.[60] An industry effort was also initiated in late 2010 to establish a small wind turbine installers certification led by the North American Board of Certified Energy Practitioners. The program, however, garnered modest participation in its first year. In 2011, the American Wind Energy Association stated that "[f]uture attempts to build a broad cadre of certified wind installers may depend on greater consumer education, refining testing eligibility, policy interventions and industry support such as reduced group insurance premiums."[61]

By the end of the first decade into the twenty-first century, the small wind industry had experienced its biggest expansion. By 2009, the American Wind Energy Association estimated that 100 megawatts of wind-electric power were generated by almost 100,000 small wind turbines, which were rated 100 kilowatts or less, and approximately half of that 100 megawatts was the result of those machines installed between 2007 and 2009. In 2009 alone, approximately 9800 small wind turbines valued at $82.4 million were sold in the United States for a total rated output of 20.3 megawatts, which equated to a 15 percent growth in rated output over 2008.[62] In addition to the myriad state and federal government incentive programs now available for small wind turbine purchases, homeowners, farms, and small businesses across the country were increasingly drawn to this renewable energy source as a means of shrinking their monthly electric bills. Utility-generated electricity prices had risen 6 percent to 9.44 cents per kilowatt hour between 2006 and 2009.[63] Michael Bergey, president of Bergey Windpower, told *Time* magazine in a 2010 interview that "[t]his is the best business environment we've ever seen, so I'd have to say that these are exciting times."[64] By 2011, the American Wind Energy Association tabulated that approximately 7800 small wind turbines valued at $115 million and with a cumulative rated output of 19 megawatts were installed, increasing the collective number of machines operating in the United States to 151,300, for a total capacity of 198 megawatts.[65] In addition to the 39 percent federal tax credit, thirty-nine state governments in 2011 provided approximately 700 small wind turbine purchasers various rebates, grants, tax credits, and low-interest loans in the amount of more than $38 million, compared to $30 million in 2010 and $35.6 million over the entire 2001–2009 period. The machines that benefited most from these incentives during 2011 were generally 10 kilowatts or larger and grid-connected, while the national average-size small wind turbines that sold that year were 2.6 kilowatts, with numerous off-grid applications.[66] Despite all the available incentives, small wind turbines in the United States still cost about $5430 per kilowatt or about 2.5 times more than the cost per kilowatt of the giant megawatt turbines of the wind farms in 2011. Homeowners also continued to encounter restrictive zoning laws for erecting these machines on their properties.[67] The small wind industry's biggest emerging competitor at the start of the new decade, however, was another renewable energy source—solar.

The first tests using sunlight to produce electricity were conducted by French scientist Edmond Becquerel in 1839. This research was furthered by English electrical engineer Willoughby Smith in 1873 who discovered the photoconductivity of selenium when sunlight struck the material, and, in 1883, Charles Fritts of New York made the first solar cell by applying a thin layer of gold to selenium. In 1954, Bell Laboratories scientists Calvin S. Fuller, Gerald L. Pearson, and Daryl M. Chapin manufactured the first solar cells made from silicon, which was 6 percent more efficient than selenium at converting sunlight to electricity. But at a cost of $2000 per kilowatt of electricity generated during peak sun exposure, photovoltaic solar applications remained limited to powering satellites and other space technologies.[68] While solar energy research continued through the 1950s and 1960s, it was not until the early 1970s in the aftermath of the Arab Oil Embargo that the US government through legislation and funding accelerated the technology's development for residential and industrial applications. The Department of Energy concentrated this research within the newly formed Solar Energy Research Institute in 1977. The Energy Tax Act of 1978 included a residential energy investment tax credit of 30 percent on the first $2000 paid for

qualifying solar systems and 20 percent, or $2500, on a maximum $8000 investment. By 1980, solar cells generated electricity at $8 to $15 per peak watt, with an industry goal to lower the cost per peak watt to 70 cents by 1986.[69] While the investment tax credit for residential wind turbines expired in 1985, it was extended for commercial solar installations in the Tax Reform Act of 1986 and the Energy Policy Act of 1992. Investment tax credits for residential solar installations were reinstated by the Energy Policy Act of 2005 with a $2000 cap, which was lifted three years later by the Emergency Economic Stabilization Act. The American Recovery and Reinvestment Act of 2009 also expanded funding mechanisms for solar systems.[70] In addition to various state incentives for residential solar energy systems, the persistent federal investment tax credits and research funding since 1985, as well as billions of dollars in manufacturing investments from venture capitalists, were viewed by the small wind turbine industry as providing an unfair cost advantage to photovoltaic (PV) solar manufacturers and installers.[71] The American Wind Energy Association stated:

> In 2008, PV manufacturers generally focused on reducing costs, as the technology had become largely commoditized. Today, many manufacturers are looking for new ways to differentiate their technologies or company brands. Small wind systems have not yet reached a stage of commoditization, though 99% of all small wind turbines sold in 2009—and throughout the industry's history—have been of a single general configuration (horizontal axis, mounted on a tower).[72]

The price per solar PV module dropped to $1.85 to $2.25 per watt in 2009 from $3.50 to $4.00 per watt the year before, helping to catapult residential installations in the United States from 78 megawatts for 2008 to 156 megawatts for 2009. The average installation cost for a photovoltaic system also fell by 10 percent in 2009. These factors led the Solar Energy Industries Association to declare for 2010 and the rest of the decade that "PV is getting ready to go big."[73]

Residential photovoltaic solar systems have setups similar to those of small wind systems in that the electricity from the panels starts as DC power, which can either be stored directly in batteries for off-grid use or sent through an inverter for conversion to 60 hertz, 120-volt AC power suitable for use in grid-tied electric systems. Unlike small wind turbines, however, photovoltaic solar panels have no moving parts and offer low physical profiles when attached to roofs or ground mounts. There is no need for tall towers. These technological characteristics have generally contributed to residential solar systems winning quicker approvals for installations from local zoning authorities. While the average-size residential photovoltaic solar system in 2007 was 4.5 kilowatts, the system sizes were rapidly increasing since they were only limited by the size of the roof and overall exposure to the sun. The more roof surface available with a southerly exposure to the sunlight, the more significant the electrical output from the panels. The industry standard for a solar panel's size became 65 inches long by 39

Craig Toepfer of Chelsea, Michigan, uses photovoltaic solar panels to supplement power to his home and charge the batteries of his electric vehicles. Courtesy Craig Toepfer, Chelsea, MI.

Bergey Windpower's wind turbine test yard at its factory in Norman, Oklahoma. Photograph by author.

inches wide. Including mounting hardware, a solar panel generally weighs between 2 and 4 pounds per square foot, which is an acceptable weight for most home and commercial building roofs. By 2005, it was not uncommon to find homeowners with wind turbines of 10 kilowatts or smaller adding solar panels to their systems, since the two technologies could share the same electrical infrastructure.[74]

By 2011, the top three small wind turbine manufacturers in the United States in terms of sales were Northern Power, Southwest Windpower, and Bergey Windpower. In addition to the increased competition from photovoltaic solar systems, these companies still found it challenging to reduce their wind turbine production expenses. The average cost to install a wind turbine of 100 kilowatts or less in the United States in 2011 increased 11 percent to $6040 per kilowatt over 2010.[75] State and federal tax incentives, while still important to the small wind industry's well-being, continued to encounter administrative hurdles and resistance from public utilities to grid connections. The industry also received a public relations setback in 2011 due to several widespread turbine failures, resulting in a handful of states with generous incentives, such as Alaska, California, Nevada, New Jersey, Ohio, and Wisconsin, either to curtail or to suspend those programs. California, which was considered the largest small wind turbine market at the time, froze its incentive program for almost year.[76] Overall, the small wind sector struggled to regain its footing in the depressed US economy. The American Wind Energy Association noted that "[t] he extended weak economy affected all sectors, but residential [wind] applications were hit hardest because homeowners remained reluctant to invest in their properties, especially in the more populated states."[77] By 2012, US domestic sales of off-grid and residential-size wind turbines, which were by now part of a wider portfolio known as "distributed wind" or wind energy systems connected to the customer side of the meter for onsite usage or directly connected to the local distribution grid, had dropped to about 3700 turbines (less than 100 kilowatts) or half the number sold the year before.[78]

One of the most stunning exits by a small wind turbine manufacturer involved Southwest Wind-

Southwest Windpower's charismatic founders Andy Kruse (left), vice president of sales and marketing, and David Calley, chief executive officer, in 2004. Courtesy Andy Kruse, Boulder, CO.

A Southwest Windpower factory worker in Flagstaff, Arizona, assembled air turbine nacelles in March 2010. Courtesy Brent Summerville, Boone, NC.

power, which without notice on February 20, 2013, closed its operations. The company claimed to have sold more than 170,000 wind turbines across 120 countries since 1987 and had a sales presence in 88 countries. While its sales consisted mostly of the 12-volt Air microturbines, which were installed for off-grid applications such as remote cabins, telecom transmitters, oil platforms, traffic signs, and sailboats, Southwest Windpower sold more than 8000 Skystreams worldwide between 2006 and 2012.[79] In the eyes of its installers and customers, as well as competitors, the company appeared to be in an enviable position with its $10 million investment in 2009 from General Electric to innovate its turbines and an arrangement reached with Home Depot in 2011 to sell and install the Skystream 3.7.[80] However, internally Southwest Windpower was experiencing difficulties with its finances and manufacturing supply chain. The company's two founders, David Calley and Andy Kruse, had departed by 2010 and early 2012, respectively, to seek new business opportunities.[81] Southwest Windpower ended production of its Whisper turbines by 2012 and sold off its successful Air microturbine line to Lakewood, Colorado-based aerospace company Primus Metals in January 2013, which established Primus Wind Power.[82] Nearly four months after Southwest Windpower's closure, XZERES Corporation of Wilsonville, Oregon, acquired the Skystream 3.7 manufacturing assets, including tooling and fixtures, production line and testing equipment, along with five patents and five more pending in the United States and overseas, and the plans to build the next-generation Skystream. "The acquisition of these assets and supporting IP (intellectual property) perfectly complements our existing world-class 10 kW system, as we continue to build XZERES into a global renewable energy company," said Frank Greco, XZERES's chief executive officer, in a July 9, 2013, press release. (Greco previously served as chief executive officer of Southwest Windpower until 2010.)[83] The downside to the acquisition was that the warranties for those Skystreams sold before Southwest Windpower's closure were no longer valid, but XZERES said it would provide support and maintenance services to those wind turbine owners.[84]

With photovoltaic solar power's expanding presence in both off-grid and grid-tied markets, other small wind turbine companies found it difficult to compete on price, and those with unsuccessful manufacturing operations or subpar machines left the industry. This exodus was further exacerbated

Primus Wind Power took over the highly successful Air microturbine line after Southwest Windpower closed in 2012. The 12-volt wind turbine used for off-grid application continues to be the most sold machine on the market. Courtesy Primus Wind Power, Lakewood, CO.

in January 2015 with the commencement of the Internal Revenue Service's requirement that small turbines be certified under the American Wind Energy Association Small Wind Turbine Performance and Safety Standard 9.1–2009 or the International Electrotechnical Commission 61400–1, 61400–12 (power) and 61400–11 (acoustics) to qualify for federal tax incentives.[85] The Department of Energy noted that by 2016 the number of small wind turbine manufacturers—both foreign and domestic—reporting sales in the United States rapidly declined from thirty-one in 2012 to twelve (nine of which were US headquartered).[86] US small wind turbine sales (less than 100 kilowatts) similarly dropped from about 2700 units, valued at $36 million and with a collective output of 5.6 megawatts, in 2013 to about 1695 units, valued at $21 million and with a collective output of 4.3 megawatts by 2015. California, New York, and

Minnesota continued to lead the way with their favorable small wind incentive programs on top of the federal tax breaks.[87] The total number of small turbines operating in the United States by 2016, an estimated 77,200, still made up less than 1 percent of the country's total operating wind energy capacity, which remained dominated by utility-scale wind farms with about 42,430 turbines.[88] Most surviving US small wind turbine manufacturers relied on export markets, especially to Japan, the United Kingdom, and Italy, which at the time offered the most generous feed-in-tariffs in the world to help stay afloat.[89] Consistently among the top five small US wind turbine providers since the start of the 2000s, based on sales, were Northern Power Systems with its 100-kilowatt machines, Bergey Windpower and its 10-kilowatt machine, and the former Southwest Windpower, whose turbines were now split between XZERES Corporation with its 2.1-kilowatt (former Skystream 3.7), 10-kilowatt, and 50-kilowatt machines, and Primus with its 160-kilowatt and 400-kilowatt Air models. By 2016, Primus was a top global seller of microturbines for off-grid applications, including remote cabins,

sailboats, oil and gas equipment, telecom installations, scientific instrumentation, and militaries.[90]

Those companies that still produced small wind turbines in the United States after 2010 continued to seek out design and manufacturing process improvements to maintain their competitiveness in the renewable energy market. The US Department of Energy's Wind Energy Technologies Office and the National Renewable Energy Laboratory in 2012 initiated the Distributed Wind Competitiveness Improvement Project (CIP) with the goal of assisting the fledgling industry to "facilitate the development of next-generation, U.S. manufactured small and mid-size wind turbine technology."[91] The program annually awarded cost-shared subcontracts to individual companies, with a focus on component improvements and systems optimization, manufacturing process improvements, prototype testing, certification testing, and type certification. CIP project subcontracts generally lasted eighteen to twenty-four months. By 2018, the National Renewable Energy Laboratory had overseen the award of twenty-eight subcontracts to fifteen companies, valued at $6.3 million, over six annual rounds, and

After five years of research and development in partnership with the US Department of Energy, Bergey Windpower introduced its 15-kilowatt Excel 15 turbine to the commercial market in early 2019. Courtesy Michael Bergey, Norman, OK.

in 2019 awarded another round of contracts to US small wind industry participants in the amount of $2 million.[92] Companies that received CIP subcontracts between 2013 and 2018 were Bergey Windpower, Endurance Wind Power, Integrid, Northern Power Systems, Pecos Wind Power, Pika Energy, Primus Windpower, Rock Concrete, Sonsight, Star Wind Turbines, Urban Green Energy, Ventura Wind, Wetzel Engineering, Windurance, and XZERES.[93] One of the biggest beneficiaries of the CIP program was Bergey Windpower, which received four subcontracts between 2013 and 2017 that contributed to the company's development and successful launch in early 2019 of its new 15-kilowatt Bergey Excel 15 turbine and tower system, a replacement for its twenty-year-old 10-kilowatt Bergey Excel 10 machine. The Excel 15 achieved a levelized cost of energy of 13 cents per kilowatt-hour, or nearly half that of the Excel 10.[94]

Learning a lesson from the photovoltaic solar industry's successful rooftop leasing program, the small wind industry also began working with third-party finance and leasing companies to establish similar arrangements with homeowners, farmers, and institutions. The customers of these arrangements agreed to host a wind turbine on their property and make monthly payments toward the turbine's installation, operation, and maintenance costs. These arrangements eliminated the risk of directly owning and operating a wind turbine and allowed the property owner to benefit from lower monthly utility bills through power purchase agreements. Small wind turbine manufacturers also benefited from the increased sales of machines through these arrangements with the leasing companies.[95] As these leasing arrangements gained traction after 2010, private equity firms began taking notice and made investments with the providers of these services in recent years. In January 2016, Brooklyn, New York-based United Wind, a distributed wind leasing company, started in 2013 by former managers of Talco Electronics and Wind Analytics, secured $200 million in project equity capital from Forum Equity Partners to expand its turbine lease program. In January 2017, the firm announced a purchase agreement with Bergey Windpower to acquire one hundred 10-kilowatt turbines, which it planned to erect throughout central and western New York to benefit from the state's generous small wind incentives. United Wind has since extended these arrangements to farm operators in Colorado and Kansas and has installed grid-tied turbines with rated outputs of 50 to 100 kilowatts.[96] "Farmers like these turbines. They look at them as any other

United Wind, a distributed wind leasing company, installed this 50-kilowatt Endurance wind turbine at Double A Vineyards in Fredonia, New York, in June 2014. Courtesy United Wind, Brooklyn, NY.

farm equipment," said Russell Tencer, founder and chief executive officer of United Wind.[97] Between 2014 and 2018, United Wind installed seventy 10-kilowatt Bergey turbines, six 50-kilowatt Endurance turbines, and three 100-kilowatt turbines from Northern Power Systems.[98] Another firm, North Findley, Ohio-based One Energy Enterprises, owns and erects larger wind turbines and sells power to commercial and industrial customers at twenty-year fixed rates through power purchase agreements. In December 2017, One Energy secured $80 million in financing from Prudential Capital Group to expand its offering of utility-scale, 1.5-megawatt wind turbines across Ohio.[99]

Small wind turbines in the 100-kilowatt range have remained a key component for electric-power generation in remote geographies where transmission lines from large public utilities still do not reach. In Alaska, isolated villages once dependent solely on diesel generators during the past three decades have supplemented this costly power source with these wind turbines. Unless a continuous wind source is available, however, most remote power generation applications increasingly rely on a combination of wind turbines and photovoltaic solar panels to produce year-round electricity. Off-grid power suppliers began exploring the combination of wind and solar, or "hybrid" systems, during the early 1980s. The Jacobs Wind Energy Company, then part of Eden Prairie, Minnesota-based Earth Energy Systems Inc., at that time designed a hybrid power plant for remote applica-

By 2018, Primus Wind Power estimated that 99 percent of its micro-wind turbines sold were combined with solar systems, like this railroad signal in southern California. Courtesy Primus Wind Power, Lakewood, CO.

tions which included a 10-kilowatt wind turbine, a 10-kilowatt engine generator, up to six photovoltaic solar panels of 4.8 kilowatts, an electrical control center, a battery, and a DC/AC inverter. The engine generator, battery, and electronics were kept inside a standard 20-foot-long ocean freight container, on top of which were mounted the solar panels. The container was positioned at the base of the wind turbine tower. It is unknown how many of these systems were deployed before Earth Energy Systems closed its wind energy business in 1986.[100]

The deployment of hybrid systems, which combine both wind and solar energy, as well as battery storage, increased in the 1990s; these systems are now common for today's remote electric-power applications.[101] By 2018, Primus Wind Power estimated that 99 percent of its micro-wind turbines sold were combined with solar systems. Ken Kotalik, the company's director of global sales and operations, explained that the two renewable technologies operate "seamlessly," with the wind turbines and solar panels charging batteries by day. During the nighttime, and if there is a breeze, the wind turbine will continue to charge the batteries.[102] On a larger scale, Juhl Energy, a pioneering community wind project developer based in Chanhassen, Minnesota, completed a first-of-its-kind utility-scale, hybrid power installation in 2017 for Otter Tail Power Company, a Red Falls, Minnesota, utility, that integrated two General Electric 2.3-megawatt wind turbines with a 1-megawatt solar array. The system included General Electric's Wind Integrated Solar Energy (WiSE) technology platform, which allowed electricity generated from both the wind turbines and solar array to pass through the same controller on its way to the utility substation. Earlier hybrid systems used separate electric controllers for wind and solar to feed the grid. In November 2018, Juhl Energy completed the installation of a 2-megawatt solar-wind hybrid system for Lake Region Electric Cooperative in Pelican Rapids, Minnesota, which includes a 2.7-megawatt General Electric wind turbine and half-a-megawatt solar array, which feeds into the grid through the WiSE controller. The company has plans for more of these hybrid projects.[103] "Distributed generation will play a major role in the implementation of renewable energy in the U.S.

electrical market in the years to come," Juhl Energy founder Dan Juhl said.[104]

Another promising area for small wind turbines was their inclusion in microgrids. The US Department of Energy's National Renewable Energy Laboratory has defined microgrids, which had been in development since the start of the 2000s, as a singularly controlled group of "interconnected" electric-power generation resources that can disconnect from the grid and operate independently.[105] Microgrids have been set up in recent years in urban, industrial, and rural areas. Electric power generation specialists see microgrids easing the operational strain on and costs for expanding already aging public utility infrastructures. Through renewable energy sources, microgrids allow for the generation and use of electricity locally onsite, a process also known as "islanding."[106] Meanwhile, the price of solar panels continued to fall ($2.70 per watt for direct current to residential systems and $1.83 per watt for direct current to commercial systems by the first quarter of 2018) largely due to improved manufacturing processes, continued government support for this form of renewable energy, and increased volumes of inexpensive Chinese imports, further facilitating the development of microgrids.[107] Similarly, battery storage technologies have improved in terms of their efficiency and are becoming increasingly cost effective, which ensures that microgrids can have electricity available for use around the clock.[108] "Our analysis shows that solar-plus-battery systems will reach grid parity—for growing numbers of customers in certain geographies, especially those with high retail electricity prices—well within the 30-year period by which utilities capitalize major power assets. Millions of customers, commercial earlier than residential, representing billions of dollars in utility revenues will find themselves in a position to cost effectively defect from the grid if they so choose," concluded the authors of the 2014 report, *The Economics of Grid Defection*.[109] Small wind turbine manufacturers are eager to play an increasing role in the development of microgrids, particularly since the performance of solar panels is diminished on cloudy days and nonfunctioning at night. During those times, small turbines could

Niagara Wind & Solar, Inc., installed this second 100-kilowatt Northern Power Systems turbine at Miller Sonshine Acres in Corfu, New York, in 2016. Courtesy Padma Kasthurirangan, Amherst, NY.

take over the electric-power generating function for the microgrid.[110]

The US Department of Energy's Office of Energy Efficiency and Renewable Energy foresees the biggest usage of distributed wind in rural America, with both large and small turbines connected on the customer side of the meter to power grids or at off-grid sites to generate electricity for local use. In a 2018 announcement establishing the Wind Innovations for Rural Economic Development (WIRED) program, the department noted that many rural locations throughout the country, with their quality wind resources and high public utility rates, were ideally suited for wind turbines.[111] Many rural communities and businesses receive their electric power from small, not-for-profit cooperatives. According to the National Rural Electric Cooperative Association, cooperatives deliver electricity to over 19 million customers in 2500 of the 3141 counties.[112] WIRED, which received up to $6.1 million in funding at the start of 2019, will foster the development of wind energy technology that can be combined with other distributed energy resources, such as photovoltaic solar arrays, and the simplification of distributed energy project development through standard applications and technical assistance.[113] The program's proponents acknowledged that rural cooperatives might be cautious at first about embracing distributed wind

and its associated technological and performance risks within their systems. Over time, however, they will realize the benefits of wind energy and endorse ways to include this power source among their customer service options. Through WIRED, the US Department of Energy in the long term foresees rural cooperatives applying a range of distributed wind applications within their electric transmission infrastructures. "Hybrid wind-solar-storage systems, microgrids, beneficial electrification, and commercial and industrial applications represent potential high-value opportunities for distributed wind for rural economic development," the department said at the conclusion of a December 2018 WIRED workshop.[114]

US manufacturers of wind turbines of less than 100 kilowatts, however, have struggled to increase their market share against cheaper solar photovoltaic technology. In 2018, the Department of Energy reported on the 50.5 megawatts of distributed wind that were installed in the United States during 2018; it found that only 1.5 megawatts were derived from small wind turbines, or 2661 units, which was a drop from the 1.7 megawatts of small wind turbines, or 3269 units, installed in 2017. The department also reported five small wind turbine manufacturers and three importers operating in the United States during 2018.[115] The longtime manufacturer of 100-kilowatt wind

Chava Wind has developed a three-bladed, Darrieus-troposkein rotor, capable of generating 25 kilowatts in a 25-mph wind. Courtesy Chava Wind, Homestead, FL.

turbines, Barre, Vermont-based Northern Power Systems, announced the closure of its operations in April 2019 after struggling in recent years to compete with larger wind turbine manufacturers.[116] Despite the challenges to small wind, the business has continued to attract new entrants, as well as the occasional return of industry veterans from the 1970s and 1980s. Their endeavors are encouraged both by ongoing research support from the federal government and by the many homeowners, businesses, farms, and communities that are increasingly looking to reduce their dependence on costly grid-tied electricity. These mostly entrepreneurial enterprises are enthusiastic about the prospects of introducing new lines of small wind turbines to the market. Some of these companies have reintro-

duced designs that were earlier abandoned for their lack of efficiency. In 2014, Hagen Ruff, founder and chief executive officer of Homestead, Florida-based Chava Wind began designing a vertical-axis wind turbine that included a three-bladed, Darrieus-troposkein rotor. The curved blades, which are made from a carbon composite, present a rotor diameter of 31.5 feet. The belt-driven induction generator is rated to produce 25 kilowatts of electric power in a 25-mph wind and is erected on a nearly 100-foot-tall monopole tower. In 2019, Chava Wind was completing prototype testing of its Windleaf 2500.[117] Ken Visser, an aerospace engineering professor at Clarkson University, co-founded a company in Potsdam, New York, to build a wind turbine that uses a "ducted" rotor design. Visser and his students believe they have solved an efficiency problem with past ducted rotor designs by placing the rotor closer to the aft of the duct where the wind exits. A test unit, which consisted of a four-bladed rotor with a 9.8-foot diameter inside a 12-foot-diameter duct, was erected in 2018 on the roof of Clarkson University's Technology Advancement Center. Ducted Turbines International's blades are composed of lightweight carbon fiber. The turbine is rated to produce 3.5 kilowatts of electricity in a 25-mph wind.[118] In East Dorset, Vermont, Jason Day, owner of Star Wind Turbines, has introduced a downwind turbine with six blades, reminiscent in appearance to those machines built by Pennsylvanian Terrance Mehrkam in the late 1970s and early 1980s, and in five model sizes ranging from the smallest with a rotor diameter of 28 feet and 5 kilowatts of output to the largest with a rotor diameter of 72 feet and 45 kilowatts of output.[119] Carter Wind Energy of Wichita Falls, Texas, announced plans to reenter the wind turbine market with its "two bladed, downwind, teetering, bearing less rotor hub design," which was successfully introduced in the mid-1970s, for between 100 kilowatt and 1-megawatt distributed wind applications.[120] In addition to test results, these wind turbine manufacturers equally tout efficient shipping in standard-size, 40-foot-long marine containers and towers that can be raised or lowered hydraulically, dispensing with the traditional and costly requirement of cranes to lift them in place.

However, before customers can benefit from federal and state incentive programs, most of these new wind turbines must successfully complete and obtain their third-party performance certifications. Citing the advice of James Jarvis, founder of data logger manufacturer APRS World, to prevent the recycling of past failed machine designs, Paul Gipe, the longtime wind industry historian and consultant, warned new entrants of the wind industry and their prospective customers in his 2016 book, *Wind Energy for the Rest of Us*: "Test early, test often. Test everything independently and together. Test more than one unit. Whether a purported manufacturer of a new wind turbine has met these requirements should be the first question. . . . If they haven't, they don't have a product. They have a dream. They may have a drawing—or a fancy website—but they don't have a product."[121]

For 2020 and beyond, small wind turbines will remain an important component in the US renewable energy portfolio, but their collective output will continue to be significantly dwarfed by large, multimegawatt turbines deployed by the country's wind farm operators. US wind farms are projected to include nearly 100 gigawatts of installed capacity by the start of this century's third decade.[122] Turbines used for wind farm applications also continue to increase in both size and output. Vestas in 2019 began supplying turbines to a US wind farm developer, which were the most powerful onshore units at 3.6 megawatts of output and the tallest with a combined tower and blade tip height of 590 feet installed in the United States. These turbines are two to three times larger in terms of output than those installed in US wind farms ten years earlier.[123] In March 2019, GE Renewable Energy erected its largest onshore wind turbine for testing at Wieringermeer, Netherlands. Called the Cypress, the turbine has a rated output of 5.3 megawatts.[124] Even larger wind turbines are in production for offshore wind farms, including GE Renewable Energy's Haliade-X turbine, which was first installed on land in the Netherlands in January 2019 for testing and is rated to generate 12 megawatts or enough electricity for 16,000 homes. The Haliade-X has a 720-foot-diameter rotor and, from the base of

A water-pumping Aermotor windmill is dwarfed by these 2.5-megawatt GE wind turbines with their 116-meter rotors at the Persimmon Creek 1 wind farm near Woodward, Oklahoma, on July 11, 2017. Courtesy Herman Drees, Thousand Oaks, CA.

the tower to the tip of its blades, stands 853 feet tall or just under 200 feet from matching the height of the radio tower on top of New York City's Chrysler Building.[125] Both GE Renewable Energy's largest onshore and offshore wind turbines are expected to be eclipsed in size and output by competing manufacturers in the next five to ten years. A report released in 2022 by the National Renewable Energy Laboratory forecasted that by 2035 onshore wind turbines will increase their electric power generation from an average 2.5 megawatts in 2018 to 5.5 megawatts, while offshore wind turbines will have an output of up to 17 megawatts compared to 4.4 megawatts in 2018. Subsequently, onshore wind turbine hub heights are expected to reach 425 feet by 2035 compared to 330 feet in 2018, with rotor diameters for these machines increasing to 575 feet from 385 feet over the same period. Offshore wind turbine hub heights will reach a hub height of 495 feet in 2035 from 295 feet in 2018, with their rotor diameters expanding to 820 feet compared to 435 feet in 2018, according to the laboratory's researchers.[126] The size of these giant turbines and their encroachment on rural communities across the United States, as well as their control and ownership by large power companies, continues to stir up public backlash and increased resistance to their installation.[127] On April 16, 2019, the *Milwaukee Journal Sentinel* published an article with the dramatic headline, "Wisconsin Wind Turbine Project Pits Brother against Brother, Clean Energy against Rural Vistas," about a proposed wind farm near Jefferson, Wisconsin.[128] This emerging discontent regarding wind farms could result in an increased acceptance of residential-scale wind turbines for localized power generation in the decades ahead.

Glossary of Terms

Courtesy of the US Department of Energy's Office of Energy Efficiency and Renewable Energy and the Distributed Wind Energy Association.

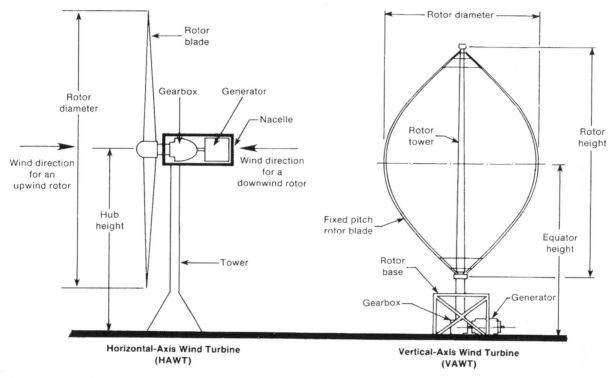

Wind turbine configurations. From *Five Year Research Plan 1985–1990[,] Wind Energy Technology: Generating Power from the Wind[,] January 1985*, Washington, DC (1985).

Airfoil—The shape of the blade cross section, which for most modern horizontal-axis wind turbines is designed to enhance the lift and improve turbine performance.

Alternator—An electric generator for producing alternating current. See also Generator.

Ampere-hour—A unit for the quantity of electricity obtained by integrating current flow in amperes over the time in hours for its flow; used as a measure of battery capacity.

Anemometer—A device to measure the wind speed.

Availability—A measure of the ability of a wind turbine to make power, regardless of environmental conditions. Generally defined as the time in a period when a turbine is able to make power, expressed as a percentage.

Average wind speed—The mean wind speed over a specified period of time.

Behind-the-meter/behind-the-fence generation—An electrical generating system connected on the user's side of a utility meter, primarily for energy usage on site instead of for sale to energy retailers. See also Net metering/net billing.

Betz limit—The maximum power coefficient (Cp) of a theoretically perfect wind turbine equal to 16/27 (59.3%) as proven by German physicist Albert Betz in 1919. This is the maximum amount of power that can be captured from the wind. In reality, this limit is never achieved because of aerodynamic drag, electrical losses, and mechanical inefficiencies.

Blades—The aerodynamic surface that catches the wind. See also *Airfoil; Rotor.*

Brake—Various systems used to stop the rotor from turning.

Certification—A process by which small wind turbines (100 kW and under) can be certified by an independent certification body to meet or exceed the performance and durability requirements of the American Wind Energy Association (AWEA) Standard.

Converter—See *Inverter.*

Corrosivity—A measure of oxidation and/or material degradation.

Cut-in wind speed—The wind speed at which a wind turbine begins to generate electricity.

Cut-out wind speed—The wind speed at which a wind turbine ceases to generate electricity.

Direct drive—A blade and generator configuration in which the blades are connected directly to the electrical generating device so that one revolution of the rotor equates to one revolution of the electrical generating device.

Distributed generation—Energy generation projects in which electrical energy is generated primarily for on-site consumption. The term is applied for wind, solar, and nonrenewable energy.

Downwind—On the opposite side from the direction from which the wind blows.

Drag—An aerodynamic force that acts in the direction of the airstream flowing over an airfoil.

Dual metering—Buying electricity from the utility and selling it to the utility with two different energy rates, typically retail (buying) and wholesale (selling).

Electric cost adjustment—An energy charge (dollars per kilowatt-hour) on a utility bill in addition to the standard rate in the tariff, which is associated with extra costs to purchase fuel, control emissions, construct transmission upgrades, and so on. These various costs may be itemized or rolled into one electric cost adjustment rate. Sometimes referred to as fuel cost adjustment.

Electric utility company—A company that engages in the generation, transmission, and distribution of electricity for sale, generally in a regulated market. Electric utilities may be investor owned, publicly owned, cooperatives, or nationalized entities.

Energy curve—A diagram showing the annual energy production at different average wind speeds.

Energy production—Energy is power exerted over time. Energy production is hence the energy produced in a specific period of time. Electrical energy is generally measured in kilowatt-hours (kWh).

Environmental conditions—Of or pertaining to the ambient state of the environment. See also *Corrosivity; Humidity; Temperature; Wind.*

Flagging—The deformation of local vegetation toward one direction, indicating the prevailing wind direction and relative strength.

Furling—A passive protection for the turbine in which the rotor folds up or around the tail vane.

Gearbox—A compact, enclosed unit of gears or the like for the purpose of transferring force between machines or mechanisms, often with changes of torque and speed. In wind turbines, gearboxes are used to increase the low rotational speed of the turbine rotor to a higher speed required by many electrical generators.

Generator—A machine that converts mechanical energy to electricity. The mechanical power for an electric generator is usually obtained from a rotating shaft. In a wind turbine, the mechanical power comes from the wind causing the blades on a rotor to rotate. See also *Alternator; Blades; Rotor; Stator.*

Geographic information system (GIS) software—GIS software is used for managing map-based information and data. It may also be used to visualize the relationships between terrain, wind data, land-use boundaries, obstacles, and potential wind turbine locations.

Governor—A device used to limit the revolutions per minute (RPM) of the rotor. Limiting RPM serves to reduce centrifugal forces acting on the wind turbine and rotor as well as to limit the electrical output of the generating device. Governors can be electrical, also known as "dynamic braking," or mechanical. Mechanical governors can be "passive"—using springs to pitch the blades out of their ideal orientation, or an offset rotor that pitches out of the wind—or "active"—electrically or hydraulically pitching blades out of their ideal orientation.

Grid—The utility distribution system. The network that connects electricity generators to electricity users.

Grid-connected—Small wind energy systems that are connected to the electricity distribution system. These often require a power-conditioning unit that makes the turbine output electrically compatible with the utility grid. See also *Inverter*.

Gross annual energy production—The amount of annual energy (usually in kilowatt-hours) estimated for a given wind turbine at a given location, before adjusting for losses. See also *Net annual energy production*.

Guyline—A guyline (or guy wire) supports guyed towers, which are the least expensive way to support a wind turbine. Guyed towers can consist of lattice sections, pipe, or tubing. Because the guy radius must be one-half to three-quarters of the tower height, guyed towers require more space to accommodate them than monopole or self-standing lattice towers.

Horizontal-axis wind turbine (HAWT)— A wind turbine with a rotor axis that lies in or close to a horizontal plane; often called a "propeller-style" wind turbine.

Hub—That component of a wind turbine to which the blades are affixed. See also *Blades*; *Rotor*.

Hub height—The distance from the foundation to which the tower is attached to the center of the hub of a HAWT.

Humidity—A measure of moisture content in the air.

Induction generator—An asynchronous AC motor designed for use as a generator. It generates electricity by being spun faster than the motor's standard "synchronous" speed. It must be connected to an already-powered circuit to function (i.e., the grid), but it does not require an inverter to produce grid-ready electricity.

Interconnection standards—The technical and procedural process by which a customer connects an electricity-generating device to the grid. Such standards include the technical and contractual terms that system owners and utilities must abide by. State public utility commissions typically establish standards for interconnection to the distribution grid, while the Federal Energy Regulatory Commission (FERC) establishes standards for interconnection to the transmission grid. Although many states have adopted interconnection standards, the standards of some states apply only to investor-owned utilities and not to municipal utilities or electric cooperatives.

Intermittency—Stopping or ceasing for a time; alternately ceasing and beginning again. Wind and solar resources are described as intermittent because they change without regard to people's needs or wants.

International Electrotechnical Commission (IEC)— The international wind-industry standards body.

Inverter—A device that converts direct current (DC) to alternating current (AC).

kW—Kilowatt, a measure of power for electrical current (1000 watts).

kWh—Kilowatt-hour, a measure of energy equal to the use of 1 kilowatt in 1 hour.

Lattice—A structure of crossed wooden or metal strips usually arranged to form a diagonal pattern of open spaces between the strips. Lattice towers, either guyed or free-standing, are often used to support small wind turbines.

Lift—An aerodynamic force that acts at right angles to the air stream flowing over an airfoil.

Micrositing—A resource assessment tool used to determine the exact position of one or more wind turbines on a parcel of land to optimize power production.

Microturbine—A very small wind turbine, usually under a 1000-watt rating, which is appropriate for small energy needs (e.g., for cabins, campers, sailboats, very small communication stations, or other small off-grid loads).

Monopole—A free-standing type of tower that is essentially a tube, often tapered.

MW—Megawatt, a measure of power (1 million watts).

Nacelle—The body of a propeller-type wind turbine, containing the gearbox, generator, blade hub, and other parts.

Nameplate capacity—The power capacity of a generating device that is typically affixed to the generating device. Nameplate capacity typically, but not necessarily, represents the maximum continuous power output of the generating device.

Net annual energy production—The amount of annual energy (usually in kilowatt-hours) produced or estimated for a given wind turbine at a given location, after subtracting losses from the gross annual energy production. A variety of losses may be estimated for obstacle wind shadows, turbulence, turbine wake effects, turbine availability, high-wind hysteresis effects, electrical efficiency, blade icing, blade soiling and surface degradation, idling parasitic losses, control errors, low temperature shutdown, utility system maintenance, and other issues specific to a given turbine installation.

Net metering/net billing—For electric customers who generate their own electricity, net metering allows for the flow of electricity both to and from the customer. When a customer's generation exceeds the customer's use, electricity from the customer flows back to the grid, offsetting electricity consumed by the customer at a different time during the same billing cycle. In effect, the customer uses excess generation to offset electricity that the customer otherwise would have to purchase at the utility's full retail rate. Net metering is required by law in most US states, but state policies vary widely. See also Behind-the-meter/behind-the-fence generation.

Noise—Generally defined as unwanted sound. Sound power is measured in decibels (dB). Building and planning authorities often regulate sound power levels from facilities. See also *Sound*.

O & M costs—Operation and maintenance costs.

Obstruction—A general term for any significant object that would disturb wind flow passing through a turbine rotor. Most common examples are homes, buildings, trees, silos, and fences. Topographical features, such as hills or cliffs that might also affect wind flow, are not called obstructions.

Off-grid—Energy-generating systems that are not interconnected directly into an electrical grid. Energy produced in these systems is often used for battery charging.

Overall height—The total height of a wind turbine from its base at grade to its uppermost extent. See also *Total height*.

Peak demand—The maximum electricity consumption level (in kilowatts) reached during the month or billing period, usually for a 15- or 30-minute duration. The definition of peak demand may vary by electric utility. This is a simplified definition of a complex topic.

Peak power—The maximum instantaneous power that can be produced by a power-generating system or consumed by a load. Peak power may be significantly higher than average power.

Permitting—The process of obtaining legal permission to build a project, potentially from a number of government agencies, but primarily from the local building department (i.e., the city, county, or state). During this process, a set of project plans is submitted for review to ensure that the project meets local requirements for safety, sound, aesthetics, setbacks, engineering, and completeness. The permitting agency typically inspects the project at various milestones for adherence to the plans and building safety standards.

Power coefficient—The ratio of the power extracted by a wind turbine to the power available in the wind stream.

Power curve—A chart showing a wind turbine's power output across a range of wind speeds.

Prevailing wind—The most common direction or directions from which the wind comes at a site. Prevailing wind usually refers to the amount of time the wind blows from that particular direction but may also refer to the direction in which the wind with the greatest power density comes.

PUC—Public Utility Commission, a state agency that regulates utilities. In some areas it is known as the Public Service Commission (PSC).

PURPA—Public Utility Regulatory Policies Act (1978), 16 U.S.C. § 2601.18 CFR §292; refers to small generator utility-connection rules.

Rated output capacity—The output power of a wind machine operating at the rated wind speed.

Rated wind speed—The lowest wind speed at which the rated output power of a wind turbine is produced.

Rotor—The rotating part of a wind turbine, including either the blades and blade assembly or the rotating portion of a generator.

Rotor diameter—The diameter of the circle swept by the rotor.

Rotor speed—The revolutions per minute of the wind turbine rotor.

Setback—In zoning parlance, the distance required between a structure and another structure, property line, utility easement, or other demarcation.

Site assessment—The act of evaluating a site to determine a favorable location for a wind turbine, which includes assessing the expected wind resource and potential turbine performance at that location.

Small wind turbine—A wind turbine that has a rating of up to 100 kilowatts and is typically installed near the point of electric usage, such as near homes, businesses, remote villages, and other kinds of buildings.

Sound—Pressure waves occurring at a frequency in the audible range of human hearing that are registered as sensory input by the ear. See also *Noise*.

Startup wind speed—The wind speed at which a wind turbine rotor will begin to spin. See also *Cut-in wind speed*.

Stator—The stationary part of a rotary machine or device, especially a generator or motor. It is most especially related to the collection of stationary parts in its magnetic circuits. The stator and rotor interact to generate electricity in a generator and to turn the driveshaft in a motor.

Swept area—The area swept by the turbine rotor, $A = \pi R2$, where R is the radius of the rotor. See also *Rotor diameter*.

Tariff—An official schedule of rates or charges from a utility, usually with different rate schedules by customer classification (e.g., residential, commercial, industrial, farm, or other designation) and/or a service or meter rating for the customer.

Temperature—A measure of thermal energy.

Tip-speed ratio—The speed at the tip of the rotor blade as it moves through the air divided by the wind velocity. This is typically a design characteristic of the turbine.

Total height—The height of the wind system from the top of the foundation to which the tower is attached to the tip of a blade extended upward. See also *Overall height*.

Tower—A structure designed to support a wind turbine at a substantial height above grade in a wind flow. Typical types include monopole, guyed lattice, and self-supporting lattice designs.

Turbulence—The changes in wind speed and direction, frequently caused by obstacles and/or terrain and vegetation.

Upwind—On the same side as the direction from which the wind is blowing—windward.

Upwind rotor—A horizontal-axis wind turbine whose propeller is located upwind of the tower; a wind turbine with an architecture such that the wind flow passes through the propeller prior to flowing past the tower.

Vertical-axis wind turbine (VAWT)— A wind turbine whose rotor spins about a vertical or near-vertical axis.

Wind—The movement of an air mass.

Wind farm—A group of wind turbines, often owned and maintained by one company. Also known as a wind power plant.

Wind rose—A visual means of representing the frequency with which the wind blows from different directions.

Wind shadow—A turbulent and/or low-wind-speed region downwind of (behind) an object such as a building, tower, or trees.

Wind shear—The difference in wind speed and direction over a relatively short distance in the atmosphere.

Wind turbine—A mechanical device that converts kinetic energy in the wind into electrical energy.

Yaw—The movement of the tower top turbine that allows the turbine to stay faced into the wind.

Zoning—Most land has been delegated to various zones by a region's local government and building department officials (at the city, county, or state level [occasionally]). The zones control types of land use, such as agricultural, residential, commercial, and industrial, and include subcategories. Each type of zoning carries its own specific permitting restrictions, such as building height and property line offsets (required separation distance).

Notes

Chapter 1

1. F. Stuart Chapin, "How We Waste Our Coal," *Popular Electricity Magazine* 6, no. 4 (August 1913): 366.

2. "The Wasted Wind," *Chambers's Journal* 1, no. 6 (December 1897–November 1898): 247.

3. Hans E. Wulff, *The Traditional Crafts of Persia: Their Development, Technology, and Influence on Eastern and Western Civilization* (Cambridge, MA: MIT Press, 1966), 284–85; Michael Harverson, *Persian Windmills*, Biblioteca Molinologica 10 (Sprang Capelle, The Netherlands: International Molinological Society, 1991); T. Lindsay Baker, "An Overview of Horizontal Windmills," *Windmillers' Gazette* 18, no. 1 (Winter 1999): 2.

4. Joseph Needham, *Science and Civilization in China*, Vol. 4, *Physics and Physical Technology*; pt. 2, *Mechanical Engineering* (London: Cambridge University Press, 1965), 556–60; Hong-Sen Yang, *Reconstruction Designs of Lost Ancient Chinese Machinery*, History of Mechanism and Machine Science 3 (Dordrecht, The Netherlands: Springer, 2007), 85; Baichun Zhang, "Ancient Chinese Windmills" (paper presented at the Third International Symposium on History of Machines and Mechanisms, National Cheng Keng University, Tainan, Taiwan, November 11–14, 2008), accessed February 19, 2014, http://link.springer.com/chapter/10.1007%2F978-1-4020-9485-9_15.

5. R. Pernoud, *Die Kreuzzuge in Augenseugenberichten* (Dusseldorf, Germany: Karl Rauch, 1961), 363; Edward J. Kealey, *Harvesting the Air: Windmill Pioneers in Twelfth-Century England* (Berkeley: University of California Press, 1987), 41–46.

6. Suzanne Beedell, *Windmills* (New York: Charles Scribner's Sons, 1979), 13–14.

7. J. Smeaton, "An Experimental Enquiry Concerning the Natural Powers of Water and Wind to Turn Mills, and Other Machines, Depending on a Circular Motion," *Philosophical Transactions of the Royal Society* 51 (1759): 138–74.

8. Stanley Freese, *Windmills and Millwrighting* (London: Cambridge University Press, 1957), 8; Wilfried Nijs and Frans Brouwers, "Wieksystemen," Windmill course, Levende Molens, Aartselaar, Belgium, 2011–2012.

9. William John MacQuorn Rankine, *A Manual of the Steam Engine and Other Prime Movers* (London: Charles Griffin and Company, 1866), xx–xxiii; Stephen Roper, *A Catechism of High Pressure or Non-Condensing Steam Engines* (Philadelphia: Edward Meeks, 1893), 206; Roy Gregory, *The Industrial Windmill in Britain* (West Sussex, UK: Phillimore & Company, 2005), 136.

10. Derek Ogden and Anne Burke, "The Windmill at Flowerdew Hundred," *Old Mill News* 6, no. 1 (January 1978): 4–5; Bob Norberg, "Fort Ross Shows Off New Russian-Built Windmill," *Press Democrat* (Santa Rosa, CA), October 20, 2012; Sandy Jones Birkland, "Replica Russian Windmill Gifted to Fort Ross: A Symbol of Peace Linking Two Nations," *Old Mill News* 40, no. 4 (Fall 2012): 15–17.

11. William P. Quinn, *The Saltwater of Historic Cape Cod: A Record of the Nineteenth Century Economic Boom in Barnstable County* (Orleans, MA: Parnassus, 1993), 15–17; Albert Cook Church, "The Padanaram Salt Works," *New England Magazine* 41, no. 2 (October 1909): 489–92; John R. Watkins, "A Common Crystal," *Strand Magazine* (London) 17, no. 98 (February 1899): 175; William Marks and Charles Coleman, *The History of Wind-Power on Martha's Vineyard* (n.p.: National Association of Wind-Power Resources, 1981), 11; Christopher Gillis, "Sea Breezes to Salt," *Windmillers' Gazette* 32, no. 3 (Summer 2013): 9–10; Pam Winterbauer, "A Short History of Hayward," accessed March 9, 2017, http://activerain.com/blogsview/475061/a-short-history-of-hayward.

12. T. Lindsay Baker, *A Field Guide to American Windmills* (Norman: University of Oklahoma Press, 1985), 18–19, 89–91; Everett Dick, "Water: A Frontier Problem," *Nebraska History* 49, no. 3 (Autumn 1968):

242; Christopher C. Gillis, *Still Turning*[:] *A History of Aermotor Windmills* (College Station: Texas A&M University Press, 2015), 14–16.

13. Harold Cook, "Water Stop," *Railroad Magazine* 66, no. 6 (October 1955): 14; T. Lindsay Baker, "Windmills and Railroad Water Systems," *Windmillers' Gazette* 28, no. 2 (Spring 2009): 2–5; Samuel Porcello, "Windmills and Railroads: A Successful Partnership," *Windmillers' Gazette* 35, no. 3 (Summer 2016): 6–7.

14. Gillis, *Still Turning*, 18.

15. Patent No. 11,629, Daniel Halladay, US Department of Commerce, Patent Office, Washington, DC; T. Lindsay Baker, "A Product History of the US Wind Engine and Pump Company," *Windmillers' Gazette* 2, no. 1 (Winter 1983): 5–6; T. Lindsay Baker, "Windmills with Variable-Pitch Blades," *Windmillers' Gazette* 8, no. 4 (Autumn 1989): 2–3; T. Lindsay Baker, "Large-Diameter Halladay Standard Windmills," *Windmillers' Gazette* 30, no. 2 (Spring 2011): 2–6.

16. F. G. Hobart, "History of the 'Eclipse' Windmill and Its Production," *Windmillers' Gazette* 1, no. 1 (Winter 1982): 6–9; T. Lindsay Baker, "Hard Times and Hard Feelings: The Untold Story of the Early 'Eclipse' Windmills," *Windmillers' Gazette* 19, no. 3 (Summer 2000): 2–3.

17. "Tustin's Improved Adjustable Windmill," *Mining and Scientific Press* (San Francisco) 16, no. 16 (April 18, 1868): 241; Roger S. Manning, "The Windmill in California," *Journal of the West* 14, no. 3 (July 1975): 33; T. Lindsay Baker, "Andrew J. Corcoran: Maker of America's Premiere Wooden-Wheel Windmills," *Windmillers' Gazette* 23, no. 2 (Spring 2004): 3–4.

18. R. L. Ardrey, *American Agricultural Implements: A Review of Invention and Development in the Agricultural Implement Industry of the United States* (Chicago: privately printed, 1894): 142–43; T. Lindsay Baker, "Iron Turbine," *Windmillers' Gazette* 7, no. 1 (Winter 1988): 2–4.

19. *Still Turning*, 31–35; *A Field Guide to American Windmills*, 33–36; T. Lindsay Baker, "Steel Windmills versus Wooden: A War of Words," *Windmillers' Gazette* 7, no. 1 (Winter 1988): 8–9; T. Lindsay Baker, "Pioneer Metal Windmills of the Plains: The Kirkwood Iron Wind Engines," *Windmillers' Gazette* 14, no. 1 (Winter 1995): 2–6.

20. F. L. Dole, "History and Development of the Windmill," *Export Implement Age* (Philadelphia) 14, no. 5 (August 1906): 28; *Still Turning*, 42–43; *A Field Guide to American Windmills*, 37–39.

21. T. Lindsay Baker, "Power Windmills," *Windmillers' Gazette* 6, no. 1 (Winter 1987): 3–7; T. Lindsay Baker, "Halladay & Wheeler's Patent Windmill," *Windmillers' Gazette* 6, no. 1 (Winter 1987): 8–10; T. Lindsay Baker, "'Every Farmer His Own Miller:' The Use of Power Windmills," *Windmillers' Gazette* 22, no. 4 (Autumn

2003): 2–7; T. Lindsay Baker, "Power Aermotor Windmills," *Windmillers' Gazette* 30, no. 3 (Summer 2011): 2–7.

22. A. M. Tanner, "The Electrical Utilization of Water and Wind Power First Proposed by Nollet in the Year 1840," *The Electrical World* (New York) 19, no. 15 (April 9, 1892): 242.

23. Glen E. Swanson, "The Great Eclipse of 1878 and Thomas Edison's Wind Turbine," *Windmillers' Gazette* 36, no. 3 (Summer 2017): 6–8; "Deposition of Thomas A. Edison," *Electric Railway Company of the United States v. The Jamaica and Brooklyn Road Company*, 497–99 (ED NY 1893), accessed April 3, 2017, http://edison. rutgers.edu/NamesSearch/glocpage.php?gloc=QE001&; "Thomas A. Edison Testimony," *Edison v. Siemens v. Field*, 74, 99 (US Patent Office, 1881), accessed April 3, 2017, http://edison.rutgers.edu/NamesSearch/glocpage. php?gloc=QD001&.

24. A. E. Dolbear, "Moses Gerrish Farmer," *Proceedings of the American Academy of Arts and Sciences* 29 (May 1893–May 1894): 415–16, accessed March 28, 2017, http://www.jstor.org/stable/20020568.

25. Glen E. Swanson, "History of Moses Farmer Windmill Patent Model" (research notes, Grand Valley State University, Allendale, MI, March 10, 2017); Glen E. Swanson, email to author, March 31, 2017.

26. Sir William Thomson, "Address to the Mathematical and Physical Science Section of the British Association," *The Chemical News and Journal of Physical Science* (London) 44, no. 1138 (September 16, 1881): 136.

27. Ibid.

28. [A. R. Wolff], "Are Windmills Expensive Prime Movers?" *The American Engineer* (Chicago) 2, no. 11 (November 1881): 209.

29. Alfred B. [sic] Wolf[f], "Windmills for Generating Electricity," *The Engineer* (London) 65 (February 3, 1888): 88; Alfred R. Wolff, *The Windmill as A Prime Mover*, 2nd ed. (New York: John Wiley & Sons, 1900), 133.

30. James Blyth, "On a New Form of Windmill," *Report of the Sixty-Second Meeting of the British Association for the Advancement of Science held at Edinburgh in August 1892* (London: John Murray, 1893): 869.

31. Ibid.

32. Ibid., 870.

33. James Carlill, "Chapter IX[;] Wind Power," ed. Harold Cox, *The Edinburgh Review or Critical Journal* (Norwich, UK) 228, no. 466 (October 1918): 345–46; "Professor Blyth on Wind Power for Electric Lighting," *The Electrician* (London) 21, no. 522 (May 18, 1888): 38; "The Use of Wind Power," *The Electrical Engineer* (London) 13, no. 2 (January 12, 1894): 36.

34. "Windmill for Producing Electric Light at Cape de la Heve," *Scientific American Supplement* 28, no. 709 (August 3, 1889): 11327.

35. "Generating Electricity with Windmills," *The Implement and Machinery Review* (London) 13, no. 150 (October 1, 1887): 9888.

36. "Windmill for Producing Electric Light at Cape de la Heve."

37. Ibid.

38. Etienne Rogier, "The First Wind Generator in France, 1887," *Windmillers' Gazette* 38, no. 2 (Spring 2019): 4–5.

39. "Wind Power for Electric Lighting," *The Farm Implement News* (Chicago) 12, no. 2 (February 1890): 30–31; T. B. Hannaford, "To the Editor[:] Mr. Hannaford's Lighthouse," *New Zealand Herald* 27, no. 8232 (April 17, 1890); "Hannaford's Light," *Timespanner*[:] *A Journey through Avondale, Auckland and New Zealand History*, April 19, 2009, accessed March 21, 2017, http://time spanner.blogspot.com/2009/04/hannafords-light.html.

40. "The Early Work of Mr. Brush," *Gas Power* (St. Joseph, MI) 4, no. 6 (November 1906): 20; Jeffrey La Favre, "Charles Brush and the Arc Light," 1998, accessed March 20, 2017, http://www.lafavre.us/brush/brushbio. htm; "Charles F. Brush: 1849–1929," The Brush Foundation, accessed March 20, 2017, http://fdnweb.org/brush/brush-history/; Peter Tocco, "The Night They Turned the Lights on in Wabash," *Indiana Magazine of History* 95, no. 4 (December 1999): 352–61; Henry Schroeder, "History of Electric Light," *Smithsonian Miscellaneous Collections* 76, no. 2 (August 15, 1923): xii; California Electric Light Company, "The Brush Electric Light" (Advertisement), *The Pacific Rural Press* (September 9, 1882): 196; Patent Nos. 219,208; 260,652 and 312,184, Charles F. Brush, US Patent Office, Washington, DC.

41. Charles F. Brush, "Laboratory Notebook, May 1880–January 1883," Box 5, Series 2: Laboratory Notes, 1880–1929, Folder 24, Charles F. Brush, Sr. papers in the Special Collections Research Center, Kelvin Smith Library, Case Western Reserve University, transcription provided to author by Glen E. Swanson, instructor, Grand Valley State University, Allendale, MI, and Charles F. Brush researcher, June 23, 2017.

42. Charles F. Brush, patent application, "A System and Apparatus for Charging Secondary Batteries," US Patent Office (January 5, 1886): 1–17, Box 9, Series 3, Folder 6, File 00905, Charles F. Brush, Sr. papers in the Special Collections Research Center, Kelvin Smith Library, Case Western Reserve University, photocopy provided to author by Glen E. Swanson, instructor, Grand Valley State University, Allendale, MI, and Charles F. Brush researcher, March 4, 2015. For reasons unknown, Brush never obtained a patent from his application.

43. "Mr. Brush's Windmill," *Cleveland Plain Dealer*, July 11, 1887. Glen E. Swanson, "The Brush Windmill Timeline[:] 1885–1956," 1–9, email to author, March 4, 2015.

44. "The Brush Windmill Timeline[:] 1885–1956," 1–2.

45. "Mr. Brush's Windmill Dynamo," *Scientific American* 63, no. 25 (December 20, 1890): 389; "The Electric Value of Wind Power," *Scientific American* 93, no. 21 (November 18, 1905): 394–95.

46. "Famous Wind Wheel Passes," *Cleveland Plain Dealer*, July 9, 1907; "The Brush Windmill Timeline."

47. "Mr. Brush's Windmill Dynamo."

48. Therese Quistgaard, "The Experimental Windmills at Askov 1891–1903," in *Wind Power—The Danish Way*, ed. Benny Christensen (Vejen, Denmark: The Poul la Cour Foundation, 2009), 12–13; "Den Sociale Vision," Poul la Cour Museum, accessed April 2, 2019, http://www.poullacour.dk/dansk/vision.htm.

49. "The Experimental Windmills at Askov 1891–1903," 13–15; "Electrical Notes. Central Station Driven by Electric Power," *Scientific American Supplement* 56, no. 1459 (December 19, 1903): 23386.

50. "The Experimental Windmills at Askov 1891–1903," 15–16.

51. "Engineering Notes. In a Lecture Recently Delivered at Copenhagen," *Scientific American Supplement* 46, no. 1185 (September 17, 1898): 18997.

52. Ibid.

53. Jytte Thorndahl, "Electricity and Wind Power for the Rural Areas 1903–1915," in *Wind Power—The Danish Way*, ed. Benny Christensen (Vejen, Denmark: The Poul la Cour Foundation, 2009), 18–23.

54. "Electric Notes. The Problem of Using Wind Power," *Scientific American Supplement* 57, no. 1469 (February 27, 1904): 23546.

55. Patent No. 452,546, James M. Mitchell, US Patent Office, Washington, DC.

56. Ibid.

57. Patent No. 457,657, James M. Mitchell, US Patent Office, Washington, DC.

58. Patent No. 452,546.

59. "The Wind Mill Industry," *The Farm Implement News* (Chicago) 15, no. 49 (December 6, 1894): 20.

60. Ibid.

61. W. O. A., "The Storage of Wind Power," *Scientific American* 49, no. 2 (July 14, 1883): 17; "Storing Wind Power," *Scientific American* 49, no. 15 (October 13, 1883): 229; Francis B. Crocker, *Electric Lighting*[:] *A Practical Exposition of the Art for the Use of Engineers, Students, and Others Interested in the Installation or Operation of Electrical Plants*[,] *Vol. 1*[,] *The Generating Plant* (New York: D. Van Nostrand Company, 1896): 223; "Electricity from Wind," *Muskogee Phoenix* (Muskogee, Indian Territory), April 14, 1898.

62. "The Lewis Electric Car Lighting System," *The Electrical Engineer* (New York) 17, no. 298 (January 17, 1894): 43–44; "Windmills for Electric Lighting," *The Electrical World* (New York) 23, no. 5 (February 3, 1894):

157–58; "The Lewis Train Electric Lighting System," *The Electrical World* (New York) 24, no. 4 (July 28, 1894): 85–86; Lt. I. N. Lewis, "Generating Electricity by Wind Mills," *The Farm Implement News* (Chicago) 16, no. 34 (August 22, 1895): 21.

63. "The Lewis System of Electric Lighting by Windmills," *The Electrical Engineer* (New York) 17, no. 300 (January 31, 1894): 86–87.

64. "Generating Electricity by Wind Mills," 21, 24; "The Windmill Electric Lighting Plant at Marblehead Neck, Mass.," *The Electrical Engineer* (New York) 18, no. 342 (November 21, 1894): 412–13; "Successful Electric Wind Mill Plant," *The Farm Implement News* (Chicago) 16, no. 11 (March 14, 1895): 17.

65. Charles J. Jager Company, *Illustrated Catalogue of Windmills, Tanks and Pumps as Applied to Water Supply Systems, also Windmills Adapted for Power* (Boston: Charles J. Jager Company, ca. 1895), 93–94. (In the Library and Archive of the Museum of the Great Plains, Lawton, OK).

66. "An Electric Light Plant Operated by a Wind Mill," *The Farm Implement News* (Chicago) 18, no. 3 (January 21, 1897): 46; Windmill Light & Power Company, Walpole, MA, "Do Your Own Electric Lighting" (advertisement), *Munsey's Magazine* (New York) 17, no. 4 (July 1897): n.p.; "The Windmill's Electric Generating Services," *The Implement and Machinery Review* (London) 24, no. 282 (October 1, 1898): 23270.

67. Franklin L. Pope, "An Isolated Electric-Lighting Plant Has Recently Been Installed in London Which Is Being Successfully Operated by an American Windmill," *Engineering Magazine* 3, no. 3 (June 1892): 399–400.

68. "Mr. John Wallis Titt," *The Implement and Machinery Review* (London) 21, no. 244 (August 1, 1895): 19408.

69. T. Lindsay Baker, "Windmills with Variable-Pitch Blades," *Windmillers' Gazette* 8, no. 4 (Autumn 1989): 6–7.

70. "Wind and Electricity: Extensive Improvements at Boyle Hall, West Ardsley," *The Implement and Machinery Review* (London) 24, no. 287 (March 2, 1899): 23774; "Mr. J. Wallis Titt's New Wind Engine at Boyle Hall: A Remarkably Economical and Automatic Motor," *The Implement and Machinery Review* (London) 24, no. 288 (April 1, 1899): 23879–80; "Electric Lighting by Wind Power," *The Farm Implement News* (Chicago) 20, no. 17 (April 27, 1899): 16.

Chapter 2

1. "Electric Signs Add Much to the Brilliance and Attractiveness of the Streets of Cities and Villages," *Popular Electricity* 1, no. 10 (February 1909): 619; "The Lighting of Show Windows," *Popular Electricity* 1, no. 12 (April 1909): 735–38; "All Day Service in Small Cities," *Popular Electricity* 2, no. 1 (May 1909): 23; "Brilliant Street Lighting in Warren [, Ohio]," *Popular Electricity* 4, no. 5 (September 1911): 432–33.

2. "Electricity Supply for Small Communities," *Popular Electricity Magazine* 6, no. 3 (July 1913): 258.

3. "A City Electrical," *Popular Electricity* 1, no. 2 (June 1908): 77; "Lights Enough to Encircle the Globe," *Popular Electricity* 1, no. 2 (June 1908): 91.

4. Edwin J. Houston, *Electricity in Every-Day Life*, vol. 1 (New York: P. F. Collier & Son, 1905), 1.

5. Elizabeth H. Callahan, "Electricity in the Household[:] More Comfort in the Home," *Popular Electricity* 1, no. 2 (June 1908): 110–12; "A Motor for the Sewing Machine," *Popular Electricity* 1, no. 2 (June 1908): 115; "An Electric Laundry Washer," *Popular Electricity* 1, no. 2 (June 1908): 115; "The Electric Motor in the Home," *Popular Electricity* 1, no. 4 (August 1908): 250–51; "Electricity in the Household[:] Equipment of an Electric Kitchen," *Popular Electricity* 1, no. 5 (September 1908): 310–14; "Electricity in the Household[:] The Latest Electric Household Conveniences," *Popular Electricity* 1, no. 6 (October 1908): 382–84; "Electricity in the Household[:] Sanitary Refrigeration for the Home," *Popular Electricity* 1, no. 7 (November 1908): 446–47; "Electricity in the Household[:] The Wonders of 'Electric Shop,'" *Popular Electricity* 1, no. 12 (April 1909): 780–85; "Electric Light Companies and the People," *Popular Electricity* 2, no. 9 (January 1910): 594; "Electric Light the Safest in the Home," *Popular Electricity* 1, no. 10 (February 1909): 647; "Electricity, the Busiest Worker," *Popular Electricity* 2, no. 7 (November 1909): 462–64; "An Electrical Fan the Year Around," *Popular Electricity* 3, no. 1 (May 1910): 53; "Electricity in the Household[:] Speaking of Wash Day," *Popular Electricity* 3, no. 2 (June 1910): 144–45; "A Little Motor for the Home," *Popular Electricity* 3, no. 7 (November 1910): 639–40.

6. G. A. Fleming, "The New Servant Girl," *Popular Electricity* 1, no. 2 (June 1908): 114; Ruth Schwartz Cowan, *More Work for Mother* (New York: Basic Books, 1983), 91–94.

7. Varney V. Shumaker, "The House without a Chimney," *Popular Electricity* 3, no. 10 (February 1911): 926–28.

8. William Keily, "Electricity in the Household[:] Is Electricity Light Too Dear for Modest Purses?" *Popular Electricity* 2, no. 1 (May 1909): 42–45.

9. Edmund Searles, "What Electricity Offers as a Life Work," *Popular Electricity* 3, no. 10 (February 1911): n.p.; "Functions and Methods of Trade Schools," *Popular Electricity* 4, no. 8 (December 1911): 699–704.

10. "Hazardous Line Construction," *Popular Electricity* 2, no. 12 (April 1910): 782; Waldon Fawcet[t], "The Pathfinders of the Wires," *Popular Electricity* 5, no. 5 (September 1912): 428–31; "Linking the Links of the Long Distance 'Phone," *Popular Electricity Magazine* 5,

no. 7 (November 1912): 670–72; R. D. Coombs, *Pole And Tower Lines For Electric Power Transmission* (New York: McGraw-Hill Book Company, 1916): 76–77.

11. Waldon Fawcett, "Electrical Invention and a Larger Patent Office," *Popular Electricity* 5, no. 7 (November 1912): 630–32.

12. "Electric Farm Lighting," *Popular Electricity* 5, no. 5 (September 1912): 431.

13. Guy E. Tripp, *Electric Development as an Aid to Agriculture* (New York: Knickerbocker Press, 1926), 39;

14. P. J. O'Gara, "How a Farmer Built His Own Electric Power Plant," *Popular Electricity* 1, no. 7 (November 1908): 432–36; Warren H. Miller, "Fifty Kilowatts of Water Power," *Popular Electricity* 3, no. 7 (November 1910): 618–21; George F. Worts, "Running a Farm by the Power of a Brook," *Popular Electricity* 5, no. 10 (February 1913): 1058–59.

15. "The Story of an Electric Farm," *Popular Electricity* 4, no. 4 (August 1911): 289–97.

16. Frederick Irving Anderson, *Electricity for the Farm* (New York: MacMillan Company, 1915), 218–20.

17. National Electric Light Association, *National Electric Light Association Thirty-Sixth Convention[,] Commercial Sessions[,] Papers, Reports and Discussions[,] Chicago, Ill.[,] June 2–6, 1913* (New York: James Kempster Printing Company, 1913): 161.

18. Andrey A. Potter, *Farm Motors* (New York: McGraw-Hill Book Company, 1917), 265.

19. "Electricity on the Farm," *Popular Electricity* 1, no. 4 (August 1908): 216–20; "The Farmer's Light and Power," *Popular Electricity* 3, no. 5 (September 1910): 373–75; "Electricity—The Farm Hand," *Popular Electricity* 3, no. 12 (April 1911): 1056–61; "Electricity Enhances Farm Values," *Popular Electricity* 4, no. 3 (July 1911): 230–31; "A Pennsylvania Farm Electrified," *Popular Electricity* 5, no. 2 (June 1912): 138–39.

20. "An Inexpensive Farm Electric Plant," *Popular Electricity* 4, no. 8 (December 1911): 709.

21. T. A. Boyd, *Professional Amateur[:] The Biography of Charles Franklin Kettering* (New York: E. P. Dutton & Company, 1957), 92–94; Delco-Light Company, *The Delco-Light Story* (Dayton, OH: Delco-Light Company, 1922), 6–9 (in author's files); Norman H. Schneider, *Low Voltage Electric Lighting with the Storage Battery* (New York: Spon & Chamberlain, [ca. 1920]), 64–66; The Domestic Engineering Company, Dayton, OH, "Delco-Light[:] Electricity for Every Farm" (advertisement), *The Saturday Evening Post* (Philadelphia) 189, no. 11 (September 9, 1916): 60–61.

22. Westinghouse Electric & Manufacturing Company, East Pittsburgh, PA, "Everything Electrical for the Farm" (advertisement), *Farm Light and Power Year Book* (New York, 1922): 7–11; Allis-Chalmers Manufacturing Company, Milwaukee, WI, "A Complete Power Plant furnishing Light and Power to Small Communities or Large Farms" (advertisement), *Farm Light and Power Year Book* (New York, 1922): 13; Fairbanks, Morse & Company, Chicago, "What Our Selling Franchise Means to You" (advertisement), *Farm Light and Power Year Book* (New York, 1922): 15; Phelps Light & Power Company, Rock Island, IL, "Phelps Pioneers The Way" (advertisement), *Farm Light and Power Year Book* (New York, 1922): 27; Western Electric Company, New York, "Western Electric Power & Light" (advertisement), *Farm Light and Power Year Book* (New York, 1922): 35.

23. R. C. Cosgrove, "Developing the Farm Electric Industry," *Farm Light and Power* (New York) 3, no. 10 (June 15, 1923): 10.

24. Kohler Company, *The Principle and the Proof* (Kohler, WI: Kohler Company, 1924), 29 (in author's files); Kohler Company, *And It's the Thriftiest Thing on the Place* (Kohler, WI: Kohler Company, 1925), brochure (in author's files).

25. "Are You Ready for State and County Fairs?" *Farm Light and Power* (New York) 2, no. 12 (August 15, 1922): 12; Mike Walin, "Value of Handling a Diversified Line," *Farm Light and Power* (New York) 4, no. 6 (February 15, 1924): 193–94, 213–15.

26. Jerry Dare, "Isolated Lighting Plants," *Everyday Engineering Magazine* (New York) 9, no. 2 (May 1920): 105.

27. "Why the Farmer Buys a Plant," *Farm Light and Power* (New York) 3, no. 8 (April 14, 1923): 24; Enoch Lundquist, "Selling Electrical Appliances to the Farmer," *Farm Light and Power* (New York) 3, no. 11 (July 15, 1923): 19, 29, 31.

28. "How Are We Going to Sell the Farm Women?" *Farm Light and Power* (New York) 4, no. 2 (October 23, 1923): 51–52.

29. "Good Openings for Wind Engines," *The Implement and Machinery Review* (London) 32, no. 378 (October 2, 1906): 662–63; "An Interesting Wind Turbine Electric Plant Has Recently Been Installed at Buckenhill, Bromyard, by Messrs. J. G. Childs & Company, Ltd., of Willesden-green, London," *The Implement and Machinery Review* (London) 34, no. 416 (December 1, 1909): 982; J. G. Childs & Company, Ltd., London, "Electricity from the Wind" (advertisement), *The Implement and Machinery Review* (London), 35, no. 410 (June 1, 1909): 248; "With a Little Coaxing, the Wind Can Be and Is Made Very Serviceable as a Power Generator," *The Implement and Machinery Review* (London) 35, no. 411 (July 2, 1909): 375; "Wind Power Electric Plants," *Popular Electricity* 2, no. 7 (November 1909): 451.

30. "German Wind Power Plant," *Popular Electricity* 6, no. 7 (November 1913): 799.

31. Jytte Thorndahl, "Electricity and Wind Power for the Rural Areas 1903–1915," in *Wind Power—The Danish Way*, ed. Benny Christensen (Vejen, Denmark: The Poul la Cour Foundation, 2009), 18.

32. "The Future of the Windmill," *Scientific American* 108, no. 4 (April 5, 1913): 309; "Making Electricity by Wind Power," *Popular Electricity* 2, no. 9 (January 1910): 583–84; F. E. Powell, *Windmills and Wind Motors*[:] *How to Build and Run Them* (New York: Spon & Chamberlain, 1918), 65–78.

33. "Domestic Electric Light Plant Driven by a Windmill," *Scientific American* 96, no. 22 (June 1, 1907): 448.

34. Putnam A. Bates, "Farm Electric Lighting by Wind Power," *Scientific American* 107, no. 13 (September 28, 1912): 262; "Windmill Farm Lighting Plant," *Popular Electricity Magazine* 5, no. 7 (November 1912): 655–56; J. F. Forrest, "A Practical Windmill Electric Plant," *Popular Electricity Magazine* 6, no. 1 (May 1913): 58–59.

35. P. C. Day, "The Winds of the United States and Their Economic Uses," *Yearbook of the United States Department of Agriculture* (Washington DC: US Department of Agriculture, 1911), 349–50.

36. "Current from the Wind," *Popular Electricity* 4, no. 5 (September 1911): 429; Patent No. 996,334, Caryl D. Haskins, US Patent Office, Washington, DC.

37. "Wind Power Plants," *Farm Light and Power Year Book*[:] *Dealers' Catalog and Service* (New York: Farm Light and Power Publishing Company, 1922), 41.

38. "Ordinary Windmill Will Not Do," *Farm Light and Power Year Book*[:] *Dealers' Catalog and Service* (New York: Farm Light and Power Publishing Company, 1922), 42.

39. Perkins Corporation, Mishawaka, IN, *Electricity from the Wind* (Mishawaka, IN: Perkins Corporation, [ca. 1922]), 1–12, brochure (in the *Windmillers' Gazette* research files); Perkins Corporation, Mishawaka, IN, *Aerolectric Light and Power from the Wind*[,] *Electricity*[,] *a City Luxury for Every Farm*, Form 1001, (Mishawaka, IN: Perkins Corporation, [ca. 1922]), 1–6, folder (in the *Windmillers' Gazette* research files); Perkins Corporation Mishawaka, IN, typewritten letter to W. E. Holt, Minooka, IL (Electricity from the Wind questionnaire and envelope), April 11, 1922 (in the *Windmillers' Gazette* research files); Perkins Corporation Mishawaka, IN, *Electricity from the Wind* (Mishawaka, IN: Perkins Corporation, 1923), folder with front printed as letterhead stationary (in the *Windmillers' Gazette* research files).

40. Perkins Corporation, Mishawaka, IN, typewritten letter to W. E. Holt, Minooka, IL, April 27, 1922 (in the *Windmillers' Gazette* research files).

41. *Electricity from the Wind*, ca. 1922, 5.

42. "American Farm Light and Power Plants Purchased in India," *Farm Light and Power* (New York) 3, no. 7 (March 15, 1923): 18.

43. "Henry Ford Buys Aerolectric Plant," *Farm Light and Power* (New York) 4, no. 2 (October 15, 1923): 58.

44. Christopher C. Gillis, *Still Turning*[:] *A History*

Aermotor Windmills (College Station: Texas A&M University Press, 2015), 104–5.

45. Aermotor Company, Chicago, *Electric Aermotor*, Publication E. A. 1. (Chicago: Aermotor Company, [ca. 1920]), folder (in the *Windmillers' Gazette* research files).

46. "Model of Windmill Electric Plant," *Popular Electricity Magazine* 6, no. 1 (May 1913): 57.

47. Charles B. Hayward, "A Wind Electric Plant with Novel Features," *Farm Light and Power* (New York) 2, no. 12 (August 15, 1922): 20–21, 43; K. J. T. Ekblaw, "Wind-Driven Power Plants," *The Farm Journal* (Philadelphia) 46, no. 8 (August 1922): 6–7; "Wind Electric Company[,] Minneapolis, Minn.," *Farm Light and Power*[:] *Dealers' Service Book* (New York: Farm Light and Power Publishing Company [ca. 1923]), 54; "Design of 'Aerolite' Plant," *Farm Light and Power*[:] *Dealers' Service Book* (New York: Farm Light and Power Publishing Company, [ca. 1923]), 135.

48. A. H. Steenrod, "Fritchle Wind Electric System for Windmills," *Farm Light and Power* (New York) 2, no. 10 (June 15, 1922): 29–30, 38; "Battery, Switchboard and Resistance of a Wind-Driven Plant," *Farm Light and Power* (New York) 2, no. 12 (August 15, 1922): 32; "Details of Fritchle Attachable Unit," *Farm Light and Power*[:] *Dealers' Service Book* (New York: Farm Light and Power Publishing Company, [ca. 1923]), 133; T. Lindsay Baker, "Wind Electric News: The Papers of Oliver P. Fritchle," *Windmillers' Gazette* 10, no. 4 (Autumn 1991): 9–10.

49. "Fritchle Wind Electric System for Windmills," 29.

50. "Woodmanse Mfg. Co.[,] Freeport, Ill.," *Farm Light and Power*[:] *Dealers' Service Book* (New York: Farm Light and Power Publishing Company, [ca. 1923]), 54; Woodmanse Manufacturing Company, Freeport, IL, *Light, Power & Water from the Wind* (Freeport, IL: Woodmanse Manufacturing Company, [ca. 1922]), 1–16, brochure (in the *Windmillers' Gazette* research files); Woodmanse Manufacturing Company, Freeport, IL, *The Fritchle Wind-Electric* (Freeport, IL: Woodmanse Manufacturing Company, [ca. 1922]), folder (in the *Windmillers' Gazette* research files); Woodmanse Manufacturing Company, Freeport, IL, *Price List No. 1 Covering The Fritchle Wind-Electric System* (Freeport, IL: Woodmanse Manufacturing Company, 1922), folder (in the *Windmillers' Gazette* research files).

51. "Ordinary Windmill Will Not Do"; Christopher Gillis, *Windpower* (Atglen, PA: Schiffer Publishing Ltd., 2008), 27; "Aerodynamic Wind Mills," *Scientific American* 140, no. 6 (June 1929): 525.

52. Greville Bathe, *Horizontal Windmills, Draft Mills and Similar Air-Flow Engines* (Philadelphia, privately printed, 1948), 19–23; T. Lindsay Baker, *A Field Guide to American Windmills* (Norman: University of Oklahoma Press, 1985), 13–15, 29–32; T. Lindsay Baker, "D. H. Bausman's Pennsylvania-Made Windmills," *Windmillers' Gazette* 6, no. 3 (Summer 1987): 6–7; T. Lindsay Baker,

"An Overview of Horizontal Windmills," *Windmillers' Gazette* 18, no. 1 (Winter 1999): 2–5; Christopher Gillis, "D. H. Bausman: A Pennsylvania Windmill Maker," *Windmillers' Gazette* 36, no. 1 (Winter 2017): 2–3.

53. "The Eustis Turbine Windmill," *Popular Electricity Magazine* 5, no. 10 (February 1913): 1033.

54. "An Overview of Horizontal Windmills," 6.

55. "Windmill Supplies Light," *Popular Science* 103, no. 6 (December 1923): 69.

56. Manning Engineering and Sales Company, Elkhart, IN, *Wind Driven Electric Light, Power and Irrigation Plants* (Elkhart, IN: Manning Engineering & Sales Company, [ca. 1923]), folder (in the Kregel Windmill Company Papers, Nebraska State Historical Society, Lincoln, Nebraska).

57. Twiford Corporation, Chicago Heights, IL, *Twiford Wind Motors with Electrical & Irrigation Equipment Especially Designed for Farm Lighting, Water Systems, and the Operation of All Household Electrical Appliances. Also for Power Driven Machinery Such as Feed Grinders, Corn Shellers, Circular Saws, Drilling Machines, Iron and Wood Lathers, Etc.* (Chicago Heights, IL: Twiford Corporation, [ca. 1924]), 1–31, brochure (in the *Windmillers' Gazette* research files).

58. Sigurd J. Savonius, *The Wind-Rotor in Theory and Practice* (Helsingfors, Finland: Savonius & Company, [ca. 1925]), 11–12, 17–19.

59. Ibid., 39.

60. Patent Nos. 1,697,574 and 1,766,765, S. J. Savonius, US Patent Office, Washington, DC.; Patent No. 11121, Sigurd J. Savonius, National Board of Patents and Registration of Finland. Savonius sought patents for his wing rotor design in other countries between 1925 and 1928, including France (1925), Greece (1925), Austria (1926), Denmark (1927); and Canada (1928), (in the files of Etienne Rogier, Toulouse, France).

61. "An Overview of Horizontal Windmills," 7; Patent No. 1,835,018, G. J. M. Darrieus, US Patent Office, Washington, DC; Etienne Rogier, "Georges Darrieus and the 'Egg-Beater' Vertical-Axis Wind Turbines," *Windmillers' Gazette* 23, no. 1 (Winter 2004): 12–14.

62. R. G. Kloeffler and E. L. Sitz, "Electric Energy From Winds," *Kansas State College Bulletin* (Manhattan, KS) 30, no. 9 (September 1, 1946): 15.

63. Patent No. 1,786,057, Elisha N. Fales, US Patent Office, Washington, DC.

64. Ibid.

65. Ibid.

66. Mike Werst, owner of Wincharger.com, telephone interview by author, May 8, 2017.

67. Perkins Corporation, South Bend, IN, *The New Perkins Aeroelectric* (South Bend, IN: Perkins Corporation, [ca. 1924]), folder, sent with letter dated December 9, 1924 (in the *Windmillers' Gazette* research files); "A New Windmill Electricity Generating Plant," *The Imple-ment and Machinery Review* (London) 51, no. 601 (May 1, 1925): 70.

68. Aerodyne Corporation, Minneapolis, *Electricity from Wind!* (Minneapolis: Aerodyne Corporation, [ca. 1930]), folder (in author's files); Karen Klinkenberg, "Radisson Farm & Its Influences[:] Historical Notes on the Farm, the Hotel and the Road," *Blaine Historical Society* (Blaine, MN) (December 2010), 1–2, accessed May 13, 2017, http://www.blainehistory.org/Radisson_Farm/Radisson_Farm_Article.pdf.

69. A. R. Grundel, sales manager, Herbert E. Bucklen Corporation, Elkhart, IN, letter to James F. Smith, M. D., Hayes Store, VA, November 5, 1927, in untitled notebook consisting of mimeographed descriptive data, photostatic copies of photographs, and order forms (in the Windmill Manufacturers' Trade Literature Collection, Panhandle-Plains Historical Museum, Canyon, TX); Mike Werst, "Herbert E. Bucklen Corporation and Its Propeller-Driven Wind Plants," *Windmillers' Gazette* 37, no. 1 (Winter 2018): 2–3.

70. Patent No. 1,792,212, Herbert E. Bucklen and Harlie O. Putt, US Patent Office, Washington, DC; Herbert E. Bucklen Corporation, Elkhart, IN, *Light and Power at the Cost of Pumping Water by Wind-Mill* (Elkhart, IN: Herbert E. Bucklen Corporation, [ca. 1927]), brochure (in author's files); "Herbert E. Bucklen Corporation and Its Propeller-Driven Wind Plants," 3.

71. "Electricity from the Wind," *Farm Mechanics* 17, no. 4 (August 1927): 25–27; "Herbert E. Bucklen Corporation and Its Propeller-Driven Wind Plants," 3–4.

72. Herbert E. Bucklen Corporation, Elkhart, IN, *HEBCO All Service Wind Electrics Price List 101-F[,] Aug. 5, 1928* (Elkhart, IN: Herbert E. Bucklen Corporation, 1928), folder (in the *Windmillers' Gazette* research files); "Herbert E. Bucklen Corporation and Its Propeller-Driven Wind Plants," 4.

73. Authur Van Vlissingen, "Riches from the Wind[:] How Two Boy Mechanics Started a Brand-New Industry," *Popular Science Monthly* 132, no. 5 (May 1938): 54–55, 116–17; Wincharger Corporation, Sioux City, IA, *There's Power in the Air* (Sioux City, IA: Wincharger Corporation, [ca. 1937]), 3, folder (in author's files); Mick Sagrillo, "How It All Began," *Home Power*, no. 27 (February-March 1992): 16. For background on Fridtjof Nansen's early use of a ship-mounted wind-electric generator, see Etienne Rogier and T. Lindsay Baker, "The Polar Wind Machines: The Wind Generator on the *Fram* during Nansen's 1893–96 Arctic Expedition," *Windmillers' Gazette* 17, no. 2 (Spring 1998): 7–11.

74. Robert W. Righter, "Reaping the Wind[:] The Jacobs Brothers, Montana's Pioneer 'Windsmiths,'" *Montana*[:] *The Magazine of Western History* 46, no. 4 (Winter 1996): 40.

75. Letter from Mr. and Mrs. A. J. Jacobs, Dawson County, MT, "The Electricity Equipped Farm Home,"

The Dakota Farmer (Aberdeen, SD) 43, no. 5 (March 1, 1923): 1.

76. Righter, "Reaping the Wind[:] The Jacobs Brothers, Montana's Pioneer 'Windsmiths,'" 40.

77. Ibid., 40–41.

78. Marcellus L. Jacobs, Jacobs Wind Electric Company, Fort Myers, FL, "Experience with Jacobs Wind-Driven Electric Generating Plant, 1931–1957" (paper presented at the Wind Energy Conversion Systems Workshop, Washington, DC, June 11–13, 1973), 155.

79. Ibid.

80. Jacobs Wind Electric Company, Minneapolis, *The Jacobs Super Automatic Wind Electric Plant* (Minneapolis: Jacobs Wind Electric Company, [ca. 1940]), handbill (in author's files); Jacobs Wind Electric Company, Minneapolis, *Electricity from the Air*[:] *The Super Automatic Wind Electric Plant with Fly Ball Governor Control* (Minneapolis: Jacobs Wind Electric Company, [ca. 1940]), brochure (in author's files).

81. Paul R. Jacobs, Corcoran, MN, email to author, May 9, 2017.

82. Righter, "Reaping the Wind[:] The Jacobs Brothers, Montana's Pioneer 'Windsmiths,'" 41.

83. Ibid, 42.

84. "Experience with Jacobs Wind-Driven Electric Generating Plant, 1931–1957," 156.

85. Ibid.

86. Richard Evelyn Byrd, *Discovery*[:] *The Story of the Second Byrd Antarctic Expedition* (New York: G. P. Putnam's Sons, 1935): 12, 195–97; Melville Bell Grosvenor, "Admiral of the Ends of the Earth," *The National Geographic Magazine* (Washington, DC) 112, no. 1 (July 1957): 36–48; Jacobs Wind Electric Company, Minneapolis, *With Byrd in Little America* (Minneapolis: Jacobs Wind Electric Company, [ca. 1940]), handbill (in author's files). The Jacobs machine was reported to remain in operable condition and all three blades intact through subsequent expeditions until 1955 when it was removed from the tower.

87. M. L. Jacobs, secretary-treasurer, Jacobs Wind Electric Company, Minneapolis, letter to Elmer Kreidman, Fort Atkinson, WI, August 24, 1938, typewritten letter (in author's files); Jacobs Wind Electric Company, Minneapolis, *The Jacobs Twin Motor Electric* (Minneapolis: Jacobs Wind Electric Company, 1938), brochure (in author's files); Jacobs Wind Electric Company, Minneapolis, *Twin Motor Electric*[,] *Automatic Proven Speed Control, Charging Control, Voltage Control*[,] *Price List*[,] *Effective February 1, 1938* (Minneapolis: Jacobs Wind Electric Company, 1938), handbill (in author's files).

88. "Experience with Jacobs Wind-Driven Electric Generating Plant, 1931–1957," 156–57.

89. Paul R. Jacobs, Corcoran, MN, email to author, May 9, 2017.

90. Perkins Corporation, South Bend, IN, *The Guiding Star of the Air Mail Service* (South Bend, IN: Perkins Corporation, [ca. 1926]), folder (in Windmillers' Gazette research files).

91. *Light and Power at the Cost of Pumping Water by Wind-Mill*; Herbert E. Bucklen Corporation, Elkhart, IN, *Modern Conveniences Are Yours*[,] *Just Snap on the Switch* (Elkhart, IN: Herbert E. Bucklen Corporation, [ca. 1932]), brochure (in author's files); Herbert E. Bucklen Corporation, Elkhart, IN, *What the United States Government and Others Think and Say about HEBCO Wind-Electrics*, Publication Form 129-A (Elkhart, IN: Herbert E. Bucklen Corporation, [ca. 1927]), circular (in author's files).

92. "Wind-Driven Electric Plant at Arecibo Light Station, Porto Rico," *Lighthouse Service Bulletin* (Washington, DC) 3, no. 54 (June 1, 1928): 247.

93. "Wind-driven Electric Generator at Kalae Light Station, Hawaii," *Lighthouse Service Bulletin* (Washington, DC) 4, no. 1 (January 2, 1930): 6; "Centennial Improvements at Point Lookout, Md.," *Lighthouse Service Bulletin* (Washington, DC) 4 no. 4 (April 1, 1930): 17.

94. C. A. Carlisle Jr., "How to Sell Wind Driven Light and Power Plants," *Farm Light and Power* (New York) 2, no. 8 (April 15, 1922): 28, 40; Coyne Electrical School, Chicago, *Electricity*[,] *Gateway To Opportunity*[,] *Coyne Electrical School*[,] *Established 1899*[,] *Chicago*[,] *World's Greatest Electrical Center* (Chicago: Coyne Electrical School, [ca. 1940]), 32–33, brochure (in author's files); Coyne Electrical School, Chicago, *I Will Include An . . . Extra Radio Course* (Chicago: Coyne Electrical School, [ca. 1940]), folder (in author's files).

95. Jacobs Wind Electric Company, Minneapolis, *Introduction to Wind Electric Generation*[:] *Background (1920–1970)* (Minneapolis: Jacobs Wind Electric Company, 1980), 1–2, handbill (in author's files).

96. Peerless Battery Manufacturing Company, Lincoln, NE, *Let the Wind Do Your Work*[,] *the System Without a Peer*[,] *Peerless Battery Manufacturing Company*[,] *Lincoln*[,] *Nebraska* (Lincoln, NE: Peerless Battery Manufacturing Company, [ca. 1920[), folder, (in the Kregel Windmill Company Papers, Nebraska State Historical Society, Lincoln, NE); National Batteries, St. Paul, MN, *Isolated Lighting Plant Batteries* (St. Paul, MN: National Batteries, [ca. 1930]), brochure (in author's files); The Electric Storage Battery Company, Philadelphia, *The Exide-Hyray Battery* (Philadelphia: The Electric Storage Battery Company, 1921), 1–2, brochure with price sheets (in author's files); Universal Battery Company, Chicago, *The New Universal Guide for Lengthening Battery Life* (Chicago: Universal Battery Company, 1932): 1–24, brochure (in author's files); Wind-Power Light Company, Newton, IA, *End Your Battery Troubles With the Battery Made for Submarines and Supertrains*[:]

Get a Practically New Edison All Steel Alkaline Battery and . . . You're Through! (Newton, IA: Wind-Power Light Company, [ca. 1935]), brochure (in author's files).

97. National Manufacturing Company, Lincoln, NE, *Hang This Card Up on the Wall Near Your Light Plant, Storage Batteries on in Work Shop* (Lincoln, NE: National Manufacturing Company, [ca. 1935]), handbill with cover letter (in author's files); Sheldon & Sartor Wind Electric Manufacturing Company, Nehawka, NE, *Specially Designed Wind Electric Propeller Hubs Made to Fit Our Aluminum Alloy Propellers* (Nehawka, NE: Sheldon & Sartor Wind Electric Manufacturing Company, [ca. 1935]), handbill (in author's files); Sheldon & Sartor Wind-Electric Company, Nehawka, NE, *Wind Electrics and Aluminum Propellers!* (Nehawka, NE: Sheldon & Sartor Wind-Electric Company, [ca. 1935]), handbill (in author's files).

98. E. A. McCardell Sr., *The History of Winpower Mfg. Company* (*A Story of Diversification*) (Newton, IA: n.p., ca. 1980), n.p. (in the Jasper County Historical Museum, Newton, IA); Ed McCardell, "History of Winpower Mfg. Company," in *A History of Newton, Iowa*, ed. Larry Ray Hurto (Wolfe City, TX: Henington Publishing Company, 1982), 335–36.

99. J. C. Fleming Jr., sales manager, Bucklen-Perkins Aerolectric, Inc., Elkhart, IN, to Henry Pechanec, Timken, KS, January 15, 1930, typewritten letter (in the *Windmillers' Gazette* research files); M. C. Metz, sales department, Universal Battery Company, Chicago, to Ernst Fuelling, Decatur, IN, August 15, 1935, typewritten letter (in author's files).

100. "Riches from the Wind," 117.

101. Sears, Roebuck and Company, Atlanta, *The Latest Merchandise for Spring and Summer 1939* (Atlanta: Sears, Roebuck and Company, 1939), 724 (in author's files); Montgomery Ward & Company, Chicago, *Wards Powerlite Electric Light and Power Plants*. Publication 2–15–38 (Chicago: Montgomery Ward & Company, 1938), 10–11 (in author's files).

102. Ruralite Engineering Company, Sioux City, IA, *Plenty of Dependable Electric Light and Power* (Sioux City, IA: Ruralite Engineering Company, [ca. 1938]), folder (in the *Windmillers' Gazette* research files); Ruralite Engineering Company, Sioux City, IA, *Ruralite Wind Chargers 1938 Models* (Sioux City, IA: Ruralite Engineering Company, 1938), brochure (in author's files).

103. Craig Toepfer, *The Hybrid Electric Home* (Atglen, PA: Schiffer Publishing Ltd., 2010), 67–68.

104. *A Report on the Use of Windmills for the Generation of Electricity*, Oxford University, Institute of Agricultural Engineering, Bulletin No. 1 (Oxford, UK: The Clarendon Press, 1926), 1–63 (in the F. Hal Higgins Agricultural History Collection, University Library, University of California at Davis); "Windmill Genera-tion Plants. A Detailed Report on Their Practicability and Economy," *The Implement and Machinery Review* (London) 51, no. 611 (March 1, 1926), 1220–24.

105. C. A. Cameron Brown, *Windmills for the Generation of Electricity*, 2nd ed. (Oxford, UK: Institute for Research in Agricultural Engineering, University of Oxford, 1933), 9.

106. Etablissement des Aéromoteurs Cyclone, Margny-lès-Compiègne, France, *Aéromoteurs "Cyclone" pour L'Élévation de L'Eau et al. Production de L'Électricité*[,] *Pompes*[,] *Pompes a Bras—Pompes a Moteurs*[,] *Pompes D'Irrigation—Réservoirs* (Margny-les-Compiègne, France: Établts des Aéromoteurs "Cyclone," [ca. 1928]), 1–20, brochure (in the files of Etienne Rogier. Toulouse, France); Etienne Rogier, Toulouse, France, email message to author, April 18, 2017.

107. James Alston & Sons Pty. Ltd., South Melbourne, Australia, *"Alston" Patent Electric Generating Windmill* (South Melbourne, Australia: James Alston & Sons Pty. Ltd., [ca. 1931]), brochure (in *The Windmill Journal* research files, Morawa, Australia).

108. Helen Walter, "Hannan Bros Ltd.," *The Windmill Journal* (Morawa, Australia) 14, no. 3 (September 2015): 9–11; "Events in the History of Dunlite," http://www.pearen.ca/dunlite/History-3.pdf, accessed on August 21, 2016; Helen Walter, "Saunders' Break-of-Gauge and Engineering Company Ltd., Saunders Engineering Company, Ltd. & Speedy Windmill and Pump Company," *The Windmill Journal* (Morawa, Australia) 11, no. 4 (December 2012): 8–9.

109. *Windmills for the Generation of Electricity*, 9.

110. C. Hersholdt, Agrico Manufacturing Company, Ltd., Copenhagen, Denmark, typewritten letter to Kregel Windmill Co[mpany], Nebraska City, NE, February 2, 1925 (in the Kregel Windmill Company Papers, Nebraska State Historical Society, Lincoln, NE).

111. Mike Werst, "Pre-Rural Electrification Wind Generators—Speed Governing Mechanisms," *Windmillers' Gazette* 35, no. 1 (Winter 2016): 2.

112. Werst, "Pre-Rural Electrification Wind Generators—Speed Governing Mechanisms," 2–3; Wincharger Corporation, Sioux City, IA, *Giant 32 Volt Wincharger*[,] *Factory Direct to You*[,] *$69.95* (Sioux City, IA: Wincharger Corporation, [ca. 1936]), brochure (in author's files); Wincharger Corporation, Sioux City, IA, *Every Farm Home Can Now Enjoy "Big City" Radio Reception* (Sioux City, IA: Wincharger Corporation, [ca. 1937[), brochure (in author's files); Wincharger Corporation, Sioux City, IA, *Light Your Farm for 50¢ a Year*[,] *Power Operating Cost* (Sioux City, IA: Wincharger Corporation, [ca. 1937]), brochure (in author's files); Wincharger Corporation, Sioux City, IA, *Enjoy Your Own Electric Light & Power*[,] *Pay Only 10¢ A Year Power Operation Cost*, Form No. 839 (Sioux City, IA:

Wincharger Corporation, [ca. 1938]), brochure (in author's files); Wincharger Corporation, Sioux City, IA, *Parts Price List for 6-Volt Winchargers*[,] *Covers Models '36, '37, '38* (Sioux City, IA: Wincharger Corporation, 1938), brochure (in author's files).

113. Werst, "Pre-Rural Electrification Wind Generators—Speed Governing Mechanisms," 3; Jacobs Wind Electric Company, Minneapolis, *Jacobs*[:] *The Standard of Comparison* (Minneapolis: Jacobs Wind Electric Company, [ca. 1935]), brochure (in author's files); M. L. Jacobs, general manager, Jacobs Wind Electric Company, Minneapolis, to A. Clyde Eide, Columbus, OH, April 15, 1955, typewritten letter, photocopy (in the *Windmillers' Gazette* research files).

114. Werst, "Pre-Rural Electrification Wind Generators—Speed Governing Mechanisms," 3–4; *The History of Winpower Mfg. Company.*

115. Werst, "Pre-Rural Electrification Wind Generators—Speed Governing Mechanisms," 4; Parris-Dunn Corporation, Clarinda, IA, *Free Power for Radio and Lights with the Hy-Tower* (Clarinda, IA: Parris-Dunn Corporation, [ca. 1940]), brochure (in author's files); Parris-Dunn Corporation, Clarinda, IA, *32 Volt Free Lite*[:] *World's Most Complete Line of Direct-Drive Wind Electric Power Plants* (Clarinda, IA: Parris Dunn Corporation, [ca. 1940]), brochure (in author's files).

116. H. F. McColly and Foster Buck, *Homemade Six-Volt Wind Electric Plants*, Circular 58 (Fargo, ND: Agricultural Experimental Station, North Dakota Agricultural College, 1935), 2.

117. Arnold Benson, *Plans for Construction of a Small Wind-Electric Plant for Oklahoma Farms*, Publication No. 33 (Stillwater: Oklahoma Agricultural and Mechanical College), 8, no. 1 (June 1937): 5–8, 17, 23–24, 31.

118. Harley Freeman, "Wind-Driven Power Plants You Can Build," *Popular Mechanics* 57, no. 6 (June 1932): 1043–45; C. A. Crowley, "Wind-Driven Generator Charges Batteries," *Popular Mechanics* 70, no. 1 (July 1938): 146–51.

119. LeJay Manufacturing Company, Minneapolis, *LeJay 6 and 32 Volt Slow Speed Wind Plants Catalog No. 12* (Minneapolis: LeJay Manufacturing Company, [ca. 1935]), 1–16, brochure (in author's files); LeJay Manufacturing Company, Minneapolis, *LeJay 6 and 32 Volt Slow Speed Wind Plants* (Minneapolis: LeJay Manufacturing Company, [ca. 1935]), 1–32, brochure (in author's files); LeJay Manufacturing Company, Minneapolis, *LeJay Slow Speed Wind Plants Catalog No. 15* (Minneapolis: LeJay Manufacturing Company, 1936), 1–32, brochure (in author's files); LeJay Manufacturing Company, Minneapolis, *LeJay Slow Speed Wind Plants Catalog No. 16* (Minneapolis: LeJay Manufacturing Company, 1936), 1–32, brochure (in author's files); LeJay Manufacturing Company, Minneapolis, *LeJay Slow Speed Wind Plants*

Catalog No. 17 (Minneapolis: LeJay Manufacturing, 1936), 1–32, brochure (in author's files).

120. Klinsick Mechanical Shop, Optima, OK, *The Wind is Free*[,] *Why Not Use It!* (Optima, OK: Klinsick Mechanical Shop, [ca. 1935]), brochure (in the files of Norman Marks, Geneva, NE); Klinsick Mechanical Shop, Optima, OK, *Price List* (Optima, OK: Klinsick Mechanical Shop, [ca. 1935]), loose-leaf page, (in the files of Norman Marks, Geneva, NE).

121. Fred W. Hawthorn and Robert W. Hawthorn, *Idlewild Farm, A Century of Progress* (Lake Mills, IA: Graphic Publishing Company, 1976), 56–57.

122. Ibid., 57.

123. Harold Huxford Beaty, "Wind Electric Plants" (master's thesis, Iowa State College, Ames, 1941), 77–78, accessed February 2, 2017, http://lib.dr.iastate.edu/cgi/viewcontent.cgi?article=18340&context=rtd.

124. "Windmills Protect Pipelines from Corrosion," *Agricultural Engineering* 18, no. 7 (July 1937): 295.

125. Jacobs Wind Electric Company, Minneapolis, *Pipeline Cathodic Protection Plants* (Minneapolis: Jacobs Wind Electric Company, [ca. 1950]), handbill (in the files of Paul R. Jacobs, Corcoran, MN); Jacobs Wind Electric Company, Minneapolis, *Pipe Line Protection with the New Jacobs System* (Minneapolis: Jacobs Wind Electric Company, [ca. 1950]), handbill (in the files of Paul R. Jacobs, Corcoran, MN); Jacobs Wind Electric Company, Minneapolis, *Jacobs Wind Electric*[,] *Pipe Line Cathodic Protection* (Minneapolis: Jacobs Wind Electric Company, [ca. 1950]), handbill (in the files of Paul R. Jacobs, Corcoran, MN); *The History of Winpower Mfg. Company.*

126. Paul R. Jacobs, interview by author, Minneapolis, September 30, 2016; M. L. Jacobs, general manager, Jacobs Wind Electric Company, Minneapolis, "The Use of Wind-Driven Generators as an External Source of Protective Currents" (paper presented at Cathodic Protection: A Symposium by the Electrochemical Society and the National Association of Corrosion Engineers, Pittsburgh, PA, December 1947), 77–79.

127. Ernest Greenwood, *Aladdin, U. S. A.* (New York: Harper & Brothers Publishers, 1928), 52.

128. J. Roland Hamilton, *Using Electricity on the Farm* (Englewood Cliffs, NJ: Prentice-Hall, Inc., 1959), 5.

129. "Home Wired for 20 Years[,] He Finally Gets Current," *Rural Electrification News* 2, no. 11 (July 1937): 13; "He Waited 10 Years For It," *Rural Electrification News* 3, no. 1 (September 1937): 11–12.

130. "Three Years of REA," *Rural Electrification News* 3, no. 9 (May 1938): 4; Harry Slattery, *Rural America Lights Up* (Washington, DC: National Home Library Foundation, 1940), 27–37; Sam H. Schurr, Calvin C. Burwell, Warren D. Devine, Jr., and Sidney Sonenblum, *Electricity in the American Economy: Agent of Technological Progress*

(Westport, CT: Greenwood Publishing, 1990), 234; National Rural Electric Cooperative Association, *The Next Greatest Thing*, ed. Richard A. Pence (Silver Spring, MD: McArdle Printing Company, 1984), 81.

131. *Idlewild Farm, A Century of Progress*, 57.

132. *The History of Winpower Mfg. Company*.

133. Jacobs Wind Electric Company, Minneapolis, *Navy Special Jacobs Engine Electric Plant* (Minneapolis: Jacobs Wind Electric Company, [ca. 1943]), handbill (in files of Paul R. Jacobs, Corcoran, MN); Paul R. Jacobs, Corcoran, MN, email to author, May 9, 2017.

134. Wincharger Corporation, Sioux City, IA, *Souvenir Booklet of the Army-Navy "E" Presentation to the Employees and Management of Wincharger Corporation* (Sioux City, IA: Wincharger Corporation, 1943), 6–7, brochure (in author's files).

135. Parris-Dunn Associates, Clarinda, IA, *Manual of Arms for the Victory TraineRifle* (Clarinda, IA: Parris-Dunn Associates, [ca. 1943]), brochure (in author's files); "Parris-Dunn Manufacturing Co.," Windcharger.org, accessed May 11, 2017, http://www.windcharger.org/Wind_Charger/Parris-Dunn_Corp..html.

136. "Aerodynamic Wind Mills"; "New Schemes for Harnessing the Winds," *Popular Science* 135, no. 2 (August 1939): 100–01; Palmer Cosslett Putnam, *Power from the Wind* (New York: Van Nostrand Reinhold Company, 1948), 98–108; Christopher Gillis, *Offshore Windpower* (Atglen, PA: Schiffer Publishing Ltd., 2011), 28–34.

137. *Power from the Wind*, 1–14, 109–33; "Mountain-Top Windmill to Feed Vermont Electric Lines," *Popular Science* 139, no. 1 (July 1941): 115–17; Grant H. Voaden, "The Smith-Putnam Wind Turbine . . . A Step Forward in Aero-Electric Power Research," *Turbine Topics* (York, PA) 1, no. 3 (June 1943): 3–8; John B. Wilbur, "The Smith-Putnam Wind Turbine Project," *Journal of the Boston Society of Civil Engineers* 29 (July 1942): 211–28; Lawrence M. Howard, "Power from the Winds," *Vermont Life* 10, no. 2 (Winter 1955–1956): 51–55; Beauchamp E. Smith, "Smith-Putnam Wind Turbine Experiment" (paper presented at the Wind Energy Conversion Systems Workshop, Washington, DC, June 11–13, 1973), 5–7.

138. *Offshore Windpower*, 34–35; Jytte Thorndahl and Benny Christensen, "Time for Survival and Development 1920–1945," in *Wind Power—The Danish Way*, ed. Benny Christensen (Vejen, Denmark: The Poul la Cour Foundation, 2009), 36–38.

139. E. W. Golding, *The Generation of Electricity by Wind Power* (New York: Philosophical Library, 1956), 3–4.

140. J. P. Schaenzer, *Rural Electrification* (Milwaukee, WI: The Bruce Publishing Company, 1948), 4–5; D. Clayton Brown, *Electricity for Rural America*[:] *The Fight for the REA* (Westport, CT: Greenwood Press, 1980), 112–13.

141. Delwin Bass, Torrington, WY, handwritten letter to author, May 26, 2017.

142. Ibid.

143. Arlo Jacobson, "Air-Electric Machine Units Cut Weeds—Flab," *Des Moines Sunday Register*, December 6, 1970; L. A. Christensen, Air-Electric Machine Company, Lohrville, IA, letter to Robert J. Hayes, Muscatine, IA, January 5, 1971, typewritten letter (in author's files); Mike Werst and John Killam, "Air-Electric Machine Company: The Early Years," *Windmillers' Gazette* 38, no. 1 (Winter 2019): 4–5; Parker-McCrory Manufacturing Company, Kansas City, MO, *The Modern Fencing Method*[,] *Parmak Electric Fencer with the Amazing Flux Diverter and Dry Weather Intensifier* (Kansas City, MO: Parker-McCrory Manufacturing Company, 1939), brochure (in author's files).

144. Jacobs Wind Electric Company, Minneapolis, *Jacobs Air Way*[,] *The New Model 35* (Minneapolis: Jacobs Wind Electric Company, [ca. 1949]), brochure (in author's files); M. L. Jacobs, general manager, Jacobs Wind Electric Company, Minneapolis, letter to L. L. Romersheuser, Hayes Center, NE, January 9, 1953, typewritten letter (in author's files); M. L. Jacobs, president, Jacobs Wind Electric Company, Fort Myers, FL, typewritten letter to Jim Martin, Xenia, OH, October 6, 1977, typewritten letter (in author's files).

145. Donald Marier, "Marcellus Jacobs: 1903–1985," *Alternative Sources of Energy* (Milaca, MN), no. 75 (September/October 1985): 6; "About Queen for a Day," Queen for a Day, accessed June 7, 2017, http://queenforaday.com/about.php.

146. Jacobs Wind Electric Company, Minneapolis, *New 4th Generation of Jacobs Wind Systems* (Minneapolis: Jacobs Wind Electric Company, 2009), handbill (in author's files).

Chapter 3

1. John Noble Wilford, "Nation's Energy Crisis: Is Unbridled Growth Indispensable to the Good Life?" *The New York Times*, July 8, 1971; Ford Foundation, *A Time to Choose*[:] *America's Energy Future* (Cambridge, MA: Ballinger Publishing Company, 1974), 5–6, 181; J. R. Johnson, "An Engineer's Perspective of Our Energy Dilemma," *Alternative Sources of Energy* no. 21 (June 1976): 11; Victor Klassen, "The Energy Crisis in Historical Perspective," *Alternative Sources of Energy* no. 12 (October–November 1973): 21–22.

2. "A Holiday is Given in Effort to Reduce Los Angeles Smog," *The New York Times*, July 27, 1973; Jim Dwyer, "Remembering a City Where the Smog Could Kill," *The New York Times*, February 28, 2017.

3. Peter Kihss, "Worried Drivers Swamp Stations Selling Gasoline," *The New York Times*, February 5, 1974.

4. "The New Highway Guerrillas," *Time* 102, no. 25 (December 17, 1973): 33; "Nixon Approves Limit of 55 M.P.H.," *The New York Times*, January 3, 1974.

5. "The Squeeze Hits the Purse," *Newsweek* (December 3, 1973): 92; Edward Cowan, "A Year of Costly Oil and Abrupt Changes," *The New York Times*, October 13, 1974; "The Fuel Crisis—Nixon Acts," *Newsweek* (December 3, 1973): 24–25.

6. James Ridgeway, *The Last Play*[:] *The Struggle to Monopolize the World's Energy Resources* (New York: The New American Library, 1973), 6–7; Larry Bogart, "The Most Important Thing in the World," *Alternative Sources of Energy* no. 12 (October-November 1973): 15–16; Skip Laitner, "The Case Against Nuclear Power: An Overview[,] Part I," *Alternative Sources of Energy* no. 23 (December 1976): 22–24; Douglas LaFollette, "Guest Editorial[:] The Economic Myth of Nuclear Power," *Alternative Sources of Energy* no. 30 (February 1978): 2–3.

7. Ralph Keyes, "Learning to Love the Energy Crisis," *Newsweek* (December 3, 1973): 17.

8. Ralph Borsodi, *This Ugly Civilization* (New York: Simon and Schuster, 1929), n.p., accessed June 5, 2019, https://soilandhealth.org/wp-content/uploads/0303 critic/030302borsodi.ugly/030302borsodi.ch15.html.

9. Helen and Scott Nearing, *Living the Good Life* (Harborside, ME: Social Science Institute, 1954), 6.

10. Carolyn Kimsey, "The Plowboy Interview: Dr. Ralph Borsodi," *The Mother Earth News* no. 26 (March 1974): 9–10.

11. David Gumpert, "The New Pioneers: Eliot and Sue Coleman find 'Homesteading' Is Satisfying Way of Life," *The Mother Earth News* no. 11 (September 1971), 38–39.

12. Ibid., 41.

13. Nicolas Wade, "The New Alchemy Institute: Search for an Alternative Agriculture," *Science* 187, no. 4178 (February 28, 1975): 727–29; Wade Green, "The New Alchemists," *The New York Times*, August 8, 1976; Gary Hirshberg, Earle Barnhart, Tyrone Cashman, and Bill Sheperdson, "Wind & Solar at New Alchemy," *Wind Power Digest* no. 11 (Winter 1977–1978): 16–25; Michael Gery, "The New Alchemists: Creating a Gentle Science to Heal the Earth," *New Roots* (Greenfield, MA) (November–December 1979): 20–26; Gary Hirshberg, *The New Alchemy Water Pumping Windmill Book* (Andover, MA: Brick House Publishing Company, 1982), 52–94.

14. "Wind & Solar at New Alchemy," 16.

15. "Power!," *The Mother Earth News* no. 6 (November 1970): 60–61.

16. "The Plowboy Interview: Dr. Ralph Borsodi," 12.

17. Henry Clews, "Henry Clews' Miraculous Wind-Powered Homestead," *The Mother Earth News* no. 18 (November 1972): 25.

18. Henry Clews, *Electric Power from the Wind* (East Holden, ME: Solar Wind Company, 1973), 1–29 (in author's files); Henry M. Clews, *Solar Wind Progress Report* (East Holden, ME: Solar Wind Company, 1973), 1 (in author's files).

19. *Solar Wind Progress Report*, 1–2

20. Ibid., 2.

21. Henry M. Clews, "Wind Power Systems for Individual Applications" (paper presented at the Wind Energy Conversion Systems Workshop, Washington, DC, June 11–13, 1973), 165–69.

22. *Solar Wind Progress Report*, 2.

23. Solar Wind Company, East Holden, ME, *Wind Generators Currently in Production Throughout the World* (East Holden, ME: Solar Wind Company, 1973), type-written letter (in author's files).

24. Solar Wind Company, East Holden, ME, *Solar Wind Company*[,] *System and Component Price List*[,] *January 1974* (East Holden, ME: Solar Wind Company, 1974), 1–4, brochure (in author's files).

25. Len Buckwalter, "Watts from the Wind," *Mechanix Illustrated* 71, no. 562 (March 1975): 40.

26. Joe Carter, "Earl Rich's Big Windmill: Wind Power in N.H.," *Wind Power Digest* no. 7 (December 1976): 27–30; Earle Rich, interview by author, Amherst, NH, August 26, 2016.

27. Hans Meyer, "Wind Generators: Here's an Advanced Design You Can Build," *Popular Science* 201, no. 5 (November 1972): 103–5, 142.

28. Jim Sencenbaugh, "I Built a Wind Charger for $400!" *The Mother Earth News* no. 20 (March 1973): 32–36.

29. Ibid., 32, 36.

30. Alan Altman, "'O$_2$ Powered Delight' Plans," *Alternate Sources of Energy* no. 14 (May 1974): 5.

31. "A Talk with Jim Sencenbaugh," *Wind Power Digest* 1, no. 6 (September 1976): 5.

32. Harry Kolbe, "Build Your Own Budget Wind-charger," *Mechanix Illustrated* 75, no. 597 (February 1978): 57–60; *Mechanix Illustrated Budget Windcharger* (Wayne, NJ: CBS Publications, 1978), 1–15.

33. Ken Smith, ".75 KW Windcycle," *Alternative Sources of Energy* no. 14 (May 1974): 2–4.

34. George Helmholz and Larry Wheat, "22 Ft Wind Generator," *Alternative Sources of Energy* no. 18 (July 1975): 22–23.

35. Michael A. Hackleman, *Wind and Windspinners* (Culver City, CA: Peace Press Printing and Publishing, 1977), 119, 123–24; Michael Hackleman, "The Savonius Super Rotor!" *The Mother Earth News* no. 26 (March 1974): 78–80; Michael Hackleman, "More about the S-Rotor," *The Mother Earth News* no. 27 (May 1974): 39.

36. E. F. Lindsley, "Wind Power: How New Technology is Harnessing an Age-Old Energy Source," *Popular Science* 205, no. 1 (July 1974): 56–57.

37. Don Marier, "Some Notes on Windmills," *Alternative Sources of Energy* no. 13 (February 1974): 28.

38. Michael Hackleman, *The Homebuilt, Wind-Generated Electricity Handbook* (Culver City, CA: Peace Press, 1975), 8–18.

39. Jack Park and Dick Schwind, *Wind Power for Farms, Homes & Small Industry* (Mountain View, CA: Nielsen Engineering & Research, 1978), 6–12.

40. Craig Toepfer, Chelsea, MI, telephone interview by author, August 14, 2017.

41. James B. DeKorne, "The Answer Is Blowin' in the Wind," *The Mother Earth News* no. 24 (November 1973): 70.

42. Paul Biorn, "Recovering a Jacobs Wind Electric," *Wind Power Digest* 1, no. 3 (December 1975): 20.

43. Dermot McGuigan, *Harnessing the Wind for Home Energy* (Charlotte, VT: Garden Way Publishing, 1978), 84–86.

44. Coulson Wind Electric, Polk City, IA, *New Parts and Price List as of April 1, 1977* (Polk City, IA: Coulson Wind Electric, 1977), double-sided typewritten handbill (in author's files).

45. Donald Marier, "Reaching Up, Reaching Out," in "Wind Energy & Wood Heating," special issue, *Alternative Sources of Energy* no. 41 (January-February 1980): 13.

46. Craig Toepfer, Chelsea, MI, email message to author, February 5, 2016.

47. North Wind Power Company, Warren, VT, *North Wind Power Co[.]* (Warren, VT: North Wind Power Company, [ca. 1979]), 6–8, brochure (in author's files); *Harnessing the Wind for Home Energy*, 99–100; Donald Mayer, Waitsfield, VT, telephone interview by author, April 24, 2015.

48. *Wind Power for Farms, Homes & Small Industry*, 6–12.

49. Joe Carter, "At Home with Martin Jopp[,] Wind Power Pioneer," *Wind Power Digest* no. 8 (Spring 1977): 6–8.

50. Ken Meter, "Interview with Martin Jopp," *Alternative Sources of Energy* no. 28 (October 1977): 27.

51. Don Marier, "Some Notes on Windmills," *Alternative Sources of Energy* no. 12 (October-November 1973): 2–3; Martin Jopp, "Martin Answers[:] Questions and Answers on Windmills and Electricity," *Alternative Sources of Energy* no. 18 (July 1975): 21.

52. "Staff Notes," *Alternative Sources of Energy* no. 28 (October 1977): 3; Martin Jopp, "Martin Answers," *Alternative Sources of Energy* no. 28 (October 1977): 46.

53. "Feedback: Marcellus Jacobs," *Wind Power Digest* no. 9 (Summer 1977): 5.

54. "This Special Issue of *Alternative Sources of Energy Magazine* is Dedicated to the Memory of Martin Jopp," in "Wind/Hydropower," special issue, *Alternative Sources of Energy* no. 46 (November-December 1980): 5.

55. Volta Torrey, *Wind-Catchers: American Windmills of Yesterday and Tomorrow* (Brattleboro, VT: The Stephen Greene Press, 1976), 192.

56. US Energy Information Administration, *Total Energy[,] Annual Energy Review[,] Average Retail Prices of Electricity, 1960–2011*, September 2012, accessed July 25, 2017, https://www.eia.gov/totalenergy/data/annual/showtext.php?t=ptb0810.

57. *Wind-Catchers*, 192.

58. Robert N. Meroney, "Wind Power Applications in Rural and Remote Areas" (paper presented at the ASCE Specialty Conference on Solar Energy and Remote Sensing, Fort Collins, CO, July 20–21, 1976), 6.

59. Roger Hamilton, "Can We Harness The Wind?" *National Geographic* 148, no. 6 (December 1975): 815–16; Roger Bergerson, "40 Years with Jacobs' Gems," *Public Power* 37, no. 5 (September-October 1979): 19.

60. *The Mother Earth News Handbook of Homemade Power* (New York: Bantam Books, 1974), 126–203; William Randolph, "Power to the People (Wincharger Style)," *Wind Power Digest* no. 8 (Spring 1977): 10–14.

61. Jim Erdman, interview by author, Custer, WI, June 17, 2016.

62. Ibid.

63. Jim Erdman, "Letters[:] 32 Volt Bulbs," *Alternative Sources of Energy* no. 21 (June 1976): 44.

64. Jim Erdman interview.

65. Ibid.

66. Ibid.

67. Jan Erdman, interview by author, Custer, WI, June 17, 2016.

68. Jim Erdman interview.

69. Diane Burnside, "Sound of the Wind," *Wind Power Digest* 1, no. 6 (September 1976): 28–30.

70. "Obituaries[:] Elliott John Bayly," *Steamboat Pilot & Today* (Steamboat Springs, CO), January 28, 2012.

71. Tim Rounds, "PCR Employs Windpower," *The Clock* (Plymouth State College, Plymouth, NH), November 30, 1978; "At State College Wind Power used for Radio Station," *The Herald* (Provo, UT), December 7, 1978; "In Sync[,] Up and Coming in Broadcast Technology[,] It's a Breeze," *Broadcasting* (Washington, DC) 48 (January 8, 1979): 42; David Price, "Harnessing the Wind," *New Hampshire Magazine* 8, no. 1 (October 1979): 7–8.

72. David B. Price, director, Office of Public Relations, Plymouth State College, Plymouth, NH, letter to Paul Shulins, manager, WPCR Radio, Plymouth State College, Plymouth, NH, January 30, 1979, typewritten letter (in the files of Paul Shulins, Boston, MA).

73. Jeffrey N. Tellis, president, The Intercollegiate Broadcasting System, Inc., Vails Gate, NY, letter to Paul Shulins, general manager, Radio Station WPCR, Plymouth State College, Plymouth, NH, January 12, 1979, typewritten letter (in the files of Paul Shulins, Boston,

MA); Paul Shulins, Boston, MA, telephone interview by author, October 5, 2015.

74. Travis L. Price III, interview by author, Washington, DC, July 29, 2016; Energy Task Force, *Windmill Power for City People*[:] *A Documentation of the First Urban Wind Energy System* (Washington, DC: US Government Printing Office, 1977), 1–3; Kenneth Herts, "Energy Task Force[:] Rebuilding the Inner City with Affordable Sources of Energy," *Living Alternatives Magazine* 1, no. 6 (February 1980): 19; Joyce Maynard, "519 East 11th St.: Neighbors Rebuild Hopes," *The New York Times*, July 8, 1976.

75. *Windmill Power For City People*, 3–6.

76. US Office of Consumer Affairs, *People Power: What Communities are Doing to Counter Inflation*, Report No. ED193164, edited by Mary S. Gordon (Washington, DC: US Office of Consumer Affairs, 1980), 220–21.

77. Robert Mcg. Thomas, "11th St. Tenants with Windmill and Con Edison," *The New York Times*, November 13, 1976.

78. Price interview; *Windmill Power For City People*, 34–37; *People Power*, 219; Public Utility Act of 1978, Public Law 95–617, 95th Cong., 1st sess. (November 9, 1978), 3137–67.

79. Joe Carter and Ted Finch, "Wind & Solar at East 11th," *Wind Power Digest* no. 9 (Summer 1977): 12–17; "The 11th Street Movement!" *Alternative Sources of Energy* no. 30 (February 1978): 31.

80. Mary M. Christianson, "Alternative Technology in Low Income New York: Energy Task Force's First Four Years" (paper presented at The First Conference on Community Renewable Energy Systems, University of Colorado, Boulder, August 20–21, 1979), 85–86.

81. Josh Weil, "The Wind Farmers of East 11th Street," *The New York Times*, August 3, 2008; Shayla Love, "The Almost Forgotten Story of the 1970s East Village Windmill," *The Gothamist*, September 29, 2014, accessed September 30, 2014, http://gothamist.com/2014/09/29/east_village_windmill_nyc.php#photo-1.

82. "A Talk With Jim Sencenbaugh," 6.

83. Ibid.

84. Frank B. Edwards, "Windsteading Into the Eighties," *Harrowsmith* (Ontario, Canada) 4, no. 22 (September 1979): 30.

Chapter 4

1. [Craig Toepfer,] "American Wind Energy Association" (Detroit: American Wind Energy Association, [ca. 1975]), 2, typewritten (in the files of Craig Toepfer, Chelsea, MI).

2. "The American Wind Energy Association," *The Mother Earth News* no. 30 (November 1974): 16; Allan

O'Shea, Copemish, MI, telephone interview by author, September 6, 2017.

3. "The American Wind Energy Association"; O'Shea interview.

4. Allan O'Shea, Environmental Energies, Inc., Detroit, American Wind Energy Association meeting invitation, August 5, 1974, typewritten (in author's files).

5. O'Shea interview.

6. *American Wind Energy Association*, 2.

7. Ibid., 2–3.

8. "American Wind Energy Association;" *American Wind Energy Association*, 4–6.

9. Daniel Mullendore, "Letter to the Members[,] April 15, 1975," *American Wind Energy Association Newsletter* (Detroit) (August 1975): 13–14.

10. *American Wind Energy Association*, 3.

11. "American Wind Energy Conference: April 10–11, 1975," *American Wind Energy Association Newsletter* (Detroit) (ca. March 1975): n.p.

12. "Membership Application," *American Wind Energy Association Newsletter* (Detroit) (June 1976): n.p.

13. Allan O'Shea, "Memo from the President," *American Wind Energy Association Newsletter* (Detroit) (August 1975): 2.

14. Wind Energy Society of America, Pasadena, CA, *Wind Energy Society of America*[:] *The Challenge*[,] *Goals* (Pasadena, CA: Wind Energy Society of America, [ca. 1975]), handbill (in the files of Craig Toepfer, Chelsea, MI).

15. Ibid.

16. Wind Energy Society of America, "Annual Meeting of the Wind Energy Society of America, August 2, 1975," *Wind Energy Society of America Newsletter* (Pasadena, CA), no. 8 ([August] 1975): 1.

17. Wind Energy Society of America, "WESA Status," *Wind Energy Society of America Newsletter* (Pasadena, CA), no. 10 ([August]1976): 1.

18. Judith Woelke, "Minutes," *AWEA Newsletter* (Detroit) (April 1976): 3–6; Don Mayer, "The Development of the AWEA," *AWEA Windletter* (Washington, DC) (February 1978): 6.

19. "The Federal Wind Energy Program," *AWEA Newsletter* (Detroit) (September 1976): 10–11.

20. "The Development of the AWEA," 6.

21. "Fall Meeting Set[:] Warren, VT chosen as Wind Exposition Site," *AWEA Newsletter* (Detroit) (September 1976): 3–4; Woody Stoddard, "Commentary and Review[:] American Wind Energy Association Annual Meeting and Exposition," *Alternative Sources of Energy* no. 23 (December 1976): 31.

22. Frank R. Eldridge, "A Preliminary Federal Commercialization Plan for WEC's," *AWEA Newsletter* (Bristol, IN) (Spring 1977): 14–19.

23. Herman M. Drees, "Note from the Editor," *Wind Technology Journal* 1, no. 1 (Spring 1977): 2; "Wind

Journal Arrives," *AWEA Newsletter* (Bristol, IN) (Spring 1977): 13.

24. Paul N. Vosburgh, *Commercial Applications of Wind Power* (New York: Van Nostrand Reinhold Company, 1983), 7.

25. "The Development of the AWEA," 6; Richard Katzenberg, "President's Report," *AWEA Newsletter* (Bristol, IN) (Summer 1977): 5–6; "AWEA Research," *AWEA Windletter* (Washington, DC) (February 1978): 10.

26. "The Development of the AWEA," 6.

27. Richard Katzenberg, interview by author, Amherst, NH, August, 26, 2016.

28. Richard Katzenberg, "Quarterly Report," *AWEA Newsletter* (Bristol, IN) (Spring 1977): 23.

29. George Tennyson, "Test and Product Standards for WECS," *AWEA Newsletter* (Bristol, IN) (Summer 1977): 4.

30. Ibid.

31. Ibid.

32. Ibid.

33. "AWEA Research," 10.

34. "More on Small Wind Generators," *AWEA Windletter* (Washington, DC) (February 1978): 9; "DOE issues Three Contracts for 1-KW Systems," *AWEA Windletter* (Washington, DC) (June 1978): 3.

35. Herman Drees, Westlake Village, CA, telephone interview by author, September 18, 2017.

36. Jay Carter, Jr., letter to the editor, *AWEA Windletter* (Washington, DC) (April 1980): 6.

37. "1978 Board of Directors," *AWEA Windletter* (Washington, DC) (June 1978): 4.

38. "AWEA Board Developments," *AWEA Windletter* (Washington, DC) (June 1978): 7.

39. "The AWEA Establishes an Office in Washington, D.C.," *AWEA Windletter* (Washington, DC) (June 1978): 3.

40. Mike Evans, "Interview with Rick Katzenberg," *Wind Power Digest* no. 13 (Fall 1978): 5.

41. Christopher Gillis, *Windpower* (Atglen, PA: Schiffer Publishing Ltd., 2008), 33.

42. Paul Gipe, "PURPA: A New Law Helps Make Small-Scale Power Production Profitable," *Sierra* 66 (November/December 1981): 54.

43. Ibid.

44. "A Special Report[:] Section 2[,] P.U. R. P. A. and Small Systems," *Wind Power Digest* no. 22 (Spring 1981): 13.

45. "A Special Report[:] Section 2[,] P.U. R. P. A. and Small Systems," 14–23; Don Bain, "Small-Scale Wind and the Public Utility Regulatory Policies Act," *Wind Power Digest* no. 17 (Fall 1979): 5–6; Mike Evans, "Understanding the P.U. R. P. A. Puzzle," *Wind Power Digest* no. 24 (Summer 1982): 16–17.

46. Energy Tax Act of 1978, Public Law 95–618, 95th Cong., 1st sess., (November 9, 1978), 3175.

47. Rick Katzenberg, "Where Are the Dallas Cowgirls When We Need Them?" *Wind Power Digest* no. 16 (Summer 1979): 5.

48. Ibid., 5–7. A quad is a unit of energy, which in terms of electricity equates to 293.07 terawatt-hours.

49. Thomas Gray, interview by Gregory L. Sharrow, June 19, 2008, interview 2008.202, transcript, NRG Research Project, Vermont Folklife Center, Middlebury, VT.

50. Tom Gray, AWEA standards development manager, Memo to All Interested Parties, July 14, 1980, typewritten letter (in author's files); Performance Subcommittee, American Wind Energy Association, *W. E. C. S. Performance Standards Development Program*[:] *Initial Outline* (Washington, DC: American Wind Energy Association, 1980), 2; Interim Standards Subcommittee, American Wind Energy Association, *Draft Working Document*[,] *Interim Standards Subcommittee*[,] *AWEA Standards Program*[,] *February 7, 1981* (Washington, DC: American Wind Energy Association, 1981), n.p.

51. Electric Power Subsystems Subcommittee, American Wind Energy Association, *Draft Working Document*[,] *Electric Power Subsystems Subcommittee*[,] *AWEA Standards Program*[,] *January 27, 1981*[,] *SWECS Electrical System Interconnect Technical Guidelines* (Washington, DC: American Wind Energy Association, 1981), n.p.; Terminology Standards Subcommittee, American Wind Energy Association, *Draft Working Document*[,] *Terminology Standards Subcommittee*[,] *AWEA Standards Program*[,] *February 26, 1981* (Washington, DC: American Wind Energy Association, 1981), n.p.

52. Gray interview.

53. T. O. Gray, "Wind Power and the Utilities: A Wind Industry Perspective" (paper presented at the Rural Electric Wind Energy Workshop Proceedings, Boulder, CO, June 1–3, 1982), 227–28.

54. Ibid., 228.

55. Gray interview.

56. "Amarillo Wind Expo '82[:] The Wind Industry Has Matured," *Wind Power Digest* no. 25 (Fall 1982): 48, 52.

57. Ibid, 52.

58. *Federal Energy Regulatory Commission v. Mississippi*, 102 S. Ct. 2126 (1982); *American Paper Institute v. American Electric Power Service Corporation*, 103 S. Ct. 1921 (1983); Jim Hartwig, "PURPA Prevails," *Soft Energy Notes* (San Francisco) 5 (September–October 1982): 106; Peyton G. Bowman, III, Carl W. Ulrich, et al., "Report of the Committee on Cogeneration and Small Power Production Facilities," *Energy Law Journal* 4, no. 2 (1983): 279–82; Charles A. Tievsky, "Full Electric Utility Ownership of PURPA Qualifying Cogeneration Facilities: Trouble Down the Line?" *Washington University Journal of Urban and Contemporary Law* 27 (January

1984): 330–33; Sharon P. Gross, "Energy Conservation—A New Factor in Certain Utility Rate Regulation," *Natural Resources Journal* 24 (Spring 1984): 487–96.

59. Donald Marier, "The Consumer Wind Market: What Will the Future Hold," *Alternative Sources of Energy* no. 61 (May-June 1983): 4.

60. Michael Lotker, "Making the Most of Federal Tax Laws: A New Way to Look at WECS Development," *Alternative Sources of Energy* no. 63 (September-October 1983): 42–43.

61. Ibid.

62. C. P. Gilmore, "What's New?[:] Blowing in the Wind," *Popular Science* 222, no. 4 (April 1983): 77.

63. "AWEA Joins Peers to Lobby for Tax Credit Extension," *World Wind* (Ottawa, Canada) 1, no. 2 (January 1985): 4.

64. Ibid.

65. Melinda Welsh, "Tax Credits Update," *Windpower Monthly* (Knebel, Denmark) 1, no. 6 (June 1985): 14.

66. Melinda Welsh, "Future for Tax Credits Darkens," *Windpower Monthly* (Knebel, Denmark) 1, no. 7 (July 1985): 4.

67. Melinda Welsh, "Congress Not Likely to Take Decision on Wind in '85," *Windpower Monthly* (Knebel, Denmark) 1, no. 9 (September 1985): 8.

68. Jim Schefter, "New Harvest of Energy from Wind Farms," *Popular Science* 222, no. 1 (January 1983): 58–61.

69. Janet L. Hopson, "They're Harvesting a New Cash Crop in California Hills," *Smithsonian* 13, no. 8 (November 1982): 123.

70. "Looking in on Altamont's Wind Farms . . . by Car or on a Tour," *Sunset* 174 (April 1985): 78.

71. Gillis, *Windpower*, 55.

72. Robert W. Righter, *Wind Energy in America*[:] *A History* (Norman: University of Oklahoma Press, 1996), 209.

73. Ellen Paris, "The Great Windmill Tax Dodge," *Forbes* 133 (March 12, 1984): 40.

74. The Great Windmill Tax Dodge"; Ellen Paris, "Palm Springs and the Wind People," *Forbes* 135 (June 3, 1985): 170–71.

75. Elizabeth Willson, "Fayette Manufacturing: Is Wind Power Gone with the Tax Credits?" *High Technology* 6, no. 1 (January 1986): 17.

76. "A Special Report: U. S. Wind Energy After Seven Years," *AWEA Wind Energy Weekly* (Washington, DC) 7, no. 300 (April 24, 1988), n.p.

77. Gillis, *Windpower*, 34.

78. "AWEA Briefs Hill Staff on Wind and Developing World," *AWEA Wind Energy Weekly* (Washington, DC) 7, no. 310 (July 5, 1988): n.p.; Michael L. S. Bergey, president and CEO, Bergey Windpower Company, interview by author, Norman, OK, October 23, 2015.

79. "Atlantic Orient Plans Changes to Enertech E44 Wind Turbine," *AWEA Wind Energy Weekly* (Washington, DC) 10, no. 463 (1991): n.p.

80. "Northern Power Advanced Turbine Would Cut Power Costs by 46%," *AWEA Wind Energy Weekly* (Washington, DC) 10, no. 464 (1991): n.p.

81. Andy Kruse, Boulder, CO, telephone interview by author, December 13, 2017.

82. Gray interview.

83. "Wind Can Help Avert Greenhouse Effect, AWEA Says," *AWEA Wind Energy Weekly* (Washington, DC) 7, no. 311 (July 11, 1988): n.p.

84. Ibid.

85. "FERC Deregulation Ignores Environmental Costs: AWEA," *AWEA Wind Energy Weekly* (Washington, DC) 7, no. 314 (July 30, 1988): n.p.

86. "AWEA Members to Fund Legislative Push," *AWEA Wind Energy Weekly* (Washington, DC) 7, no. 327 (November 6, 1988): n.p.; "New U. S. Administration May Bring Changes," *AWEA Wind Energy Weekly* (Washington DC) 7, no. 328 (November 13, 1988): n.p.

87. Gray interview.

88. Randall S. Swisher, Silver Spring, MD, email to author, July 20, 2018.

89. Ibid.

90. Gray interview.

91. Swisher, email to author, July 20, 2018.

92. Ibid.

93. Randall S. Swisher, interview by author, Silver Spring, MD, October 11, 2017.

94. Randy Swisher, interview by Gregory L. Sharrow, August 14, 2008, interview 2008.2038, transcript, NRG Research Project, Vermont Folklife Center, Middlebury, VT.

95. Ibid.

96. Ibid.

97. "AWEA Reports to Members on Legislative Push," *AWEA Wind Energy Weekly* (Washington, DC) 8, no. 343 (March 12, 1989): n.p.

98. "AWEA Asks Key Senators to Mull Production Incentives," *AWEA Wind Energy Weekly* (Washington, DC) 8, no. 348 (April 17, 1989): n.p.

99. Energy Policy Act of 1992, Public Law 102–486, 102nd Cong. (October 24, 1992), 2969–70.

100. Gillis, *Windpower*, 89; "Competition Favors Renewables, says NIMO Vice President," *AWEA Wind Energy Weekly* (Washington, DC) 13, no. 594 (April 25, 1994): n.p.; "Utilities Upbeat about Windpower Prospects," *AWEA Wind Energy Weekly* (Washington, DC) 13, no. 597 (May 23, 1994): n.p.

101. Mike Wiser, "Growing Tall Turbines," *Iowa Farmer Today* (Cedar Rapids, IA), November 18, 2011.

102. Nancy Rader, Scott Sklar, and Randy Swisher, To: Reviewers, (Re: Concepts for Federal Renewable Energy Policies/Programs), August 3, 1990, typewritten letter

(in files of Nancy Rader, Berkeley, CA); Nancy Rader, "A Review and Critique of Environmental Quantification in Electric Utility Planning" (term paper, University of California at Berkeley, 1992), 1–2; "Swisher Finds NARUC Meeting 'Encouraging' for Wind," *AWEA Wind Energy Weekly* (Washington, DC) 14, no. 647 (May 22, 1995): n.p.; Nancy A. Rader and Richard B. Norgaard, "Efficiency and Sustainability in Restructured Electric Markets: The Renewables Portfolio Standard," *The Electricity Journal* 9, no. 6 (July 1996): 37–49; Nancy Rader and Scott Hempling, *The Renewables Portfolio Standard*[:] *A Practical Guide* (Washington, DC: National Association of Regulatory Utility Commissioners, 2001), 1–1—1–5; Nancy A. Rader, executive director, California Wind Energy Association, Berkeley, CA, email to author, March 16, 2018.

103. American Wind Energy Association, Washington, DC, *News from the American Wind Energy Association*[,] *for Immediate Release*[,] *AWEA Hails CPUC Support for Renewables*[,] *Restructuring Proposals Disagree on Market Structure, But Converge on Renewables Requirements*, May 24, 1995 (in files of Randall S. Swisher, Silver Spring, MD).

104. Ibid.

105. Randall Swisher and Kevin Porter, "Renewable Policy Lessons from the US: The Need for Consistent and Stable Policies," in *Renewable Energy Policy and Politics: A Handbook for Decision-Making*, ed. Karl Mallon (London: Earthscan, 2006): 191; Nancy A. Rader, executive director, California Wind Energy Association, Berkeley, CA, email to author, April 2, 2018.

106. Swisher interview, October 11, 2017.

107. "Growing Tall Turbines."

108. "Texas Passes Law for Big Renewable Energy Portfolio," *Windpower Monthly* (London), July 1, 1999, accessed March 18, 2018, http://www.windpowermonthly.com/article/955103/texas-passes-law-big-renewable-energy-portfolio; National Association of Regulatory Utility Commissioners Subcommittee on Renewable Energy, "State Activities[:] Texas," *State Renewable Energy News* (Washington, DC) 8, no. 2 (Summer 1999): 4; Ryan Wiser and Ole Langniss, "The Renewable Porfolio Standard in Texas: An Early Assessment," *Energy Policy* 31, no. 6 (May 2003): 527–35; Paul Komor, *Renewable Energy Policy* (New York: Diebold Institute for Public Policy Studies, 2004), 165–66; Kate Galbraith and Asher Price, *The Great Texas Wind Rush* (Austin, TX: University of Texas Press, 2013), 4.

109. Gillis, *Windpower*, 55.

110. "'Thousand Cuts' Cause Drop in Kenetech's Earnings," *AWEA Wind Energy Weekly* (Washington, DC) 14, no. 659 (August 14, 1995): n.p.; "Kenetech Board Struggles to Resolve Financial Troubles," *AWEA Wind Energy Weekly* (Washington, DC) 14, no. 676 (December 11, 1995): n.p.; American Wind Energy Association, Washington, DC, *Statement of Randall Swisher*[,] *Executive Director, American Wind Energy Association on Kenetech Windpower's Filing for Bankruptcy Protection*, May 30, 1996 (in the files of Randall S. Swisher, Silver Spring, MD).

111. *Statement of Randall Swisher.*

112. Gillis, *Windpower*, 58.

113. "NAE, Micon to Install Largest Single Turbine in Midwest," *AWEA Wind Energy Weekly* (Washington, DC) 14, no. 678 (December 25, 1995): n.p.

114. American Wind Energy Association, Washington, DC, *Urgent Memorandum*[,] *to: Wind Energy Supporters*[,] *from: Jared Barlage, AWEA*[,] *Subject: Massive Cuts to the Federal Wind Program Budget*, June 12, 1995 (in files of Randall S. Swisher, Silver Spring, MD); American Wind Energy Association, Washington, DC, *News from the American Wind Energy Association*[,] *Massive Budget Cuts Threaten Job Losses and Economic Downturn in Many American Communities*[,] *Proposed Cuts in Renewable Energy Run Contrary to Public Opinion*, June 13, 1995 (in files of Randall S. Swisher, Silver Spring, MD); American Wind Energy Association, Washington, DC, *News from the American Wind Energy Association*[,] *PURPA Crucial to Development of Fully Competitive Electricity Market, Says AWEA*, June 6, 1995 (in files of Randall S. Swisher, Silver Spring, MD); American Wind Energy Association, Washington, DC, *Legislative Alert from the American Wind Energy Association*[,] *The Fiscal Year (FY) 1996 U. S. Federal Wind Energy Research and Development (R&D) Budget Is in Its Final Stage as the House and Senate Prepare to Iron Out the Differences in Their Respective Energy and Water Appropriations Bills in Just a Few Short Weeks,* August 29, 1995 (in the files of Randall S. Swisher, Silver Spring, MD); American Wind Energy Association, Washington, DC, *Federal Legislative Update*[,] *Wind Energy Research and Development (R&D) Budget*, September 22, 1995 (in the files of Randall S. Swisher, Silver Spring, MD); "National Renewable Energy Laboratory to Trim Staff," *AWEA Wind Energy Weekly* (Washington, DC) 14, no. 673 (November 20, 1995): n.p.; American Wind Energy Association, Washington, DC, *CPUC Restructuring Decision Crucial to Wind and Other Renewables Says AWEA*[,] *Some California Wind Plants Already Being Dismantled, Industry Notes*, December 14, 1995 (in the files of Randall S. Swisher, Silver Spring, MD); "CPUC Restructuring Decision Includes Portfolio Standard[,[Order Calls for Partial Transition to Direct Access by 1998, Complete Phase-In by 2003," *AWEA Wind Energy Weekly* (Washington, DC) 14, no. 678 (December 25, 1995): n.p.; "House Panel Explores PURPA Reform, Industry Structure[,] Karas Testifies That PURPA Has Not Yet Accomplished All of Its Goals," *AWEA Wind Energy*

Weekly (Washington, DC) 15, no. 683 (February 5, 1996): n.p.; American Wind Energy Association, Washington, DC, *News from the American Wind Energy Association*[,] *AWEA Calls California Utilities' Renewables Portfolio Standard Proposal 'Step In Right Direction'*[,] *However, Changes Needed in Implementation Details*, May 7, 1996 (in files of Randall S. Swisher, Silver Spring, MD); "Legislative Alert[:] Make Your Voice Heard on the Renewable Energy Research Budget!" *AWEA Wind Energy Weekly* (Washington, DC) 15, no. 685 (February 19, 1996): n.p.; "Restructuring Good for Environment, Utility Executive Contends," *AWEA Wind Energy Weekly* (Washington, DC) 15, no. 692 (April 8, 1996): n.p.; "Slash R&D? 'Historic Blunder,' Say DOE Executives[,] Renewable R&D Vital to Nation's Future: DOE Officials," *AWEA Wind Energy Weekly* (Washington, DC) 15, no. 693 (April 15, 1996): n.p.; American Wind Energy Association, Washington, DC, *News from the American Wind Energy Association*[,] *AWEA Blasts House Committee Plan to Slash Federal Wind Research*, April 19, 1996 (in files of Randall S. Swisher, Silver Spring, MD); American Wind Energy Association, Washington, DC, *AWEA Cheers Senate Agreement to Restore $23.7 Million to Renewable Energy Budgets*, July 31, 1996 (in the files of Randall S. Swisher, Silver Spring, MD).

115. Randy Swisher, executive director, American Wind Energy Association, Washington, DC, letter to membership, (Subject: Taking Action to Stop Repeal of the Wind Production Tax Credit), September 26, 1995 (in the files of Randall S. Swisher, Silver Spring, MD); American Wind Energy Association, Washington, DC, *AWEA Legislative Alert*[,] *Senator James Jeffords (R-Vermont) Has Authored a Letter to Senator William Roth (R-Delaware), Chairman of the Senate Finance Committee, Asking That the Production Tax Credit for Wind Be Maintained as it Exists in Current Law*, October 5, 1995 (in files of Randall S. Swisher, Silver Spring, MD); American Wind Energy Association, Washington, DC, *Legislative Update from the American Wind Energy Association*[,] *To: AWEA Board, Legislative Committee and Supporters*[,] *from: Randall Swisher and Jared Barlage*[,] *Re: Status Report on Wind Energy Production Tax Credit*, October 12, 1995 (in the files of Randall S. Swisher, Silver Spring, MD); American Wind Energy Association, Washington, DC, *News from the American Wind Energy Association*[,] *Roth Preserves Wind Energy Production Tax Credit*[,] *AWEA Praises Recommendation to Leave Credit Intact, Maintaining Tax Equity*, October 17, 1995 (in files of Randall S. Swisher, Silver Spring, MD); American Wind Energy Association, Washington, DC, *Legislative Update from the American Wind Energy Association*[,] *Wind Energy Production Tax Update*, October 31, 1995 (in files of Randall S. Swisher, Silver Spring, MD); American Wind Energy Association, Washington, DC, *Legislative Update from the American*

Wind Energy Association[,] *Wind Credit Remains in Law*[,] *from: Mike Marvin and Jared Barlage, AWEA*, November 13, 1995 (in the files of Randall S. Swisher, Silver Spring, MD); American Wind Energy Association, Washington, DC, *Legislative Update* [,] *Wind Energy Production Tax Credit*[,] *from: Michael Marvin and Jared Barlage, AWEA*, November 21, 1995 (in the files of Randall S. Swisher, Silver Spring, MD); American Wind Energy Association, Washington, DC, *Legislative Update from the American Wind Energy Association*[,] *Production Tax Credit Update*[;] *from: Mike Marvin and Jared Barlage, AWEA*, December 8, 1995 (in the files of Randall S. Swisher, Silver Spring, MD).

116. Mitchell Ward, "The PTC and Wind Energy: Restructuring the Production Tax Credit as a More Effective Incentive," *Houston Business and Tax Law Journal* 11, no. 2 (2011): 464–65; Gillis, *Windpower*, 89.

117. American Wind Energy Association, Washington, DC, *Boom: 2003 Close to Best Ever Year for New Wind Installations; Bust: Expiration of Key Incentives Lowers Hopes for 2004*, January 22, 2004 (in author's files).

118. American Wind Energy Association, Washington, DC, *First Quarter Market Report: Wind Industry Trade Group Sees Little to No Growth in 2004, Following Near-Record Expansion in 2003*, May 12, 2004 (in author's files).

119. American Wind Energy Association, Washington, DC, *Second Quarter Market Report: American Wind Industry Needs Consistent Business Environment to Bring Wind Power's Promise to the Country*, August 10, 2004 (in author's files).

120. American Wind Energy Association, Washington, DC, *Energy Bill Extends Wind Power Incentive Through 2007*, July 29, 2005 (in author's files).

121. Gillis, *Windpower*, 90.

122. "Northern Power to Design Large-Scale Polar Turbine," *AWEA Wind Energy Weekly* (Washington, DC) 14, no. 634 (February 13, 1995): n.p.

123. "Atlantic Orient Plans Wind-Diesel Project in Alaska," *AWEA Wind Energy Weekly* (Washington, DC) 14, no. 637 (March 6, 1995): n.p.

124. "Bergey Ships Nine Wind Pumping Systems to Indonesia," *AWEA Wind Energy Weekly* (Washington, DC) 14, no. 674 (November 27, 1995): n.p.; "Bergey Turbines in Brazil under World Bank Project," *AWEA Wind Energy Weekly* (Washington, DC) 15, no. 679 (January 8, 1996): n.p.

125. Sara Schaeffer Munoz, "A Novel Way to Reduce Home Energy Bills," *The Wall Street Journal*, August 15, 2006; Jennifer Alsever, "A Nice Stiff Breeze, and a Nice Little Power Bill," *The New York Times*, April 1, 2007.

126. "Business Energy Investment Tax Credit (ITC)," US Department of Energy, accessed April 2, 2017, https://www.energy.gov/savings/business-energy-investment-tax-credit-itc.

127. Keith R. Knox, "Strategies & Warnings for Wind Generator Buyers," *Wind Power Digest* no. 24 (Summer 1982): 54–56; "Who's Mill Is This?" *Wind Power Digest* no. 25 (Fall 1982): 16; Brad Haugh and Carol Hoerig, "Fortune!!![:] In Wind Energy, 40 Percent Tax Break, Mr. Hoyler," *Wind Power Digest* no. 25 (Fall 1982): 16–17; "Hoyler's Response," Wind Power Digest no. 25 (Fall 1982): 17.

128. Paul Gipe, "New Federal Subsidies Distort the US Small Wind Market," Wind-Works, November 14, 2008, accessed July 4, 2019, http://www.wind-works.org/cms/index.php?id=64&tx_ttnews%5Btt_news%5D=54&cHash=69ab65d9bf5142d330e8b7b70be45c40.

129. "A Nice Stiff Breeze, and a Nice Little Power Bill"; "Small-Scale Wind Turbines Catching On," *The Associated Press*, May 17, 2009; Kevin Spear, "Lake County Man's Wind Turbine Could Help Feed Florida's Power Grid," *Orlando Sentinel* (Orlando, FL), September 29, 2009.

130. Office of the Governor, Commonwealth of Pennsylvania, Harrisburg, PA, *Governor Rendell Making Small Wind Energy Systems Available to Local Governments*[,] *15 Small-Scale Systems to Place Alternative Energy Source in Public View*, April 7, 2006 (in author's files).

131. American Wind Energy Association, *AWEA Small Wind Turbine Global Market Study 2007* (Washington, DC: American Wind Energy Association, 2007), 2–3.

132. American Wind Energy Association, *AWEA Small Wind Turbine Global Market Study: 2009* (Washington, DC: American Wind Energy Association, 2009), 3, 11.

133. Ibid., 3.

134. Lot Tan, "Small, Cheaper Turbine Might Work for Homes," *The Columbus Dispatch* (Columbus, OH), September 14, 2008; Alexander Haislip, "Startup Raises $2 Million for Small-Scale Wind Turbines," *Reuters*, May 12, 2009; "Green Energy Technologies Introduces Wind-Cube," *North American Windpower* (Southbury, CT), May 14, 2009; Shelley Widhalm, "Rooftop Wind Turbine Seen as Green Energy Starter Kit," *Loveland Reporter-Herald* (Loveland, CO), May 16, 2009; "Startup Green Energy Tech Installs First Small-Wind Concentrators," *Reuters*, June 3, 2009; "Inside a Small Wind-Turbine Beta Test," *Green Tech* (Boston), July 22, 2009; Chris Atchison, "Turbine developer feels the wind at his back," *The Globe and Mail* (Ontario, Canada), October 20, 2009.

135. Michael L. S. Bergey, chairman, Performance Subcommittee, American Wind Energy Association, *Performance Rating Document* (Washington, DC: American Wind Energy Association, [1980]), 1–3.

136. "ANSI Approves AWEA as Standards-Making Body," *AWEA Wind Energy Weekly* (Washington, DC) 14, no. 639 (March 20, 1995): n.p.

137. Paul Gipe, "Comments on a Proposed Performance Standard for Small Wind Turbines," Wind-Works, May 25, 2000, accessed July 4, 2019, http://www.wind-works.org/cms/index.php?id=64&tx_ttnews%5Btt_news%5D=54&cHash=69ab65d9bf5142d330e8b7b70be45c40.

138. American Wind Energy Association, *AWEA Small Wind Turbine Performance and Safety Standard*[,] *AWEA 9.1–2009* (Washington, DC: American Wind Energy Association, 2009), II-IV.

139. *AWEA Small Wind Turbine Performance and Safety Standard*[,] *AWEA 9.1–2009*, 1–6; Colin Miner, "Standards for Small-Scale Wind Power," *The New York Times*, August 28, 2009; Larry Sherwood, president and CEO, Interstate Renewable Energy Council Inc., Latham, NY, telephone interview by author, March 23, 2018.

140. *AWEA Small Wind Turbine Performance and Safety Standard*[,] *AWEA 9.1–2009*, 2.

141. Matthew Gagne, "How Long Will It Take to Certify a Turbine?" Small Wind Certification Council, Brea, CA, accessed March 6, 2018, http://smallwindcertification.org.

142. "Interview with Ron Stimmel: AWEA Small Wind Advocate," *Small Wind Newsletter* (Interstate Renewable Energy Council, Latham, NY), no. 38 (July 2009), n.p.

143. "Applications—Small Turbines," Small Wind Certification Council, accessed May 14, 2018, http://smallwindcertification.org/for-applicants/small-turbines/application-process-small/applications/; "AWEA 9.1 Standard Testing and Certification," Intertek, accessed May 14, 2018, http://www.intertek.com/wind/awea-standard/.

144. Michael L. S. Bergey, president and CEO, Bergey Windpower Company, Norman, OK, email message to author, December 28, 2017.

145. Ibid.

146. Ibid.

147. "About Us," Small Wind Conference, LLC, accessed October 26, 2019, http://smallwindconference.com/?page_id=2. The last Small Wind Conference was held at Stevens Point, Wisconsin, in June 2018.

148. "DOE Seeks Comment on Distributed Energy Generation Research Plans," *AWEA Wind Energy Weekly* (Washington, DC) 11, no. 499 (May 18, 1992): n.p.; Distributed Wind Energy Association, *Distributed Wind Energy Association* (Durango, CO: Distributed Wind Energy Association, [2010]), 1–4.

149. Mike Bergey, "An Overview of DWEA[:] New National Trade Association for Distributed Wind" (presentation at the Sixth Annual Small Wind Conference, Stevens Point, WI, June 15–16, 2010), 5–8, 11–12.

150. Bergey email; Jennifer Jenkins, Flagstaff, AZ, telephone interview by author, January 18, 2018. In July 2018, Jenkins joined the American Wind Energy Association to lead its distributed wind program. See American Wind Energy Association, Washington, DC, *AWEA Hires Jennifer Jenkins to Launch National Distributed Wind Program*, July 24 2018.

151. "American Wind Power and AWEA Grow by Leaps and Bounds," American Wind Energy Association, accessed June 18, 2019, https://www.awea.org/wind-101/history-of-wind/2000s.

152. Bergey email.

153. Distributed Wind Energy Association, *DWEA 2012 Annual Report* (Durango, CO: Distributed Wind Energy Association, 2013), 2.

154. Ibid.

155. *DWEA 2012 Annual Report*, 2; Jenkins interview.

156. *DWEA 2012 Annual Report*, 1.

157. Distributed Wind Energy Association, *DWEA Distributed Wind Vision—2015–2030[:] Strategies to Reach 30 GW of "Behind-the-Meter" Wind Generation by 2030[,] March 2015* (Durango, CO: Distributed Wind Energy Association, 2015), 6–10.

158. Ibid., 20.

159. Ibid., 21.

160. Ibid., 22–23.

161. National Institute of Standards and Technology, US Department of Commerce, Washington, DC, *SMART Wind Consortium: Developing a Consensus Based Sustainable Manufacturing, Advanced Research and Technology Roadmap for Distributed Wind*, May 6, 2014.

162. Jennifer Jenkins, Heather Rhoads-Weaver, Trudy Forsyth, Brent Summerville, Ruth Baranowski and Britton Rife, *SMART Wind Roadmap: A Consensus-Based, Shared Vision[,] Sustainable Manufacturing, Advanced Research & Technology Action Plan for Distributed Wind* (Durango, CO: Distributed Wind Energy Association, 2016), 1, 18.

163. Ibid., 2–5.

164. Ibid., VII.

165. Ibid., C-2—C-9.

166. Ibid., 44–49.

167. Jenkins interview; Padma Kasthurirangan, president, Buffalo Renewables, Niagara Falls, NY, "Welcome and Opening Remarks" (presentation at Distributed Wind 2018, Washington, DC, February 28, 2018).

168. Bipartisan Budget Act of 2018, H. Res. 1892, 115th Cong., 2nd sess. (February 9, 2018): H1892–87; Distributed Wind Energy Association, Durango, CO, *DWEA Lauds New ITC Legislation Supporting Distributed Wind Power*, February 9, 2018.

169. American Wind Energy Association, Washington, DC, *Wind Power Closes 2017 Strong, Lifting the American Economy*, January 30, 2018; US Geological Survey, Reston, VA, *Mapping the Nation's Wind Turbines*, May 16, 2018.

170. Mike Bergey, president and CEO, Bergey Windpower, "Leadership Panel: State of the Industry" (presentation at Distributed Wind 2018, Washington, DC, February 28, 2018).

Chapter 5

1. Patricia Buffer, *Beyond the Buildings at the Place Called 'Rocky Flats'* (Washington, DC: Office of Legacy Management, US Department of Energy, 2003), 3.

2. "Rocky Flats: Testing Small-Scale Wind Systems," *Wind Power Digest* no 7 (December 1976): 39; Darrell Dodge, "To Catch the Wind," *Rockwell News, Rocky Flats Plant* (Boulder, CO) 7, no. 8 (June 20, 1980): 3.

3. Dodge, "To Catch the Wind," 3.

4. Louis Divone, "Overview of WECS Program" (paper presented at Second Workshop on Wind Energy Conversion Systems, Washington, DC, June 9–11, 1975), 5–6.

5. "The Federal Wind Program," *Wind Power Digest* no. 13 (Fall 1978): 42–43.

6. "Rocky Flats: Testing Small-Scale Wind Systems."

7. National Aeronautics and Space Administration, Lewis Research Center, Cleveland, OH, *Wind Energy Systems: A Non-Pollutive, Non-Depletable Energy*, December 1973 (in author's files).

8. Donald J. Vargo, *Wind Energy Developments in the 20th Century* (Cleveland, OH: Lewis Research Center, National Aeronautics and Space Administration, 1975), 7–8.

9. Solar Energy Research, Development, and Demonstration Act of 1974, Public Law 93–473, 88 Stat. 1431 (1974), codified at 42 U. S. C. 5551, et seq. (1988).

10. "1970s: Oil Prices Skyrocket & AWEA Emerges," American Wind Energy Association, accessed June 18, 2019, https://www.awea.org/wind-101/history-of-wind/1970s.

11. Samuel Walters, "Power from Wind," *Mechanical Engineering* 96, no. 4 (April 1974): 55; Daniel M. Simmons, *Wind Power* (Park Ridge, NJ: Noyes Data Corporation, 1975), 138.

12. Joseph M. Savino, "Introduction to Wind Energy Conversion Systems Technology" (paper presented at Second Workshop on Wind Energy Conversion Systems. Washington, DC, June 9–11, 1975), 200.

13. Ronald L. Thomas, "Introduction to Large System Design" (paper presented at Second Workshop on Wind Energy Conversion Systems, Washington, DC, June 9–11, 1975), 13.

14. Vargo, *Wind Energy Developments,* 5–6.

15. Ibid.

16. Richard L. Puthoff, "100 KW Experimental Wind Turbine Generator Project" (paper presented at Second Workshop on Wind Energy Conversion Systems, Washington, DC, June 9–11, 1975), 21–25.

17. James L. Schefter, *Capturing Energy from the Wind* (Washington, DC: Scientific and Technical Information Branch, National Aeronautics and Space Administration, 1982), 54–67; "Three-Machine Windfarm Starts Up, Then Shuts Down for Repairs," *Solar Times* 3, no. 7 (July

1981): 1; "Electricity-Producing Windmills Generate Hope," *The Standard-Times* (New Bedford, MA), September 5, 1979; Christopher Gillis, *Windpower* (Atglen, PA: Schiffer Publishing Ltd., 2008), 45–50.

18. "Small Wind Systems 1979[,] SWECS Representatives Meet in Boulder," *Wind Power Digest*, no. 15 (Spring 1979): 27–28.

19. Dodge, "To Catch the Wind," 3.

20. Wind Systems Branch, Division of Solar Technology, US Department of Energy, Washington, DC, *Technical and Management Support for the Development of Wind Systems for Farm, Remote and Rural Use[,] Annual Report for the Period October 1976—September 1977[,] Wind Systems Program[,] Rocky Flats Plant[,] Rockwell International[,] October 1977*, Contract E(20–2)-3533 (Washington, DC: US Department of Energy, 1977), 12–15.

21. Ibid., 15–22.

22. *Technical and Management Support for the Development of Wind Systems for Farm, Remote and Rural Use*, 22; Wind Systems Branch, Division of Solar Technology, US Department of Energy, Washington, DC, *A Guide to Commercially Available Wind Machines[,] April 3, 1978* (Washington, DC: US Department of Energy, 1978), 38.

23. *Technical and Management Support for the Development of Wind Systems for Farm, Remote and Rural Use*, 22; *A Guide to Commercially Available Wind Machines*, 49.

24. *Technical and Management Support for the Development of Wind Systems for Farm, Remote and Rural Use*, 22; *A Guide to Commercially Available Wind Machines*, 54–55, 65.

25. *Technical and Management Support for the Development of Wind Systems for Farm, Remote and Rural Use*, 22; *A Guide to Commercially Available Wind Machines*, 77.

26. *Technical and Management Support for the Development of Wind Systems for Farm, Remote and Rural Use*, 22; *A Guide to Commercially Available Wind Machines*, 72, 76.

27. Michael Evans, "Rocky Flats Small Wind System Tests[,] Manufacturers Prove Your Specs!," *Wind Power Digest*, no. 24 (Summer 1982): 37–42.

28. *Technical and Management Support for the Development of Wind Systems for Farm, Remote and Rural Use*, 22; *A Guide to Commercially Available Wind Machines*, 44, 67.

29. *Technical and Management Support for the Development of Wind Systems for Farm, Remote and Rural Use*, 25–32.

30. "Rocky Flats: Testing Small-Scale Wind Systems," 39.

31. "Small Is Beautiful," *New Scientist* 73, no.1042 (March 10, 1977): 574; *Technical and Management Support for the Development of Wind Systems for Farm, Remote and Rural Use*, B-1; Warren S. Bollmeier II, Kaneohe, HI, telephone interview by author, June 12, 2016.

32. "Small Is beautiful."

33. "$60 Million U. S. Program Helps Accelerate Use of Wind Power," *Canadian Renewable Energy News* (March 1979): 14.

34. Jonathan Hodgkin, Essex Junction, VT, telephone interview by author, January 25, 2016.

35. *Technical and Management Support for the Development of Wind Systems for Farm, Remote and Rural Use*, 35–36.

36. Ibid., 36, 38–47.

37. "Rocky Flats Small Wind System Tests[,] Manufacturers Prove Your Specs!," 40.

38. Ibid., 43.

39. Warren S. Bollmeier, II, Kaneohe, HI, email to author, August 1, 2019.

40. "Rocky Flats: Testing Small-Scale Wind Systems," 41.

41. *Technical and Management Support for the Development of Wind Systems for Farm, Remote and Rural Use*, 48–50.

42. *Technical and Management Support for the Development of Wind Systems for Farm, Remote and Rural Use*, 55–59; T. A. Murrin, "UTRC 8 KW Wind Turbine Development" (paper presented at Small Wind Turbine Systems 1981, Boulder, CO, May 12–14, 1981), 68–76; V. Daniel Hunt, *Windpower[:] A Handbook on Wind Energy Conversion Systems* (New York: Van Nostrand Reinhold Company, 1981), 389–92; "DOE Contracts Awarded," *Wind Power Digest* no. 11 (Winter 1977–1978): 26–27.

43. *Technical and Management Support for the Development of Wind Systems for Farm, Remote and Rural Use*, 50.

44. Ibid.

45. Ibid., 51.

46. Hunt, *Windpower*, 394; B. A. Goodale, "Development and Testing of the Kaman 40 KW Wind Turbine Generator" (paper presented at Small Wind Turbine Systems 1981, Boulder, CO, May 12–14, 1981), 82–92.

47. Christopher C. Gillis, *Still Turning[:] A History of Aermotor Windmills* (College Station: Texas A&M University Press, 2015), 157–158; Robert D. McConnell, "Giromill Overview" (paper presented at Wind Energy Innovative Systems Conference, Colorado Springs, CO, May 23–25, 1979), 1–12; John Anderson, "The 40 KW Giromill" (paper presented at Small Wind Turbine Systems 1981, Boulder, CO, May 12–14, 1981), 93–105.

48. *Technical and Management Support for the Development of Wind Systems for Farm, Remote and Rural Use*, 60.

49. Ibid., 60.

50. Hunt, *Windpower*, 385, 387–88; "DOE Contracts Awarded."

51. Hunt, *Windpower*, 385, 387–88.

52. Ibid.

53. *Technical and Management Support for the Development of Wind Systems for Farm, Remote and Rural Use*, A-1.

54. *Still Turning*, 158–59.

55. Windworks, Mukwonago, WI, *Windworks—An Energy Systems Corporation • R&D • Product Engineering • Manufacturing* (Mukwonago, WI: Windworks, Inc., [ca. 1980]), brochure (in author's files).

56. Enertech, Norwich, VT, *Enertech 1500*[:] *A Beautiful Way to Save Electricity* (Norwich, VT: Enertech, 1979), brochure (in *Windmillers' Gazette* research files, Rio Vista, TX); Enertech, Norwich, VT, *Enertech E44* (Norwich, VT: Enertech, [ca. 1982]), 1, 3 (in author's files).

57. North Wind Power Company, Warren, VT, *Reliable Power from the Wind*[:] *Cost Effective for Remote Applications* (Warren, VT: North Wind Power Company, 1980), 1–8, brochure (in author's files); North Wind Power Company, Moretown, VT, *North Wind Power Co.* (Moretown, VT: North Wind Power Company, ca. 1982), brochure (in author's files).

58. I. B. Allen, "Status of SWECS Testing Programs at the Rocky Flats Small Wind Systems Test Center" (paper presented at Small Wind Turbine Systems 1981, Boulder, CO, May 12–14), 53.

59. Ibid., 53–54.

60. Ibid., 54–56.

61. Dodge, "To Catch the Wind," 3–4; "Small Wind Systems 1979," 29.

62. Dodge, "To Catch the Wind," 4.

63. Allen, "Status of SWECS Testing Programs," 56–57; Chris Gillis, "Shake, Rattle and Roll," *American Shipper* (Jacksonville, FL) 52, no. 11 (November 2010): 60.

64. *Technical and Management Support for the Development of Wind Systems for Farm, Remote and Rural Use*, 65–67; William E. Heronemus, "The University of Massachusetts Wind Furnace Project: A Summary Statement, June 1975" (paper presented at Second Workshop on Wind Energy Conversion Systems, Washington, DC, June 9–11, 1975), 404–07; William Heronomous [sic.], "The U. Mass. Wind Furnace Program," *Wind Power Digest* 1, no. 7 (December 1976): 34–38; Mary Prince and Woody Stoddard, "University of Massachusetts Solar Wind House," *Alternative Sources of Energy* no. 26 (June 1977): 26–27; D. J. De Renzo, ed., *Wind Power*[:] *Recent Developments* (Park Ridge NJ: Noyes Data Corporation, 1979), 22, 308–11.

65. *Technical and Management Support for the Development of Wind Systems for Farm, Remote and Rural Use*, 67–68; "New Design for Windmill on Display," *Osceola Sun* (Kissimmee, FL), February 20, 1974; E. F. Lindsley, "Wind Power: How New Technology Is Harnessing an Age-Old Energy Source," *Popular Science* 205, no. 1 (July 1974): 54–55, 124; Simmons, *Wind Power*, 108.

66. Rockwell International, Rocky Flats Wind Systems Program, Golden, CO, *Commercially Available Small Wind Systems and Equipment*[,] *March 31, 1981*[,] *A Checklist Prepared by the Rocky Flats Wind Systems Program* (Golden, CO: Rockwell International, 1981), 1–21.

67. Thomas G. Bolle, *Financial Problems Facing the Manufacturers of Small Wind Energy Conversion Systems*[,] *Final Report*[,] *November 1979* (Washington, DC: JDB & Company, 1979), i.

68. W. S. Bollmeier, C. P. Butterfield, R. P. Dingo, D. M. Dodge, A. C. Hansen, D. C. Shepherd, and J. L. Tangler, *Small Wind Systems Technology Assessment*[:] *State of the Art and Near Term Goals*[,] *February 1980* (Golden, CO: Rockwell International, 1980), i.

69. Ibid., 42.

70. Allen, "Status of SWECS Testing Programs," 51.

71. Dodge, "To Catch the Wind," 3.

72. Allen, "Status of SWECS Testing Programs," 51.

73. Dodge, "To Catch the Wind," 4.

74. *Technical and Management Support for the Development of Wind Systems for Farm, Remote and Rural Use*, 81–82.

75. Dodge, "To Catch the Wind," 4.

76. Gillis, *Windpower*, 50; R. H. Braasch, "Power Generation with a Vertical-Axis Wind Turbine" (paper presented at Second Workshop on Wind Energy Conversion Systems, Washington, DC, June 9–11, 1975), 417–19.

77. Gillis, *Windpower*, 50; P. C. Klimas, "Darrieus Wind Turbine Program at Sandia Laboratories" (paper presented at Wind Energy Innovative Systems Conference, Colorado Springs, CO, May 23–25, 1979), 59.

78. Gillis, *Windpower*, 50; Klimas, 59; US Department of Energy, Division of Solar Technology, *Federal Wind Energy Program*[,] *Program Summary*[,] *January 1978* (Washington, DC: US Department of Energy, 1978), 4.

79. Gillis, *Windpower*, 50; Klimas, 61, 63, 66.

80. W. Ken Bell, "Wind Energy Information at Sandia Laboratories" (paper presented at Panel on Information Dissemination for Wind Energy, Albuquerque, NM, August 2–3, 1979), 31.

81. Gillis, *Windpower*, 50; Herbert J. Sutherland, Dale E. Berg, and Thomas D. Ashwill, *Sandia Report*[:] *A Retrospective of VAWT Technology* (Albuquerque, NM: Sandia National Laboratories, 2012), 9.

82. US Department of Agriculture, Agricultural Research Service, *Agriculture Information Bulletin Number 446*[,] *Wind Power Research for Agriculture* (Washington, DC: US Department of Agriculture, 1981), 4–5; Joe Carter, "USDA's Wind Testing Program," *Wind Power Digest* no. 8 (Spring 1977): 38.

83. US Department of Agriculture, Agricultural Research Service, Southern Great Plains Research Center, *Irrigation Pumping with Wind Energy* (Bushland, TX: US Department of Agriculture, 1978), 1; Paul Gipe, "USDA: Defining Wind's Role in Agriculture[,] Applied Wind Energy Research Continues," *Wind Power Digest* no. 18 (Winter 1979–1980): 14.

84. Carter, 38–39; *Irrigation Pumping with Wind Energy*, 2–4.

85. US Department of Agriculture, Agricultural Research Service, *Development of Rural and Remote Applications of Wind-Generated Energy*[,] *Summary of Expected Requests for Proposal*[,] *July 1977—June 1978* (Beltsville, MD: US Department of Agriculture, 1977), 1–4, with cover letter; Carter, 38–40; "USDA: Defining Wind's Role in Agriculture," 14–16; *Agriculture Information Bulletin Number 446*, 6–11.

86. R. Nolan Clark, "Applications Testing by USDA—An Overview" (paper presented at Small Wind Turbine Systems 1981, Boulder, CO, May 12–14, 1981), n.p.

87. "USDA: Defining Wind's Role in Agriculture," 16, 19; Mike Evans, "Wind in Rural America," *Wind Power Digest,* no. 25 (Fall 1982): 31–32.

88. Vaughn Nelson, Kenneth Starcher, and William Pinkerton, "AEI-Wind Test Center" (paper presented at Wind Energy Expo '86 and National Conference, Cambridge, MA, September 1–3, 1986), 42–51.

89. "Applications Testing by USDA—An Overview"; L. H. Soderholm, "Non-Interconnected Applications of Wind Energy" (paper presented at Rural Electric Wind Energy Workshop, Boulder, CO, June 1–3, 1982), 285–89.

90. Soderholm, "Non-Interconnected Applications," 289–90.

91. US Department of Agriculture and Alternative Energy Institute, *Wind Energy Research*[:] *System Evaluation* (Bushland, TX: US Department of Agriculture, Agricultural Research Service, [ca. 1983]), n.p., brochure (in author's files).

92. Gillis, *Windpower*, 51; Sutherland, 16–17; "Huge Windmill with One Blade is Planned," *The New York Times*, April 8, 1986.

93. National Academy of Sciences, *Establishment of a Solar Energy Research Institute* (Washington, DC: National Academy of Sciences, 1975), 3–4.

94. Pat Weis, "The Technical Information Dissemination Program for Wind Energy at SERI" (paper presented at Panel on Information Dissemination for Wind Energy, Albuquerque, NM, August 2–3, 1979), 8–9.

95. US Department of Energy, *Wind Energy Systems Program Summary*[,] *Fiscal Years 1981 and 1982* (Washington, DC: US Department of Energy, 1983), 17–18.

96. Ibid., 19–20.

97. Bob Irwin, "The Gas Crunch of '79," *AutoWeek* (May 14, 1979): 2; "Coping: Showing the Way to a Gas-Short Future," *AutoWeek* (May 14, 1979): 2; Tyler Priest, "The Dilemmas of Oil Empire," *Journal of American History* 99, no. 1 (June 2012): 236–51.

98. Donald Janson, "Radiation is Released in Accident at Nuclear Plant in Pennsylvania," *The New York Times*, March 29, 1979; "Backgrounder on the Three Mile Island Accident," U. S. Nuclear Regulatory Commission, accessed February 10, 2018, https://www.nrc.gov/reading-rm/doc-collections/fact-sheets/3mile-isle.html.

99. Adam Clymer, "Gas Shortage Spurs Carter Decline in Poll," *The New York Times*, July 13, 1979; Jimmy Carter, "Energy and the National Goals, July 15, 1979," in *Public Papers of the Presidents of the United States: Jimmy Carter, 1979, Book II-June 23 to December 31, 1979*, 1235 (Washington, DC: Government Printing Office, 1979).

100. "Energy and the National Goals," 1239.

101. Wind Energy Systems Act of 1980, Public Law 96–345, 96th Cong. (September 8, 1980), 1139–40.

102. *Wind Energy Systems Program Summary*[,] *Fiscal Years 1981 and 1982*, 7.

103. David Biello, "Where Did the Carter White House's Solar Panels Go?" *Scientific American*, August 6, 2010, accessed December 24, 2017, http://www.scientificamerican.com/article/carter-white-house-solar-panel-array.

104. "Plans for '83[:] Rocky Flats," *Wind Power Digest* no. 25 (Fall 1982): 27–28.

105. "New Center Formed[,] DOE Merges SERI and Rocky Flats," *World Wind* (Ottawa, Canada) 1, no. 2 (January 1985): 1.

106. US Department of Energy, Wind Energy Technology Division, *Five Year Research Plan 1985–1990*[,] *Wind Energy Technology: Generating Power from the Wind*[,] *January 1985* (Washington, DC: US Department of Energy, 1985), 24, 27; Daniel F. Ancona, "Federal Wind Energy Research Programs: Current Status and Future Plans" (paper presented at Wind Energy Expo '84 and National Conference, Pasadena, CA, September 24–26, 1984), 263.

107. Gillis, *Windpower*, 49.

108. Andrew R. Trenka, letter to the editor, "Update: Rocky Flats/SERI Merger," *Wind Power Digest* no. 27 (1985): 5.

109. Ibid.

110. *Five Year Research Plan 1985–1990*, 5.

111. Vaughn Nelson, *Wind Energy and Wind Turbines* (Canyon, TX: Alternative Energy Institute, West Texas A&M University, [ca. 1995]), 134.

112. Ibid., 134, 143–44.

113. Ibid., 134.

114. Biello, "Where Did the Carter White House's Solar Panels Go?"

115. Nelson, *Wind Energy and Wind Turbines*, 144.

116. George H. W. Bush, "Remarks on Designating the Solar Energy Research Institute as the National Renewable Energy Laboratory, September 16, 1991," in *Public Papers of the Presidents of the United States: George Bush, 1991, Book II-July 1 to December 31, 1991*, 1155–56 (Washington, DC: US Government Printing Office, 1992).

117. Nelson, *Wind Energy and Wind Turbines*, 144.

118. Amy Royden, "US Climate Change Policy under

President Clinton: A Look Back," *Golden Gate University Law Review* 32, no. 4 (January 2002): 416–20.

119. National Renewable Energy Laboratory, Golden, CO, *National Renewable Energy Laboratory to Reduce Staff*, November 14, 1995; National Renewable Energy Laboratory, Golden, CO, *NREL Funding Reductions to Further Impact Lab's Work Force*, December 22, 1995.

120. Royden, "US Climate Change Policy under President Clinton," 449–50.

121. National Renewable Energy Laboratory, Golden, CO, *Companies Selected for Small Wind Turbine Project*, November 27, 1996.

122. Office of Energy Efficiency and Renewable Energy, US Department of Energy, *Wind Powering America*[,] *America's Wind Power . . . A National Resource* (Washington, DC: US Department of Energy, 2001), brochure; Office Energy Efficiency and Renewable Energy, US Department of Energy, *Wind and Water Power Program*[,] *The Wind Powering America Initiative* (Washington, DC: US Department of Energy, 2011), handbill.

123. National Renewable Energy Laboratory, *National Renewable Energy Laboratory*[:] *Ten-Year Site Plan*[,] *FY2007-FY2018* (Washington, DC: Office of Energy Efficiency and Renewable Energy, US Department of Energy, 2006), 1–1, 1–2, 5.3–1.

124. Ibid., 5.3–1.

125. Ibid.

126. R. Nolan Clark and Brian Vick, "Determining the Proper Motor Size for Two Wind Turbines Used in Water Pumping" (paper presented at the 1995 American Society of Mechanical Engineers SED, Houston, TX, January 29—February 1, 1995), 65–72; Brian D. Vick, R. Nolan Clark, and Saiful Molla, "Performance of a 10 Kilowatt Wind-Electric Water Pumping System for Irrigating Crops" (paper presented at Windpower 1997, Austin, TX, June 15–18, 1997), 523–31; Brian D. Vick and R. Nolan Clark, "Ten Years of Testing a 10 Kilowatt Wind-Electric System for Small Scale Irrigation" (presented at the 1998 American Society of Agricultural Engineers Annual International Meeting, Orlando, FL, July 12–16, 1998), 1, 7–8; Brian D. Vick, Byron Neal, and R. Nolan Clark, "Wind Powered Irrigation for Selected Crops in the Texas Panhandle and South Plains" (paper presented at Windpower 2001, Washington, DC, June 3–7, 2001), 1–10; Brian D. Vick, Byron A. Neal, and R. Nolan Clark, "Performance of a Small Wind Powered Water Pumping System" (paper presented at Windpower 2008, Houston, TX, June 1–4, 2008), 1–13.

127. Don Comis, "Wind Power Where You Want It," *Agricultural Research* (US Department of Agriculture, Washington, DC) 42, no. 6 (June 1994): 4–7; R. Nolan Clark, "Wind-Electric Water Pumping Systems for Rural Domestic and Livestock Water" (paper presented at the 5th European Wind Energy Association Conference and Exhibition, Thessaloniki-Macedonia, Greece, October 10–14, 1994), 1136–37, 1139.

128. Paul Gipe, "Windmills: Still Primed for Pumping Water," *Independent Energy* 19, no. 6 (July –August 1989): 64; Michael L. S. Bergey, "Wind-Electric Pumping Systems for Communities" (paper presented at the 1st International Symposium on Safe Drinking Water in Small Systems, Washington, DC, May 10–13, 1998), n.p.

129. Brian D. Vick, R. Nolan Clark, and Shitao Ling, "One and a Half Years of Field Testing a Wind-Electric System for Watering Cattle in the Texas Panhandle" (paper presented at Windpower 1999, Burlington, VT, June 20–23, 1999), n.p.; Brian D. Vick, email to author, January 17, 2018.

130. Brian D. Vick and R. Nolan Clark, "Testing of a 2-Kilowatt Wind-Electric System for Water Pumping" (paper presented at Windpower 2000, Palm Springs, CA, April 30-May 4, 2000), n.p.

131. Brian D. Vick and R. Nolan Clark, "Performance and Economic Comparison of a Mechanical Windmill to a Wind-Electric Water Pumping System" (paper presented at the 1997 ASAE Annual International Meeting, Minneapolis, MN, August 10–13, 1997), 6.

132. Brett G. Ziter, "Electric Wind Pumping for Meeting Off-Grid Community Water Demands," *Guelph Engineering Journal* (Ottawa, Canada) 2 (2009): 15.

133. Brian D. Vick and Byron A. Neal, "Analysis of Off-Grid Hybrid Wind Turbine/Solar PV Water Pumping Systems," *Solar Energy* 86 (2012): 1206.

134. Brian D. Vick, R. Nolan Clark, and David Carr, "Analysis of Wind Farm Efficiency Produced in the United States" (paper presented at Windpower 2007, Los Angeles, June 3–6, 2007), 1–13.

135. Donny R. Cagle, Anthony D. May, Brian D. Vick, and Adam J. Holman, "Evaluation of Airfoils for Small Wind Turbines" (paper presented at Windpower 2007, Los Angeles, June 3–6, 2007), 6.

136. Ibid., 2, 6.

137. Brian D. Vick, email to author, January 10, 2018.

138. M. Jureczko, M. Pawlak and A. Mężyk, "Optimisation of Wind Turbine Blades," *Journal of Materials Processing Technology* 167 (2005): 467–68; Jose R. Zayas and Wesley D. Johnson, *Sandia Report*[:] *3X-100 Blade Field Test* (Albuquerque, NM: Sandia National Laboratories, 2008), 13, 17–18, 39–40, 63; Karen Wood, "Wind Turbine Blades: Glass vs. Carbon Fiber," *Composites-World*, May 31, 2012, accessed January 14, 2018; https://www.compositesworld.com/articles/wind-turbine-blades-glass-vs-carbon-fiber; R. Nolan Clark, Amarillo, TX, telephone interview by author, January 2, 2018.

139. Vick, email, January 10, 2018; Brian Vick and Sylvia Broneske, "Effect of Blade Flutter and Electrical Loading on Small Wind Turbine Noise," *Renewable Energy* 50 (2013): 1044–52; Brian D. Vick and Tim A. Moss, "Adding Concentrated Solar Power Plants to Wind

Farms to Achieve a Good Utility Electrical Load Match," *Solar Energy* 92 (2013): 298–312.

140. Vick, email, January 10, 2018; David Brauer, acting laboratory director, Conservation and Production Research Laboratory, Agricultural Research Service, US Department of Agriculture, Bushland, TX, email to author, January 2, 2018.

141. Alice C. Orrell, Nikolas F. Foster, Scott L. Morris, and Juliet S. Homer, *2016 Distributed Wind Market Report* (Richland, WA: Pacific Northwest National Laboratory, 2017), 13–14.

142. Sandia National Laboratories, Albuquerque, NM, *Advanced Wind Energy Projects Test Facility Moving to Texas Tech University*, July 27, 2011; Chris Cook, Texas Tech University, Lubbock, TX, *Advanced Wind Energy Test Facility Makes a Move*, July 27, 2011.

143. *Advanced Wind Energy Projects Test Facility Moving to Texas Tech University*; *Advanced Wind Energy Test Facility Makes a Move*.

144. Brian Naughton, "Wind Plant Optimization[:] Sandia Wake Imaging System Successfully Deployed at the SWiFT Facility," *Wind & Water Power Newsletter* (Sandia National Laboratories) (July-August 2015), accessed February 3, 2018, http://content.govdelivery.com/accounts/USDOESNLEC/bulletins/114227e.

145. Karin Slyker, Texas Tech University, Lubbock, TX, *Texas Tech Commissions New Wind Research Facility*, July 9, 2013.

146. Ibid.

147. Jonathan R. White, Wind Energy Technologies Department, Sandia National Laboratories, Albuquerque, NM, email to author, February 8, 2018.

148. Chris Lo, "Vestas's Multi-Rotor Wind Turbine: Are Four Rotors Better Than One?" *Power Technology* (October 17, 2016), accessed February 10, 2018, http://www.power-technology.com/features/featurevestass-multi-rotor-wind-turbine-are-four-rotors-better-than-one-5001819/.

149. White, email.

150. US Department of Energy, Office of Energy Efficiency and Renewable Energy, Solar Energy Technologies Office, Washington, DC, *The SunShot Initiative*, 2017, accessed January 31, 2018, https://energy.gov/eere/solar/sunshot-initiative.

151. Jose Zayas, "DOE's Competitiveness Improvement Projects Are Delivering Substantial Cost Reductions in the U. S. Distributed Wind Industry," *North American Clean Energy* (July-August 2017), accessed January 31, 2018, http://www.nacleanenergy.com/articles/27632/doe-s-competitiveness-improvement-projects-are-delivering-substantial-cost-reductions-in-the-u-s-distributed-wind-industry; US Department of Energy, Office of Energy Efficiency and Renewable Energy, Washington, DC, *Energy Department Helps Manufacturers of Small and Mid-Size Wind Turbines*

Meet Certification Requirements, October 1, 2015, accessed January 23, 2018, https://energy.gov/eere/articles/energy-department-helps-manufacturers-small-and-mid-size-wind-turbines-meet.

152. Patrick Gilman, modeling and analysis program manager, Wind Energy Technologies Office, US Department of Energy, Golden, CO, telephone interview by author, January 24, 2018.

153. Gilman interview; National Renewable Energy Laboratory, Golden, CO, *NREL Announces New Distributed Wind Competitiveness Improvement Project Contracts*, November 1, 2017, accessed January 31, 2018, https://www.nrel.gov/news/program/2017/nrel-announces-new-distributed-wind-competitiveness-improvement-project-contracts.html.

154. US Department of Energy, Office of Energy Efficiency and Renewable Energy, Wind and Water Power Technologies Office, Washington, DC, *Distributed Wind Competitiveness Improvement Project* (Washington, DC: US Department of Energy, 2016), handbill (in author's files).

155. US Department of Energy, Office of Energy Efficiency and Renewable Energy, Washington, DC, *Pika Energy Develops Innovative Manufacturing Process and Lowers Production Cost Under DOE Competitiveness Improvement Project*, September 16, 2015, accessed January 23, 2018, https://energy.gov/eere/wind/articles/pika-energy-develops-innovative-manufacturing-process-and-lowers-production-cost; US Department of Energy, Office of Energy Efficiency and Renewable Energy, Washington, DC, *Wind Turbine Showcased in Energy Department Headquarters*, February 26, 2016, accessed January 23, 2018, https://energy.gov/eere/articles/wind-turbine-showcased-energy-department-headquarters.

156. US Department of Energy, Office of Energy Efficiency and Renewable Energy, Washington, DC, *EERE Success Story—Distributed Wind Competitiveness Improvement Project* (CIP) *Partner Delivers Next-Generation Wind Turbine for On-Site Power*, April 11, 2017, accessed January 23, 2018, https://energy.gov/eere/success-stories/articles/eere-success-story-distributed-wind-competitiveness-improvement.

157. *NREL Announces New Distributed Wind Competitiveness Improvement Project Contracts*.

158. *2016 Distributed Wind Market Report*, 13.

Chapter 6

1. Steve Weichelt, "The Plowboy Interview[:] Marcellus Jacobs," *The Mother Earth News* no. 24 (November 1973): 58;

2. *Production of Power by Means of Wind-Driven Generator: Hearing before the Committee on Interior and Insular Affairs*, HR 4286, 82nd Cong., 1st sess., September 19, 1951, 40–41.

3. Robert W. Righter, "Reaping the Wind[:] The Jacobs Brothers, Montana's Pioneer 'Windsmiths,'" *Montana[:] The Magazine of Western History* 46, no. 4 (Winter 1996): 45–46.

4. Paul R. Jacobs, owner of the Jacobs Wind Electric Company, interview by author, Minneapolis, MN, September 30, 2016.

5. Paul R. Jacobs interview; Patent Nos. 3,326,005 and 3,733,830, Marcellus L. Jacobs, US Patent and Trade Office, Washington, DC.

6. Paul R. Jacobs, Corcoran, MN, email message to author, February 5, 2017.

7. Paul R. Jacobs, Corcoran, MN, email message to author, January 29, 2017.

8. Ibid.

9. Ibid.

10. Ibid.

11. Ibid.

12. Ibid.

13. Ibid.

14. Ibid.

15. Ibid.

16. Ibid.

17. "The Plowboy Interview[:] Marcellus Jacobs," 52–58.

18. Ibid., 52.

19. Ibid., 58.

20. Paul R. Jacobs, Corcoran, MN, email to author, January 26, 2017.

21. Paul R. Jacobs, Corcoran, MN, email to author, January 24, 2017.

22. Senator Lee Metcalf, *From the Office of Senator Lee Metcalf*[,] *News Release*, October 16, 1974 (in the files of Paul R. Jacobs, Corcoran, MN).

23. Senator Metcalf of Montana, speaking on Wind Power, on August 7, 1974, 93rd Cong., 2nd sess., *Congressional Record* 120, pt. 20: 27081–27322.

24. Roger Bergerson, "40 Years with Jacobs' Gems," *Public Power* 37, no. 5 (September–October 1979): 19.

25. Email to author, February 5, 2017.

26. M. L. Jacobs, letter of the editor, "Marcellus Jacobs on Raising a Tower," *Alternative Sources of Energy* no. 29 (December 1977): 11–13.

27. "The American Wind Energy Association," *The Mother Earth News* no. 30 (November 1974): 16. Paul Jacobs ended the Jacobs Wind Electric Company's membership to AWEA in the late 1990s after the association reorganized its fees to support the interests of large wind turbine manufacturers and wind farm developers.

28. "Reaping the Wind," 47.

29. Paul R. Jacobs, Corcoran, MN, email to author, January 31, 2017.

30. Paul R. Jacobs, Corcoran, MN, email to author, January 26, 2017.

31. Ibid.

32. M. L. Jacobs, president, Jacobs Wind Electric Company, Fort Myers, FL, letter to T. J. Healy, manager, Wind Systems Program, Rocky Flats [Plant], Golden, CO, July 29, 1977; M. L. Jacobs, president, Jacobs Wind Electric Company, Fort Myers, FL, letter to Terry J. Healy, Rockwell International, Atomics International Division, Rocky Flats Plant, Golden, CO, February 17, 1978; Terry J. Healy, manager, Rockwell International, Wind Systems Program, Atomics International Division, Rocky Flats Plant, Golden, CO, letter to M. L. Jacobs, president, Jacobs Wind Electric Company, Fort Myers, FL, March 17, 1978; M. L. Jacobs, president, Jacobs Wind Electric Company, Fort Myers, FL, letter to Terry J. Healy, Rockwell International, Atomics International Division, Rocky Flats Plant, Golden, CO, March 17, 1978; M. L. Jacobs, president, Jacobs Wind Electric Company, Fort Myers, FL, letter to T. J. Healy, Manager, Wind Systems Program, Rocky Flats Plant, Energy Systems Group, Rockwell International, Golden, CO (includes an eleven-page paragraph-by-paragraph breakdown by the Jacobs Wind Electric Company calling out the errors in the 1979 Rocky Flats report), July 12, 1979; M. L. Jacobs, president, Jacobs Wind Electric Company, Fort Myers, FL, letter to R. Anderson, chairman of the board, Rockwell International, Pittsburgh, PA, July 13, 1979; R. O. Williams Jr., vice president and general manager, Rockwell International, Rocky Flats Plant, Energy Systems Group, Golden, CO, letter to M. L. Jacobs, president, Jacobs Wind Electric Company, Fort Myers, FL, August 2, 1979; R. O. Williams, Jr., vice president and general manager, Rockwell International, Rocky Flats Plant, Energy Systems Group, Golden, CO, letter to M. L. Jacobs, president, Jacobs Wind Electric Company, Fort Myers, FL, August 14, 1979; M. L. Jacobs, president, Jacobs Wind Electric Company, Fort Myers, FL, letter to R. O. Williams, Jr., vice president and general manager, Rocky Flats Plant, Energy Systems Group, Rockwell International, Golden, CO, August 23, 1979; M. L. Jacobs, [president], Jacobs Wind Electric Company, Fort Myers, FL, letter to Robert Anderson, chairman of the board, Rockwell International, Pittsburgh, PA, August 24, 1979 (all of the above in the files of Paul R. Jacobs, Corcoran, MN).

33. Paul R. Jacobs, Corcoran, MN, email to author, February 3, 2017.

34. Paul R. Jacobs, Corcoran, MN, email to author, October 2, 2016.

35. Jacobs Wind Electric Company, Fort Myers, FL, *Important Features of the New Jacobs 8 KVA Wind Energy System* (Fort Myers, FL: Jacobs Wind Electric Company, 1979), handbill (in the files of Paul R. Jacobs, Corcoran, MN).

36. Jacobs Wind Electric Company, Fort Myers, FL, *New Jacobs Wind Energy System Summary Information* (Fort Myers, FL: Jacobs Wind Electric Company, 1979), handbill (in the files of Paul R. Jacobs, Corcoran, MN).

37. Patent Nos. 3,891,347; 4,059,771; 4,068,131; 4,088,420; 4,228,361; 4,228,363; and 4,297,075, Marcellus L. Jacobs and Paul R. Jacobs, US Patent and Trade Office, Washington, DC.

38. Jacobs Wind Electric Company, Fort Myers, FL, *Preliminary Report from Suppliers for Components of New Wind Electric Plant*[:] *26 March 1979* (Fort Myers, FL: Jacobs Wind Electric Company, 1979), typewritten document (in the files of Paul R. Jacobs, Corcoran, MN).

39. Ibid.

40. Paul R. Jacobs, Corcoran, MN, email to author, February 3, 2017.

41. *Preliminary Report from Suppliers for Components of New Wind Electric Plant.*

42. Jacobs Wind Electric Company, Fort Myers, FL, and Control Data Corporation Minneapolis, MN, *Statement of Understanding by and between Marcellus L. Jacobs Doing Business as Jacobs Wind Electric Co., Inc. and Control Data Corporation* (Minneapolis, MN: Jacobs Wind Electric Company and Control Data Corporation, 1979), 1–4 (in the files of Paul R. Jacobs, Corcoran, MN); E. E. Strickland, Control Data Corporation, Minneapolis, MN, Interoffice Memorandum (Jacobs Wind Generators), March 20, 1979 (in the files of Paul R. Jacobs, Corcoran, MN); "Jacobs Unveils New Machine," *Wind Power Digest* no. 17 (Fall 1979): 18; Elizabeth Holland, "Jacobs Is Back in the Wind Machine Business," *Solar Age* 5, no. 2 (February 1980): 18.

43. Paul R. Jacobs, Corcoran, MN, email to author, January 31, 2017.

44. Ibid.

45. Ibid.

46. Paul R. Jacobs, Corcoran, MN, email to author, January 29, 2017.

47. Paul R. Jacobs, Corcoran, MN, email to author, January 31, 2017.

48. M. L. Jacobs, Jacobs Wind Electric Company, Plymouth, MN, Notice to Dealers[:] Product Improvements, May 8, 1981 (Plymouth, MN: Jacobs Wind Electric Company, 1981), 1–2, handbill (in the files of Paul R. Jacobs, Corcoran, MN).

49. Paul R. Jacobs, Corcoran, MN, email to author, January 31, 2017.

50. Shoshana Hoose, "Wind Power Propels Family's Trade," *The Minneapolis Star*, January 7, 1982.

51. Nancy L. Roberts, "Wind Power[:] Time Has Come, Again, for This Energy Pioneer," *The Dispatch* (Minneapolis), May 4, 1981.

52. Jon Naar, *The New Wind Power* (New York: Penguin Books, 1982), 76–78.

53. Paul R. Jacobs, Corcoran, MN, email to author, January 29, 2017.

54. Chuck Raasch, "Wind Power on the Brink of Revival," *USA Today*, April 5, 1983.

55. "CDC Taking Charge of Jacobs Wind Electric," *Wind Energy Report* 6, no. 1 (January 14, 1983): 7.

56. Paul R. Jacobs, Corcoran, MN, email to author, February 6, 2017.

57. "Jacobs Wind Farm Being Built in Hawaii," *Wind Energy Report* 6, no. 1 (January 14, 1983): 7.

58. Renewable Energy Ventures, Inc., Washington, DC, *December 21, 1982*[,] *Joint Venture to Develop Wind Farms*[:] *A Business Plan*[,] *Prepared for Control Data Corporation, Minneapolis, Minnesota 55440* (Washington, DC: Renewable Energy Ventures, 1982), I (in the files of Paul R. Jacobs, Corcoran, MN).

59. "CDC Taking Charge of Jacobs Wind Electric."

60. Donald Marier, "Conference Report: The CALWEA Annual Meeting[,] Setting the Record Straight on Jacobs," *Alternative Sources of Energy*, no. 61 (May–June 1983): 31.

61. Paul R. Jacobs, Corcoran, MN, email to author, February 6, 2017.

62. *December 21, 1982*[,] *Joint Venture to Develop Wind Farms*[:] *A Business Plan*, 14.

63. Paul R. Jacobs, Corcoran, MN, email to author, February 6, 2017.

64. Jim Schefter, "New Harvest of Energy from Wind Farms," *Popular Science* 222, no. 1 (January 1983): 60.

65. Renewable Energy Ventures, Inc., Encino, CA, *REV Wind Power Partners 1981–1*[,] *Preview Product Profile*[,] *Summary* (Encino, CA: Renewable Energy Ventures, Inc. 1984), 1–6 (in the files of Paul R. Jacobs, Corcoran, MN); Renewable Energy Ventures, Inc., Encino, CA, *Renewable Energy Ventures, Inc.*[,] *A Subsidiary of Earth Energy Systems, Inc.* (Encino, CA: Renewable Energy Ventures, Inc., 1985), brochure (in the files of Paul R. Jacobs, Corcoran, MN).

66. Earth Energy Systems Inc., Eden Prairie, MN, letter To: Winco Accounts, 1985, 1 leaf (in the files of Paul R. Jacobs, Corcoran, MN); Earth Energy Systems Inc., Eden Prairie, MN, letter To: All Winco Dealers, 1985, 1 leaf (in the files of Paul R. Jacobs, Corcoran, MN); Larry N. Stoiaken, "Jacobs Finalizes Dyna Technology Acquisition: Unveils New Name—Earth Energy Systems Inc.," *Alternative Sources of Energy*, no. 72 (March–April 1985): 53; Ann Anderson, "Plymouth Company Pioneers Hybrid Power Systems," *Plymouth Post* (Plymouth, MN), May 17, 1984; Earth Energy Systems Inc., Eden, Prairie, MN, *Hybrid Power Plant*[:] *Class A up to 10 KW Continuous* (Eden Prairie, MN: Earth Energy Systems, [ca. 1985]), brochure (in the files of Craig Toepfer, Chelsea, MI); Earth Energy Systems Inc., Eden Prairie, MN, *Bringing Energy to the World* (Eden Prairie, MN: Earth Energy Systems, 1985), brochure (in files of Craig Toepfer, Chelsea, MI); Earth Energy Systems, Inc., Eden Prairie, MN, *Class A Hybrid Power Plant*[,] *Wind/Solar/Engine/Battery*[:] *Energy-Minder Theory of Operation* (Eden Prairie, MN: Earth Energy Systems Inc., [ca. 1985]), handout (in the files of Craig Toepfer, Chelsea,

MI); Earth Energy Systems, Inc., Eden Prairie, MN, *Hybrid Power Plant*[,] *Wind/Solar/Engine/Battery*[:] *Wincharger 1000 Theory of Operation* (Eden Prairie, MN: Earth Energy Systems Inc., [ca. 1985]), handout (in the files of Craig Toepfer, Chelsea, MI).

67. Paul R. Jacobs, Corcoran, MN, email to author, February 6, 2017.

68. Ibid.

69. Jacobs Wind Electric Company, Plymouth, MN, *Consulting Report Index*[,] *Rev. 6/17/85* (Plymouth, MN: Jacobs Wind Electric Company, 1985), typewritten index, 1 leaf; Jacobs Wind Electric Company, Plymouth, MN, *Master List*[,] *Possible Recipients of M. L. Jacobs Consulting Reports* (Plymouth, MN: Jacobs Wind Electric Company, 1984), typewritten, 1 leaf; M. L. Jacobs, Plymouth, MN, *February 12, 1985*[,] *Consulting Report No. 14*[:] *Follow-up Report on the Governor Hub Nut Problems at the Jacoby-Kerr Project In Palm Springs, California* (Plymouth, MN: Jacobs Wind Electric Company, 1985), typewritten memorandum; M. L. Jacobs, Plymouth, MN, *February 12, 1985*[,] *Consulting Report No. 15*[:] *Governor Blade Shaft Zerk Fittings—Buck Ranch Observations* (Plymouth, MN: Jacobs Wind Electric Company, 1985), typewritten memorandum; M. L. Jacobs, Plymouth, MN, *February 14, 1985*[,] *Consulting Report No. 16*[:] *Tower Manufacturer Mistakes Found at the Buck Ranch Site in Palm Springs, California* (Plymouth, MN: Jacobs Wind Electric Company, 1985), typewritten memorandum; M. L. Jacobs, Plymouth, MN, *February 19, 1985*[,] *Consulting Report No. 18*[:] *Tower Angle Brace Rod Breakage* (Plymouth, MN: Jacobs Wind Electric Company, 1985), typewritten memorandum; M. L. Jacobs, Plymouth, MN, *February 21, 1985*[,] *Consulting Report No. 19*[:] *A. I. Tower Corner Rail Cracking (Failure) at Palm Springs* (Plymouth, MN: Jacobs Wind Electric Company, 1985), typewritten memorandum; M. L. Jacobs, *March 29, 1985*[,] *Consulting Report No. 24*[:] *Reply to R. E. Fletcher's Letter of February 27, 1985: 20 KW Governor Lubrication (Copy of Letter Attached)* (Plymouth, MN: Jacobs Wind Electric Company, 1985), typewritten memorandum; M. L. Jacobs, Plymouth, MN, *May 10, 1985*[,] *Consulting Report No. 28*[:] *Installation and Maintenance Problems with the "X" Braced Angle Iron Tower at Palm Springs* (Plymouth, MN: Jacobs Wind Electric Company, 1985), typewritten memorandum; M. L. Jacobs, Plymouth, MN, *May 10, 1985*[,] *Consulting Report No. 29*[:] *Heavier Control Spring for Dual Fold Tail Vane* (Plymouth, MN: Jacobs Wind Electric Company, 1985), typewritten memorandum; M. L. Jacobs, Plymouth, MN, *May 31, 1985*[,] *Consulting Report No. 32*[:] *Further Comments on Governor Hub Lubrication and Reference to Consulting Report No. 24* (Plymouth, MN: Jacobs Wind Electric Company, 1985), typewritten memorandum; M. L. Jacobs, Plymouth, MN, *June 13, 1985*[,] *Consulting Report No. 33*[:] *New Tower Girt to Improve Present Windfarm Towers* (Plymouth, MN: Jacobs Wind Electric Company, 1985), typewritten memorandum with drawing (all of the above in the files of Paul R. Jacobs, Corcoran, MN).

70. *February 12, 1985*[,] *Consulting Report No. 14*.

71. Paul R. Jacobs, Corcoran, MN, email to author, January 29, 2017.

72. Paul R. Jacobs, Corcoran, MN, emails to author, February 3, 2017 and February 6, 2017; Minnesota Department of Health, Section of Vital Statistics, St. Paul, MN, *Certificate of Death*[,] *Marcellus Luther Jacobs*, July 23, 1985 (in the files of Paul R. Jacobs, Corcoran, MN); Donald Marier, "Marcellus Jacobs: 1903–1985," *Alternative Sources of Energy* no. 75 (September-October 1985): 6.

73. Claude W. Eggleton, "In Memorial . . . Marcellus L. Jacobs (1903–1985)," *Wind Power Digest* no. 27 (1985): 3.

74. David Phelps and Steve Gross, "Against the Wind: CDC's $90 Million Failure," *Minneapolis Star and Tribune*, December 9, 1986; *Renewable Energy Ventures, Inc.*[,] *a Subsidiary of Earth Energy Systems, Inc.*; Paul R. Jacobs, Corcoran, MN, email to author, February 7, 2017.

75. Paul R. Jacobs, Corcoran, MN, email to author, February 22, 2017.

76. Paul R. Jacobs, Corcoran, MN, emails to author, February 7, 2017 and February 22, 2017; *Charles Casserly, Trustee v. Shearson Lehman Hutton, Inc.*, 4–90–125, 1–16 (4th Div. D Minn., 1991).

77. James Walker, "Tide Gate Eyed to Ease Pollution," *Tampa Tribune* (Tampa, FL), April 5, 1975; Tampa Port Authority, Tampa, FL, *Permit For Work In The Navigable Waters Of The Hillsborough County Port District* (Tampa, FL, July 9, 1980) (in the files of Paul R. Jacobs, Corcoran, MN); Steve Piacente, "Tide Gate Could Improve Bay Water Quality," *Tampa Tribune* (Tampa, FL), May 22, 1980; Karen Wolfson, "Tide Gates Ready to Flush Stagnant Waterway," *Tampa Tribune* (Tampa, FL), December 23, 1982; C. J. Barkett, "Infringement—Exclusive Federal Court Jurisdiction over Patent Infringement Suits Does Not Preclude State Court from Exercising Jurisdiction over Due Process Takings and Conversion Claims Brought by Patent Holder against State," *The United States Law Week* 62 (October 26, 1993): 2241; *Jacobs Wind Electric Co., Inc. v. Thomas R. Shahady*, 05–16685, 1–2 (17th Cir., Fla. 2010); *Jacobs Wind Electric Co., Inc. v. Thomas R. Shahady*, 05–16685 CACE 21, 1–2 (17th Cir., Fla. 2010); Julie Kay, "Adorno Lawsuit Was Under Radar until Paychecks Bounced," *Daily Business Review* (Miami-Dade, FL), December 10, 2010.

78. Mark B. Dayton, commissioner, Minnesota Department of Energy and Economic Development, St. Paul, MN, letter to Paul Jacobs, April 17, 1986 (in the files of Paul R. Jacobs, Corcoran, MN); Robert D. Sykes, associate professor and principal investigator, and Lance M. Neckar, professor and department head, University of Minnesota,

Department of Landscape Architecture, College of Design, Minneapolis, MN, letter to Paul R. Jacobs, June 2, 2009 (in the files of Paul R. Jacobs, Corcoran, MN).

79. Patent No. 4,909,703, Paul R. Jacobs, US Patent and Trade Office, Washington, DC.

80. Paul R. Jacobs, Corcoran, MN, email to author, February 18, 2017.

81. Ibid.

82. REPCO, Bismarck, ND, *Articles of Association of REPCO-Rural Energy Producers Electric Power Cooperative* (Bismarck, ND: REPCO, 1993), handout (in the files of Paul R. Jacobs, Corcoran, MN); REPCO, *News Release*[,] *New Crop Coop Formed*, January 3, 1994 (in the files of Paul R. Jacobs, Corcoran, MN); REPCO, Bismarck, ND, *The Origins of REPCO* (Bismarck, ND: REPCO, 1994), handout (in the files of Paul R. Jacobs, Corcoran, MN); REPCO, Bismarck, ND, *The Case for Action Now in Establishing a Dispersed Wind Rec* (Bismarck, ND: REPCO, 1994), handout (in the files of Paul R. Jacobs, Corcoran, MN); REPCO, Bismarck, ND, *The REPCO Mission* (Bismarck, ND: REPCO, 1994), handout (in the files of Paul R. Jacobs, Corcoran, MN).

83. John Farrell, "Why Aren't Rural Electric Cooperatives Champions of Local Clean Power?" *Renewable Energy World.Com* (Tulsa, OK) December 3, 2014; Paul R. Jacobs, Corcoran, MN, email to author, February 18, 2017.

Chapter 7

1. Karl H. Bergey, Jr., founder, Bergey Windpower Company, interview by author, Norman, OK, October 23, 2015.

2. "1920s Aviation[:] The Dawn of Air Travel and Air Freight," 1920–30.com, accessed April 23, 2016, http://1920–30.com; Patricia Trenner, "10 All-Time Great Pilots," *Air & Space Magazine* 17, no. 6 (March 2003): 67–72; "Aviation: From Sand Dunes to Sonic Booms[:] Aviation Pioneers," National Park Service, accessed April 23, 2016, https://www.nps.gov/nr/travel/aviation/pioneers.htm.

3. Karl Bergey interview.

4. Ibid.

5. Evan T. Towne, "Mr. Model Engine—Bill Brown," Craftsmanship Museum, accessed April 23, 2016, http://www.craftmanshipmuseum.com/BrownJr.htm; "William (Bill) L. Brown IV[,] May 30, 1911—January 8, 2003," State College Radio Control Club (Bellefonte, PA), accessed April 23, 2016, http://scrc-club.com/Archive%20files/Bill_Brown/billbrwn.htm; Bill Mohrbacher, "History of Model Engines," *Model Aviation* (February 2012), accessed April 23, 2016, http:///www.modelaviation.com/enginehistory.

6. David J. Peery, *Aircraft Structures* (New York: McGraw-Hill Book Company, 1949), v.

7. Karl Bergey interview.

8. Ibid.

9. Karl H. Bergey, Jr., typewritten letter to author, February 15, 2017.

10. Karl Bergey interview.

11. Karl Bergey, letter to author.

12. Karl Bergey interview.

13. Karl H. Bergey, "Wind Power Demonstration and Siting Problems" (paper presented at Wind Energy Conversion Systems Workshop, Washington, DC, June 11–13, 1973), 41; University of Oklahoma, *OU Urban Car* (Norman, OK: University of Oklahoma, [ca. 1973]), handout (in author's files).

14. "Wind Power Demonstration and Siting Problems"; *OU Urban Car.*

15. "Wind Power Demonstration and Siting Problems"; *OU Urban Car.*

16. Karl Bergey interview.

17. K. H. Bergey, "Feasibility of Wind Power Generation for Central Oklahoma" (Norman, OK: University of Oklahoma, 1971), 16.

18. "Wind Power Demonstration and Siting Problems," 41; Stephen Kidd and Doug Garr, "Can We Harness Pollution-free Electric Power from Windmills?" *Popular Science* 201, no. 5 (November 1972): 71.

19. "Wind Power Demonstration and Siting Problems," 41.

20. Ibid., 42.

21. Ibid., 43–44.

22. Karl H. Bergey, "Wind Power Potential for the United States," *AWARE Magazine* (Madison, WI) no. 49 (October 1974): 2.

23. "Wind Power Demonstration and Siting Problems," 42.

24. Ibid., 44.

25. Michael L. S. Bergey, president, Bergey Windpower Company, interview by author, Norman, OK, October 23, 2015.

26. Mike Bergey, "Wind Power in Oklahoma," *Wind Power Digest* no. 13 (Fall 1978): 8.

27. Ibid.

28. Mike Bergey, "The Vertical Axis, Articulated Blade, Wind Turbine" (paper presented at the National Conference, American Wind Energy Association, Amarillo, TX, March 1–5, 1978), 13–17.

29. R. V. Brulle, *Feasibility Investigation of the Giromill for Generation of Electrical Power*[,] *Volume I—Executive Summary*[,] *Final Report for the Period April 1975—April 1976* (Prepared for the US Energy Research and Development Administration, Division of Solar Energy) (St. Louis: McDonnell Aircraft Company, 1977), 1–7, 21; R. V. Brulle, *Feasibility Investigation of the Giromill for Generation of Electric Power*[,] *Volume II—Technical Discussion*[,] *Final Report for the Period April 1975—April 1976* (Prepared for the US Energy

Research and Development Administration, Division of Solar Energy) (St. Louis: McDonnell Aircraft Company, 1977), 2–3, 12–20, 59–63; W. A. Moran, *Giromill Tunnel Test and Analysis*[,] *Volume 2. Technical Discussion*[,] *Final Report for the Period July 1976—October 1977* (Prepared for the US Energy Research and Development Administration, Division of Solar Energy) (St. Louis: McDonnell Aircraft Company, 1977), 1, 82; Robert V. Brulle, *Engineering the Space Age: Rocket Scientist Remembers* (Maxwell Air Force Base, AL: Air University Press, 2008), 220; Christopher C. Gillis, *Still Turning*[:] *A History of Aermotor Windmills* (College Station: Texas A&M University Press, 2015), 157–58.

30. "The Vertical Axis, Articulated Blade, Wind Turbine," 18–22; "Wind Power in Oklahoma," 8–10.

31. Michael Bergey interview.

32. Ibid.

33. "The Vertical Axis, Articulated Blade, Wind Turbine," 21–22.

34. Ibid.

35. Michael Bergey interview.

36. Christopher Gillis, *Windpower* (Atglen, PA: Schiffer Publishing Ltd., 2008), 33; Jeff Donn, "Town has been Testbed for Wind Power," *Associated Press*, June 14, 2004.

37. Michael Bergey interview.

38. "Wind Power in Oklahoma," 10.

39. Michael Bergey interview.

40. "Wind Power in Oklahoma," 8.

41. Gillis, *Windpower*, 33.

42. Michael Bergey interview.

43. Michael Bergey interview; Patent No. 4,291,235, Karl H. Bergey Jr., and Michael L. S. Bergey, US Patent Office, Washington, DC.; Karl H. Bergey, "Development of High-Performance, High-Reliability Windpower Generators" (paper presented at the American Wind Energy Association National Conference, Summer 1980, Pittsburgh, PA, June 8–11, 1980), 101–05; Ed Jacobs, "Wind Generator has Variable-Pitched Blades with Fixed-Blade Simplicity," *Popular Science* 220, no. 6 (June 1982): 114.

44. Michael Bergey interview; "Wind Generator Has Variable-Pitched Blades with Fixed-Blade Simplicity."

45. Michael Bergey interview.

46. Ibid.

47. Donald Marier, "The Bergey Windpower Corp.: A Profile," *Alternative Sources of Energy* no. 61 (May–June 1983): 20.

48. Ibid.

49. Bergey Windpower Company, Norman, OK, *BWC 1000*[,] *Bergey Windpower* (Norman, OK: Bergey Windpower Company, [ca. 1982]), folder (in author's files).

50. *BWC 1000*[,] *Bergey Windpower*; Bergey Windpower Company, Norman, OK, *Bergey Windpower* (Norman, OK: Bergey Windpower Company, [ca. 1982]), handbill (in *Windmillers' Gazette* research files, Rio Vista, TX); Bergey Windpower Company, Norman, OK, *Annual Energy Output Curve*[,] *Revision 2-3-82* (Norman, OK: Bergey Windpower Company, 1982), handbill (in *Windmillers' Gazette* research files, Rio Vista, TX); Bergey Windpower Company, Norman, OK, *Bergey Windpower Powersync Inverter Data Sheets* (Bergey Wind Power Company, 1982), 3 leaves (in *Windmillers' Gazette* research files, Rio Vista, TX).

51. *Bergey Windpower.*

52. Ibid.

53. Michael Bergey interview.

54. Ibid.

55. Ibid.

56. Mike Bergey, "Wind Energy Potpourri" (paper presented at Rural Electric Wind Energy Workshop of the National Rural Electric Cooperative Association, Boulder, CO, June 1–3, 1982), 261–62.

57. Ibid., 262.

58. Michael Bergey interview.

59. E. F. Lindsley, "33 Windmills You Can Buy Now!" *Popular Science* 221, no. 1 (July 1982): 52–55.

60. David Bumke, "Success Usually Doesn't Come Easy. People Who Win with Wind Power," *Rodale's New Shelter* 4, no. 4 (April 1983): 50.

61. Ibid.

62. Gillis, *Windpower*, 33.

63. Michael Bergey interview.

64. "Wind Energy Potpourri," 265.

65. Michael Bergey interview.

66. "The Bergey Windpower Corp.: A Profile," 21; "New Bergey Unit Coming," *Wind Power Digest*, no. 25 (Fall 1982): 33.

67. Michael Bergey interview.

68. "The Bergey Windpower Corp.: A Profile," 21; "New Bergey Unit Coming."

69. "The Bergey Windpower Corp.: A Profile," 21; "New Bergey Unit Coming."

70. Michael Bergey interview.

71. Larry N. Stoiaken, "Going International[:] Bergey Approached by Japanese and Chinese for License Production of BWC 1000," *Alternative Sources of Energy*, no. 63 (September-October 1983): 25.

72. Ibid.

73. Ibid.

74. Karl H. Bergey interview.

75. Mike Bergey, "Performance Testing and Rating Standards for Wind Energy Conversion Systems" (paper presented at the National Conference, Summer 1980, American Wind Energy Association, Pittsburgh, PA, June 8–11, 1980), 163–67.

76. Ibid.

77. Ibid.

78. Ibid.

79. David Brooks, "Remembering the World's First Wind Farm—in New Hampshire," Granite Geek, February 24, 2016, accessed January 22, 2017, http://granitegeek.concordmonitor.com/2016/02/24/remembering-worlds-first-wind-farm-new-hampshire/; "World's First Windfarm," *Farm Show Magazine* 5, no. 5 (1981): 18.

80. Karl H. Bergey, "Windfarm Strategies: The Economics of Scale" (paper presented at Wind Energy Expo '83 and National Conference, San Francisco, CA, October 17–19, 1983), 63, 67,70–71.

81. Ibid, 71.

82. "Success Usually Doesn't Come Easy," 50.

83. "Windpower '85 a Success[,] Over 650 Go to San Francisco," *Alternative Sources of Energy* no. 76 (November-December 1985): 50.

84. Gillis, *Windpower*, 34.

85. "Town has been Testbed for Wind Power"; Larry Sherwood, "Interview: Mike Bergey of Bergey Windpower," *Small Wind Energy News* (Interstate Renewable Energy Council, Latham, NY), November 5, 2013, accessed November 12, 2016, http://www.irecusa.org/2013/11/interview-mike-bergey-of-bergey-windpower/.

86. Michael Bergey interview.

87. Ibid.

88. *Still Turning*, 192; Don Comis, "Wind Power Where You Want It," *Agricultural Research* (Washington, DC) 42, no. 6 (June 1994): 4–7; R. Nolan Clark, "Wind-Electric Water Pumping Systems for Rural Domestic and Livestock Water" (paper presented at the 5th European Wind Energy Association Conference and Exhibition, Thessaloniki-Macedonia, Greece, October 10–14, 1994), 1136–37, 1139.

89. "About Bergey[:] Company Background and Customers," Bergey Windpower Company, accessed February 26, 2007, http://www.bergey.com/About_BWC.htm; Bergey Windpower, Norman, OK, *The BWC 1500 Wind Turbine System* (Norman, OK: Bergey Windpower, 1994), handbill (in author's files).

90. *Still Turning*, 192–93; R. Nolan Clark and Brian D. Vick, "Chapter 13[:] Livestock Watering with Renewable Energy," in *Agriculture as a Producer and Consumer of Energy*, ed. J. Outlaw, K. J. Collins and J. A. Duffield (Wallingford, United Kingdom: CABI Publishing, 2005), 232–38; Paul Gipe, "Windmills: Still Primed for Pumping Water," *Independent Energy* 19, no. 6 (July-August 1989): 64; Michael L. S. Bergey, "Wind-Electric Pumping Systems for Communities" (paper presented at the 1st International Symposium on Safe Drinking Water in Small Systems, Washington, DC, May 10–13, 1998), accessed February 19, 2014, http://bergey.com/wind-school/articles/wind-electric-pumping-systems-for-communities-2; Brett G. Ziter, "Electric Wind Pumping for Meeting Off-Grid Community Water Demands," *Guelph Engineering Journal* (Ontario, Canada) 2 (2009): 15.

91. Ros Davidson, "Remote Area Potential Proving Attractive," *Wind Power Monthly* (Knebel, Denmark) 6, no. 9 (September 1990): 18.

92. Art Lilley and Mike Bergey, "Lessons Learned: Deployment of Wind Turbines in Indonesia" (paper presented at Windpower '94, Minneapolis, MN, May 10–13, 1994), 705, 711.

93. David Corbus and Mike Bergey, "Costa de Cocos 11-KW Wind-Diesel Hybrid System" (paper presented at Windpower '97, Austin, TX, June 15–18, 1997), 191, 195–96.

94. Holly Davis, Steve Drouilhet, Larry Flowers, Michael Bergey, and Michael Brandemuehl, "Wind-Electric Ice Making for Villages in the Developing World" (paper presented at Windpower '94, Minneapolis, MN, May 10–13, 1994), 635–36, 641.

95. "Windmills: Still Primed for Pumping Water," 64.

96. "Bergey Wind Turbines—Quality-Performance-Value" [advertisement], *Home Power* (Ashland, OR), no. 57 (February-March 1997), 61; Bergey Windpower, Norman, OK, *BWC XL.1[,] 1 kW Class Wind Turbine* (Norman, OK: Bergey WIndpower, [ca. 2000]), handbill (in author's files); *The BWC 1500 Wind Turbine System*.

97. "Windmills: Still Primed for Pumping Water," 64.

98. *Windpower*, 34.

99. American Wind Energy Association, Small Wind Turbine Committee, *The U. S. Small Wind Turbine Industry Roadmap[:] A 20-Year Industry Plan for Small Wind Turbine Technology* (Washington, DC: American Wind Energy Association, 2002), 7.

100. Ibid., 3.

101. Ibid., 7.

102. Ibid., 8–9.

103. "Town has been Testbed for Wind Power"; Bergey Windpower, Norman, OK, *BWC Excel[,] 10 KW Class Wind Turbine* (Norman, OK: Bergey Windpower, [ca. 2000]), handbill (in author's files).

104. Bergey Windpower, Norman, OK, *BWC XL.50[,] 50 KW Class Wind Turbine* (Norman, OK: Bergey Windpower, [ca. 2000]), handbill (in author's files).

105. "Business Energy Investment Tax Credit (ITC)," US Department of Energy, accessed April 2, 2017, https://energy.gov/savings/business-energy-investment-tax-credit-itc.

106. "The Bergey Excel 6 Wind Turbine," Bergey Windpower Company, accessed January 27, 2017, http://bergey.com/products/wind-turbines/6kw-bergey-excel.

107. Michael Bergey interview.

108. Distributed Wind Energy Association, *SMART Wind Roadmap* (Durango, CO: Distributed Wind Energy Association, 2016), iv.

109. "DWEA Committees," Distributed Wind Energy Association, accessed January 31, 2017, http://distributedwind.org/home/committees/; "DWEA Member Directory," Distributed Wind Energy Association,

accessed January 31, 2017, http://distributedwind.org/dwea-members/.

110. "Interview: Mike Bergey of Bergey Windpower."

111. Michael Bergey interview.

112. Katherine Ling, "Small Turbines Get Second Wind from Climate Regs," *E&E News* (Washington, DC), August 17, 2015; United Wind, Brooklyn, NY, *United Wind & Bergey WindPower Close Landmark Purchase of 100 Distributed-Scale Wind Turbines*, January 20, 2017 (in author's files).

113. *United Wind & Bergey WindPower Close Landmark Purchase of 100 Distributed-Scale Wind Turbines*.

114. US Department of Energy, Office of Energy Efficiency and Renewable Energy, Washington, DC, *Energy Department's Competitiveness Improvement Project Hits First Certification Milestone; Opens Next Round of Applications*, February 28, 2018 (in author's files); Jose Zayas, "DOE's Competitiveness Improvement Projects Are Delivering Substantial Cost Reductions in the U. S. Distributed Wind Industry," *North American Clean Energy* (July-August 2017), accessed January 31, 2018. http://www.nacleanenergy.com/articles/27632/doe-s-competitiveness-improvement-projects-are-delivering-substantial-cost-reductions-in-the-u-s-distributed-wind-industry; Patrick Gilman, "Enabling Wind to Contribute to a Distributed Energy Future—The New DOE Distributed Wind Plan" (presentation at Distributed Wind 2019, Washington, DC, February 28, 2019); US Department of Energy, Office of Energy Efficiency and Renewable Energy, Golden, CO, *Distributed Wind Competitiveness Improvement Project* (Golden, CO: National Renewable Energy Laboratory, 2019), handbill (in author's files).

115. *Distributed Wind Competitiveness Improvement Project*; Ian Baring-Gould, "Latest CIP Round" (presentation at Distributed Wind 2019, Washington, DC, February 28, 2019); Bergey Windpower, Norman, OK, *Bergey Excel 15*[:] *Advanced Technology—Superior Economics* (Norman, OK: Bergey Windpower, [ca. 2019]), handbill (in author's files).

116. Michael Bergey, "Leadership Panel: State of the Industry" (presentation at Distributed Wind 2019, Washington, DC, February 28, 2019).

Chapter 8

1. Winco, Division of Dyna Technology, Inc., Sioux City, IA, *Wincharger*[:] *12 Volt Wind Electric Battery Charger* (Sioux City, IA: Winco, Division of Dyna Technology, Inc., ca. 1975), handbill (in author's files).

2. Daniel M. Simmons, *Wind Power* (Park Ridge, NJ: Noyes Data Corporation, 1975), 253–54; Volta Torrey, *Wind-Catchers*[:] *American Windmills of Yesterday and Tomorrow* (Brattleboro, VT: The Stephen Greene Press, 1976), 193–96.

3. "Events in the History of Dunlite," http://www.pearen.ca/dunlite/History-3.pdf, accessed on August 21, 2016; M. Schaufelberger, *37 Years Experience with Elektro Windmills* (Winterthur, Switzerland: Elektro GmbH, [ca. 1985]), 1–4; Etienne Rogier, Toulouse, France, "AEROWATT, c'est l'administration," email to author, August 5, 2018; Torrey, 195–96.

4. Simmons, *Wind Power*, 254.

5. Ed Harper, "Problems with a 6 KW Elektro," *Wind Power Digest* 1, no. 5 (June 1976): 14.

6. Syverson Consulting, *Dunlite*[,] *Brushless*[,] *Wind Driven Power Plants* (North Mankato, MN: Syverson Consulting, 1975), brochure (in the files of Bob McBroom, Holton, KS).

7. "Problems with a 6 KW Elektro," 11–14.

8. M. Schaufelberger, Elektro GmbH, Winterhur, Switzerland, letter to Henry Clews, Solar Wind, East Holden, ME, "Fixation of Propeller," January 22, 1975, typewritten letter (in personal files of Leander Nichols).

9. Joe Carter, "Testing an Elektro-Gemini Wind System," *Wind Power Digest* 1, no. 6 (September 1976): 14–15.

10. Paul Gipe, "Wind Catchers[:] Machines that Turn Wind into Electricity," *New Roots* (Greenfield, MA), no. 18 (Harvest 1981): 12.

11. Christopher Flavin, "Wind Power: A Turning Point," *Worldwatch Paper 45* (Washington, DC: Worldwatch Institute, 1981), 35; Rockwell International, Rocky Flats Wind Systems Program, *Commercially Available Small Wind Systems and Equipment*[,] *March 31, 1981* (Golden, CO: Rockwell International, 1981), 6–8.

12. Dakota Wind & Sun Ltd., Aberdeen, SD, *Common Questions and General Information*[,] *April 1978* (Aberdeen, SD: Dakota Wind & Sun Ltd., 1978), folder (in the *Windmillers' Gazette* research files, Rio Vista, TX); Dakota Wind & Sun Ltd., Aberdeen, SD, *Common Questions and General Information*[,] *September 1978* (Aberdeen, SD: Dakota Wind & Sun Ltd., 1978), folder (in the *Windmillers' Gazette* research files, Rio Vista, TX); Dakota Wind & Sun Ltd., Aberdeen, SD, *Greetings from the Factory. February 19, 1979* (Aberdeen, SD: Dakota Wind & Sun Ltd., 1979), 5 leaves (in the *Windmillers' Gazette* research files, Rio Vista, TX); Dakota Wind & Sun Ltd., Aberdeen, SD, *Greetings from the Factory! April 4, 1979* (Aberdeen, SD: Dakota Wind & Sun Ltd., 1979), 7 leaves (in the *Windmillers' Gazette* research files, Rio Vista, TX); Dakota Wind & Sun Ltd., Aberdeen, SD, *Greetings from the Factory! May 18, 1979* (Aberdeen, SD: Dakota Wind & Sun Ltd., 1979), 7 leaves (in the *Windmillers' Gazette* research files, Rio Vista, TX); Paul Gipe, "Dakota Sun & Wind[:] Making a Good Machine Better," *Wind Power Digest* no. 15 (Spring 1979): 13–14; Paul Gipe, "Wind Activists in the Commonwealth," *Wind Power Digest* no. 9 (Summer 1977): 32–35; Independent

Energy Systems, Inc., Fairview, PA, *IES Independent Energy Systems Inc. Integrated Utility Free Energy Systems* (Fairview, PA: Independent Energy Systems, Inc., [ca. 1977]), brochure (in the *Windmillers' Gazette* research files, Rio Vista, TX); Independent Energy Systems, Inc., Fairview, PA, *IES Independent Energy Systems, Inc. Catalog #216* (Fairview, PA: Independent Energy Systems, Inc., 1979), 1–4, brochure (in the *Windmillers' Gazette* research files, Rio Vista, TX).

13. Larry N. Stoiaken, "The Small Wind Energy Conversion System Market: Will 1984 be 'The Year of the SWECS'?" *Alternative Sources of Energy* no. 63 (September-October 1983): 10–17, 20–23.

14. Larry N. Stoiaken, "1984 SWECS Shakeouts," *Alternative Sources of Energy* no. 73 (May–June 1985): 57.

15. Christopher Flavin, "Electricity for a Developing World: New Directions," *Worldwatch Paper 70* (Washington, DC: Worldwatch Institute, 1986), 5–9, 52–57; "Small Wind Systems Continue to Find Markets," *Alternative Sources of Energy* no. 71 (January–February 1985): 55–56; Christopher Gillis, *Windpower* (Atglen, PA: Schiffer Publishing Ltd., 2008), 33.

16. Gillis, *Windpower*, 34.

17. A. Betz, "The Maximum of the Theoretically Possible Exploitation of Wind by Means of a Wind Motor," *Wind Engineering* 37, no. 4 (2013): 441–46 (Translation of *Das Maximum der theoretisch möglichen Ausnützung des Windes durch Windmotoren*, Zeitschrift für das gesamte Turbinenwesen, Heft 26, Seiten 307 bis 309, 1920, by H. Hamman, J. Thayer, and A. P. Schaffarczyk, University of Applied Sciences, Kiel, Germany).

18. Walter Döring, *A Vision Becomes Reality*[:] *Wie die Windreich AG das Vermächtnis des Windpioniers Ulrich Hütter in die tat Umsetzt* (Wolfshlugen, Germany: Windreich AG, [ca. 2012]), 84–87.

19. Peter M. Moretti and Louis V. Divone, "Modern Windmills," *Scientific American* 254, no. 6 (June 1986): 114–15.

20. Döring, *A Vision Becomes Reality*, 84–87.

21. Ibid., 90.

22. T. E. Sweeney, *The Princeton Windmill Program*[,] *AMS Report No. 1093*[,] *March, 1973* (Princeton, NJ: Princeton University, Department of Aerospace and Mechanical Sciences, 1973), 2.

23. *The Princeton Windmill Program*, 2–7; Stephen Kidd and Doug Garr, "Can We Harness Pollution-Free Electric Power From Windmills?" *Popular Science* 201, no. 5 (November 1972): 70–72; Stephen Kidd, "Windmills and Other Things," *Princeton Alumni Weekly* 73, no. 24 (April 24, 1973): 12–13; T. E. Sweeney and W. B. Nixon, "An Introduction to the Princeton Sailwing Windmill" (paper presented at the Wind Energy Conversion Systems Workshop, Washington, DC, June 11–13, 1973), 70–72; T. E. Sweeney, "Sailwing Windmill Technology" (paper presented at the Second Workshop on Wind Energy Conversion Systems, Washington, DC, June 9–11, 1975), 224–28.

24. Frank Farwell, "New Design Spurs a Resurgence of Windmills," *The New York Times*, February 3, 1980.

25. Hans Meyer, "The Use of Paper Honeycomb for Prototype Blade Construction for Small to Medium-Sized Wind Driven Generators" (paper presented at the Wind Energy Conversion Systems Workshop, Washington, DC, July 11–13, 1973), 73–74.

26. Hans Meyer, "Wind Generators: Here's An Advanced Design You Can Build," *Popular Science* 201, no. 5 (November 1972): 103–05, 142.

27. "The Use of Paper Honeycomb for Prototype Blade Construction for Small to Medium-Sized Wind Driven Generators," 74.

28. "Wind Generators: Here's an Advanced Design You Can Build," 103.

29. "New Design for Windmill on Display," *Osceola Sun* (Kissimmee, FL), February 20, 1974; Patent No. 3,942,839A, US Patent and Trademark Office, Washington, DC; American Wind Turbine, St. Cloud, FL, *Dear Wind Power Enthusiast* (St. Cloud, FL: American Wind Turbine, [ca. 1975]), handbill, 3 leaves (in the *Windmillers' Gazette* research files, Rio Vista, TX).

30. E. F. Lindsley, "Wind Power: How New Technology Is Harnessing an Age-Old Energy Source," *Popular Science* 205, no. 1 (July 1974): 54–55.

31. "American Wind Turbine Revisited," *Wind Power Digest* no. 2 (Summer 1975): 13.

32. "AEA Introduces Amerenalt Wind Turbine," *Wind Power Digest* no. 2 (Summer 1975): 31–32; American Energy Alternatives Inc., Boulder, CO, *Amerenalt Series Energy Conversion Systems* (Boulder, CO: American Energy Alternatives Inc., 1975), brochure (in the *Windmillers' Gazette* research files, Rio Vista, TX).

33. Chalk Wind Systems, St. Cloud, FL, *Modern Technology Comes to an Old Idea* (St. Cloud, FL: Chalk Wind Systems, [ca. 1979]), folder (in the *Windmillers' Gazette* research files, Rio Vista, TX).

34. "Tom Chalk's American Wind Turbine," *Wind Power Digest* no. 21 (Winter 1980–1981): 19.

35. E. F. Lindsley, "33 Windmills You Can Buy Now!" *Popular Science* 221, no. 1 (July 1982): 52–55.

36. "Wind Catchers[:] Machines that Turn Wind into Electricity," 13.

37. David Boyt, "Use of Laminated Wood in Wind Turbine Blades" (paper presented at the Wind Energy Expo '82 and National Conference, Amarillo, TX, October 24–27, 1982), 174–76; Roger Anderson, "Pultruded Fiberglass Reinforced Plastic (FRP) Wind Turbine Blades" (paper presented at the Wind Energy Expo '82 and National Conference, Amarillo, TX, October 24–27, 1982), 177–81; Paul Gipe, "Wind in America:

The Changing Landscape," *Solar Age* 8, no. 10 (October 1983): 33.

38. "33 Windmills You Can Buy Now!" 55.

39. Ed Jacobs, "Wind Generator Has Variable-Pitched Blades with Fixed-Blade Simplicity," *Popular Science* 220, no. 6 (June 1982): 114; Bergey Windpower Company, Norman, OK, *Windpower! BWC 1000 Series* (Norman, OK: Bergey Windpower Company, [ca. 1983]), brochure (in author's files); Bergey Windpower Company, Norman, OK, *Windpower! BWC EXCEL Series* (Norman, OK: Bergey Windpower Company, [ca. 1983]), brochure (in author's files); Bergey Windpower Company, Norman, OK, *BWC EXCEL*[,] *Bergey Windpower Announces the BWC EXCEL* (Norman, OK: Bergey Windpower Company, 1983), handbill (in author's files); Bergey Windpower Company, Norman, OK, *BWC EXCEL Series*[,] *Bergey Wind Energy Systems* (Norman, OK: Bergey Windpower Company, 1983), handbill (in author's files); Jacobs Wind Electric Company, Minneapolis, MN, *Jacobs Air Way*[,] *The New Model 35* (Minneapolis, MN: Jacobs Wind Electric Company, [ca. 1949]), brochure (in author's files); Jacobs Wind Electric Company, Fort Myers, FL, *Important Features of the New Jacobs 8 KVA Wind Energy System* (Fort Myers, FL: Jacobs Wind Electric Company, 1979), handbill (in author's files).

40. Parris-Dunn Corporation, Clarinda, IA, *Direct Drive 6–12–32–110 Volts Wind Electric Light & Power Plants* (Clarinda, IA: Parris-Dunn Corporation, [ca. 1940]), handbill (in the files of Mike Werst, Manor, TX); Don Mayer, "North Wind's 1 KW High Reliability WTG Program" (paper presented at the American Wind Energy Association National Conference, Amarillo, TX, March 1–5, 1978), 50–52; North Wind Power Company, Warrant, VT, *The 2 KW High Reliability Wind Turbine Generator* (Warren, VT: North Wind Power Company, 1979), handbill, 5 leaves (in the *Windmillers' Gazette* research files, Rio Vista, TX); "Designing a High Reliability Wind Machine—How One Company Did It," *Wind Power Digest* no. 25 (Fall 1982): 6–13; North Wind Power Company, Warren, VT, *North Wind Power Co., Inc.* (Warren, VT: North Wind Power Company, [ca. 1980]), brochure (in author's files); North Wind Power Company, Warren, VT, *Reliable Power from the Wind* (Warren, VT: North Wind Power Company, [ca. 1980]), brochure (in author's files); North Wind Power Company, Moretown, VT, *Reliable Power from the Wind* (Moretown, VT: North Wind Power Company, 1981), brochure (in author's files); Don Mayer, Waitsfield, VT, telephone interview by author, April 24, 2015.

41. *A Vision Becomes Reality*, 17, 102.

42. "The UMass Wind Turbine WF-1, A Retrospective," University of Massachusetts Wind Energy Center, accessed September 20, 2018, https://www.umass.edu/windenergy/about/history/windturbinewf1; Duane Cromack and Michael Edds, "Operational Aspects

of the UMass Wind Turbine" (paper presented at the American Wind Energy Association National Conference, Amarillo, TX, March 1–5, 1978), 62–72; Alvin Duskin, "An Introduction to United States Wind Power" (paper presented at the First Conference on Community Renewable Energy Systems, Boulder, CO, August 20–21, 1979), 143–44; Ellen Perley Frank, "Breaking the Energy Impasse[:] A Small Massachusetts Company Puts Windpower in the Marketplace," *New Roots* (Greenfield, MA) no. 7 (September-October 1979): 30–32; Mary Podevin, "Promoting Large-Scale Wind[:] A Conversation with Wind Farm Entrepreneur Alvin Duskin," *Wind Power Digest* no. 18 (Winter 1979–1980): 40–41; Woody Stoddard, "Appreciation[:] The Life and Work of Bill Heronemus, Wind Engineering Pioneer," *Wind Engineering* 26, no. 5 (2002): 338.

43. Jay Carter Jr., president, CEO and principal design engineer, Carter Aviation Technologies, interview by author, Wichita Falls, TX, October 24, 2015; "J. Carter's Development of Fiberglass Technology," *Wind Power Digest* no. 17 (Fall 1979): 40–44; Jay Carter, Jr., "Design and Development of the Carter 25 Generator" (paper presented at the American Wind Energy Association National Conference, Summer 1980, Pittsburgh, PA, June 8–11, 1980), 88–89; Jay Carter Enteprises, Inc., Burkburnett, TX, *Carter Wind Generator Model 25 General Description* (Burkburnett, TX: Jay Carter Enterprises, Inc., [ca. 1981]), brochure (in the files of Bob McBroom, Holton, KS).

44. "Windmill Dynamo Created," *Lubbock Avalanche-Journal* (Lubbock, TX), March 12, 1975; "Zephyr Unveils 7.5 KW Generator," *Wind Power Digest* no. 2 (Summer 1975): 29; Joe Carter, "A Visit with Zephyr Wind Dynamo," *Wind Power Digest* no. 7 (December 1976): 14–17; Zephyr Wind Dynamo Company, Brunswick, ME, *Zephyr Wind Dynamo Company* (Brunswick, ME: Zephyr Wind Dynamo Company, [ca. 1976]), brochure (in the files of Earle Rich, Mont Vernon, NH).

45. "The Kedco 1200," *Wind Power Digest* no. 2 (Summer 1975): 14; Kedco, Inc., Inglewood, CA, *Harnessing Electricity from the Wind*[,] *Kedco 1200* (Inglewood, CA: Kedco, Inc., [ca. 1975]), handbill (in the *Windmillers' Gazette* research files, Rio Vista, TX).

46. "Wind in America: The Changing Landscape," 33; E. F. Lindsley, "Advanced Design Puts New Twists in Windmill Generator," *Popular Science* 215, no. 4 (October 1979): 84–85.

47. Clint Coleman and Wayne Hunnicutt, "Rotor and Control System Design for the 8 KW, 10 Meter Diameter, DOE/Windworks WECS" (paper presented at the American Wind Energy Association National Conference, Spring 1979, San Francisco, CA, April 16–19, 1979), 135–39; Windworks, Inc., Mukwonago, WI, *Windworks* (Mukwonago, WI: Windworks, Inc., [ca. 1980]), brochure (in the files of Craig Toepfer, Chelsea,

MI); H. Meyer, "The Development of a 10 Meter Wind Turbine Generator" (paper presented at the 5th Annual Energy-Sources Technology Conference and Exhibition, New Orleans, LA, March 7–10, 1982), 107–11; "Windworks/Wisconsin Power Merger," *Wind Power Digest*, no. 24 (Summer 1982): 18–19; David Stoeffler, "WPL Hopes to Get Boost from Wind," *Wisconsin State Journal* (Madison, WI), June 17, 1982; Wisconsin Power & Light Company, Madison, WI, *Quarterly Report*[,] *A Report to Investors* 9, no. 2 (August 1982): 2–3 (in the files of Craig Toepfer, Chelsea, MI); "Wind Conversion," *Daily Globe* (Dodge City, KS), October 27, 1982; Wisconsin Power & Light Company, "The Winds of Change Blow Strong and Sure for Windworks," *Concepts* (Madison, WI) 9, no. 4 (Fall 1983): 2–9; John Torinus Jr., "Windworks Finds Home with Utility," *Milwaukee Sentinel*, September 27, 1983; "Bridge Power," *San Francisco Chronicle*, September 10, 1983; "Charging Windmills," *The Tribune* (Oakland, CA), September 11, 1983.

48. Jon Naar, *The New Wind Power* (New York: Penguin Books, 1982), 186–90; William Drake, "The Enertech 1500 Watt Wind Turbine" (paper presented at the American Wind Energy Association National Conference, Fall 78, Cape Cod, MA, September 25–27, 1978), 59–63; Enertech Corporation, Norwich, VT, *Enertech 1500*[,] *A Beautiful Way to Save Electricity* (Norwich, VT: Enertech Corporation, 1979), brochure (in the *Windmillers' Gazette* research files, Rio Vista, TX); Enertech Corporation, Norwich, VT, *Enertech 1800*[,] *A Beautiful Way to Save Electricity* (Norwich, VT: Enertech Corporation, 1981), brochure (in the *Windmillers' Gazette* research files, Rio Vista, TX); Enertech Corporation, Norwich, VT, *Enertech E44* (Norwich, VT: Enertech Corporation, [ca. 1986]), brochure (in the *Windmillers' Gazette* research files, Rio Vista, TX).

49. Fayette Manufacturing Corporation, Fayette, CA, *Fayette Manufacturing Corporation* (Fayette, CA: Fayette Manufacturing Corporation, [ca. 1985]), brochure (in the files of Earle Rich, Mont Vernon, NH); "Wind in America: The Changing Landscape," 33.

50. WhirlWind Power Company, Denver, CO, *Finally, a Cost-Effective Wind System* (Denver, CO: WhirlWind Power Company, [ca. 1980]), folder (in the *Windmillers' Gazette* research files, Rio Vista, TX); WhirlWind Power Company, Denver, CO, *All New for 1982! Economical, Reliable Wind Electric Systems for Home, Business and Remote Power Needs. WhirlWind Series 3000*[,] *3 KW Wind Generators* (Denver CO: WhirlWind Power Company, 1982), handbill (in the *Windmillers' Gazette* research files, Rio Vista, TX); WhirlWind Power Company, Duluth, MN, *WhirlWind Power Company*[,] *Catalog of Wind-Powered Generators and Accessories* (Duluth, MN: WhirlWind Power Company, [ca. 1983]), brochure (in the *Windmillers' Gazette* research files, Rio Vista, TX); WhirlWind Power Company, Duluth, MN, *WhirlWind*

Power Company[,] *Wind Powered Generators*[,] *2 Kilowatt—Series W2*[,] *4 Kilowatt—Series W4*[,] *10 Kilowatt—Series W10* (Duluth, MN: WhirlWind Power Company, [ca. 1987]), brochure (in the *Windmillers' Gazette* research files, Rio Vista, TX).

51. Bruce A. Goodale, "Development of a 40 KW Horizontal Axis Wind Turbine Generator" (paper presented at the American Wind Energy Association National Conference, Spring 1979, San Francisco, CA, April 16–19, 1979), 142–43; Kaman Aerospace Corporation, Bloomfield, CT, *Kaman Wind Energy Systems* (Bloomfield, CT: Kaman Aerospace Corporation, 1980), handbill (in the *Windmillers' Gazette* research files, Rio Vista, TX).

52. Kenneth Speiser, manager, advanced systems, Grumman Energy Systems, Ronkonkoma, NY, letter to B. McBroom, Holton, KS, October 12, 1976, typewritten (in the files of Bob McBroom, Holton, KS); Grumman Energy Systems, Ronkonkoma, NY, *Grumman Windstream 25* (Ronkonkoma, NY: Grumman Energy Systems, 1976), handbill (in the files of Bob McBroom, Holton, KS).

53. M. C. Cheney, "UTRC 8 KW Wind Turbine" (paper presented at the American Wind Energy Association, National Conference, Spring 1979, San Francisco, CA, April 16–19, 1979), 156–85; Herman M. Drees, Thousand Oaks, CA, (wind turbine innovations in the 1970s and 1980s), email to author, September 2, 2018.

54. *The New Wind Power*, 91–92; Energy Sciences Incorporated, Boulder, CO, *The New Generation in Wind Technology* (Boulder, CO: Energy Sciences Incorporated, [ca. 1982]), brochure (in author's files); Thomas E. Jewett, "Design of the ESI-80/200" (paper presented at the Wind Energy Expo '83 and National Conference, San Francisco, CA, October 17–19, 1983), 1–20; C. P. Butterfield and W. D. Musial, "EPRI/ESI Test Program" (paper presented at Windpower '85, San Francisco, CA, August 27–30, 1985), 109–16.

55. Jeanne McDermott, "Harnessing Urban Winds[:] The Bronx Frontier Organizes a Unique Wind-Assisted Composting Project in the Big Apple," *Wind Power Digest* no. 18 (Winter 1979–1980), 7–12; Ted Finch, "Field Testing of a Commercial Wind Energy Conversion System" (paper presented at the 5th Annual Energy-Sources Technology Conference and Exhibition, New Orleans, LA, March 7–10, 1982), 247–55.

56. *The New Wind Power*, 203–06; Paul Gipe, "Wind Power in PA: Terry Mehrkam's Wind Machine," *Wind Power Digest* no. 9 (Summer 1977): 27–31; Paul Gipe, "The Mehrkams Re-visited[:] A 3 KW Unit This Time," *Wind Power Digest* no. 10 (Fall 1977): 13–17; Paul Bruce Gipe, "Tomorrow's Wind Turbines Being Built Today in Pennsylvania," *Alternative Sources of Energy* no. 29 (December 1977): 5–10; Paul Gipe, "Clowning Around at Dorny [sic.] Park," *Wind Power Digest* no. 15 (Spring 1979): 39–43; Paul Gipe, "Mehrkam's Windmills—He's

Overpowering NASA," *Popular Science* 214, no. 4 (April 1979): 82–83; Energy Development Company, Hamburg, PA, *Energy of the Future Available Today* (Hamburg, PA: Energy Development Company, [ca. 1980]), brochure (in the files of Bob McBroom, Holton, KS); Paul Gipe, "Photos of Mehrkham [sic.] Turbines by Paul Gipe," Wind-Works, May 8, 2013, accessed September 29, 2018, http://www.wind-works.org/cms/index.php?id=516.

57. Jytte Thorndahl, "Johannes Juul and the Birth of Modern Wind Turbines," in *Wind Power The Danish Way*, ed. Benny Christensen (Askov, Denmark: Poul la Cour Foundation, 2009), 43–44.

58. Thorndahl, "Johannes Juul," 45; Marshall Merriam, "The Gedser Mill Re-vitalized," *Wind Power Digest* no. 11 (Winter 1977–1978): 34–38; Mogens Johansson, *Present Condition of the Gedser Wind Turbine and a Cost Estimate of Refurbishing the Turbine for Test Purposes* (Lyngby, Denmark: Danske Elvaerkers Forenings Udredningsafdeling, 1977), 1–5.

59. Allen P. Spaulding Jr., "MP 1–200 Technical Description" (paper presented at the American Wind Energy Association National Conference, Fall 78, Cape Cod, MA, September 25–27, 1978), 53–58; WTG Energy Systems Incorporated, Buffalo, NY, *MP 1–200 General Description* (Buffalo, NY: WTG Energy Systems Incorporated, [ca. 1979]), 1–6, handbill (in the files of Earle Rich, Mont Vernon, NH); William R. Lowstuter Jr., "Wind Supplies Much of Cuttyhunk Island's Electric Power," *Electrical Consultant* 59, no. 5 (September-October 1979): 26–30.

60. *Generation on the Wind*, directed by David Vassar (David Vassar, 1979), accessed October 6, 2019, https://www.imdb.com/title/tt0079201/; National Archives Theater, Washington, DC, *Generation of the Wind*[,] *1979* (Washington, DC: National Archives Theater, 2018), handbill (in author's files).

61. *The New Wind Power*, 197.

62. William Sheperdson, "W. T. G. Energy Systems: 200 KW for Cuttyhunk Island," *Wind Power Digest* no. 10 (Fall 1977): 6–11; *The New Wind Power*, 197–200; WTG Energy Systems Inc., Buffalo, NY, *MP-20 Detailed Schematic*[,] *WTG Energy Systems Inc.* (Buffalo, NY: WTG Energy Systems Inc., [ca. 1980]), handbill, 4 leaves (in the *Windmillers' Gazette* research files, Rio Vista, TX).

63. "California May Allocate 15 Mil for State Wind Program," *Wind Power Digest* no. 11 (Winter 1977–1978): 29; Lee Johnson and Marshall Merriam, "Small Groups, Big Windmills," *Wind Power Digest* no. 11 (Winter 1977–1978): 30–31; "3 MW Wind Turbine Slated for Southern California," *Wind Power Digest* no. 12 (Spring 1978): 26; "California Utility to Test 3-MW Wind Turbine," *Wind Energy Report* (July 1978): 1–2; Charles Schachle and Robert L. Scheffler, "The Southern California Edison 3 MW Wind Turbine Generator (WTG) Demonstration Project" (paper presented

at the American Wind Energy Association National Conference, Fall 78, Cape Cod, MA, September 25–27, 1978), 78–80; Charles Schachle and Robert L. Scheffler, "Southern California Edison's 3MW Wind Demonstration Project," *Wind Power Digest* no. 16 (Summer 1979): 32–34.

64. Jay Carter, Sr., "Jay Carter Enterprises. Inc[.] Experience" (paper presented at the Rural Electric Wind Energy Workshop, Boulder, CO, June 1–3, 1982), 267.

65. Joe Carter, "New Machines . . . Wind Power Systems' Storm Master," *Wind Power Digest* no. 13 (Fall 1978): 34–35; Ed Salter, "Flexible Rotor Technology and the 'Storm Master' Wind Turbine[:] A Brief History—1977 to 2009," Scribd, accessed October 9, 2018, https://www.scribd.com/document/134004067/Storm-Master-Rotor-Tech-4–09; Paul Gipe, "Photos of StormMaster," Wind-Works, May 9, 2013, accessed October 9, 2018, http://www.wind-works.org/cms/index.php?id=512.

66. Drees, email to author, September 2, 2018.

67. Hans E. Wulff, *The Traditional Crafts of Persia: Their Development, Technology, and Influence on Eastern and Western Civilization* (Cambridge, MA: MIT Press, 1966), 284–85; Michael Haverson, *Persian Windmills*. Biblioteca Molinologica 10 (Sprang Capelle, Netherlands: International Molinological Society, 1991); T. Lindsay Baker, "An Overview of Horizontal Windmills," *Windmillers' Gazette* 18, no. 1 (Winter 1999): 2.

68. Joseph Needham, *Science and Civilization in China*, vol. 4, *Physics and Physical Technology, Pt. II: Mechanical Engineering* (London: Cambridge University Press, 1965), 556–60; Hong-Sen Yang, *Reconstruction Designs of Lost Ancient Chinese Machinery*, History of Mechanism and Machine Science 3 (Dordrecht, Netherlands: Springer, 2007), 85; Baichun Zhang, "Ancient Chinese Windmills" (paper presented at the Third International Symposium on History of Machines and Mechanisms, National Cheng Keng University, Tainen, Taiwan, November 11–14, 2008), 203–14, accessed February 19, 2014, http://linkspringer.com/chapter/10.1007%2F978-1-4020-9485-9_15.

69. J. Brownlee Davidson and Leon Wilson Chase, *Farm Machinery and Farm Motors* (New York: Orange Judd Company, 1910), 299–302; Edward Charles Murphy, *The Windmill: Its Efficiency and Economic Use*[,] *Part II*, US Department of the Interior, Geological Survey, Water-Supply and Irrigation Paper No. 42 (Washington, DC: Government Printing Office, 1901), 125–27; T. Lindsay Baker, *A Field Guide to American Windmills* (Norman, OK: University of Oklahoma Press, 1985), 13–15; Christopher Gillis, "D. H. Bausman: A Pennsylvania Windmill Maker," *Windmillers' Gazette* 36, no. 1 (Winter 2017): 2–4.

70. M. H. Simonds and A. Bodek, *Performance Test of a Savonius Rotor. Technical Report No. T10*[,] *January*

1964 (Quebec, Canada: Brace Research Institute, McGill University, 1964), 4; W. Vance, "Vertical Axis Wind Rotors—Status and Potential" (paper presented at the Wind Energy Conversion Systems Workshop, Washington, DC, June 11–13, 1973), 96–102; Michael Hackleman, *Wind and Windspinners* (Culver City, CA: Peace Press Printing and Publishing, 1977), 66–94; Paul Gipe, "VAWT's[:] A Brief Introduction," *Wind Power Digest* no. 19 (Spring 1980): 9–10.

71. National Research Council Canada, National Aeronautical Establishment, *Laboratory Technical Report*[,] *LTR-LA-74*[,] *Preliminary Tests of a High S peed Vertical Axis Windmill Model*, by P. South and R. S. Rangi (Ottawa, Canada: National Research Council Canada, 1971), 4–7; Peter South and Raj Rangi, "The Performance and Economics of the Vertical-Axis Wind Turbine Developed at the National Research Council, Ottawa, Canada" (paper presented at Annual Meeting of the Pacific Northwest Region of the American Society of Agricultural Engineers, Calgary, Canada, October 10–12, 1973), 2–9.

72. "The Performance and Economics of the Vertical-Axis Wind Turbine," 7.

73. Ibid., 9.

74. Richard Stepler, "Eggbeater Windmill Is Self-Starting, Cheaper to Build," *Popular Science* 206, no. 5 (May 1975): 74–76; Joe Carter, "VAWT Research at Sandia Labs," *Wind Power Digest* no. 9 (Summer 1977): 42–45; Barbara Francis, "Sandia Darrieus Technology Advances," *Wind Power Digest* no. 19 (Spring 1980): 27–29.

75. Coy Harris, executive director, American Windmill Museum, Lubbock, TX, email to author, February 9, 2016; Herman M. Drees, Thousand Oaks, CA (vertical-axis wind turbine developments in the 1970s and 1980s), email to author, October 19, 2018.

76. Dominion Aluminum Fabricating Limited, Mississauga, Canada, *Wind Turbines* (Mississauga, Canada: Dominion Aluminum Fabricating Limited, 1975), folder (in the *Windmillers' Gazette* research files, Rio Vista, TX); "DAF Darrius [sic.] Marketed," *Wind Power Digest* no. 5 (Summer 1976): 41; Paul Gipe, "Wind and Utilities[:] Alcoa Markets VAWT's," *Wind Power Digest* no. 17 (Fall 1979): 29–32; Paul Gipe, "Alcoa Soon to Dominate VAWT Market," *Wind Power Digest* no. 19 (Spring 1980): 20–25; Joe Szostak, "DAF," *Wind Power Digest* no. 19 (Spring 1980): 30–32; Joe Szostak, Brian Toller, and Doug Whiteway, "NRC[:] Rebirth of the Darrieus Rotor," *Wind Power Digest* no. 19 (Spring 1980): 33–35; Howard R. Kutcher, "Alcoa Vertical Axis Wind Turbines" (paper presented at the American Wind Energy Association National Conference, Summer 1980, Pittsburgh, PA, June 8–11, 1980), 13–15; Commonwealth Electric Company, Boston, MA, *Wind to Watts* (Boston, MA: Commonwealth Electric Company, [ca.

1981]), folder (in the *Windmillers' Gazette* research files, Rio Vista, TX); *The New Wind Power*, 108–10.

77. "NH Utility Purchasing Experimental VAWT," *Wind Energy Report* (July 1978): 5; "Darrieus-Silo Demonstration Project Now Underway," *Wind Power Digest* no. 14 (Winter 1978): 18–19; William Sheperdson, "Dynergy[:] One Entrepreneur's Experience in Wind," *Wind Power Digest* no. 19 (Spring 1980): 13–18.

78. Paul N. Vosburgh, "VAWTPOWER 185, Forecast Vertical Axis Wind Turbines" (paper presented at the Wind Energy Expo '82 and National Conference, Amarillo, TX, October 24–27, 1982), 55–61; Paul N. Vosburgh, *Commercial Applications of Wind Power* (New York: Van Nostrand Reinhold Company, 1983), 68; William J. Broad, "Small, Cheap Windmills Outdo High Tech," *The New York Times*, August 14, 1984. Paul Gipe, "Alcoa Darrieus VAWT," Wind-Works, n.d., accessed October 28, 2018, http://www.wind-works.org/cms/index.php?id=504.

79. "FloWind Dedication," *Wind Power Digest* no. 24 (Summer 1982): 61; Michael Cornwall, "FloWind Begins Operation of Turbine in Washington State," *Sun*Up Energy News Digest* (Yucca Valley, CA) (September 1982): 34; FloWind Corporation, Kent, WA, *FloWind Model 120 Vertical Axis Wind Turbine* (Kent, WA: FloWind Corporation, [ca. 1985]), folder (in author's files); FloWind Corporation, *FloWind 170 Vertical Axis Wind Turbine* (Kent, WA: FloWind Corporation, [ca. 1985]), folder (in author's files); FloWind Corporation, Kent, WA, *FloWind Corporation*[:] *An Integrated Wind Energy Company* (Kent, WA: FloWind Corporation, [ca. 1985]), folder (in author's files); Drees, email to author, October 19, 2018.

80. Niels Sønder, "Darrieus Windmills: Past. President—and Future?" *Windpower Monthly* (Knebel, Denmark) 1, no. 12 (December 1985): 18–19.

81. Dana Hornig, "Energy Out of Thin Air: The Windmills of Marstons Mills," *The Register* (Cape Cod, MA), December 16, 1976; "Interview with Herman Drees[,] President—Pinson Energy Corp.," *Wind Power Digest* no. 7 (December 1976), 5–13; Bill Smith, "Happily, This Windmill Does More Than Spin—It Works for a Living," *Sunday Cape Cod Times* (Hyannis, MA), May 22, 1977; Pinson Energy Corp., Marstons Mills, MA, *The Cycloturbine Vertical Axis Wind Turbine* (Marstons Mills, MA: Pinson Energy Corp., 1977), brochure (in author's files); Richard B. Noll, Herman M. Drees, and Lea Nichols, "Development of the 1–2 KW Cycloturbine" (paper presented at the American Wind Energy Association National Conference, Amarillo, TX, March 1–5, 1978), 40–49; "DOE Contract Announced," *Wind Power Digest*, no. 12 (Spring 1978): 28; H. M. Drees, "The Cycloturbine and Its Potential for Broad Application" (paper presented at the Second International Symposium on Wind Energy Systems, Amsterdam, the Netherlands,

October 3–6, 1978); E7–81—E7–88; Herman Meijer Drees, "The Cycloturbine and Its Potential for Broad Application" (paper presented at the American Wind Energy Association National Conference, Fall 78, Cape Cod, MA, September 25–27, 1978), 64–68; Bob Dunham, "Pinson[:] An Alternate Solution," *Wind Power Digest* no. 19 (Spring 1980): 50–52; Pinson Energy Corp., *Power from the Wind, A Renewable Energy Source* (Marstons Mills, MA: Pinson Energy Corp., [ca. 1980]), folder (in author's files); R. B. Noll, N. D. Ham, H. M. Drees, and L. B. Nichol, *ASI/Pinson 1 Kilowatt High Reliability Wind System Development[,] Phase I—Design and Analysis[,] Technical Report[,] March 1982* (prepared for Rockwell International Corporation, Energy Systems Group, Rocky Flats Plant, Wind Systems Program, Golden, CO, as part of the US Department of Energy, Wind Energy Technology Division, Federal Wind Energy Program) (Burlington, MA: Aerospace Systems Inc., 1982), 1–1—1–8; *The New Wind Power*, 98–102; Herman M. Drees, Thousand Oaks, CA, telephone interview by author, April 30, 2016; Herman M. Drees, interview by author, Westlake Village, CA, August 25, 2017; Drees, email to author, 0ctober 19, 2018.

82. Drees, email to author, October 19, 2018.

83. Christopher C. Gillis, *Still Turning[:] A History of Aermotor Windmills* (College Station: Texas A&M University Press, 2015), 157–59; Robert V. Brulle and Harold C. Larsen, "Giromill (Cyclogiro Windmill) Investigation for Generation of Electrical Power" (paper presented at the Second Workshop on Wind Energy Conversion Systems, Washington, DC, June 9–11,1975), 452–60; Robert V. Brulle, *Engineering the Space Age: A Rocket Scientist Remembers* (Maxwell Air Force Base, AL: Air University Press, 2008), 211–27; Robert D. McConnell, "Giromill Overview" (paper presented at the Wind Energy Innovative Systems Conference, Colorado Springs, CO, May 23–25, 1979), 1–2, 10; John W. Anderson, Robert V. Brulle, Edwin B. Birchfield, and William D. Duwe, *Development of a 40 KW Giromill[,] Phase I[,] Volume II—Design and Analysis* (prepared for Rockwell International Corporation, Energy Systems Group, Rocky Flats Plant, Wind Systems Program, Golden, CO, as part of the US Department of Energy, Division of Distributed Solar Technology, Federal Wind Energy Program) (St. Louis, MO: McDonnell Aircraft Company, 1979), 1, 22–23, 109, 229; Valley Industries, Inc., "Giromill Begins Performance Tests," *Pipeline* (St. Louis) (June 1980): 1; "Giromill Testing Underway," *Wind Power Digest* no. 21 (Winter 1980–1981): 10–12.

84. *Engineering the Space Age: A Rocket Scientist Remembers*, 227.

85. Jack Park and Dick Schwind, *Wind Power for Farms, Homes & Small Industry* (Washington, DC: US Department of Energy, Division of Solar Technology, Federal Wind Energy Program, 1978), 5–44; US

Department of Energy, *Home Wind Power* (Charlotte, VT: Garden Way Publishing, 1981), 94.

86. Alan W. Wilkerson, president, Gemini Company, Thiensville, WI, *Synchronous Inversion for Wind Power Utilization* (Thiensville, WI: Gemini Company, 1975), 1–12, typewritten (in the files of Craig Toepfer, Chelsea, MI); Alan W. Wilkerson, president, Gemini Company, Thiensville, WI, *Synchronous Inversion Techniques for Utilization of Waste Energy* (Thiensville, WI: Gemini Company, 1976), 1–21, typewritten (in the files of Craig Toepfer, Chelsea, MI); Patent Nos. 3,946,242; 4,059,772; 4,200,833; and 4,366,388, U. S. Patent and Trademark Office, Washington, DC; Hans Meyer, Windworks Inc., Mukwonago, WI, *Synchronous Inversion[:] Concept & Application* (Mukwonago, WI: Windworks Inc., 1976), 1–12, typewritten (in the files of Craig Toepfer, Chelsea, MI); Windworks Inc., Mukwonago, WI, *Gemini Synchronous Inverter Systems* (Mukwonago, WI: Windworks Inc., 1978), brochure (in the files of Craig Toepfer, Chelsea, MI); Windependence Electric Company, Ann Arbor, MI, *Windependence Electric Co.* (Ann Arbor, MI: Windependence Electric Company, [ca. 1976]), brochure (in the files of Craig Toepfer, Chelsea, MI); *The New Wind Power*, 94–96; Solomon S. Kagin, *Buyers Guide to Wind Power* (Santa Rosa, CA: Real Gas & Electric Company, 1978), 11; "The Acheval Wind Electronic Co.[:] A Profile," *Alternative Sources of Energy* no. 63 (September-October 1983): 44–46; Donald W. Bingley, "Wind System Utility Intertie with Inverters/Frequency Chargers" (paper presented at the American Wind Energy Association Wind Energy Expo '82 and National Conference, Amarillo, TX, October 24–27, 1982), 159–61; Alan W. Wilkerson, president, Gemini Company, Thiensville, WI, "License Agreement" (typewritten document conferring patent usage to Omnion Power Engineering, Mukwonago, WI), January 4, 1986 (in the files of Sherry Jones, Cedarburg, WI); Sherry Jones, owner, Gemini Controls, Cedarburg, WI, email to author, January 23, 2017.

87. Living Energy Consultants Ltd., Colorado Springs, CO, *Something about Inverters and DWS Multi-Verters* (Colorado Springs, CO: Living Energy Consultants Ltd., 1980), handbill (in the *Windmillers' Gazette* research files, Rio Vista, TX).

88. Henry S. Ruess, letter to the editor, *The New York Times*, June 27, 1976; "Windependence," in "Wind/Hydropower," ed. Don Marier, supplement, *Alternative Sources of Energy* no. 46 (November–December 1980): 50–51.

89. Craig Toepfer, Chelsea, MI, email to author, October 18, 2018.

90. *Synchronous Inversion[:] Concept & Application*, 12; E. F. Lindsley, "New Inverter Gives Wind Power Without Batteries," *Popular Science* 207, no. 4 (October 1975): 52.

91. Leander Nichols, "Wind-Driven Alternators," *Wind Power Digest* no. 9 (Spring 1977): 34–37; Bob Kirchner, *Something About Generators & Alternators* (Colorado Springs, CO: Living Energy Consultants Ltd., [ca. 1979]), handbill (in the *Windmillers' Gazette* research files, Rio Vista, TX); Benjamin Bell, "Synchronous Wind Machines," *Solar Age* 8, no. 10 (October 1983): 36–41; "Wind in America: The Changing Landscape," 33–34.

92. Palmer Cosslett Putnam, *Power From The Wind* (New York: D. Van Nostrand Company, 1948), 1, 109–14.

93. P. W. Carlin, A. S. Laxson, and E. B. Muljadi, "The History and State of the Art of Variable-Speed Wind Turbine Technology," *Wind Energy* 6 (2003): 136–37; Drees, email to author, September 2, 2018.

94. John J. Pullen, "Will It Pay You to Put up a Windmill?" *Country Journal* 7, no. 7 (July 1980): 74–75; *Enertech 1500*[,] *A Beautiful Way to Save Electricity*; *Enertech 1800*[,] *A Beautiful Way to Save Electricity*; *Enertech E44*.

95. *The New Wind Power*, 89; *Carter Wind Generator Model 25 General Description*; Carter Wind Systems, Inc., Burkburnett, TX, *Carter Model 250* (Burkburnett, TX: Carter Wind Systems, Inc., [ca. 1983]), brochure (in the files of Bob McBroom, Holton, KS); Carter Wind Systems, Inc., Burkburnett, TX, *Carter Wind Systems Model 25 Generator—July 1983* (Burkburnett, TX: Carter Wind Systems, Inc., 1983), handbill (in the files of Bob McBroom, Holton, KS); Carter Wind Systems, Inc., *Model 200 Generator Specifications* (Burkburnett, TX: Carter Wind Systems, Inc., [ca. 1983]), handbill (in the files of Bob McBroom, Holton, KS); *The New Generation in Wind Technology*.

96. "The History and State of the Art of Variable-Speed Wind Turbine Technology," 137; Drees, email to author, September 2, 2018.

97. T. Lindsay Baker, "Windmill Towers and Their Infinite Variety," *Windmillers' Gazette* 3, no. 1 (Winter 1984): 3–6; T. Lindsay Baker, "How to Build a Wooden Windmill Tower," *Windmillers' Gazette* 6, no. 4 (Autumn 1987): 6–8; T. Lindsay Baker, "Aermotor Windmill Towers," *Windmillers' Gazette* 11, no. 3 (Summer 1992): 2–14; T. Lindsay Baker, "'Turning Fair to the Wind'[:] An Occupational Vocabulary of Windmilling," *Windmillers' Gazette* 15, no. 4 (Autumn 1996): 7; T. Lindsay Baker, *A Field Guide to American Windmills* (Norman, OK: University of Oklahoma Press, 1985), 93–94.

98. "Aermotor Windmill Towers," 2–4; *A Field Guide to American Windmills*, 94; *Still Turning*, 49–52.

99. Wincharger Corporation, Sioux City, IA, *Every Farm Home Can Now Enjoy "Big City" Radio Reception* (Sioux City, IA: Wincharger Corporation, [ca. 1937]), brochure (in author's files); Wincharger Corporation, Sioux City, IA, *Free Electricity for Radio and Lights*[.] *Brought to the Farm by Wincharger* (Sioux City, IA: Wincharger Corporation, [ca. 1935]), brochure (in author's files); Parris-Dunn Corporation, Clarinda, IA, *Instruction Manual for Model 206—Six Volt* [and] *Model 212—Twelve Volt*[,] *Direct-Drive*[,] *Slip-the-Wind Power Plants* (Clarinda, IA: Parris-Dunn Corporation, [ca. 1935]), 3–4 (in the files of Mike Werst, Manor, TX).

100. *Free Electricity for Radio and Lights*[.] *Brought to the Farm by Wincharger*; Parris-Dunn Corporation, Clarinda, IA, *Hy-Tower*[,] *The World's Best 6-Volt Charger*[,] *Free Power for Radio and Lights with the Hy-Tower* (Clarinda, IA: Parris-Dunn Corporation, [ca. 1935]), brochure (in author's files); LeJay Manufacturing Company, Minneapolis, MN, *LeJay Slow Speed Wind Plants Catalog No. 18A* (Minneapolis, MN: LeJay Manufacturing Company, [ca. 1937]), brochure (in author's files).

101. Wincharger Corporation, Sioux City, IA, *Light Your Farm for 50¢ a Year*[,] *Power Operating Cost* (Sioux City, IA: Wincharger Corporation, [ca. 1937]), brochure (in author's files); Wincharger Corporation, Sioux City, IA, *Giant 32 Volt Wincharger*[,] *Factory Direct to You*[,] *$69.95* (Sioux City, IA: Wincharger Corporation, 1936), brochure (in author's files); Wincharger Corporation, Sioux City, IA, *Self-Supporting Wincharger Tower* (Sioux City, IA: Wincharger Corporation, [ca. 1950]), handbill (in author's files); Wincharger Corporation, Sioux City, IA, *New and Stronger Wincharger Tower* (Sioux City, IA: Wincharger Corporation, [ca. 1950]), handbill (in author's files); Wincharger Corporation, Sioux City, IA, *Here's That Popular, Low-Priced Steel Antenna Tower Made by Wincharger* (Sioux City, IA: Wincharger Corporation, [ca. 1953]), brochure (in author's files).

102. *Wind Power for Farms, Homes & Small Industry*, 7-2—7-3; Gillis, *Windpower*, 34–35; Bruce H. Bailey, *New York State Wind Energy Handbook* (Albany, NY: New York State Energy Office, 1982), 25–26.

103. Ulrich Hütter, "Some Extemporaneous Comments on Our Experiences with Towers for Wind Generators" (paper presented at the Wind Energy Conversion Systems Workshop, Washington, DC, June 11–13, 1973), 206–07.

104. *Wind Power for Farms, Homes & Small Industry*, 5-45—5-48; *Home Wind Power*, 95–97.

105. [Robert Kirschner], *Something About Towers* (Colorado Springs, CO: Living Energy Consultants Ltd., 1980), handbill (in the *Windmillers' Gazette* research files, Rio Vista, TX).

106. *Wind Power for Farms, Homes & Small Industry*, 5-45—5-48; *Home Wind Power*, 95–97; *Something About Towers*.

107. Natural Power Tower Inc., New Boston, NH, *Octahedron Module Tower* (New Boston, NH: Natural Power Inc., 1976), handbill (in author's files); Natural Power Tower Inc., New Boston, NH, *Octahedron Module Tower* (New Boston, NH: Natural Power Inc., 1977),

handbill (in author's files); Hannah Martin, "Buckminster Fuller's Geodesic Dome and Other Forward-Looking Architecture," *Architectural Digest*, February 12, 2016, accessed September 9, 2018, https://www.architectural digest.com/gallery/buckminster-fuller-architecture/all.

108. *Octahedron Module Tower*, 1977; Earle Rich, Mont Vernon, NH, telephone interview by author, August 30, 2018.

109. Samuel M. King and Michael Duffy, "Analysis and Testing of a Tower Structure for Application to Wind Turbine Systems" (paper presented at the American Wind Energy Association National Conference, Amarillo, TX, March 1–5, 1978), 97–106; Herman M. Drees, Thousand Oaks, CA, email to author, October 17, 2018.

110. Earle Rich, interview by author, Amherst, NH, August 26, 2016.

111. Rohn Manufacturing Company Peoria, IL, *"This Is Your Line"*[,] *TV and Communication Towers* (Peoria, IL: Rohn Manufacturing Company, 1959), brochure (in author's files); Unarco-Rohn, Division of Unarco Industries, Inc., Peoria, IL, *Rohn* (Peoria, IL: Unarco-Rohn, [ca. 1978]), handbill (in author's files); "Company History," Rohn Products, accessed October 4, 2015, http://www.rohnnet.com/rohn-company-history.

112. Charles A. Wright, "Charles A. Wright of Rohn Manufacturing" [ca. 1974], one-page, typewritten (in author's files); "Comments by Charlie Wright," *Wind Power Digest* no. 2 (Summer 1975): 25–26.

113. "Charles A. Wright of Rohn Manufacturing."

114. C. A. Wright, letter to Nancy Horning, American Wind Energy Association, August 29, 1974, two pages, typewritten (in author's files); Philip W. Metcalfe, Brimfield, IL, telephone interview by author, November 16, 2015.

115. Unarco-Rohn, Division of Unarco Industries, Inc., Peoria, IL, *Rohn 25G Tower* (Peoria, IL: Unarco-Rohn, 1978), brochure (in author's files); Unarco-Rohn, Division of Unarco Industries, Inc., Peoria, IL, *Rohn No. 45G Communication Tower* (Peoria, IL: Unarco-Rohn, 1979), brochure (in author's files); Unarco-Rohn, Division of Unarco Industries, Inc., Peoria, IL, *Rohn SSV Self-Supporting Communication Towers* (Peoria, IL: Unarco-Rohn, 1978), brochure (in author's files); Clean Energy Products, Seattle, WA, *Wind Power Systems and Components* (Seattle, WA: Clean Energy Products, 1979), brochure (in author's files).

116. Philip W. Metcalfe, Unarco-Rohn, Division of Unarco Industries, Inc., *To: R. A. Kleine*[,] *Market Survey and Forecast for Wind Generator Towers*[,] *February 20, 1980* (Peoria, IL: Unarco-Rohn, 1980), 1–3, typewritten memorandum (in author's files).

117. Ibid., 3.

118. Metcalfe interview.

119. Henry Clews, "Electricity from the Wind: Ready-Made Units," in *Producing Your Own Power*[:]

How to Make Nature's Energy Sources Work for You, ed. Carol Hupping Stoner (Emmaus, PA: Rodale Press, Inc., 1974), 18–22; Harry L. Wegley and William T. Pennell, "Siting Small Wind Machines" (paper presented at the American Wind Energy Association National Conference, Fall 78, Cape Cod, MA, September 25–27, 1978), 110–26; Ralph Wolfe and Peter Clegg, *Home Energy for the Eighties* (Charlotte, VT: Garden Way Publishing, 1979), 138; Jeanne McDermott, "How to Make a Wind-Site Analysis," *Popular Science* 217, no. 1 (July 1980): 100–02; *Home Wind Power*, 35–45; Harry L. Wegley, James V. Ramsdell, Montie M. Orgill, and Ron L. Drake, *A Siting Handbook for Small Wind Energy Conversion Systems*[,] *PNL-2521 Rev 1*[,] *March 1980* (prepared for the US Department of Energy, Washington, DC, as part of the Pacific Northwest Laboratory, Battelle Memorial Institute) (Richland, WA: Pacific Northwest Laboratory, 1980), 5–34; Lisa Cohn, "Prospecting for Wind Pits Man Against Machine," *Windpower Monthly* (Knebel, Denmark) 15, no. 6 (June 1999): 42–43.

120. Robert B. Keller, "Problems Relating to Small Wind Energy Conversion Systems as Seen by a Dealer/Installer" (paper presented at the Small Wind Turbine Systems 1981, Boulder, CO, May 12–14, 1981), 407.

121. *A Siting Handbook for Small Wind Energy Conversion Systems*, 1.

122. *A Siting Handbook for Small Wind Energy Conversion Systems*, 35–52; "Anemometers and Recorders," in "The Solar Age Wind Products Supplement," *Solar Age Magazine* (1980), 6.

123. Natural Power Inc., New Boston, NH, *Questions That Predicate the Future Direction of the Company* (New Boston, NH: Natural Power Inc., 1975), 1–2, typewritten memorandum (in the files of Earle Rich, Mont Vernon, NH); "Review of Hardware," *Wind Power Digest* no. 6 (September 1976): 44–46; Natural Power Inc., New Boston, NH, *Specialists in Instrumentation and Controls for Alternate Energy Industry* (New Boston, NH: Natural Power Inc., 1976), handbill (in author's files); Natural Power Inc., New Boston, NH, *Wind Survey Techniques Using Natural Power Inc. Recording Anemometers* (New Boston, NH: Natural Power Inc., 1976), brochure (in author's files); Natural Power Inc., New Boston, NH, *Wind Energy Monitor* (New Boston, NH: Natural Power Inc., 1976), handbill (in author's files); Natural Power Inc., New Boston, NH, *Wind Data Compiler* (New Boston, NH: Natural Power Inc., 1976), handbill (in author's files); Natural Power Inc., New Boston, NH, *Data Accumulator* (New Boston, NH: Natural Power Inc., 1976), handbill (in author's files); Natural Power Inc., New Boston, NH, *Price List* (New Boston, NH: Natural Power Inc., 1976), handbill (in author's files); Natural Power Inc., New Boston, NH, *About Natural Power Inc.* (New Boston, NH: Natural Power Inc., 1977), handbill (in the *Windmillers' Gazette* research files, Rio Vista,

TX); Natural Power Inc., New Boston, NH, *Wind and Solar Energy Products* (New Boston, NH: Natural Power Inc., 1977), brochure (in the files of Richard Katzenberg, Amherst, NH); Charles Puckette, Natural Power Inc., *To Rick Katzenberg[,] 1979 Marketing Plan[,] November 20, 1978* (New Boston, NH: Natural Power Inc., 1978), 1–8, typewritten memorandum (in the files of Earle Rich, Mont Vernon, NH);

124. Natural Power Inc., New Boston, NH, *Anemometer—Recorder System* (New Boston, NH: Natural Power Inc., 1976), handbill (in author's files).

125. Pete Maule, "Oregon Anemometer Loan Program," *Wind Power Digest* no. 20 (Summer 1980): 36–41; Paul Gipe, "Adopt an Anemometer—A Unique Anemometer Loan Program" (paper presented at the Wind Energy Expo '82 and National Conference, Amarillo, TX, October 24–27, 1982), 79.

126. Natural Power Inc., *Price List*, 1976.

127. Natural Power Inc., New Boston, NH, *Tachometer* (New Boston, NH: Natural Power Inc., 1976), handbill (in author's files).

128. Natural Power Inc., New Boston, NH, *Wind and Solar Energy Products* (New Boston, NH: Natural Power Inc., 1984), catalog (in author's files); Richard Katzenberg, interview by author, Amherst, NH, August 26, 2016.

129. NRG Systems, Charlotte, VT, *NRG Systems Logger 9000[,] an Advanced Real-time Serial Wind Data Logger* (Charlotte, VT: NRG Systems, 1987), folder (in the *Windmillers' Gazette* research files, Rio Vista, TX).

130. "Lone Star Wind[:] An Overview of Current Wind Energy Activities in Texas," *Wind Power Digest* no. 17 (Fall 1979): 36–40; "New Dawn in the City," *Wind Power Digest* no. 20 (Summer 1980): 18; Coy Harris, executive director, American Windmill Museum, Lubbock, TX, email to author, October 23, 2018.

131. "Windkraft zum Anfassen," Mühlenheider Wind Power Museum, accessed November 9, 2018, http://www.muehlenheider-windkraftmuseum.de/English-Site/; "The Museum," Poul la Cour Museum, accessed November 9, 2018, http://www.poullacour.dk/engelsk/museet.htm; "State of Green," The Danish Museum of Electricity (Elmuseet), accessed November 9, 2018, https://stateof green.com/en/partners/danish-museum-of-energy-energimuseet/; "Danske vindmøller 1976–2000," Danish Wind Historical Collection (Danmarks Vindkrafthistoriske Samling), accessed November 9, 2018, http://www.vindhistorie.dk.

132. Kate Gailbrath, "Museum Shows History and Power of Wind Energy," *The New York Times*, July 30, 2011; T. Lindsay Baker, "Windmill Museums as Destinations for Heritage Tourists," *Windmillers' Gazette* 31, no. 2 (Spring 2012): 3.

133. "Windmill Museums as Destinations for Heritage Tourists"; Coy Harris, executive director, American

Windmill Museum, Lubbock, TX, email to author, January 21, 2016; Coy Harris and John Baker, interview by author, Lubbock, TX, June 6, 2018; Harris, email to author, October 23, 2018.

Chapter 9

1. American Wind Energy Association, *Wind Energy in the 80s—A Decade of Development*, March 30, 1990 (in author's files).

2. "From the Editor: Saddam Hussein's Timely Reminder," *Windpower Monthly* (Knebel, Denmark) 6, no. 9 (September 1990): 10; Patrick Lee, "Impact of the Gulf War: Crude Plunges; Gasoline Prices to Dealers Cut: Energy: The Movement Defies Predictions," *Los Angeles Times*, January 18, 1991; Ole Gunnar Austvik, "The War over the Price of Oil: Oil and the Conflict on the Persian Gulf," *International Journal of Global Energy Issues* 5, nos. 2–3-4 (1993): 136, 139–41.

3. "Total Energy[,] Annual Energy Review[,] September 2012[:] Table 8.10 Average Retail Prices of Electricity, 1960–2011 (Cents per Kilowatt-hour, including Taxes)," US Energy Information Administration, September 27, 2012, accessed January 8, 2019, https://www.eia.gov/totalenergy/data/annual/showtext.php?t=ptb0810.

4. "Bush Signs Major Revision of Anti-Pollution Law," *The New York Times*, November 16, 1990; Bryan Lee, "Highlights of the Clean Air Act Amendments of 1990," *Journal of the Air & Waste Management Association* (Pittsburgh, PA) 41, no. 1 (January 1991): 16–19.

5. W. S. Bollmeier, II, The Pacific International Center for High Technology Research, Energy and Resources Division, *To: A. R. Trenka[,] Memorandum[,] Trip Report* (*Windpower '90, Washington DC, September 25–28, 1990*)[,] *October 22, 1990* (Honolulu, HI: The Pacific International Center for High Technology Research, 1990), typewritten (from files of Warren S. Bollmeier, II, Kaneohe, HI).

6. Ibid.

7. Christopher Gillis, *Windpower* (Atglen, PA: Schiffer Publishing Ltd., 2008), 89.

8. Utility Wind Interest Group, Palo Alto, CA, *The Evolving Wind Turbine* (Palo Alto, CA: Utility Wind Interest Group, 1993), brochure (from the files of Warren S. Bollmeier, II, Kaneohe, HI).

9. S. M. Hock and R. W. Thresher, "The Federal Advanced Wind Turbine Program" (paper presented at Windpower '91, Palm Springs, CA, September 24–27, 1991), 36–40; Clint (Jito) Coleman, "Northern Power's Advanced Wind Turbine Development Program" (paper presented at Windpower '91, Palm Springs, CA, September 24–27, 1991), 185–89; P. S. Hughes, R. W. Sherwin, and H. M. Clews, "Development of Atlantic Orient's 15/50 Wind Turbine" (paper presented at Windpower '91, Palm Springs, CA, September 24–27, 1991), 193–99;

R. Lynette, "Advanced 275 KW Wind Turbine" (paper presented at Windpower '91, Palm Springs, CA, September 24–27, 1991), 200–05; National Renewable Energy Laboratory, Golden, CO, "Northern Power Systems Devises Innovative Turbine Rotor," *Advanced Wind Turbines: Electricity for the 1990s and Beyond* (Golden, CO: National Renewable Energy Laboratory, 1992), brochure (from the files of Warren S. Bollmeier, II, Kaneohe, HI); National Renewable Energy Laboratory, Golden, CO, "WC-86B Wind Turbine to Show Big Jump in Performance," *Advanced Wind Turbines: Electricity for the 1990s and Beyond* (Golden, CO: National Renewable Energy Laboratory, 1992), brochure (from the files of Warren S. Bollmeier, II, Kaneohe, HI); National Renewable Energy Laboratory, Golden, CO, "New Wind Turbine Designs Target Rural, Utility Markets," *Advanced Wind Turbines: Electricity for the 1990s and Beyond* (Golden, CO: National Renewable Energy Laboratory, 1992), brochure (from the files of Warren S. Bollmeier, II, Kaneohe, HI).

10. "Northern Power's Advanced Wind Turbine Development Program," 188.

11. US Department of Energy, Solar Energy Research Institute, Golden, CO, *A Government-Industry Partnership Experience*[:] *The Cooperative Field Test Program in Wind Research* (Washington, DC: American Wind Energy Association, 1991), brochure (from the files of Warren S. Bollmeier, II, Kaneohe, HI); D. M. Dodge and W. S. Bollmeier II, *Wind-Hybrid System Tests*, US Department of Energy, National Renewable Energy Laboratory, Cooperative Field Test Program for Wind Systems[,] Final Report (Golden, CO: National Renewable Energy Laboratory, 1992), 7–2—7–7; James Tangler, *Advanced Wind Turbine Design Studies*[,] *Advanced Conceptual Study*[,] *Final Report*[,] *Atlantic Orient Corporation*[,] *Norwich, Vermont* (prepared for the US Department of Energy, National Renewable Energy Laboratory) (Golden, CO: National Renewable Energy Laboratory, 1994), vi; US Department of Energy, "Alaska to Harness Arctic Winds for Village Power," *Wind Power Today* (Washington, DC) (1997): 24–29, 31; US Department of Energy, "Technology Key to Adoption of Wind Energy in Alaska," *Wind Power Today* (Washington, DC) (1997): 30.

12. Nordtank Energy Group A/S, Balle, Denmark, *The 1.5 MW Wind Turbine of Tomorrow* (Balle, Denmark: Nordtank Energy Group, [ca. 1997]), brochure (from the files of Warren S. Bollmeier, II, Kaneohe, HI); "Nordtank, Micon Proceed with Merger Proposal," *AWEA Wind Energy Weekly* (Washington, DC) 16, no. 751 (June 9, 1997): 2; Vestas Wind System A/S, Lem, Denmark, *Annual Report 1997* (Lem, Denmark: Vestas Wind System, 1997), 10 (from the files of Warren S. Bollmeier, II, Kaneohe, HI); Vestas Wind System A/S, Lem, Denmark, *1.65 MW*[:] *Vestas V66—1.65 MW Pitchregulated* [sic.]

Wind Turbine with OptiSlip and OptiTip (Lem, Denmark: Vestas Wind System A/S, [ca. 1997]), brochure (from files of Warren S. Bollmeier, II, Kaneohe, HI); Bonus Energy A/S, Brande, Denmark, *Bonus Energy A/S* (Brande, Denmark: Bonus Energy A/S, [ca. 2000]), handbill (from the files of Warren S. Bollmeier, II, Kaneohe, HI).

13. US Department of Energy, "A New Generation of Wind Turbines on the Horizon," *Wind Power Today* (Washington, DC, 1997): 18.

14. Ibid.

15. Ibid., 19.

16. Susan Gouchoe, *Local Government and Community Programs and Incentives for Renewable Energy—National Report*[,] *North Carolina Solar Center*[,] *Industrial Extension Service*[,] *North Carolina State University*[,] *December 2000* (Raleigh, NC: Database of State Incentives for Renewable Energy, 2000), 12–13.

17. Gillis, *Windpower*, 34–35; Barbara MacMullan and Henry G. duPont, "The Economics of Utility-Interconnected Residential Wind Turbines" (paper presented at Windpower '93, San Francisco, July 12–16, 1993), 543–50; California Energy Commission, Sacramento, CA, *ABCs of Net Metering* (Sacramento, CA: California Energy Commission, 2000), brochure (in author's files); American Wind Energy Association, Washington, DC, *Is a Residential Wind System for You?* (Washington, DC: American Wind Energy Association, [ca. 1999]), handbill (from the files of Warren S. Bollmeier, II, Kaneohe, HI).

18. *A Government-Industry Partnership Experience*[:] *The Cooperative Field Test Program in Wind Research*; *Wind-Hybrid System Tests*, 7–1—7–2.

19. "A New Generation of Wind Turbines on the Horizon," 21.

20. Ibid.; Trudy L. Forsyth, "An Introduction to the Small Wind Turbine Project" (paper presented at Windpower '97, Austin, TX, June 15–18, 1997), 231–39.

21. Andy Kruse, Boulder, CO, telephone interview by author, December 13, 2017; Southwest Windpower, Flagstaff, AZ, *Air Power*[,] *New Air 403*[,] *an All New Design that Once Again Redefines How the World Looks at Wind . . .* (Flagstaff, AZ: Southwest Windpower, [ca. 1998]), brochure (in author's files).

22. Mick Sagrillo, "Home-Built Wind Generators," *AWEA WindLetter* (Washington, DC) 21, no. 5 (May 1994): 4.

23. Mick Sagrillo, "So You Want to Build a Wind Generator?" *Home Power*, no. 17 (June-July 1990): 28–30. *Home Power* ceased publishing in 2018.

24. Sagrillo, "Home-Built Wind Generators."

25. Ibid.

26. Wind Turbine Industries Corp., Prior Lake, MN, *Jacobs Wind Energy Systems 10–17.5 KW Wind Turbines* (Prior Lake, MN: Wind Turbine Industries Corp., 1986), brochure (from the files of Dan Whitehead, St. Johns, FL).

27. Daniel Whitehead, "Alternative Energy . . . or Just Plain Crazy," *Home Power*, no. 53 (June-July 1996): 6–10; Dan Whitehead, "Living with a Wind Machine," *Home Power*, no. 57 (February-March 1997): 18–21; Daniel Whitehead, St. Johns, FL, email to author, November 7, 2018.

28. Whitehead email.

29. American Wind Energy Association, Washington, DC, *Wind Energy Outlook 2000* (Washington, DC: American Wind Energy Association, 2000), brochure (from the files of Warren S. Bollmeier, II, Kaneohe, HI).

30. Ibid.

31. American Wind Energy Association, *Is a Residential Wind System for You?*

32. American Wind Energy Association, Small Wind Turbine Committee, *The U. S. Small Wind Turbine Industry Roadmap*[:] *A 20-Year Industry Plan for Small Wind Turbine Technology* (Washington, DC: American Wind Energy Association, 2002), 1–20.

33. H. Rhoads-Weaver and T. Forsyth, "Overcoming Technical and Market Barriers for Distributed Wind Applications: Reaching the Mainstream" (paper presented at the ASME 2006 International Solar Energy Conference, Denver, CO, July 8–13, 2006), 2.

34. Ibid.

35. R. Christopher Adams, *An Analysis of Wind Power Development in the Town of Hull, MA* (Hull, MA: Hull Municipal Light Plant, 2013), 6–8.

36. Jennifer L. Edwards, Ryan Wiser, Mark Bolinger, and Trudy Forsyth, *Evaluating State Markets for Residential Wind Systems: Results from an Economic and Policy Analysis Tool*[,] *Prepared for the Wind & Hydropower Technologies Program*[,] *Assistant Secretary for Energy Efficiency and Renewable Energy*[,] *U. S. Department of Energy*[,] *Paper LBNL-56344*[,] *December 2004* (Berkeley, CA: Ernest Orlando Lawrence Berkeley National Laboratory, 2004), 1–4; Sara Schaeffer Munoz, "A Novel Way to Reduce Home Energy Bills," *The Wall Street Journal*, August 15, 2006.

37. *Evaluating State Markets for Residential Wind Systems*, 4–5.

38. American Wind Energy Association, *AWEA Small Wind Turbine Global Market Study 2007* (Washington, DC: American Wind Energy Association, 2007), 2–3.

39. American Wind Energy Association, *Small Wind Turbine Global Market Study: 2009* (Washington, DC: American Wind Energy Association, 2009), 3, 11; American Wind Energy Association, *AWEA Reports 78% Growth in 2008 for U. S. Small Wind Market*, May 28, 2009 (in author's files); "US Market for Small Wind Turbines Grew 78% in 2008," *Reuters*, June 1, 2009.

40. Martin Vaughn, "Tax Report[:] Taking Credit for Energy Efficiency," *The Wall Street Journal*, December 31, 2008; Dirk Lammers, "Federal Tax Credit Boosts Home Wind Turbine Market," *Associated Press*, January

25, 2009; Ken Thomas, "Wind Energy Groups Seeking Economic Stimulus Aid," *Associated Press*, January 30, 2009; Michael A. Fletcher, "Obama Leaves D. C. to Sign Stimulus Bill[,] Renewable Energy a Focal Point in Denver as $787 Billion Effort Is Made Law," *The Washington Post*, February 18, 2009; "US Renewable Energy Industry Reacts to Stimulus Package Passage," *RenewableEnergyWorld.com*, February 18, 2009; Carolyn Starks, "More Turn to Wind Turbines," *Chicago Tribune*, April 24, 2009.

41. *Small Wind Turbine Global Market Study: 2009*, 3.

42. US Department of Energy, *20% Wind Energy by 2030*[:] *Increasing Wind Energy's Contribution to U. S. Electricity Supply*[,] *DOE/GO-102008–2567*[,] *May 2008* (Washington, DC: US Department of Energy, 2008), 55–56.

43. Andy Kruse, Boulder, CO, telephone interview by author, December 18, 2017; Southwest Windpower, Flagstaff, AZ, *Skystream 3.7 Owner's Manual* (Flagstaff, AZ: Southwest Windpower, Inc., 2006), 1–19 (in author's files); Jennifer Alsever, "A Nice Stiff Breeze, and a Nice Littler Power Bill," *The New York Times*, April 1, 2007; Renuka Rayasam, "Building This Business, the Answer's in the Wind," *U.S. News & World Report* (May 28, 2007): 64; Southwest Windpower, Flagstaff, AZ, *Skystream 3.7 by Southwest Windpower* (Flagstaff, AZ: Southwest Windpower, Inc., [ca. 2009]), brochure (in author's files); Paul Gipe, "Small Turbine Product Reviews[:] Commentary on the Skystream 3.7," Wind-Works, May 9, 2009, accessed February 7, 2019, http://www.wind-works.org/cms/index.php?id=70&tx_ttnews%5Btt_news%5D=55&cHash=71d22987868dd50c5f0b19c8047c3dea; National Renewable Energy Laboratory, Golden, CO, *Innovation*[:] *NREL Innovations Contribute to an Award-Winning Small Wind Turbine* (Golden, CO: National Renewable Energy Laboratory, 2010), handbill (in author's files).

44. "Best of What's New 2006[,] Home Tech[,] Skystream 3.7 Wind Power for Everyone," *Popular Science* 269, no. 6 (December 2006): 44; "Best Inventions of 2006[,] The Home[,] Breezy Alternative," *Time* (2006), accessed February 7, 2019, http://content.time.com/time/specials/packages/article/0,28804,1939342_1939395_1939670,00.html.

45. American Wind Energy Association, *2010 AWEA Small Wind Turbine Global Market Study* (Washington, DC: American Wind Energy Association, 2010), 18.

46. John-Laurent Tronche, "Fort Worth Energy Firm Prepares First Turbine Prototype," *Fort Worth Business Press* (Fort Worth, TX), January 19, 2009.

47. *Small Wind Turbine Global Market Study: 2009*, 11.

48. Josh Weil, "A New Spin on Wind," *Orion* 24, no. 6 (November–December 2005): 65–66.

49. Office of the Governor, Commonwealth of Pennsylvania, Harrisburg, PA, *Governor Rendell Making Small*

Wind Energy Systems Available to Local Governments[,] *15 Small-Scale Systems to Place Alternative Energy in Public View*, April 7, 2006 (in author's files); Chris Gillis, "Tapping Renewable Energies," *American Shipper* (Jacksonville, FL) 50, no. 6 (June 2008): 90–91; Kate Galbraith, "Assessing the Value of Small Wind Turbines," *The New York Times*, September 3, 2008; Debra Kahn, "San Francisco Plan Promotes Urban Wind Power," *The New York Times*, September 30, 2009.

50. Ken Belson and David W. Dunlap, "Architects and Engineers Express Doubts about Bloomberg's Windmill Proposal," *The New York Times*, August 21, 2008; Paul Kilduff, "Homeowners Can Harness the Wind," *The Chronicle* (San Francisco, CA), December 20, 2008; Richard Richtmyer, "NY Testing Wind-Energy atop Albany Skyscraper," *Associated Press*, February 2, 2009; Anne Eisenberg, "Bringing Wind Turbines to Ordinary Rooftops," *The New York Times*, February 14, 2009; Alex Wilson, "The Folly of Building-Integrated Wind," *Building Green* 18, no. 5 (April 29, 2009): n.p., accessed February 11, 2019, http://www.buildinggreen.com/feature/folly-building-integrated-wind; Shelly Widhalm, "Rooftop Wind Turbine Seen as Green Energy Starter Kit," *Loveland Reporter-Herald* (Loveland, CO), May 16, 2009; R. Nolan Clark, *Small Wind*[:] *Planning and Building Successful Installations* (Waltham, MA: Academic Press, 2014), 86; AeroVironment, Inc., Massachusetts Port Authority, and Groom Green, *Architectural Wind at Boston's Logan International Airport* (Boston, MA: Massachusetts Port Authority, 2019), handbill (in author's files).

51. Mike Bergey, president, Distributed Wind Energy Association, letter to Chris Schaffner, chair, LEED Energy & Atmosphere Technical Advisory Group, US Green Building Council, August 20, 2012 (in the files of Bergey Windpower Company, Norman, OK); Indiana Distributed Energy Alliance, "Paul Gipe Discusses Problems with LEED, USGBC and 'Bad' Wind Turbines in Indianapolis; Are Changes Needed in LEED?" July 20, 2013, accessed February 11, 2019, http://www.indianadg .net/paul-gipe-discusses-problems-with-leed-usgbc-and-bad-wind-turbines-in-indianapolis-are-changes-needed-in-leed/.

52. Rachel Barron, "They Call the Small-Wind Firm Mariah," *Green Media Tech* (Mumbai, India), April 16, 2008.

53. Melissa Block, "Big Dreams for Small Wind Turbines," *National Public Radio*, August 27, 2009.

54. A. Huskey and T. Forsyth, *NREL Small Wind Turbine Test Project: Mariah Power's Windspire Wind Turbine Test Chronology*[,] *Technical Report*[,] *NREL/ TP-500–45552*[,] *June 2009* (Golden, CO: National Renewable Energy Laboratory, 2009), 1–9.

55. Paul Gipe, "Mariah in the Running for Worst Small Turbine Install," Wind-Works, January 31, 2010, accessed July 14, 2019, http://www.wind-works.org/cms/ index.php?id=64&tx_ttnews%5Btt_news%5D=99&cH ash=9d4e873aeba4ecbaed8515276e9a44bc; Paul Gipe, "Mariah Windspire VAWT Measured Performance," Wind-Works, March 22, 2013, accessed July 14, 2019, http://www.wind-works.org/cms/index.php?id=64&tx_ ttnews%5Btt_news%5D=2295&cHash=06b4fe398a441 d737b971bb5fe47e201; Paul Gipe, "Defunct Windspire Wins Greenwashing Award," Wind-Works, June 28, 2013, accessed July 14, 2019, http://www.wind-works .org/cms/index.php?id=399&tx_ttnews%5Btt_news%5 D=2483&cHash=3b3cb9265ea9f4f24362b6e9794dbc45; Lindsay Hall, "Genoa Township Plans to Tear Down 3-Year-Old Wind Turbines," *Michigan Radio/National Public Radio*, July 16, 2013.

56. "Defunct Winspire Wins Greenwashing Award."

57. Randy Roguski, "Demand for Small-Scale Wind Turbines Growing Among Ohio Businesses," *The Plain Dealer* (Cleveland, OH), May 2, 2009; Alexander Haislip, "Startup Raises $2 Million for Small-Scale Wind Turbines," *Reuters*, May 12, 2009; "Green Energy Technologies Introduces WindCube," *North American Windpower* (Southbury, CT), May 14, 2009; "Startup Green Energy Tech Installs First Small-Wind Concentrators," *Reuters*, June 3, 2009; Paul Gipe, "Wind Cube Squarely over the Top (New Re-Branded as Wind Sphere)," Wind-Works, July 8, 2013, accessed February 11, 2019, http:// www.wind-works.org/cms/index.php?id=660&tx_ ttnews%5Btt_news%5D=149&cHash=308028ac52d4dde bd3f8b589d9e91bdb.

58. Martin LaMonica, "Inside a Small Wind-Turbine Beta Test," *Green Tech—CNET News*, July 22, 2009; Ovidiu Sandru, "$4,500 WindTronics Ultra-Efficient Low Speed Wind Turbine Available This Fall," *The Green Optimistic*, June 9, 2009; WindTronics, Muskegon, MI, *Honeywell Wind Turbine by WindTronics Now Available for Purchase*, April 21, 2011 (in author's files); Sarah Parsons, "A Personal Turbine Makes Your Rooftop into a Wind Farm," *Popular Science*, September 16, 2011, accessed September 19, 2019, https://www.popsci .com/gadgets/article/2011–09/personal-turbine-makes-your-rooftop-wind-farm/; Honeywell International Inc., Ontario, Canada, *Honeywell Wind Turbine Model WT6500 Owner's Manual* (Ontario, Canada: Honeywell International Inc., [ca. 2011]), 26 (in author's files); Paul Gipe, "Honeywell Windtronics Kaputt—Finally an End to a Sad Saga of a 'Revolutionary Roof Top Wind Turbine,'" Wind-Works, June 28, 2013, accessed July 14, 2019, http://www.wind-works.org/cms/index. php?id=674&tx_ttnews%5Btt_news%5D=2480&cHash= d3a6080425095e7cd7c8a0ce1e7085a5.

59. American Wind Energy Association, *AWEA Small Wind Turbine Performance and Safety Standard*[,] *AWEA 9.1—2009* (Washington, DC: American Wind Energy Association, 2009), II-IV; Colin Miner, "Standards for

Small-Scale Wind Power," *The New York Times*, August 28, 2009; T. Jimenez, T. Forsyth, A. Huskey, I. Mendoza, K. Sinclair, and J. Smith, "Establishment of Small Wind Regional Test Centers" (paper presented at the 2011 National Solar Conference, Raleigh, NC, May 17–21, 2011), 1–6; Larry Sherwood, president and CEO, Interstate Renewable Energy Council Inc., telephone interview by author, March 23, 2018.

60. American Wind Energy Association, *2011 U.S. Small Wind Turbine Market Report* (Washington, DC: American Wind Energy Association, 2012), 33–34.

61. Ibid., 35.

62. *2010 Small Wind Turbine Global Market Study*, 3.

63. Ibid., 14.

64. Bob Diddlebock, "Small Wind Turbines[:] Blowing Hot," Special Section[:] Global Business/Small Business, *Time* 176, no. 2 (July 12, 2010): 10.

65. *2011 U. S. Small Wind Turbine Market Report*, 4.

66. Ibid., 12.

67. Jonathan Ellis and Cody Winchester, "Sales of Wind Turbines for Home Use Are Going Strong," *USA Today*, June 29, 2011; Kate Galbraith, "Homeowners and Business Embracing Small Wind Turbines," *The New York Times*, October 2, 2011.

68. John Perlin, *The Silicon Solar Cell Turns 50* (Golden, CO: National Renewable Energy Laboratory, 2004), brochure (in author's files); "This Month in Physics History[:] April 25, 1954: Bell Labs Demonstrates the First Practical Silicon Solar Cell," *APS News* (American Physical Society, College Park, MD) 18, no. 4 (April 2009): 2; Matthew Sabas, "History of Solar Power," Institute for Energy Research, February 18, 2016, accessed February 19, 2019, https://www.instituteforenergyresearch.org/renewable/solar/history-of-solar-power/.

69. Steve Coffel, "Photovoltaics[:] Electricity from Sunshine," *Alternative Sources of Energy*, no. 42 (March–April 1980): 3.

70. "History of Solar Power."

71. *AWEA Small Wind Turbine Global Market Study 2007*, 8; Solar Energy Industries Association, *US Solar Industry Year in Review 2008* (Washington, DC: Solar Energy Industries Association, 2009), 1–5.

72. *Small Wind Turbine Global Market Study: 2009*, 25.

73. Solar Energy Industries Association, *US Solar Industry Year in Review 2009* (Washington, DC: Solar Energy Industries Association, 2010), 6; Solar Energy Industries Association, *US Solar Energy Industry Experiences Record-Breaking Growth in 2010*, May 14, 2012 (in author's files).

74. *AWEA Small Wind Turbine Global Market Study 2007*, 8.

75. *2011 U. S. Small Wind Turbine Market Report*, 9.

76. Ibid., 49.

77. Ibid.

78. A. C. Orrell, L. T. Flowers, M. N. Gagne, B. H. Pro, H. E. Rhoads-Weaver, J. O. Jenkins, K. M. Sahl, and R. E. Baranowski, *US Department of Energy[,] Energy Efficiency & Renewable Energy[,] Wind Program[,] 2012 Market Report on U. S. Wind Technologies in Distributed Applications[,] August 2013* (Richland, WA: Pacific Northwest Laboratory, 2013), 3.

79. Larry Sherwood, "What Happened to Southwest Wind Power?" *Small Wind Energy News* (Interstate Renewable Energy Council, Latham, NY), May 3, 2013, accessed November 27, 2017, http://www.irecusa.org/2013/05/what-happened-to-southwest-wind-power/; Kruse interview, December 18, 2017.

80. Southwest Windpower, Flagstaff, AZ, *$10M from GE and Current Investors Plus Federal Stimulus Incentives Propel Southwest Windpower's Expansion in Small Wind Turbines*, April 6, 2009 (in author's files); Michael Kanellos, "Big Boost for Small Wind: GE Invests in Southwest Windpower," *Greentech Media* (Boston, MA) April 6, 2009; Cyndy Cole, "Southwest Windpower gets $10M Investment," *Arizona Daily Sun*, April 7, 2009; Southwest Windpower, Flagstaff, AZ, *Power to the People: Southwest Windpower Unveils Most Efficient, Easy-to-Use Small Wind Turbine*, January 6, 2011 (in author's files); Candice Lombardi, "Home Depot's Latest Small Wind Deal," *CNET*, June 22, 2011.

81. Kruse interview, December 18, 2017.

82. "What Happened to Southwest Wind Power?"; "Primus Wind Power Acquires the AIR Wind Turbine Line from Southwest Windpower," *Cruising Outpost Magazine*, August 29, 2013; Primus Wind Power, Lakewood, CO, *Air X Owner's Manual[:] Installation * Operation * Maintenance* (Lakewood, CO: Primus Wind Power, 2013), 3 (in author's files).

83. XZERES Corporation, Wilsonville, OR, *XZERES Acquires Southwest Windpower's Skystream Product Line*, July 9, 2013 (in author's files).

84. Ibid.

85. US Internal Revenue Service, *Property Qualifying for the Energy Credit under Section 48[,] Notice 2015–4* (Washington, DC: US Internal Revenue Service, 2015), 1–8, accessed July 14, 2019, https://www.irs.gov/pub/irs-drop/n-15–04.pdf.

86. US Department of Energy, Office of Energy Efficiency and Renewable Energy, *2016 Distributed Wind Market Report* (Richland, WA: Pacific Northwest National Laboratory, 2017), 6.

87. US Department of Energy, Office of Energy Efficiency and Renewable Energy, *2015 Distributed Wind Market Report* (Richland, WA: Pacific Northwest National Laboratory, 2016), 2.

88. *2016 Distributed Wind Market Report*, 29.

89. Ibid., 9–10.

90. Ibid., 6.

91. US Department of Energy, Office of Energy Efficiency & Renewable Energy, *Energy Department's*

Competitiveness Improvement Project Hits First Certification Milestone; Opens Next Round of Applications, February 28, 2018.

92. Jose Zayas, "DOE's Competitiveness Improvement Projects Are Delivering Substantial Cost Reductions in the U.S. Distributed Wind Industry," *North American Clean Energy*, July 15, 2017, accessed January 31, 2018, http://www.nacleanenergy.com/articles/27632/doe-s-competitiveness-improvement-projects-are-delivering-substantial-cost-reductions-in-the-u-s-distributed-wind-industry; *Energy Department's Competitiveness Improvement Project Hits First Certification Milestone*; Patrick Gilman, "Enabling Wind to Contribute to a Distributed Energy Future—The New DOE Distributed Wind Plan" (presentation at Distributed Wind 2019, Washington, DC, February 28, 2019).

93. US Department of Energy, Office of Energy Efficiency and Renewable Energy, Washington, DC, *Distributed Wind Competitiveness Improvement Project* (Boulder, CO: National Renewable Energy Laboratory, 2019), handbill (in author's files).

94. *Distributed Wind Competitiveness Improvement Project*; Ian Baring-Gould, "Latest CIP Round" (presentation at Distributed Wind 2019, Washington, DC, February 28, 2019).

95. *2016 Distributed Wind Market Report*, 14–15.

96. United Wind, Brooklyn, NY, *Forum Equity Partners and United Wind Announce $200M Investment for Distributed Wind Projects*, January 6, 2016 (in author's files); United Wind, Brooklyn, NY, *United Wind & Bergey WindPower Close Landmark Purchase of 100 Distributed-Scale Wind Turbines*, January 20, 2017 (in author's files); United Wind, Brooklyn, NY, *United Wind and Smithfield Foods Announce 3MW WindLease Agreement*, March 19, 2019 (in author's files); Betsy Lillian, "Smithfield Foods Adding Wind Turbines to Colorado Hog Farms," *North American Windpower* (Southbury, CT), March 20, 2019.

97. Russell Tencer, founder and CEO, United Wind, Brooklyn, NY, telephone interview by author, November 30, 2018.

98. Ibid.

99. *2016 Distributed Wind Market Report*, 5; "What We Do," One Energy Enterprises LLC, accessed March 12, 2019, https://oneenergy.com/about-us/what-we-do/.

100. Earth Energy Systems Inc., Eden Prairie, MN, *Hybrid Power Plant[,] Class A Up to 10 kW Continuous* (Eden Prairie, MN: Earth Energy Systems Inc., ca. 1985), brochure (in the files of Craig Toepfer, Chelsea, MI).

101. Paul Gipe, *Wind Power for Home and Business* (White River Junction, VT: Chelsea Green Publishing Company, 1993), 11–13, 222–23.

102. Ken Kotalik, "General Product Overview," Primus Windpower, May 9, 2018, accessed November 29, 2018, http://www.primuswindpower.com/maintenance-service/watch-webinar/; Ken Kotalik, "Primus Wind-power: Updates and Improvements Webinar," Primus Windpower, August 27, 2018, accessed November 30, 2018, http://www.primuswindpower.com/maintenance-service/watch-webinar/; Primus Windpower, Lakewood, CO, *Air Completes Any Off Grid System* (Lakewood, CO: Primus Windpower, [ca. 2018]), brochure (in author's files); Ken Kotalik, director of global sales and operations, Primus Windpower, Lakewood, CO, telephone interview by author, November 30, 2018.

103. Betsy Lillian, "Two Worlds Collide: GE, Juhl Energy Building Hybrid Wind-Solar Project," *North American Windpower* (Southbury, CT), February 22, 2017; Frank Jossi, "Nation's First Integrated Wind and Solar Project Takes Shape in Minnesota," *Energy News Update* (London), March 2, 2017; General Electric, Boston, MA, *Juhl Energy Partners With GE Renewable Energy to Build First of its Kind Solar-Wind Hybrid Project*, November 20, 2018 (in author's files); Betsy Lillian, "GE, Juhl Energy Build Wind-Solar Hybrid Project in Minnesota," *North American Windpower* (Southbury, CT), November 21, 2018.

104. *Juhl Energy Partners With GE Renewable Energy to Build First of Its Kind Solar-Wind Hybrid Project*.

105. "Grid Modernization[:] Microgrids," National Renewable Energy Laboratory, accessed March 24, 2019, https://www.nrel.gov/grid/microgrids.html.

106. Rocky Mountain Institute, HOMER Energy, and CohnReznick Think Energy, *The Economics of Grid Defection* (Boulder, CO: Rocky Mountain Institute, 2014), 6–9, 11–17, 39; Manziel E. Velasquez, Stuart Laval, Prateek Pandey, and David Blood, "Partners Tackle How to Achieve Seamless Transfer in a Microgrid," *Solar Industry* (Southbury, CT) 10, no. 1 (February 2017): 16; Daniella Cheslow, "Pittsburgh's Microgrids Technology Could Lead the Way for Green Energy," *NPR*, November 12, 2017; Ian Baring-Gould, "MIRACL Project (Microgrids, Infrastructure Resilience, and Advanced Controls Launchpad)" (presentation at Distributed Wind 2019, Washington, DC, February 28, 2019); Betsy Lillian, "Microgrid Enables California Solar Developer to Go Off the Grid," *Solar Industry* (Southbury, CT), February 28, 2019; Andy Balaskovitz, "Microgrid Boosters Hope Michigan 'Energy District' Will Spur More Interest," *Energy News Network* (St. Paul, MN), March 13, 2019.

107. Ran Fu, David Feldman, and Robert Margolis, *U. S. Solar Photovoltaic System Cost Benchmark: Q1 2018* (Golden CO: National Renewable Energy Laboratory, 2018), 16–29, 42–45; Betsy Lillian, "National Lab: Costs Continue to Drop for Residential, Commercial PV," *Solar Industry* (Southbury, CT), December 20, 2018; Goksin Kavlak, James McNerney, and Jessika E. Trancik, "Evaluating the Causes of Cost Reduction in Photovoltaic Modules," *Energy Policy* 123 (2018): 708–9; Mark Chediak and Prashant Gopal, "California is First State to Mandate Solar Power," *Bloomberg* (in *The Frederick*

News-Post, Frederick, MD), May 12, 2018; Erin Ailworth, "Solar-Panel Makers See Daylight," *The Wall Street Journal* (weekend edition), May 12–13, 2018.

108. *The Economics of Grid Defection*, 13; Ran Fu, Timothy Remo, and Robert Margolis, *2018 U.S. Utility-Scale Photovoltaics-Plus-Energy Storage System Costs Benchmark* (Golden, CO: National Renewable Energy Laboratory, 2018), 5–7; Mike Scott, "Battery Energy Storage is a $620 Billion Opportunity as Cost Continue to Crash," *Forbes*, November 9, 2018, accessed September 1, 2019, https://www.forbes.com/sites/mikescott/2018/11/09/battery-energy-storage-is-a-1-trillion-opportunity-as-costs-continue-to-crash/#6edb23d24684.

109. *The Economics of Grid Defection*, 39.

110. Peter Lilienthal, CEO, HOMER Energy, Boulder, CO, "Advanced Hybrid System Modeling and Why Solar Microgrids Should Add Wind" (presentation at the Distributed Wind 2019, Washington, DC, February 28, 2019); HOMER Energy, Boulder, CO, *HOMER Grid*[:] *Intelligently Reduce Your Peak Power* (Boulder, CO: HOMER Energy, 2019), handbill (in author's files); HOMER Energy, Boulder, CO, *HOMER Pro*[:] *The Global Standard for Optimizing Microgrid Design* (Boulder, CO: HOMER Energy, 2019), handbill (in author's files).

111. US Department of Energy, Office of Energy Efficiency and Renewable Energy, *Workshop Report: Wind Innovations for Rural Economic Development* (*WIRED*) [,] *December 2018* (Richland, WA: Pacific Northwest National Laboratory, 2018), 1.

112. "America's Electric Cooperatives: 2017 Fact Sheet," National Rural Electric Cooperative Association, January 31, 2017, accessed March 31, 2019, https://www.electric.coop/electric-cooperative-fact-sheet/.

113. US Department of Energy, *Department of Energy Announces $28 Million in Funding for Wind Energy Research*, March 28, 2019 (in author's files).

114. *Workshop Report: Wind Innovations for Rural Economic Development* (*WIRED*), 2.

115. US Department of Energy, Office of Energy Efficiency and Renewable Energy, *2017 Distributed Wind Market Report* (Richland WA: Pacific Northwest National Laboratory, 2018), 8–9; US Department of Energy, Office of Energy Efficiency and Renewable Energy, *2018 Distributed Wind Market Report* (Richland, WA: Pacific Northwest National Laboratory, 2019), 7–9; Betsy Lillian, "DOE: U. S. Installed More Than 50 MW of Distributed Wind in 2018," *North American Windpower*, August 27, 2019.

116. Northern Power Systems Corp., Barre, VT, *Northern Power Systems Corp. Announces Disposition of Its US Service Business and Board Resignation*, April 30, 2019 (in author's files).

117. Hagen Ruff, founder and CEO, Chava Wind, "Lightning Round—New Products and Services—Part II" (presentation at the Distributed Wind 2019, Washington, DC, February 28, 2019); Chava Wind, Homestead, FL, *Chava Windleaf 2500*[:] *Chava Wind—The Art of Wind Power* (Homestead, FL: Chava Wind, [ca. 2019]), brochure (in author's files).

118. Ken Visser, co-founder, Ducted Turbines International, Potsdam, NY, "Lightning Round—New Products and Services—Part II" (presentation at the Distributed Wind 2019, Washington, DC, February 28, 2019); Melissa O'Leary, "Experimental Turbine to Bring Wind Energy to the Masses," *Composites Manufacturing* (Arlington, VA), March 8, 2019; Ducted Turbines International, Potsdam, NY, *DTI*[:] *Design Technology* (Potsdam, NY: Ducted Turbines International, 2019), handbill (in author's files); "Turbine on the Roof!" Ducted Turbines International, accessed March 26, 2019, http://ductedturbinesinternational.com/2018/10/29/turbine-on-the-roof/.

119. Jason A. Day, owner, Star Wind Turbines, East Dorset, VT, "Lightning Round—New Products and Services—Part II" (presentation at the Distributed Wind 2019, Washington, DC, February 28, 2019); Star Wind Turbines, East Dorset, VT, *Star Wind Turbines LLC* (East Dorset, VT: Star Wind Turbines, [ca. 2018]), brochure (in author's files).

120. Matthew Carter, president, Carter Wind Energy, Wichita Falls, TX, "Lightning Round—New Products and Services—Part I" (presentation at the Distributed Wind 2019, Washington, DC, February 28, 2019); "Technology," Carter Wind Energy, accessed March 26, 2019, http://www.carterwindenergy.com.

121. Paul Gipe, *Wind Energy for the Rest of Us* (Bakersfield, CA: Wind-Works.Org, 2016), 176.

122. Jack Unwin, "The Top 10 Countries in the World by Wind Energy Capacity," *Power Technology* (London), March 14, 2019.

123. Robert C. Pugh, "Wind Power Is Poised for the Future," *North American Windpower* (Southbury, CT) 14, no. 9 (October 2017): 30–32; Betsy Lillian, "Vestas Wind Installation to Be U.S.' Tallest," *North American Windpower* (Southbury, CT), December 21, 2018.

124. GE Renewable Energy, Paris, France, *GE's Largest Onshore Wind Turbine Prototype Installed and Operating in the Netherlands*, March 13, 2019 (in author's files).

125. Tomas Kellner, "A Towering Achievement: This Summer in Holland, GE will Build the World's Largest Wind Turbine," *GE Reports*, January 18, 2019; Betsy Lillian, "GE Haliade-X 12 MW Giant to Be Tested in Rotterdam," *North American Windpower* (Southbury, CT), January 21, 2019.

126. Erin Ailworth, "The Race to Build a Wind Behemoth," *The Wall Street Journal* (weekend edition), August 25–26, 2018; Jeremy Hodges, "Wind Turbines Bigger Than Jumbo Jets Seen Growing Even Larger," *Bloomberg*, January 24, 2019.

127. Dan Gearino, "AEP Cancels Nation's Largest Wind Farm: 3 Challenges Wind Catcher Faced," *Inside Climate News* (Brooklyn, NY), July 30, 2018; Donnelle Eller, "Neighbors in Eastern Iowa Fight to Bring Down Turbines—and Win," *Des Moines Register* (Des Moines, IA), November 21, 2018; Kari Lydersen, "In Wisconsin, Many Oppose Transmission Line to Bring Western Wind Power," *Energy News Network* (St. Paul, MN), January 22, 2019; Rick Daysog, "A Planned Wind Farm Is Getting Opposition from an Unusual Source: Environmentalists," *Hawaii News Now* (Honolulu, HI), February 11, 2019; Mitchell Schmidt, "Wind Turbines Haven't Been Universally Welcomed by Everyone in Iowa," *The Gazette* (Cedar Rapids, IA), February 23, 2019; Julie Buntjer, "Rural Landowners Air Opposition to Industrial Wind Farms," *Globe* (Worthington, MN), February 27, 2019; Tom Henry, "Wind Farms a Cash Cow for Communities, but Not Everyone's Sold," *The Blade* (Toledo, OH), April 7, 2019; Philipp Beiter, Joseph T. Rand, Joachim Seel, Eric Lantz, Patrick Gilman, and Ryan Wiser, "Expert Perspectives on the Wind Plant of the Future," *Wind Energy*, DOI: 10.1002/we.2735 (2022): 5.

128. Sarah Whites-Koditschek, "Wisconsin Wind Turbine Project Pits Brother against Brother, Clean Energy against Rural Vistas," *Milwaukee Journal Sentinel*, April 16, 2019.

Bibliography

Archival Materials

"1920s Aviation[:] The Dawn of Air Travel and Air Freight." 1920–30.com. Accessed April 23, 2016. http://1920–30.com.

"1970s: Oil Prices Skyrocket & AWEA Emerges." American Wind Energy Association. Accessed June 18, 2019. https://www.awea.org/wind-101/history-of-wind/1970s.

"About Bergey[:] Company Background and Customers." Bergey Windpower Company. Accessed February 26, 2007. http://www.bergey.com/About_BWC.htm.

"About Queen for a Day." Queen for a Day. Accessed June 7, 2017. http://queenforaday.com/about.php.

"About Us." Small Wind Conference, LLC. Accessed October 26, 2019. http://smallwindconference.com/?page_id=2.

AeroVironment, Inc., Massachusetts Port Authority, and Groom Green. *Architectural Wind at Boston's Logan International Airport.* Boston: Massachusetts Port Authority, 2019. Handbill. In possession of Christopher Gillis.

"America's Electric Cooperatives: 2017 Fact Sheet." National Rural Electric Cooperative Association. January 31, 2017. Accessed March 31, 2019. https://www.electric.coop/electric-cooperative-fact-sheet.

American Wind Energy Association, Washington, DC. *AWEA Cheers Senate Agreement to Restore $23.7 Million to Renewable Energy Budgets,* July 31, 1996. Available in the files of Randall S. Swisher, Silver Spring, MD.

——. *AWEA Legislative Alert[,] Senator James Jeffords (R-Vermont) Has Authored a Letter to Senator William Roth (R-Delaware), Chairman of the Senate Finance Committee, Asking That the Production Tax Credit for Wind Be Maintained as It Exists in Current Law.* October 5, 1995. Available in the files of Randall S. Swisher, Silver Spring, MD.

——. *AWEA Reports 78% Growth in 2008 for U.S. Small Wind Market,* May 28, 2009. In possession of Christopher Gillis.

——. *Boom: 2003 Close to Best Ever Year for New Wind Installations; Bust: Expiration of Key Incentives Lowers Hopes for 2004,* January 22, 2004. In possession of Christopher Gillis.

——. *CPUC Restructuring Decision Crucial to Wind and Other Renewables Says AWEA[,] Some California Wind Plants Already Being Dismantled, Industry Notes,* December 14, 1995. Available in the files of Randall S. Swisher, Silver Spring, MD.

——. *Energy Bill Extends Wind Power Incentive through 2007,* July 29, 2005. In possession of Christopher Gillis.

——. *Federal Legislative Update[,] Wind Energy Research and Development (R&D) Budget,* September 22, 1995. Available in the files of Randall S. Swisher, Silver Spring, MD.

——. *First Quarter Market Report: Wind Industry Trade Group Sees Little to No Growth in 2004, Following Near-Record Expansion in 2003,* May 12, 2004. In possession of Christopher Gillis.

——. *Is a Residential Wind System for You?* Washington, DC: American Wind Energy Association, [ca. 1999]. Handbill. Available in the files of Warren S. Bollmeier, II, Kaneohe, HI.

——. *Legislative Alert from the American Wind Energy Association[,] The Fiscal Year (FY) 1996 U.S. Federal Wind Energy Research and Development (R&D) Budget Is in Its Final Stage as the House and Senate Prepare to Iron out the Differences in Their Perspective Energy and Water Appropriation Bills in Just a Few Short Weeks,* August 29, 1995. Available in the files of Randall S. Swisher, Silver Spring, MD.

——. *Legislative Update[,] Wind Energy Production Tax Credit[,] from: Michael Marvin and Jared Barlage, AWEA.* November 21, 1995. Available in the files of Randall S. Swisher, Silver Spring, MD.

——. *Legislative Update from the American Wind Energy Association[,] Production Tax Credit Update[,] from: Mike Marvin and Jared Barlage.* December 8, 1995. Available in the files of Randall S. Swisher, Silver Spring, MD.

———. *Legislative Update from the American Wind Energy Association*[,] *To: AWEA Board, Legislative Committee and Supporters*[,] *from: Randall Swisher and Jared Barlage*[,] *Re: Status on Wind Energy Production Tax Credit.* October 12, 1995. Available in the files of Randall S. Swisher, Silver Spring, MD.

———. *Legislative Update from the American Wind Energy Association*[,] *Wind Credit Remains in Law*[,] *from: Mike Marvin and Jared Barlage, AWEA.* November 13, 1995. Available in the files of Randall S. Swisher, Silver Spring, MD.

———. *Legislative Update from the American Wind Energy Association*[,] *Wind Energy Production Tax Update,* October 31, 1995. Available in the files of Randall S. Swisher, Silver Spring, MD.

———. *News from the American Wind Energy Association*[,] *AWEA Blasts House Committee Plan to Slash Federal Wind Research,* April 19, 1996. Available in the files of Randall S. Swisher, Silver Spring, MD.

———. *News from the American Wind Energy Association*[,] *AWEA Calls California Utilities' Renewables Portfolio Standard Proposal 'Step in the Right Direction'*[,] *However, Changes Needed in Implementation Details,* May 7, 1996. Available in the files of Randall S. Swisher, Silver Spring, MD.

———. *News from the American Wind Energy Association*[,] *For Immediate Release*[,] *AWEA Hails CPUC Support for Renewables*[,] *Restructuring Proposals Disagree on Market Structure, But Converge on Renewables Requirements,* May 24, 1995. Available in the files of Randall S. Swisher, Silver Spring, MD.

———. *News from the American Wind Energy Association*[,] *Massive Budget Cuts Threaten Job Losses and Economic Downturn in Many American Communities*[,] *Proposed Cuts in Renewable Energy Run Contrary to Public Opinion,* June 13, 1995. Available in the files of Randall S. Swisher, Silver Spring, MD.

———. *News from the American Wind Energy Association*[,] *PURPA Crucial to Development of Fully Competitive Electricity Market, Says AWEA,* June 6, 1995. Available in the files of Randall S. Swisher, Silver Spring, MD.

———. *News from the American Wind Energy Association*[,] *Roth Preserves Wind Energy Production Tax Credit*[,] *AWEA Praises Recommendations to Leave Credit Intact, Maintaining Tax Equity,* October 17, 1995. Available in the files of Randall S. Swisher, Silver Spring, MD.

———. *Press Releases*[:] *AWEA Hires Jennifer Jenkins to Launch National Distributed Wind Program,* July 24, 2018. In possession of Christopher Gillis.

———. Second Quarter Market Report: American Wind Industry Needs Consistent Business Environment to Bring Wind Power's Promise to the Country, August 10, 2004. In possession of Christopher Gillis.

———. *Statement of Randall Swisher*[,] *Executive Director, American Wind Energy Association on Kenetech Windpower's Filing for Bankruptcy Protection,* May 30, 1996. Available in the files of Randall S. Swisher, Silver Spring, MD.

———. Urgent Memorandum[,] to: Wind Energy Supporters[,] from: Jared Barlage, AWEA[,] Date: June 12, 1995[,] Subject: Massive Cuts to the Federal Wind Program Budget. Typewritten letter. Available in the files of Randall S. Swisher, Silver Spring, MD.

———. *Wind Energy in the 80s—A Decade of Development,* March 30, 1990. In possession of Christopher Gillis.

———. *Wind Energy Outlook 2000.* Washington, DC: American Wind Energy Association, 2000. Brochure. Available in the files of Warren S. Bollmeier, II, Kaneohe, HI.

———. *Wind Power Closes 2017 Strong, Lifting the American Economy,* January 30, 2018. In possession of Christopher Gillis.

"American Wind Power and AWEA Grow by Leaps and Bounds." American Wind Energy Association. Accessed June 18, 2019. https://www.awea.org/wind-101/history-of-wind/2000s.

"Applications—Small Turbines." Small Wind Certification Council, Brea, CA. Accessed May 14, 2018. http://smallwindcertification.org/for-applicants/small-turbines/application-process-small/applications/.

"Aviation: From Sand Dunes to Sonic Booms[:] Aviation Pioneers." National Park Service, Accessed April 23, 2016. https://www.nps.gov/nr/travel/aviation/pioneers.htm.

"AWEA 9.1 Standard Testing and Certification." Intertek. Accessed May 14, 2018. http://www.intertek.com/wind/awea-standard/.

"Backgrounder in the Three Mile Island Accident." U.S. Nuclear Regulatory Commission. Accessed February 18, 2018. https://www.nrc.gov/reading-rm/doc-collections/fact-sheets/3mile-isle.html.

Baring-Gould, Ian, technology deployment manager, National Wind Technology Center, Office of Energy Efficiency and Renewable Energy, US Department of Energy, Golden, CO. "Latest CIP Round." Presentation at Distributed Wind 2019, Washington, DC, February 28, 2019. Notes in possession of Christopher Gillis.

———. "MIRACL Project (Microgrids, Infrastructure Resilience, and Advanced Controls Launchpad)." Presentation at Distributed Wind 2019, Washington, DC, February 28, 2019. Notes in possession of Christopher Gillis.

Bass, Delwin, Torrington, WY, to Christopher Gillis, May 26, 2017. Handwritten letter in possession of Christopher Gillis.

Beaty, Harold Huxford. "Wind Electric Plants." Master's thesis, Iowa State College, Ames, 1941. Accessed February 2, 2017. http://lib.dr.iastate.edu/cgi/viewcontent.cgi?article=18340&context=rtd.

"The Bergey Excel 6 Wind Turbine." Bergey Windpower Company. Accessed January 27, 2017. http://bergey.com/products/wind-turbines/6kw-bergey-excel.

Bergey, Karl H., Jr., founder, Bergey Windpower Company, Norman, OK, to Christopher Gillis, February 15, 2017. Typewritten letter in possession of Christopher Gillis.

Bergey, Michael L. S., president and CEO, Bergey Windpower Company, Norman, OK, to Christopher Gillis, December 28, 2017. Email in possession of Christopher Gillis.

———, Mike, president and CEO, Bergey Windpower, Norman, OK. "Leadership Panel: State of the Industry." Presentation at Distributed Wind 2018, Washington, DC, February 28, 2018. Notes in possession of Christopher Gillis.

———. "Leadership Panel: State of the Industry." Presentation at Distributed Wind 2019, Washington, DC. February 28, 2019. Notes in possession of Christopher Gillis.

———, president, Distributed Wind Energy Association, to Chris Schaffner, chair, LEED Energy & Atmosphere Technical Advisory Group, US Green Building Council, August 2012. Typewritten letter. Available in the files of Bergey Windpower Company, Norman, OK.

Bollmeier II, W. S. The Pacific International Center for High Technology Research, Energy and Resources Division. *To: A. R. Trenka*[,] *Memorandum*[,] *Trip Report* (*Windpower '90, Washington, DC, September 25–28, 1990*[,] *October 22, 1990*. Honolulu, HI: The Pacific International Center for High Technology Research, 1990. Typewritten memorandum. Available in the files of Warren S. Bollmeier, II, Kaneohe, HI.

———, Warren S., Kaneohe, HI, to Christopher Gillis, August 1, 2019. Email in possession of Christopher Gillis.

Brauer, David, acting laboratory director, Conservation and Production Research Laboratory, Agricultural Research Service, US Department of Agriculture, Bushland, TX, to Christopher Gillis, January 2, 2018. Email in possession of Christopher Gillis.

Brooks, David, "Remembering the World's First Wind Farm—in New Hampshire," Granite Geek, February 24, 2016. Accessed January 22, 2017. http://granitegeek.concordmonitor.com/2016/02/24/remembering-worlds-first-wind-farm-new-hampshire/.

Brush, Charles F. "Laboratory Notebook, May 1880—January 1883." Box 5, Series 2: Laboratory Notes, 1880–1929. Folder 24. Charles F. Brush Sr. papers. Special Collections Research Center, Kelvin Smith Library, Case Western Reserve University. Transcription provided to Christopher Gillis by Glen E. Swanson, instructor, Grand Valley State University, Allendale, MI, and Charles F. Brush, researcher.

———. Patent application. "A System and Apparatus for Charging Secondary Batteries." US Patent Office (January 5, 1886): 1–17. Box 9. Series 3. Folder 6. File 00905. Charles F. Brush Sr. papers. Special Collections Research Center, Kelvin Smith Library, Case Western Reserve University. Photocopy provided to Christopher Gillis by Glen E. Swanson, instructor, Grand Valley State University, Allendale, MI, and Charles F. Brush, researcher.

Bush, George H. W. "Remarks on Designating the Solar Energy Research Institute as the National Renewable Energy Laboratory, September 16, 1991." In *Public Papers of the Presidents of the United States: George H. W. Bush, 1991, Book II-July 1 to December 31, 1991*, 1155–56. Washington, DC: Government Printing Office, 1992.

"Business Energy Investment Tax Credit (ITC)." US Department of Energy. Accessed April 2, 2017. https://www.energy.gov/savings/business-energy-investment-tax-credit-itc.

California Energy Commission, Sacramento, CA. *ABCs of Net Metering*. Sacramento, CA: California Energy Commission, 2000. Brochure. In possession of Christopher Gillis.

Carter, Jimmy. "Energy and the National Crisis, July 15, 1979." In *Public Papers of the Presidents of the United States: Jimmy Carter, 1979, Book II-June 23 to December 31, 1979*, 1235–41. Washington, DC: Government Printing Office, 1979.

Carter, Matthew, president, Carter Wind Energy, Wichita Falls, TX. "Lightning Round—New Products and Services—Part I." Presentation at Distributed Wind 2019, Washington, DC, February 28, 2019. Notes in possession of Christopher Gillis.

"Charles F. Brush: 1849–1929." The Brush Foundation. Accessed March 20, 2017. http://fdnweb.org/brush/brush-history/.

Cheney, M. C. "UTRC 8 KW Wind Turbine." Paper presented at the American Wind Energy Association National Conference, Spring 1979, San Francisco, CA, April 16–19, 1979. In the *American Wind Energy Association National Conference, Spring 1979, Proceedings*. Edited by Vaughn Nelson. Washington, DC: American Wind Energy Association, 1979.

Christiansen, L. A., Air-Electric Machine Company, Inc., Lohrville, IA, to Robert J. Hayes, Muscatine, IA, January 5, 1971. Typewritten letter. In possession of Christopher Gillis.

Commonwealth Electric Company, Boston, MA. *Wind to Watts*. Boston, MA: Commonwealth Electric Company, [ca. 1981]. Folder. Available in the *Windmillers' Gazette* research files, Rio Vista, TX.

"Company History." Rohn Products. Accessed October 4, 2015. http://www.rohnnet.com/rohn-company-history.

Cook, Chris. Texas Tech University, Lubbock, TX. *Advanced Wind Energy Test Facility Makes a Move*, July 27, 2011. In possession of Christopher Gillis.

"Danske vindmøller 1976–2000." Danish Wind Historical Collection (Danmarks Vindkrafthistoriske Samling). Accessed November 9, 2018. http://www.vindhistorie.dk.

Day, Jason A., owner, Star Wind Turbines, East Dorset, VT. "Lightning Round—New Products and Services—Part II." Presentation at Distributed Wind 2019, Washington, DC, February 28, 2019. Notes in possession of Christopher Gillis.

Dayton, Mark B., commissioner, Minnesota Department of Energy and Economic Development, St. Paul, MN, to Paul Jacobs, April 17, 1986. Typewritten letter. Available in the files of Paul R. Jacobs, Corcoran, MN.

"Den Sociale Vision." Poul la Cour Museum. Accessed April 2, 2019. http://www.poullacour.dk/dansk/vision.htm.

"Deposition of Thomas A. Edison." *Electric Railway Company of the United States v. The Jamaica and Brooklyn Road Company*. ED NY 1893. Accessed April 3, 2017. http://edison.rutgers.edu/NamesSearch/glocpage.php?gloc=QE001&.

Distributed Wind Energy Association. *Distributed Wind Energy Association*. Durango, CO: Distributed Wind Energy Association, [2010]. In possession of Christopher Gillis.

———. *DWEA 2012 Annual Report*. Durango, CO: Distributed Wind Energy Association, 2013. In possession of Christopher Gillis.

———. *DWEA Distributed Wind Vision—2015–2030*[:] *Strategies to Reach 30 GW of "Behind-the-Meter" Wind Generation by 2030*[,] *March 2015*. Durango, CO: Distributed Wind Energy Association, 2015.

———. *DWEA Lauds New ITC Legislation Supporting Distributed Wind Power*. February 9, 2018. In possession of Christopher Gillis.

Drees, Herman M., Thousand Oaks, CA, to Christopher Gillis, September 2, 2018, October 17, 2019, and October 19, 2018. Emails in possession of Christopher Gillis.

"DWEA Committees." Distributed Wind Energy Association. Accessed January 31, 2017. http://distributedwind.org/home/committees/.

"DWEA Member Directory." Distributed Wind Energy Association. Accessed January 31, 2017. http://distributedwind.org/dwea-members/.

Earth Energy Systems Inc., Eden Prairie, MN. To: All Winco Dealers, 1985. 1 leaf. Available in the files of Paul R. Jacobs, Corcoran, MN.

———, To: Winco Accounts, 1985. 1 leaf. Available in the files of Paul R. Jacobs, Corcoran, MN.

"Events in the History of Dunlite." http://www.pearen.ca/dunlite/History-3.pdf. Accessed August 21, 2016.

Fleming, J. C., Jr., sales manager, Bucklen-Perkins Aerolectric, Inc., Elkhart, IN, to Henry Pechanec, Timken, KS, January 15, 1930. Typewritten letter. Available in *Windmillers' Gazette* research files, Rio, Vista. TX.

Gagne, Matthew. "How Long Will It Take to Certify a Turbine?" Small Wind Certification Council, Brea, CA. Accessed March 6, 2018. http://smallwindcertification.org.

GE Renewable Energy, Paris, France. *GE's Largest Onshore Wind Turbine Prototype Installed and Operating in the Netherlands*, March 13, 2019. In possession of Christopher Gillis.

General Electric, Boston, MA. *Juhl Energy Partners with GE Renewable Energy to Build First of its Kind Solar-Wind Hybrid Project*, November 20, 2018. In possession of Christopher Gillis.

Generation on the Wind. Directed by David Vassar. David Vassar, 1979. Accessed October 6, 2019. https://www.imdb.com/title/tt0079201/.

Gilman, Patrick, modeling and analysis program manager, Wind Energy Technologies Office, Energy Efficiency and Renewable Energy, US Department of Energy, Golden, CO. "Enabling Wind to Contribute to a Distributed Energy Future—The New DOE Distributed Wind Plan. Presentation at Distributed Wind 2019, Washington, DC, February 28, 2019. Notes in possession of Christopher Gillis.

Gipe, Paul. "Alcoa Darrieus VAWT," Wind-Works, N.d. Accessed October 28, 2018. http://www.wind-works.org/cms/index.php?id=504.

———. "Comments on a Proposed Performance Standard for Small Wind Turbines," Wind-Works, May 25, 2000. Accessed July 4, 2019. http://www.wind-works.org/cms/index.php?id=64&tx_ttnews%5Btt_news%5D=54&cHash=69ab65d9bf5142d330e8b7b70be45c40.

———. "Defunct Windspire Wins Greenwashing Award," Wind-Works, June 28, 2013. Accessed July 14, 2019, http://www.wind-works.org/cms/index.php?id=399&tx_ttnews%5Btt_news%5D=2483&cHash=3b3cb9265ea9f4f24362b6e9794dbc45.

———. "Honeywell Windtronics Kaputt—Finally an End to a Sad Saga of a 'Revolutionary Roof Top Wind Turbine," Wind-Works, June 28, 2013. Accessed July 14, 2019. http://www.wind-works.org/cms/index.php?id=674&tx_ttnews%5Btt_news%5D=2480&cHash=d3a6080425095e7cd7c8a0ce1e7085a5.

———. "Mariah in the Running for Worst Small Turbine Install," Wind-Works, January 31, 2010. Accessed July 14, 2019, http://www.wind-works.org/cms/index.

php?id=64&tx_ttnews%5Btt_news%5D=99&cHash=
9d4e873aeba4ecbaed8515276e9a44bc.

———. "Mariah Windspire VAWT Measured Performance," Wind-Works, March 22, 2013. Accessed July
14, 2019, http://www.wind-works.org/cms/index.
php?id=64&tx_ttnews%5Btt_news%5D=2295&cHash
=06b4fe398a441d737b971bb5fe47e201.

———. "New Federal Subsidies Distort the US Small
Wind Market," Wind-Works, November 14, 2008. Accessed July 4, 2019. http://www.wind-works.org/cms/
index.php?id=64&tx_ttnews%5Btt_news%5D=54&c
Hash=69ab65d9bf5142d330e8b7b70be45c40.

———. "Photos of Mehrkham [sic.] Turbines by Paul
Gipe," Wind-Works, May 8, 2013. Accessed September 29, 2018. http://www.wind-works.org/cms/index
.php?id=516.

———. "Photos of StormMaster," Wind-Works, May 9,
2013. Accessed October 9, 2018. http://www.wind-
works.org/cms/index.php?id=512.

———. "Small Turbine Product Reviews[:] Commentary
on the Skystream 3.7," Wind-Works, May 9, 2009. Accessed February 7, 2019. http://www.wind-works.org/
cms/index.php?id=70&tx_ttnews%5Btt_news%5D=5
5&cHash=71d22987868dd50c5f0b19c8047c3dea.

———, "Wind Cube Squarely Over the Top (New Re-
Branded as Wind Sphere," Wind-Works, July 8, 2013.
Accessed February 11, 2019. http://www.wind-works.
org/cms/index.php?id=660&tx_ttnews%5Btt_news%
5D=149&cHash=308028ac52d4ddebd3f8b589d9e91
bdb.

Gouchoe, Susan. Local Government and Community Pro-
grams and Incentives for Renewable Energy—National
Report[,] North Carolina Solar Center[,] Industrial
Extension Service[,] North Carolina State University[,]
December 2000. Raleigh, NC: Database of State Incen-
tives for Renewable Energy, 2000.

Gray, Tom, AWEA standards development manager.
Memo to All Interested Parties, July 14, 1980. Typewritten letter. In possession of Christopher Gillis.

"Grid Modernization[:] Microgrids." National Renewable
Energy Laboratory. Accessed March 24, 2019. https://
www.nrel.gov/grid/microgrids.html.

Grundel, A. R., sales manager, Herbert E. Bucklen
Corporation, Elkhart, IN, to James F. Smith, M.D.,
Hayes Store, VA, November 5, 1927. Letter in untitled
notebook consisting of mimeographed descriptive
data, photostatic copies of photographs, and order
forms. Available in Windmill Manufacturers' Trade
Literature Collection, Panhandle-Plains Historical
Museum, Canyon. TX.

"Hannaford's Light." Timespanner[:] A Journey through
Avondale, Auckland and New Zealand History. April
19, 2009. Accessed March 21, 2017. http://timespan-
ner.blogspot.com/2009/04/hannafords-light.html.

Harris, Coy, executive director, American Windmill Museum, Lubbock, TX, to Christopher Gillis, January 21,
2016; February 9, 2016; and October 23, 2018. Email
in possession of Christopher Gillis.

Healy, Terry J., manager, Rockwell International, Wind
Systems Program, Atomics International Division,
Rocky Flats Plant, Golden, CO, to M. L. Jacobs, presi-
dent, Jacobs Wind Electric Company, Fort Myers,
FL, March 17, 1978. Available in the files of Paul R.
Jacobs, Corcoran, MN.

Hersholdt, C., Agrico Manufacturing Company,
Ltd., Copenhagen, Denmark, to Kregel Windmill
Co[mpany], Nebraska City, NE, February 2, 1925.
Typewritten letter. Available in the Kregel Windmill
Company Papers, Nebraska State Historical Society,
Lincoln, NE.

Indiana Distribute Energy Alliance. "Paul Gipe Discusses Problems with LEED, USGBC and 'Bad' Wind
Turbines in Indianapolis; Are Changes Needed in
LEED?" July 20, 2013. Accessed February 11, 2019.
http://www.indianadg.net/paul-gipe-discusses-
problems-with-leed-usgbc-and-bad-wind-turbines-
in-indianapolis-are-changes-needed-in-leed/.

Jacobs, M. L., Plymouth, MN. February 12, 1985[,] Con-
sulting Report No. 14[:] Follow-up Report on the Gov-
ernor Hub Nut Problems at the Jacoby-Kerr Project in
Palm Springs, California. Plymouth, MN: Jacobs Wind
Electric Company, 1985. Typewritten memorandum.
Available in the files of Paul R. Jacobs, Corcoran, MN.

———. February 12, 1985[,] Consulting Report No. 15[:]
Governor Blade Shaft Zerk Fittings—Buck Ranch
Observations. Plymouth, MN: Jacobs Wind Electric
Company, 1985. Typewritten memorandum. Avail-
able in the files of Paul R. Jacobs, Corcoran, MN.

———. February 14, 1985[,] Consulting Report No. 16[:]
Tower Manufacturer Mistakes Found at the Buck
Ranch Site in Palm Springs, California. Plymouth,
MN: Jacobs Wind Electric Company, 1985. Typewrit-
ten memorandum. Available in the files of Paul R.
Jacobs, Corcoran, MN.

———. February 19, 1985[,] Consulting Report No. 18[:]
Tower Angle Brace Rod Breakage. Plymouth, MN:
Jacobs Wind Electric Company, 1985. Typewritten
memorandum. Available in the files of Paul R. Jacobs,
Corcoran, MN.

———. February 21, 1985[,] Consulting Report No. 19[:]
A. I. Tower Corner Rail Cracking (Failure) at Palm
Springs. Plymouth, MN: Jacobs Wind Electric Company, 1985. Typewritten memorandum. Available in
the files of Paul R. Jacobs, Corcoran, MN.

———. March 29, 1985[,] Consulting Report No. 24[:]
Reply to R. E. Fletcher's Letter of February 27, 1985: 20
KW Governor Lubrication (Copy of Letter Attached).
Plymouth, MN: Jacobs Wind Electric Company, 1985.

Typewritten memorandum. Available in the files of Paul R. Jacobs, Corcoran, MN.

———. *May 10, 1985*[,] *Consulting Report No. 28*[:] *Installation and Maintenance Problems with the "X" Braced Angle Iron Tower at Palm Springs.* Plymouth, MN: Jacobs Wind Electric Company, 1985. Typewritten memorandum. Available in the files of Paul R. Jacobs, Corcoran, MN.

———. *May 10, 1985*[,] *Consulting Report No. 29*[:] *Heavier Control Spring for Dual Fold Tail Vane.* Plymouth, MN: Jacobs Wind Electric Company, 1985. Typewritten memorandum. Available in the files of Paul R. Jacobs, Corcoran, MN.

———. *May 31, 1985*[,] *Consulting Report No. 32*[:] *Further Comments on Governor Hub Lubrication and Reference to Consulting Report No. 24.* Plymouth, MN: Jacobs Wind Electric Company, 1985. Typewritten memorandum. Available in the files of Paul R. Jacobs, Corcoran, MN.

———. *June 13, 1985*[,] *Consulting Report No. 33*[:] *New Tower Girt to Improve Present Windfarm Towers.* Plymouth, MN: Jacobs Wind Electric Company, 1985. Typewritten memorandum. Available in the files of Paul R. Jacobs, Corcoran, MN.

———, general manager, Jacobs Wind Electric Company, Minneapolis, MN, to A. Clyde Eide, Columbus, OH, April 15, 1955. Typewritten letter, photocopy. Available in the *Windmillers' Gazette* research files, Rio Vista, TX.

———, Jacobs Wind Electric Company, Minneapolis, MN, to L. L. Romersheuser, Hayes Center, NE, January 9, 1953. Typewritten letter. In possession of Christopher Gillis.

———, president, Jacobs Wind Electric Company, Fort Myers, FL, to R. Anderson, chairman of the board, Rockwell International, Pittsburgh, PA, July 13, 1979. Available in the files of Paul R. Jacobs, Corcoran, MN.

———, [president], Jacobs Wind Electric Company, Fort Myers, FL, to Richard Anderson, chairman of the board, Rockwell International, Pittsburgh, PA, August 24, 1979. Available in the files of Paul R. Jacobs, Corcoran, MN.

———, to T. J. Healy, manager, Wind Systems Program, Rocky Flats [Plant], Golden, CO, July 29, 1977. Available in the files of Paul R. Jacobs, Corcoran, MN.

———, to T. J. Healy, manager, Wind Systems Program, Rocky Flats Plant, Energy Systems Group, Rockwell International, Golden, CO, July 12, 1979. Available in the files of Paul R. Jacobs, Corcoran, MN.

———, to Terry J. Healy, Rockwell International, Atomics International Division, Rocky Flats Plant, Golden, CO, February 17, 1978. Available in the files of Paul R. Jacobs, Corcoran, MN.

———, to Terry J. Healy, Rockwell International, Atomics International Division, Rocky Flats Plant, Golden, CO, March 17, 1978. Available in the files of Paul R. Jacobs, Corcoran, MN.

———, to Jim Martin, Xenia, OH, October 6, 1977. Typewritten letter. In possession of Christopher Gillis.

———, to R. O. Williams Jr., vice president and general manager, Rocky Flats Plant, Energy Systems Group, Rockwell International, Golden, CO, August 23, 1979. Available in the files of Paul R. Jacobs, Corcoran, MN.

———, Plymouth, MN. *Notice to Dealers*[:] *Product Improvements, May 8, 1981* (Plymouth, MN: Jacobs Wind Electric Company, 1981). Handbill. Available in the files of Paul R. Jacobs, Corcoran, MN.

———, secretary-treasurer, Jacobs Wind Electric Company, Minneapolis, MN, to Elmer Kreidman, Fort Atkinson, WI, August 24, 1938. Typewritten letter. In possession of Christopher Gillis.

Jacobs, Paul R., Corcoran, MN, to Christopher Gillis, October 2, 2016; January 24, 2017; January 26, 2017; January 29, 2017; January 31, 2017; February 3, 2017; February 5, 2017; February 6, 2017; February 7, 2017; February 18, 2017; February 22, 2017; May 9, 2017. Emails in possession of Christopher Gillis.

Jacobs Wind Electric Company, Fort Myers, FL. *Preliminary Report from Suppliers for Components of New Wind Electric Plant*[,] *26 March 1979.* Fort Myers, FL: Jacobs Wind Electric Company, 1979. Typewritten document. Available in the files of Paul R. Jacobs, Corcoran, MN.

———, Plymouth, MN. *Consulting Report Index*[,] *Rev. 6/17/85.* Plymouth, MN: Jacobs Wind Electric Company, 1985. Typewritten index. I leaf. Available in the files of Paul R. Jacobs, Corcoran, MN.

———. *Master List*[,] *Possible Recipients of M. L. Jacobs Consulting Reports.* Plymouth, MN: Jacobs Wind Electric Company, 1984. Typewritten, 1 leaf. Available in the files of Paul R. Jacobs, Corcoran, MN.

———and Control Data Corporation, Minneapolis, MN. *Statement of Understanding by and between Marcellus L. Jacobs Doing Business as Jacobs Wind Electric Co. and Control Data Corporation.* Minneapolis, MN: Jacobs Wind Electric Company and Control Data Corporation, 1979. Available in the files of Paul R. Jacobs, Corcoran, MN.

Jones, Sherry, owner, Gemini Controls, Cedarburg, WI, to Christopher Gillis, January 23, 2017. Email in possession of Christopher Gillis.

Kasthurirangan, Padma, president, Buffalo Renewables, Niagara Falls, NY. "Welcome and Opening Remarks." Presentation at Distributed Wind 2018, Washington, DC, February 28, 2018. Notes in possession of Christopher Gillis.

Kirchner, Bob. *Something About Generators & Alternators.* Colorado Springs, CO: Living Energy Consultants Ltd., [ca. 1979]. Handbill. Available in the *Windmillers' Gazette* research files, Rio Vista, TX.

———, [Robert]. *Something about Towers*. Colorado Springs, CO: Living Energy Consultants Ltd., 1980. Handbill. Available in the *Windmillers' Gazette* research files, Rio Vista, TX.

Kotalik, Ken. "General Product Overview." Primus Windpower. May 9, 2018. Accessed November 29, 2018. http://www.primuswindpower.com/maintenance-service/watch-webinar/.

———. "Primus Windpower: Updates and Improvements Webinar." Primus Windpower. August 27, 2018. Accessed November 30, 2018. http://www.primuswindpower.com/maintenance-service/watch-webinar/.

La Favre, Jeffrey. "Charles Brush and the Arc Light." 1998. Accessed March 20, 2017. http://www.lafavre.us/brush/brushbio.htm.

Lilienthal, Peter, CEO, HOMER Energy, Boulder CO. "Advanced Hybrid System Modeling and Why Solar Microgrids Should Add Wind." Presentation at Distributed Wind 2019, Washington, DC, February 28, 2019. Notes in possession of Christopher Gillis.

Living Energy Consultants Ltd., Colorado Springs, CO. *Something About Inverters and DWS Multi-Verters*. Colorado Springs, CO: Living Energy Consultants Ltd., 1980. Handbill. Available in the *Windmillers' Gazette* research files, Rio Vista, TX.

Metcalf, Lee. *From the Office of Senator Lee Metcalf*[,] *News Release*, October 16, 1974. In possession of Paul R. Jacobs, Corcoran, MN.

Metcalfe, Philip W., Unarco-Rohn, Division of Unarco Industries, Inc. Peoria. IL. *To: R. A. Kleine*[,] *Market Survey and Forecast for Wind Generator Towers*[,] *February 20, 1980*. Peoria, IL: Unarco-Rohn, Division of Unarco Industries, Inc., 1980). Typewritten memorandum. In possession of Christopher Gillis.

Metz, M. C., sales department, Universal Battery Company, Chicago, to Ernst Fuelling, Decatur, IN, August 15, 1935. Typewritten letter. In possession of Christopher Gillis.

Meyer, Hans. *Synchronous Inversion*[:] *Concept & Application*. Mukwonago, WI; Windworks, Inc., 1976. Typewritten. Available in the files of Craig Toepfer, Chelsea, MI.

Minnesota Department of Health, Section of Vital Statistics, St. Paul, MN. *Certificate of Death*[,] *Marcellus Luther Jacobs, July 23, 1985*. Available in the files of Paul R. Jacobs, Corcoran, MN.

"The Museum." Poul la Cour Museum. Accessed November 9, 2018. http://www.poullacour.dk/engelsk/museet.htm.

National Aeronautics and Space Administration, Lewis Research Center, Cleveland, OH. *Wind Energy Systems: A Non-Pollutive, Non-Depletable Energy*, December 1973. In possession of Christopher Gillis.

National Archives Theater, Washington, DC. *Generation of the Wind*[,] *1979*. Washington, DC: National Archives Theather, 2018. Handbill. In possession of Christopher Gillis.

National Institute of Standards and Technology, US Department of Commerce, Washington, DC. *SMART Wind Consortium: Developing a Consensus Based Sustainable Manufacturing, Advanced Research and Technology Roadmap for Distributed Wind*. May 6, 2014. In possession of Christopher Gillis.

National Renewable Energy Laboratory, Golden, CO. *Companies Selected for Small Wind Turbine Project*, November 27, 1996. In possession of Christopher Gillis.

———. *Innovation*[:] *NREL Innovations Contribute to an Award-Winning Small Wind Turbine*. Golden, CO: National Renewable Energy Laboratory, 2010. Handbill. In possession of Christopher Gillis.

———. *National Renewable Energy Laboratory to Reduce Staff*, November 14, 1995. In possession of Christopher Gillis.

———. "New Wind Turbine Designs Target Rural, Utility Markets." *Advanced Wind Turbines: Electricity for the 1980s and Beyond*. Golden, CO: National Renewable Energy Laboratory, 1992. Brochure. Available in the files of Warren S. Bollmeier, II, Kaneohe, HI.

———. "Northern Power Systems Devises Innovative Turbine Rotor." *Advanced Wind Turbines: Electricity for the 1980s and Beyond*. Golden, CO: National Renewable Energy Laboratory, 1992. Brochure. Available in the files of Warren S. Bollmeier, II, Kaneohe, HI.

———. *NREL Announces New Distributed Wind Competitiveness Improvement Project Contracts*, November 1, 2017. Accessed January 31, 2018. https://www.nrel.gov/news/program/2017/nrel-announces-new-distributed-wind-competitiveness-improvement-project-contracts.html. In possession of Christopher Gillis.

———. *NREL Funding Reductions to Further Impact Lab's Work Force*, December 22, 1995. In possession of Christopher Gillis.

———. "WC-86B Wind Turbine to Show Big Jump in Performance." *Advanced Wind Turbines: Electricity for the 1980s and Beyond*. Golden, CO: National Renewable Energy Laboratory, 1992. Brochure. Available in the files of Warren S. Bollmeier, II, Kaneohe, HI.

Natural Power Inc., New Boston, NH. *Questions That Predicate the Future Direction of the Company*. New Boston, NH: Natural Power Inc., 1975. Typewritten memorandum. Available in the files of Earle Rich, Mont Vernon, NH.

Nijs, Wilfried, and Frans Brouwers. "Wieksystemen." Windmill course, Levende Molens, Aartselaar, Belgium, 2011–2012.

Northern Power Systems Corp., Barre, VT. *Northern Power Systems Corp. Announces Disposition of Its US Service Business and Board Resignation*, April 30, 2019. In possession of Christopher Gillis.

O'Shea, Allan, Environmental Energies, Inc., Detroit. American Wind Energy Association meeting invitation. August 5, 1974. Typewritten letter. In possession of Christopher Gillis.

Office of the Governor, Commonwealth of Pennsylvania, Harrisburg, PA. *Governor Rendell Making Small Wind Energy Systems Available to Local Governments*[,] *15 Small-Scale Systems to Place Alternative Energy Source in Public View*, April 7, 2006. In possession of Christopher Gillis.

"Parris-Dunn Manufacturing Co." Windcharger.org. Accessed May 11, 2017. http://www.windcharger.org/Wind_Charger/Parris-Dunn_Corp..html.

Perkins Corporation, Mishawaka, IN. Typewritten letter to W. E. Holt, Minooka, IL. Electricity from the Wind questionnaire and envelope. April 11, 1922. Available in *Windmillers' Gazette* research files, Rio Vista, TX.

———. Typewritten letter to W. E. Holt, Minooka, IL. April 27, 1922. Available in *Windmillers' Gazette* research files, Rio Vista, TX.

Perlin, John. *The Silicon Solar Cell Turns 50*. Golden, CO: National Renewable Energy Laboratory, 2004. Brochure. In possession of Christopher Gillis.

Price, David B., director, Office of Public Relations, Plymouth State College, Plymouth, NH. Typewritten letter to Paul Shulins, manager, WPCR Radio, Plymouth State College, NH. January 30, 1979. Available in files of Paul Shulins, Boston, MA.

Puckette, Charles. *To Rick Katzenberg*[,] *1979 Marketing Plan*[,] *November 20, 1978*. New Boston, NH: Natural Products Inc., 1978. Typewritten memorandum. Available in the files of Earle Rich, Mont Vernon, NH.

Rader, Nancy, executive director, California Wind Energy Association, Berkeley, CA, to Christopher Gillis, March 16, 2018 and April 2, 2018. Emails in possession of Christopher Gillis.

———. "A Review and Critique of Environmental Quantification in Electric Utility Planning." Term paper, University of California at Berkeley, 1992. Available in the files of Nancy Rader, Berkeley, CA.

———, Scott Sklar, and Randy Swisher, To: Reviewers, (Re: Concepts for Federal Renewable Energy Policies/Programs), August 3, 1990. Typewritten letter. Available in the files of Nancy Rader, Berkeley, CA.

Renewable Energy Ventures, Inc., Encino, CA. *REV Wind Power Partners 1981–1*[,] *Preview Product Profile*[,] *Summary*. Encino, CA: Renewable Energy Ventures, Inc., 1984. Available in the files of Paul R. Jacobs, Corcoran, MN.

Renewable Energy Ventures, Inc., Washington, DC. *December 21, 1982*[,] *Joint Venture to Develop Wind Farms*[:] *A Business Plan*[,] *Prepared for Control Data Corporation, Minneapolis, Minnesota 55440*. Washington, DC: Renewable Energy Ventures, 1982. Available in the files of Paul R. Jacobs, Corcoran, MN.

REPCO, Bismarck, ND. *News Release*[,] *New Crop Coop Formed*, January 3, 1994. In possession of Christopher Gillis.

———. *Articles of Association of REPCO-Rural Energy Producers Electric Power Cooperative*. Bismarck, ND: REPCO, 1993. Handout. Available in the files of Paul R. Jacobs, Corcoran, MN.

———. *The Case for Action Now in Establishing a Dispersed Wind Rec.* Bismarck, ND: REPCO, 1994. Handout. Available in the files of Paul R. Jacobs, Corcoran, MN.

———. *The REPCO Mission*. Bismarck, ND: REPCO, 1994. Handout. Available in the files of Paul R. Jacobs, Corcoran, MN.

Rogier, Etienne, Toulouse, France, to Christopher Gillis, April 18, 2017. Email in possession of Christopher Gillis.

———, "AEROWATT, c'est l'administration," to Christopher Gillis, August 5, 2018. Email in possession of Christopher Gillis.

Ruff, Hagen, founder and CEO, Chava Wind, Homestead, FL. "Lightning Round—New Products and Services—Part II." Presentation at Distributed Wind 2019, Washington, DC, February 28, 2019. Notes in possession of Christopher Gillis.

Sabas, Matthew, "History of Solar Power," Institute for Energy Research, February 18, 2016. Accessed February 18, 2016. https://www.instituteforenergyresearch.org/renewable/solar/history-of-solar-power/.

Sandia National Laboratories, Albuquerque, NM. *Advanced Wind Energy Projects Test Facility Moving to Texas Tech University*, July 27, 2011. In possession of Christopher Gillis.

Salter, Ed, "Flexible Rotor Technology and the 'Storm Master' Wind Turbine[:] A Brief History—1977 to 2009," Scribd. Accessed October 9, 2018. https://www.scribd.com/document/134004067/Storm-Master-Rotor-Tech-4–09.

Schaufelberger, M. Elektro GmbH, Winterhur, Switzerland. Typewritten letter to Henry Clews, Solar Wind, East Holden, ME. "Fixation of Propeller." January 22, 1975. Available in files of Leander Nichols.

Slyker, Karin, Texas Tech University, Lubbock, TX. *Texas Tech Commissions New Wind Research Center*, July 9, 2013. In possession of Christopher Gillis.

Solar Energy Industries Association, Washington, DC. *US Solar Industry Experiences Record-Breaking Growth in 2010*, May 14, 2012. In possession of Christopher Gillis.

Solar Energy Research Institute, Golden, CO. *A Government-Industry Partnership Experience*[:] *The Cooperative Field Test Program in Wind Research*. Washington, DC: American Wind Energy Association, 1991. Brochure. Available in the files of Warren S. Bollmeier, II, Kaneohe, HI.

Southwest Windpower, Flagstaff, AZ. *$10M from GE and Current Investors Plus Federal Stimulus Incentives Propel Southwest Windpower's Expansion in Small Wind Turbines*, April 6, 2009. In possession of Christopher Gillis.

———. *Power to the People: Southwest Windpower Unveils Most Efficient, Easy-to-Use Small Wind Turbine*, January 6, 2011. In possession of Christopher Gillis.

Speiser, Kenneth, manager, advanced systems, Grumman Energy Systems, Ronkonkoma, NY, to B. McBroom, Holton, KS. October 12, 1976. Typewritten letter. Available in the files of Bob McBroom, Holton, KS.

"State of Green." The Danish Museum of Electricity (Elmuseet). Accessed November 9, 2018. https://stateofgreen.com/en/partners/danish-museum-of-energy-energimuseet/.

Strickland, E. E., Control Data Corporation, Minneapolis, MN, Interoffice Memorandum (Jacobs Wind Generators), March 20, 1979. Available in the files of Paul R. Jacobs, Corcoran, MN.

Swanson, Glen E. "The Brush Windmill Timeline[:] 1885–1956," to Christopher Gillis, March 4, 2015. Email in possession of Christopher Gillis.

———, Grand Rapids, MI, to Christopher Gillis, March 31, 2017. Email in possession of Christopher Gillis.

———. "History of Moses Farmer Windmill Patent Model." Research notes, Grand Valley State University, Allendale, MI, March 10, 2017.

Swisher, Randall S., Silver Spring, MD, to Christopher Gillis, July 20, 2018. In possession of Christopher Gillis.

Swisher, Randy, executive director, American Wind Energy Association, Washington, DC, to membership, (Subject: Taking Action to Stop Repeal of the Wind Production Tax Credit), September 26, 1995. Typewritten letter. Available in the files of Randall S. Swisher, Silver Spring, MD.

Sykes, Robert D., professor and principal investigator, and Lance M. Neckar, professor and department head, University of Minnesota, Department of Landscape Architecture, College of Design, Minneapolis, MN, to Paul R. Jacobs, June 2, 2009. Typewritten letter. Available in the files of Paul R. Jacobs, Corcoran, MN.

Tampa Port Authority, Tampa, FL. *Permit for Work in the Navigable Waters of the Hillsborough County Port District*. Tampa, FL, July 9, 1980. Available in the files of Paul R. Jacobs, Corcoran, MN.

"Technology." Carter Wind Energy. Accessed March 26, 2019. http://www.carterwindenergy.com.

Tellis, Jeffrey N., president, The Intercollegiate Broadcasting System, Inc., Vails Gate, NY, to Paul Shulins, general manager, Radio Station WPCR, Plymouth State College, Plymouth, NH. January 12, 1979. Typewritten letter. Available in the files of Paul Shulins, Boston, MA.

"Thomas A. Edison Testimony." *Edison v. Siemens v. Field*. US Patent Office, 1881. Accessed April 3, 2017. http://edison.rutgers.edu/NamesSearch/glocpage.php?glo=QD001&.

[Toepfer, Craig,] "American Wind Energy Association." Detroit: American Wind Energy Association, [ca. 1975]. Typewritten handout. Available in the files of Craig Toepfer, Chelsea, MI.

Toepfer, Craig, Chelsea, MI, to Christopher Gillis, February 5, 2016, and October 18, 2018. Emails in possession of Christopher Gillis.

"Total Energy[,] Annual Energy Review[,] September 2012[:] Table 8.10 Average Retail Prices of Electricity, 1960–2011 (Cents per kilowatt-hour, including Taxes." US Energy Information Administration. September 27, 2012. Accessed January 8, 2019. https://www.eia.gov/totalenergy/data/annual/showtext.php?t=ptb0810.

Towne, Evan T. "Mr. Model Engine—Bill Brown." Craftsmanship Museum. Accessed April 23, 2016. http://www.craftmanshipmuseum.com/BrownJr.htm.

"Turbine on the Roof!" Ducted Turbines International. Accessed March 26, 2019. http://ductedturbinesinternational.com/2018/10/29/turbine-on-the-roof/.

"The UMASS Wind Turbine WF-1, A Retrospective." University of Massachusetts Wind Energy Center. Accessed September 20, 2018. https://www.umass.edu/windenergy/about/history/windturbinewf1.

United Wind, Brooklyn, NY. *Forum Equity Partners and United Wind Announce $200M Investment for Distributed Wind Projects*, January 6, 2016. In possession of Christopher Gillis.

———. *United Wind & Bergey Windpower Close Landmark Purchase of 100 Distributed-Scale Wind Turbines*, January 20, 2017. In possession of Christopher Gillis.

———. *United Wind and Smithfield Foods Announce 3MW WindLease Agreement*, March 19, 2019. In possession of Christopher Gillis.

University of Oklahoma, Norman, OK. *OU Urban Car*. Norman, OK: University of Oklahoma, [ca. 1973]. Handout. In possession of Christopher Gillis.

US Department of Agriculture and Alternative Energy Institute. *Wind Energy Research*[:] *System Evaluation*. Bushland, TX: US Department of Agriculture, Agricultural Research Service, [ca. 1983]. Brochure. In possession of Christopher Gillis.

US Department of Energy. *Department of Energy Announces $28 Million in Funding for Wind Energy Research*, March 28, 2019. In possession of Christopher Gillis.

———, Office of Energy Efficiency and Renewable Energy, Washington, DC. *Distributed Wind Competitiveness Improvement Project*. Golden, CO: National Renewable Energy Laboratory, 2019. Handbill. In possession of Christopher Gillis.

———. *EERE Success Story—Distributed Wind Competitiveness Improvement Project (CIP) Partner Delivers Next-Generation Wind Turbine for On-Site Power*, April 11, 2017. Accessed January 23, 2018. https://energy.gov/eere/success-stories/articles/eere-success-story-distributed-wind-competiveness-improvement. In possession of Christopher Gillis.

———. *Energy Department's Competitiveness Improvement Project Hits First Certification Milestone; Opens Next Round of Applications*, February 28, 2018. In possession of Christopher Gillis.

———. *Energy Department Helps Manufacturers of Small and Mid-Size Wind Turbines Meet Certification Requirements*, October 1, 2015. Accessed January 23, 2018. https://energy.gov/eere/articles/energy-department-helps-manufacturers-small-and-mid-size-wind-turbines-meet. In possession of Christopher Gillis.

———. *Pika Energy Develops Innovative Manufacturing Process and Lowers Production Cost Under DOE Competitiveness Improvement Project*, September 16, 2015. Accessed January 23, 2018. https://energy.gov/eere/wind/articles/pika-energy-develops-innovative-manufacturing-process-and-lowers-production-cost. In possession of Christopher Gillis.

———. *Wind and Water Power Program*[,] *The Wind Powering America Initiative*. Washington, DC: US Department of Energy, 2011. Handbill. In possession of Christopher Gillis.

———. *Wind Powering America*[,] *America's Wind Power . . . A National Resource*. Washington, DC: US Department of Energy, 2001. Brochure. In possession of Christopher Gillis.

———. *Wind Turbine Showcased in Energy Department Headquarters*, February 26, 2016. Accessed January 23, 2018. https://energy.gov/eere/articles/wind-turbine-showcased-energy-department-headquarters. In possession of Christopher Gillis.

———, Solar Energy Technologies Office, Washington, DC. *The SunShot Initiative*, 2017. Accessed January 31, 2018. https://energy.gov/eere/solar/sunshot-initiative.

———, Wind and Water Power Technologies Office, Washington, DC. Distributed Wind Competitiveness Improvement Project. Washington, DC: US Department of Energy, 2016. Handbill. In possession of Christopher Gillis.

US Geological Survey, Reston, VA. *Mapping the Nation's Wind Turbines*. May 16, 2018. In possession of Christopher Gillis.

Vestas Wind System A/S, Lem, Denmark, *Annual Report 1997*. Lem, Denmark: Vestas Wind System, 1997. Available in the files of Warren S. Bollmeier, II, Kaneohe, HI.

Vick, Brian D., Amarillo, TX, to Christopher Gillis, January 10, 2018, and January 17, 2018. Emails in possession of Christopher Gillis.

Visser, Ken, co-founder, Ducted Turbines International, Potsdam, NY. "Lightning Round—New Products and Services—Part II." Presentation at Distributed Wind 2019, Washington, DC, February 28, 2019. Notes in possession of Christopher Gillis.

"What We Do." One Energy Enterprises LLC. Accessed March 12, 2019. https://oneenergy.com/about-us/what-we-do/.

White, Jonathan R., Wind Energy Technologies Development, Sandia National Laboratories, Albuquerque, NM, to Christopher Gillis, February 18, 2018. Email in possession of Christopher Gillis.

Whitehead, Daniel, St. Johns, FL, to Christopher Gillis, November 7, 2018. Email in possession of Christopher Gillis.

Wilkerson, Alan W., president, Gemini Company, Thiensville, WI. "License Agreement." Typewritten document conferring patent usage to Omnion Power Engineering, Mukwonago, WI, January 4, 1986. Available in the files of Sherry Jones, Cedarburg, WI.

———. *Synchronous Inversion for Wind Power Utilization*. Thiensville, WI: Gemini Company, 1975. Typewritten. Available in the files of Craig Toepfer, Chelsea, MI.

———. *Synchronous Inversion Techniques for Utilization of Waste Energy*. Thiensville, WI: Gemini Company, 1975. Typewritten. Available in the files of Craig Toepfer, Chelsea, MI.

"William (Bill) L. Borwn IV[,] May 30, 1911—January 8, 2003." State College Radio Control Club (Bellefonte, PA). Accessed April 23, 2016. http://scrc-club.com/Archive%20files/Bill_Brown/billbrwn.htm.

Williams, R. O., Jr., vice president and general manager, Rockwell International, Rocky Flats Plant, Energy Systems Group, Golden, CO, to M. L. Jacobs, president, Jacobs Wind Electric Company, Fort Myers, FL, August 2, 1979. Typewritten memorandum. Available in the files of Paul R. Jacobs, Corcoran, MN.

———, to M. L. Jacobs, president, Jacobs Wind Electric Company, Fort Myers, FL, August 14, 1979. Typewritten memorandum. Available in the files of Paul R. Jacobs.

Wind Energy Society of America, Pasadena, CA. *Wind Energy Society of America*[:] *The Challenge*[,] *Goals*. Pasadena, CA: Wind Energy Society of America, [ca, 1975]. Handbill. Available in the files of Craig Toepfer, Chelsea. MI.

"Windkraft zum Anfassen." Mühlenheider Wind Power Museum. Accessed November 9, 2018. http://www.muehlenheider-windkraftmuseum.de/English-Site/.

WindTronics, Muskegon, MI. *Honeywell Wind Turbine by WindTronics Now Available for Purchase*, April 21, 2011. In possession of Christopher Gillis.

Winterbauer, Pam. "A Short History of Hayward." Accessed March 9, 2017. http://activerain.com/blogs-view/475061/a-short-history-of-hayward.

Wisconsin Power & Light Company, Madison, WI, "To the Owners of Wisconsin Power and Light Company," *Quarterly Report*[,] *A Report to Investors* 9, no. 2 (August 1982): 2–3. Available in the files of Craig Toepfer, Chelsea, MI.

Wright, C. A. To Nancy Horning, American Wind Energy Association. Typewritten letter. August 29, 1974.

Wright, Charles A. "Charles A. Wright of Rohn Manufacturing" [ca. 1974]. One page. Typewritten. In possession of Christopher Gillis.

XZERES Corporation, Wilsonville, OR. *XZERES Acquires Southwest Windpower's Skystream Product Line*, July 9, 2013. In possession of Christopher Gillis.

Interviews

Bergey, Karl H., Jr., founder, Bergey Windpower Company, with Christopher Gillis, Norman, OK, October 23, 2015. Handwritten notes in possession of Christopher Gillis.

Bergey, Michael L.S., president and CEO, Bergey Windpower Company, with Christopher Gillis, Norman, OK, October 23, 2015. Handwritten notes in possession of Christopher Gillis.

Bollmeier II, Warren S., Kaneohe, HI, telephone interview with Christopher Gillis, June 12, 2016. Handwritten notes in possession of Christopher Gillis.

Carter, Jay, Jr., president, CEO, and principal design engineer, Carter Aviation Technologies, with Christopher Gillis, Wichita Falls, TX, October 24, 2015. Handwritten notes in possession of Christopher Gillis.

Drees, Herman M., West Lake Village, CA, telephone interview with Christopher Gillis, April 30, 2016, and September 18, 2017. Handwritten notes in possession of Christopher Gillis.

———, with Christopher Gillis, Westlake Village, CA, October 19, 2018. Handwritten notes in possession of Christopher Gillis.

Erdman, Jan, with Christopher Gillis, Custer, WI, June 17, 2016. Handwritten notes in possession of Christopher Gillis.

Gilman, Patrick, modeling and analysis program manager, Wind Energy Technologies Office, Office of Energy Efficiency and Renewable Energy, US Department of Energy, Golden, CO, telephone interview with Christopher Gillis, January 24, 2018. Handwritten notes in possession of Christopher Gillis.

Gray, Thomas, with Gregory L. Sharrow, June 19, 2008, interview 2008.202. Transcript in files of NRG Research Project, Vermont Folklife Center, Middlebury, VT.

Harris, Coy, and John Baker, American Windmill Museum, with Christopher Gillis, Lubbock, TX, June 6, 2018. Handwritten notes in possession of Christopher Gillis.

Hodgkin, Jonathan, Essex Junction, VT, telephone interview with Christopher Gillis, January 25, 2016. Handwritten notes in possession of Christopher Gillis.

Jacobs, Paul R., owner, Jacobs Wind Electric Company, with Christopher Gillis, Minneapolis, September 30, 2016. Handwritten notes in possession of Christopher Gillis.

Jenkins, Jennifer, Flagstaff, AZ, telephone interview with Christopher Gillis, January 18, 2018. Handwritten notes in possession of Christopher Gillis.

Katzenberg, Richard, with Christopher Gillis, Amherst, NH, August 26, 2016. Handwritten notes in possession of Christopher Gillis.

Kotalik, Ken, director of global sales and operations, Primus Windpower, Lakewood, CO, telephone interview with Christopher Gillis, November 30, 2018. Handwritten notes in possession of Christopher Gillis.

Kruse, Andy, Boulder, CO, telephone interviews with Christopher Gillis, December 13, 2017, and December 18, 2017. Handwritten notes in possession of Christopher Gillis.

Mayer, Donald, Waitsfield, VT, telephone interview with Christopher Gillis, April 24, 2015. Handwritten notes in possession of Christopher Gillis.

Metcalfe, Philip W., Brimfield, IL, telephone interview with Christopher Gillis, November 16, 2015. Handwritten notes in possession of Christopher Gillis.

O'Shea, Allan, Copemish, MI, telephone interview with Christopher Gillis, September 16, 2017. Handwritten notes in possession of Christopher Gillis.

Price, Travis L., III, with Christopher Gillis, Washington, DC, July 29, 2016. Handwritten notes in possession of Christopher Gillis.

Rich, Earle, with Christopher Gillis, Amherst, NH, August 26, 2016. Handwritten notes in possession of Christopher Gillis.

———, Amherst, NH, telephone interview with Christopher Gillis, August 20, 2018. Handwritten notes in possession of Christopher Gillis.

Sherwood, Larry, president and CEO, Interstate Renewable Energy Council, Inc., Latham, NY, telephone interview with Christopher Gillis, March 23, 2018. Handwritten notes in possession of Christopher Gillis.

Shulins, Paul, Boston, MA, telephone interview with Christopher Gillis, October 5, 2015. Handwritten notes in possession of Christopher Gillis.

Swisher, Randall S., with Christopher Gillis, Silver Spring, MD, October 11, 2017. Handwritten notes in possession of Christopher Gillis.

Swisher, Randy, with Gregory L. Sharrow, August 14,

2008, interview 2008.2038. Transcript in files of RRG Research Project, Vermont Folklife Center, Middlebury, VT.

Tencer, Russell, founder and CEO, United Wind, Brooklyn, NY, telephone interview with Christopher Gillis, November 30, 2018. Handwritten notes in possession of Christopher Gillis.

Toepfer, Craig, Chelsea, MI, telephone interview with Christopher Gillis, August 14, 2017. Handwritten notes in possession of Christopher Gillis.

Werst, Mike, owner of Wincharger.com, Manor, TX, telephone interview with Christopher Gillis, May 8, 2017. Handwritten notes in possession of Christopher Gillis.

Books and Papers

Adams, R. Christopher. *An Analysis of Wind Power Development in the Town of Hull, MA*. Hull, MA: Hull Municipal Light Plant, 2013.

Allen, I. B. "Status of SWECS Testing Programs at the Rocky Flats Small Wind Systems Test Center." Paper presented at Small Wind Turbine Systems 1981, Boulder, CO, May 12–14, 1981. In *Proceedings*[:] *Small Wind Turbine Systems 1981*. Edited by the Solar Energy Research Institute. [Golden, CO]: Solar Energy Research Institute, 1981.

American Wind Energy Association. *2010 AWEA Small Wind Turbine Global Market Study*. Washington, DC: American Wind Energy Association, 2010.

———. *2011 U. S. Small Wind Turbine Market Report*. Washington, DC: American Wind Energy Association, 2012.

———. *AWEA Small Wind Turbine Global Market Study 2007*. Washington, DC: American Wind Energy Association, 2007.

———. *AWEA Small Wind Turbine Global Market Study: 2009*. Washington, DC: American Wind Energy Association, 2009.

———. *AWEA Small Wind Turbine Performance and Safety Standard*[,] *AWEA 9.1–2009*. Washington, DC: American Wind Energy Association, 2009.

———, Small Wind Turbine Committee. *The U. S. Small Wind Turbine Industry Roadmap*[:] *A 20-Year Industry Plan for Small Wind Turbine Technology*. Washington, DC: American Wind Energy Association, 2002.

———, Terminology Standards Subcommittee. *Draft Working Document*[,] *Terminology Standards Subcommittee*[,] *AWEA Standards Program*[,] *February 26, 1981*. Washington, DC: American Wind Energy Association, 1981.

Ancona, Daniel F. "Federal Wind Energy Research Program: Current Status and Future Plans." Paper presented at the Wind Energy Expo '84 and National Conference, Pasadena, CA, September 24–26, 1984.

In the *Wind Energy Expo '84 and National Conference Proceedings*. Washington, DC: American Wind Energy Association, 1984.

Anderson, Frederick Irving. *Electricity for the Farm*. New York: The Macmillan Company, 1915.

Anderson, John. "The 40 KW Giromill." Paper presented at Small Wind Turbine Systems 1981, Boulder, CO, May 12–14, 1981. In *Proceedings*[:] *Small Wind Turbine Systems 1981*. Edited by the Solar Energy Research Institute. [Golden, CO]: Solar Energy Research Institute, 1981.

Anderson, Roger. "Pultruded Fiberglass Reinforced Plastic (FRP) Wind Turbine Blades." Paper presented at Wind Energy Expo '82 and National Conference, Amarillo, TX, October 24–27, 1982. In the *Wind Energy Expo '82 and National Conference Proceedings*. Edited by Vaughn Nelson. Washington, DC: American Wind Energy Association, 1982.

Ardrey, R. L. *American Agricultural Implements: A Review of Invention and Development in the Agricultural Implement Industry of the United States*. Chicago: privately printed, 1894.

Baichun Zhang. "Ancient Chinese Windmills." Paper presented at the Third International Symposium on History of Machines and Mechanisms. National Cheng Keng University, Tainan, Taiwan. November 11–14, 2008. Accessed February 19, 2014. http://linkspringer.com/chapter/10.1007%2F978-1-4020-9485-9_15.

Bailey, Bruce H. *New York State Wind Energy Handbook*. Albany, NY: New York State Energy Office, 1982.

Baker, T. Lindsay. *A Field Guide to American Windmills*. Norman: University of Oklahoma Press, 1985.

Bathe, Greville. *Horizontal Windmills, Draft Mills and Similar Air-Flow Engines*. Philadelphia: privately printed, 1948.

Beedell, Suzanne. *Windmills*. New York: Charles Scribner's Sons, 1979.

Bell, W. Ken. "Wind Energy Information at Sandia Laboratories." Paper presented at the Panel on Information Dissemination for Wind Energy, Albuquerque, NM, August 2–3, 1979. In *Proceedings: Panel on Information Proceedings: Panel on Information Dissemination for Wind Energy*. Edited by Patricia Weis. Golden, CO: Solar Energy Research Institute, 1980.

Benson, Arnold. *Plans for Construction of a Small Wind-Electric Plant for Oklahoma Farms*. Publication No. 33. Stillwater, OK: Oklahoma Agricultural and Mechanical College, 8, no. 1 (June 1937): 1–33.

Bergey, K. H. "Feasibility of Wind Power Generation for Central Oklahoma." Typescript. Norman, OK: University of Oklahoma, 1973.

Bergey, Karl H. "Development of High-Performance, High Reliability Windpower Generators." Paper presented at the American Wind Energy Association National Conference, Summer 1980, Pittsburgh, PA,

June 8–11, 1980. In the *American Wind Energy Association National Conference, Summer 1980 Proceedings*. Edited by Vaughn Nelson. Washington, DC: American Wind Energy Association.

———. "Wind Power Demonstration and Siting Problems." Paper presented at the Wind Energy Conversion Systems Workshop, Washington, DC, June 11–13, 1973. In *Wind Energy Conversion Systems Workshop Proceedings*. Edited by Joseph M. Savino. Washington, DC: National Science Foundation, Research Applied to National Needs, and National Aeronautics and Space Administration, December 1973.

———. "Windfarm Strategies: The Economics of Scale." Paper presented at Wind Energy Expo '83 and National Conference, San Francisco, October 17–19, 1983. In *Wind Energy Expo '83 and National Conference Proceedings*. Washington, DC: American Wind Energy Association, 1983.

Bergey, Michael L. S., chairman, Performance Subcommittee, American Wind Energy Association. *Performance Rating Document*. Washington, DC: American Wind Energy Association, [1980].

———. "Wind-Electric Pumping Systems for Communities." Paper presented at the 1st International Symposium on Safe Drinking Water in Small Systems, Washington, DC, May 10–13, 1998. In *Proceedings of the 1st International Symposium on Safe Drinking Water in Small Systems*. Ann Harbor, MI: NSF International, 1999.

Bergey, Mike. "An Overview of DWEA[:] New National Trade Association for Distributed Wind. Presentation at the Sixth Annual Small Wind Conference, Stevens Point, WI, June 15–16, 2010.

———. "Performance Testing and Rating Standards for Wind Energy Conversion Systems." Paper presented at the National Conference, Summer 1980, American Wind Energy Association, Pittsburgh, PA, June 8–11, 1980. In *American Wind Energy Association National Conference, Summer 1980 Proceedings*. Edited by Vaughn Nelson. Washington, DC: American Wind Energy Association.

———. "The Vertical Axis, Articulated Blade, Wind Turbine." Paper presented at the National Conference, American Wind Energy Association, Amarillo, TX, March 1–5, 1978. In the *American Wind Energy Association National Conference Proceedings*. Edited by Vaughn Nelson. Canyon, TX: Alternative Energy Institute, West Texas State University, 1978.

———. "Wind Energy Potpourri." Paper presented at the Rural Electric Wind Energy Workshop, Boulder, CO, June 1–3, 1982. In the *Rural Electric Wind Energy Workshop Proceedings*. Washington, DC: The Energy and Environmental Cooperative Association, National Rural Electric Cooperative Association, 1982.

Bingley, Donald W. "Wind System Utility Intertie with Inverters/Frequency Changers." Paper presented at the Wind Energy Expo '82 and National Conference, Amarillo, TX, October 24–27, 1982. In the *Wind Energy Expo '82 and National Conference Proceedings*. Washington, DC: American Wind Energy Association, 1982.

Blyth, James. "On a New Form of Windmill." *Report of the Sixty-Second Meeting of the British Association for the Advancement of Science held at Edinburgh in August 1892*. London: John Murray, 1893.

Borsodi, Ralph. *This Ugly Civilization*. New York: Simon and Schuster, 1929.

Boyd, T. A. *Professional Amateur*[:] *The Biography of Charles Franklin Kettering*. New York: E. P. Dutton, 1957.

Boyt, David. "Use of Laminated Wood in Wind Turbine Blades." Paper presented at Wind Energy Expo '82 and National Conference, Amarillo, TX, October 24–27, 1982. In the *Wind Energy Expo '82 and National Conference Proceedings*. Edited by Vaughn Nelson. Washington, DC: American Wind Energy Association, 1982.

Braasch, R. H. "Power Generation with a Vertical-Axis Wind Turbine." Paper presented at Second Workshop on Wind Energy Conversion Systems, Washington, DC, June 9–11, 1975. In *Proceedings of the Second Workshop on Wind Energy Conversion Systems*. Edited by Frank R. Eldridge. [Washington, DC]: The MITRE Corporation, 1975.

Brown, C. A. Cameron. *Windmills for the Generation of Electricity*. 2nd ed. Oxford, England: Institute for Research in Agricultural Engineering, University of Oxford, 1933.

Brown, D. Clayton. *Electricity for Rural America*[:] *The Fight for the REA*. Westport, CT: Greenwood Press, 1980.

Brulle, Robert V. *Engineering the Space Age: Rocket Scientist Remembers*. Maxwell Air Force Base, AL: Air University Press, 2008.

———, and Harold C. Larsen. "Giromill (Cyclogiro Windmill) Investigation for Generation of Electrical Power." Paper presented at Second Workshop on Wind Energy Conversion Systems, Washington, DC, June 9–11, 1975. In *Proceedings of the Second Workshop on Wind Energy Conversion Systems*. Edited by Frank R. Eldridge. [Washington, DC]: The MITRE Corporation, 1975.

Buffer, Patricia. *Beyond the Buildings at the Place Called 'Rocky Flats'*. Washington, DC: Office of Legacy Management, US Department of Energy, 2003.

Butterfield, C. P., and W. D. Musial. "EPRI/ESI Test Program." Paper presented at Windpower '85, San Francisco, CA, August 27–30, 1985. In *Windpower '85 Proceedings*. Washington, DC: American Wind Energy Association, 1985.

Byrd, Richard Evelyn. *Discovery*[:] *The Story of the Second Byrd Antarctic Expedition*. New York: G. P. Putnam's Sons, 1935.

Cagle, Donny R., Anthony D. May, Brian D. Vick, and Adam J. Holman. "Evaluation of Airfoils for Small Wind Turbines." Paper presented at the Windpower 2007, Los Angeles, June 3–6, 2007. In *Windpower 2007 Proceedings*. Washington, DC: American Wind Energy Association, 2007.

Carter, Jay, Jr. "Design and Development of the Carter 25 Generator." Paper presented at the American Wind Energy Association National Conference, Pittsburgh, PA, June 8–11, 1980. In *American Wind Energy Association National Conference Proceedings*. Edited by Vaughn Nelson. Washington, DC: American Wind Energy Association, 1980.

——. "Jay Carter Enterprises, Inc[.] Experience." Paper presented at the Rural Electric Wind Energy Workshop, Boulder, CO, June 1–3, 1982. In the *Rural Electric Wind Energy Workshop Proceedings*. Washington, DC: The Energy and Environmental Cooperative Association, National Rural Electric Cooperative Association, 1982.

Christianson, Mary M. "Alternative Technology in Low Income New York: Energy Task Force's First Four Years." Paper presented at The First Conference on Community Renewable Energy Systems, University of Colorado, Boulder, CO, August 20–21, 1979. In the *Proceedings of the First Conference on Community Renewable Energy Systems*. Washington, DC: The Solar Energy Research Institute, July 1980.

Clark, R. Nolan. "Applications Testing by USDA—An Overview." Paper presented at Small Wind Turbine Systems 1981, Boulder, CO, May 12–14, 1981. In *Proceedings*[:] *Small Wind Turbine Systems 1981*. Edited by the Solar Energy Research Institute. [Golden, CO]: Solar Energy Research Institute, 1981.

——. *Small Wind*[:] *Planning and Building Successful Installations*. Waltham, MA: Academic Press, 2014.

——. "Wind-Electric Water Pumping Systems for Rural Domestic and Livestock Water." Paper presented at the 5th European Wind Energy Association Conference and Exhibition, Thessaloniki-Macedonia, Greece, October 10–14, 1994. In *Conference Proceedings*[,] *Oral Sessions*[,] *Volume II*. Edited by J. L. Tsipouridis. Thessaloniki-Macedonia, Greece: Hellenic Wind Energy Association, 1994.

——, and Brian Vick. "Chapter 13[:] Livestock Watering with Renewable Energy." In *Agriculture as a Producer and Consumer of Energy*. Edited by J. Outlaw, K. J. Collins, and J. A. Duffield. Wallingford, United Kingdom: CABI Publishing, 2005.

——. "Determining the Proper Motor Size for Two Wind Turbines Used in Water Pumping." Paper presented at American Society of Mechanical Engineers,

Solar Energy Division, Energy Sources Technology Conference and Exhibition, Houston, TX, January 29-February 1, 1995. In *Proceedings of American Society of Mechanical Engineers, Solar Energy Division*, Vol. 16. New York: American Society of Mechanical Engineers, 1995.

Clews, Henry. *Electric Power from the Wind*. East Holden, ME: Solar Wind Company, 1974.

——. "Electricity from the Wind: Ready-Made Units." In *Producing Your Own Power*[:] *How to Make Nature's Energy Sources Work for You*, edited by Carol Hupping Stoner, 13–43. Emmaus, PA: Rodale Press, 1974.

——. *Solar Wind Progress Report*. East Holden, ME: Solar Wind Company, 1973.

——. "Wind Power Systems for Individual Applications." Paper presented at the Wind Energy Conversion Systems Workshop, Washington, DC, June 11–13, 1973. In *Wind Energy Conversion Systems Workshop Proceedings*. Edited by Joseph M. Savino. Washington, DC: National Science Foundation, Research Applied to National Needs, and National Aeronautics and Space Administration, December 1973.

Coleman, Clint (Jito). "Northern Power's Advanced Wind Turbine Development Program." Paper presented at Windpower '91, Palm Springs, CA, September 24–27, 1991. In *Windpower '91 Proceedings*. Washington, DC: American Wind Energy Association, 1991.

——, Clint, and Wayne Hunnicutt. "Rotor and Control System Design for the 8 KW, 10 Meter Diameter, DOE/Windworks WECS." Paper presented the American Wind Energy Association National Conference, San Francisco, CA, April 16–19, 1979. In the *American Wind Energy Association National Conference, Spring 1979 Proceedings*. Edited by Vaughn Nelson. Washington, DC: American Wind Energy Association, 1979.

Coombs, R. D. *Pole and Tower Lines for Electric Power Transmission*. New York: McGraw-Hill Book Company, 1916.

Corbus, David, and Mike Bergey. "Costa de Cocos 11-KW Wind-Diesel Hybrid System." Paper presented at Windpower '97, Austin, TX, June 15–18, 1997. In *Windpower '97 Proceedings*. Washington, DC: American Wind Energy Association, 1997.

Cowan, Ruth Schwartz. *More Work for Mother*. New York: Basic Books, 1983.

Crocker, Francis B. *Electric Lighting*[:] *A Practical Exposition of the Art for the Use of Engineers, Students and Others Interested in the Installation or Operation of Electrical Plants*[,] *Vol. 1*[,] *The Generating Plant*. New York: D. Van Nostrand Company, 1896.

Cromack, Duane, and Michael Edds. "Operational Aspects of the UMass Wind Turbine." Paper presented at the American Wind Energy Association National Conference, Amarillo, TX, March 1–5, 1978. In

Proceedings of the American Wind Energy Association National Conference Proceedings. Edited by Vaughn Nelson. Canyon, TX: Alternative Energy Institute, West Texas State University, 1978.

Davidson, J. Brownlee, and Leon Wilson Chase. *Farm Machinery and Farm Motors.* New York: Orange Judd Company, 1910.

Davis, Holly, Steve Drouilhet, Larry Flowers, Michael Bergey, and Michael Brandemuehl. "Wind-Electric Ice Making for Villages in the Developing World." Paper presented at Windpower '94, Minneapolis, MN, May 10–13, 1994. In *Windpower '94 Proceedings.* Washington, DC: American Wind Energy Association, 1994.

Distributed Wind Energy Association. *SMART Wind Roadmap.* Durango, CO: Distributed Wind Energy Association, 2016.

Divone, Louis. "Overview of WECS Program." Paper presented at Second Workshop on Wind Energy Conversion Systems, Washington, DC, June 9–11, 1975. In *Proceedings of the Second Workshop on Wind Energy Conversion Systems.* Edited by Frank R. Eldridge. [Washington, DC]: The MITRE Corporation, 1975.

Döring, Walter. *A Vision Becomes Reality*[:] *Wie die Windreich AG das Vermächtnis des Windpioniers Ulrich Hütter in die tat Umsetzt.* Wolfshlugen, Germany: Windreich AG, [ca. 2012].

Drake, William. "The Enertech 1500 Watt Wind Turbine." Paper presented at the American Wind Energy Association National Conference, Fall 78. Cape Cod, MA, September 25–27, 1978. In the *American Wind Energy Association National Conference, Fall 78 Proceedings.* Edited by Vaughn Nelson. Washington, DC: American Wind Energy Association, 1978.

Drees, H. M. "The Cycloturbine and Its Potential for Broad Application." Paper presented at the Second International Symposium on Wind Energy Systems, Amsterdam, the Netherlands, October 3–6, 1978. In *Second International Symposium on Wind Energy Systems Proceedings.* Bedford, United Kingdom: BHRA Fluid Engineering, 1978.

Drees, Herman Meijer. "The Cycloturbine and Its Potential for Broad Application." Paper presented at the American Wind Energy Association National Conference, Fall 78. Cape Cod, MA, September 25–27, 1978. In the *American Wind Energy Association National Conference, Fall 78 Proceedings.* Edited by Vaughn Nelson. Washington, DC: American Wind Energy Association, 1978.

Duskin, Alvin. "An Introduction to United States Wind Power." Paper presented at the First Conference on Community Renewable Energy Systems, Boulder, CO, August 20–21, 1979. In *Proceedings of the First Conference on Community Renewable Energy Systems.* Edited

by Robert Odland. Washington, DC: US Department of Energy, Solar Energy Research Institute, 1980.

Electric Power Subsystems Subcommittee, American Wind Energy Association. *Draft Working Document*[,] *Electric Power Subsystems Subcommittee*[,] *AWEA Standards Program*[,] *January 27, 1981*[,] *SWECS Electrical System Interconnect Technical Guidelines.* Washington, DC: American Wind Energy Association, 1981.

Finch, Ted. "Field Testing of a Commercial Wind Energy Conversion System." Paper presented at the 5th Annual Energy-Sources Technology Conference and Exhibition, New Orleans, LA, March 7–10, 1982. In the *International Wind Energy Symposium Proceedings.* New York: American Society of Mechanical Engineers, 1982.

Flavin, Christopher. "Electricity for a Developing World: New Directions." *Worldwatch Paper 70.* Washington, DC: Worldwatch Institute, 1986.

———. "Wind Power: A Turning Point." *Worldwatch Paper 45.* Washington, DC: Worldwatch Institute, 1981.

Ford Foundation. *A Time to Choose*[:] *America's Energy Future.* Cambridge: MA: Ballinger Publishing Co., 1974.

Forsyth, Trudy L. "An Introduction to the Small Wind Turbine Project." Paper presented at Windpower '97, Austin, TX, June 15–18, 1997. In *Windpower '97 Proceedings.* Washington, DC: American Wind Energy Association, 1997.

Freese, Stanley. *Windmills and Millwrighting.* London: Cambridge University Press, 1957.

Galbraith, Kate, and Asher Price. *The Great Texas Wind Rush.* Austin: University of Texas Press, 2013.

Gillis, Christopher. *Offshore Windpower.* Atglen, PA: Schiffer Publishing Ltd., 2011.

———. *Windpower.* Atglen, PA: Schiffer Publishing Ltd., 2008.

Gillis, Christopher C. *Still Turning*[:] *A History of Aermotor Windmills.* College Station, TX: Texas A&M University Press, 2015.

Gipe, Paul. "Adopt an Anemometer—A Unique Anemometer Loan Program." Paper presented at Wind Energy Expo '82 and National Conference, Amarillo, TX, October 24–27, 1982. In the *Wind Energy Expo '82 and National Conference Proceedings.* Edited by Vaughn Nelson. Washington, DC: American Wind Energy Association, 1982.

———. *Wind Energy for the Rest of Us.* Bakersfield, CA: Wind-Works.Org, 2016.

———. *Wind Power for Home and Business.* White River Junction, VT: Chelsea Green Publishing Company, 1993.

Golding, E. W. *The Generation of Electricity by Wind Power.* New York: Philosophical Library, 1956.

Goodale, B. A. "Development and Testing of the Kaman

40 KW Wind Turbine Generator." Paper presented at Small Wind Turbine Systems 1981, Boulder, CO, May 12–14, 1981. In *Proceedings*[:] *Small Wind Turbine Systems 1981*. Edited by the Solar Energy Research Institute. [Golden, CO]: Solar Energy Research Institute, 1981.

———. "Development of a 40 KW Horizontal Axis Wind Turbine Generator." Paper presented at the American Wind Energy Association National Conference, Spring 1979, San Francisco, CA, April 16–19, 1979. In the *American Wind Energy Association National Conference, Spring 1979 Proceedings*. Edited by Vaughn Nelson. Washington, DC: American Wind Energy Association, 1979.

Gray, T. O. "Wind Power and the Utilities: A Wind Industry Perspective." Paper presented at the Rural Electric Wind Energy Workshop, Boulder, CO, June 1–3, 1982. In the *Rural Electric Wind Energy Workshop Proceedings*. Washington, DC: The Energy and Environmental Cooperative Association, National Rural Electric Cooperative Association, 1982.

Greenwood, Ernest. *Aladdin, U. S. A.* New York: Harper & Brothers Publishers, 1928.

Gregory, Roy. *The Industrial Windmill in Britain*. West Sussex: UK: Phillimore & Co., 2005.

Hackleman, Michael A. *The Homebuilt, Wind-Generated Electricity Handbook*. Culver City, CA: Peace Press, 1975.

———. *Wind and Windspinners*. Culver City, CA: Peace Press Printing and Publishing, 1977.

Hamilton, J. Roland. *Using Electricity on the Farm*. Englewood Cliffs, NJ: Prentice-Hall, 1959.

Haverson, Michael. *Persian Windmills*. Biblioteca Molinologica 10. Sprang Capelle, The Netherlands: The International Molinological Society, 1991.

Hawthorn, Fred W. and Robert W. Hawthorn. *Idlewild Farm, A Century of Progress*. Lake Mills, IA: Graphic Publishing Co., Inc., 1976.

Heronemus, William E. "The University of Massachusetts Wind Furnace Project: A Summary Statement, June 1975." Paper presented at Second Workshop on Wind Energy Conversion Systems, Washington, DC, June 9–11, 1975. In *Proceedings of the Second Workshop on Wind Energy Conversion Systems*. Edited by Frank R. Eldridge. [Washington, DC]: The MITRE Corporation, 1975.

Hirshberg, Gary. *The New Alchemy Water Pumping Windmill Book*. Andover, MA: Brick House Publishing Company, 1982.

Hock, S. M., and R. W. Thresher. "The Federal Advanced Wind Turbine Program." Paper presented at Windpower '91, Palm Springs, CA, September 24–27, 1991. In *Windpower '91 Proceedings*. Washington, DC: American Wind Energy Association, 1991.

Hong-Sen Yang. *Reconstruction Designs of Lost Ancient Chinese Machinery*. History of Mechanism and Machine Science 3. Dordrecht, The Netherlands: Springer, 2007.

Houston, Edwin J. *Electricity in Every-Day Life*. 3 vols. New York: P. F. Collier & Son, 1905.

Hughes, P. S., R. W. Sherwin, and H. M. Clews. "Development of Atlantic Orient's 15/50 Wind Turbine." Paper presented at Windpower '91, Palm Springs, CA, September 24–27, 1991. In *Windpower '91 Proceedings*. Washington, DC: American Wind Energy Association, 1991.

Hunt, V. Daniel. *Windpower*[:] *A Handbook on Wind Energy Conversion Systems*. New York: Van Nostrand Reinhold Company, 1981.

Hütter, Ulrich. "Some Extemporaneous Comments on Our Experiences with Towers for Wind Generators." Paper presented at the Wind Energy Conversion Systems Workshop, Washington, DC, June 11–13, 1973. In *Wind Energy Conversion Systems Workshop Proceedings*. Edited by Joseph M. Savino. Washington, DC: National Science Foundation, Research Applied to National Needs, and National Aeronautics and Space Administration, December 1973.

Interim Standards Subcommittee, American Wind Energy Association. *Draft Working Document*[,] *Interim Standards Subcommittee*[,] *AWEA Standards Program*[,] *February 7, 1981*. Washington, DC: American Wind Energy Association, 1981.

Jacobs, M. L. "The Use of Wind-Driven Generators as an External Source of Protective Currents." Paper presented at Cathodic Protection: A Symposium by the Electrochemical Society and the National Association of Corrosion Engineers, Pittsburgh, PA, December 1947.

Jacobs, Marcellus L. "Experience with Jacobs Wind-Driven Electric Generating Plant, 1931–1957." Paper presented at the Wind Energy Conversion Systems Workshop, Washington, DC, June 11–13, 1973. In *Wind Energy Conversion Systems Workshop Proceedings*. Edited by Joseph M. Savino. Washington, DC: National Science Foundation, Research Applied to National Needs, and National Aeronautics and Space Administration, December 1973.

Jenkins, Jennifer, Heather Rhoads-Weaver, Trudy Forsyth, Brent Summerville, Ruth Baranowski, and Britton Rife. *SMART Wind Roadmap: A Consensus-Based, Shared Vision* [,] *Sustainable Manufacturing, Advanced Research & Technology Plan for Distributed Wind*. Durango, CO: Distributed Wind Energy Association, 2016.

Jewett, Thomas E. "Design of the ESI-80/200. Paper presented at the Wind Energy Expo '83 and National Conference, San Francisco, CA, October 17–19, 1983. In *Wind Energy Expo '83 and National Conference Proceedings*. Washington, DC: American Wind Energy Association, 1983.

Jimenez, T., T. Forsyth, A. Huskey, I. Mendoza, K. Sinclair, and J. Smith. "Establishment of Small Wind Regional Test Centers." Paper presented at the 2011 National Solar Conference, Raleigh, NC, May 17–21, 2011. In the *2011 National Solar Conference Proceedings*. Boulder, CO: American Solar Energy Society, 2011.

Johansson, Mogens. *Present Condition of the Gedser Wind Turbine and a Cost Estimate of Refurbishing the Turbine for Test Purposes*. Lyngby, Denmark: Danske Elvaerkers Forenings Udredningsafdeling, 1977.

Kagin, Solomon S. *Buyers Guide to Wind Power*. Santa Rosa, CA: Real Gas & Electric Co., 1978.

Kealey, Edward J. *Harvesting the Air: Windmill Pioneers in Twelfth-Century England*. Berkeley: University of California Press, 1987.

Keller, Robert B. "Problems Relating to Small Wind Energy Conversion Systems as Seen by a Dealer/Installer." Paper presented at Small Wind Turbine Systems 1981, Boulder, CO, May 12–14, 1981. In *Proceedings[:] Small Wind Turbine Systems 1981*. Edited by the Solar Energy Research Institute. [Golden, CO]: Solar Energy Research Institute, 1981.

King, Samuel M., and Michael Duffy. "Analysis and Testing of a Tower Structure for Application to Wind Systems." Paper presented at the American Wind Energy Association National Conference, Amarillo, TX, March 1–5, 1978. In *Proceedings of the American Wind Energy Association National Conference Proceedings*. Edited by Vaughn Nelson. Canyon, TX: Alternative Energy Institute, West Texas State University, 1978.

Klimas, P. C. "Darrieus Wind Turbine Program at Sandia Laboratories." Paper presented at the Wind Energy Innovative Systems Conference, Colorado Springs, CO, May 23–25, 1979. In the *Wind Energy Innovative Systems Conference Proceedings*. Edited by Irwin E. Vas. Washington, DC: US Department of Energy, 1979.

Komor, Paul. *Renewable Energy Policy*. New York: Diebold Institute for Public Policy Studies, 2004.

Kutcher, Howard R. "Alcoa Vertical Axis Wind Turbines." Paper presented at the American Wind Energy Association National Conference, Pittsburgh, PA, June 8–11, 1980. In *American Wind Energy Association National Conference Proceedings*. Edited by Vaughn Nelson. Washington, DC: American Wind Energy Association, 1980.

Lilley, Art, and Mike Bergey. "Lessons Learned: Deployment of Wind Turbines in Indonesia." Paper presented at Windpower '94, Minneapolis, MN, May 10–13, 1994. In *Windpower '94 Proceedings*. Washington, DC: American Wind Energy Association, 1994.

Lynette, R. "Advanced 275 KW Wind Turbine." Paper presented at Windpower '91, Palm Springs, CA, September 24–27, 1991. In *Windpower '91 Proceedings*.

Washington, DC: American Wind Energy Association, 1991.

MacMullan, Barbara, and Henry G. duPont. "The Economics of Utility-Interconnected Residential Wind Turbines." Paper presented at Windpower '93, San Francisco, CA, July 12–16, 1993. In *Windpower '93 Proceedings*. Washington, DC: American Wind Energy Association, 1993.

McCardell, E. A., Sr. *The History of Winpower Mfg. Company (A Story of Diversification)*. Newton, IA: N.p., [ca. 1980]. Available in the Jasper County Historical Museum, Newton, IA.

McCardell, Ed. "History of Winpower Mfg. Company." In *A History of Newton, Iowa*. Edited by Larry Ray Hurto. Wolfe City, TX: Henington Publishing Company, 1982.

McColly, H. F., and Foster Buck. *Homemade Six-Volt Wind Electric Plants*. Circular 58. Fargo, ND: Agricultural Experimental Station, North Dakota Agricultural College, 1935.

McConnell, Robert D. "Giromill Overview." Paper presented at the Wind Energy Innovative Systems Conference, Colorado Springs, CO, May 23–25, 1979. In the *Wind Energy Innovative Systems Conference Proceedings*. Edited by Irwin E. Vas. Washington, DC: US Department of Energy, 1979.

McGuigan, Dermot. *Harnessing the Wind for Home Energy*. Charlotte, VT: Garden Way Publishing, 1978.

Marks, William, and Charles Coleman. *The History of Wind-Power on Martha's Vineyard*. N.p.: National Association of Wind-Power Resources, 1981.

Mayer, Don. "North Wind's 1 KW High Reliability WTG Program." Paper presented at the American Wind Energy Association National Conference, Amarillo, TX, March 1–5, 1978. In *American Wind Energy Association National Conference Proceedings*. Edited by Vaughn Nelson. Canyon, TX: Alternative Energy Institute, West Texas State University, 1978.

Mechanix Illustrated Budget WIndcharger. Wayne, NJ: CBS Publication, 1978.

Meroney, Robert N. "Wind Power Applications in Rural and Remote Areas." Paper presented at the ASCE Specialty Conference on Solar Energy and Remote Sensing, Fort Collins, CO, July 20–21, 1976. Washington, DC: American Society of Civil Engineers, 1976.

Meyer, H. "The Development of a 10 Meter Wind Turbine Generator." Paper presented at the 5th Annual Energy-Sources Technology Conference and Exhibition, New Orleans, LA, March 7–10, 1982. In the *International Wind Energy Symposium Proceedings*. New York: American Society of Mechanical Engineers, 1982.

Meyer, Hans. "The Use of Paper Honeycomb for Prototype Blade Construction for Small to Medium-Sized Wind Driven Generators." Paper presented at the Wind Energy Conversion Systems Workshop, Washington, DC, June 11–13, 1973. In *Wind Energy Conversion Systems Workshop Proceedings*. Edited by Joseph M. Savino. Washington, DC: National Science Foundation, Research Applied to National Needs, and National Aeronautics and Space Administration, December 1973.

The Mother Earth News Handbook. New York: Bantam Books, 1974.

Murrin, T. A. "UTRC 8 KW Wind Turbine Development." Paper presented at Small Wind Turbine Systems 1981, Boulder, CO, May 12–14, 1981. In *Proceedings*[:] *Small Wind Turbine Systems 1981*. Edited by the Solar Energy Research Institute. [Golden, CO]: Solar Energy Research Institute, 1981.

Naar, Jon. *The New Wind Power*. New York: Penguin Books, 1982.

National Electric Light Association. *National Electric Light Association Thirty-Sixth Convention*[,] *Commercial Sessions*[,] *Papers, Reports and Discussions*[,] *Chicago, Ill.*[,] *June 2–6, 1913*. New York: The James Kempster Printing Co., 1913.

National Rural Electric Cooperative. *The Next Greatest Thing*. Edited by Richard A. Pence. Silver Spring, MD: McArdle Printing Company, 1984.

Nearing, Helen, and Scott Nearing. *Living the Good Life*. Harborside, ME: Social Science Institute, 1954.

Needham, Joseph. *Science and Civilization in China*. Vol. 4. *Physics and Physical Technology, Pt. II: Mechanical Engineering*. London: Cambridge University Press, 1965.

Nelson, Vaughn. *Wind Energy and Wind Turbines*. Canyon, TX: Alternative Energy Initiative, West Texas A&M University, [ca. 1995].

———, Kenneth Starcher, and William Pinkerton. "AEI-Wind Test Center." Paper presented at the Wind Energy Expo '86 and National Conference, Cambridge, MA, September 1–3, 1986. In the *Wind Energy Expo '86 and National Conference Proceedings*. Washington, DC: American Wind Energy Association, 1986.

Noll, Richard B., Herman M. Drees, and Lea Nichols. "Development of the 1–2 KW Cycloturbine." Paper presented at the American Wind Energy Association National Conference, Amarillo, TX, March 1–5, 1978. In *American Wind Energy Association National Conference Proceedings*. Edited by Vaughn Nelson. Canyon, TX: Alternative Energy Institute, West Texas State University, 1978.

Park, Jack, and Dick Schwind. *Wind Power for Farms, Homes & Small Industry*. Mountain View, CA: Nielsen Engineering & Research, 1978.

———. *Wind Power for Farms, Homes & Small Industry*. Washington, DC: US Department of Energy, Division of Solar Technology, Federal Wind Energy Program, 1978.

Peery, David J. *Aircraft Structures*. New York: McGraw-Hill Book Company, 1949.

Performance Subcommittee, American Wind Energy Association. *W. E. C. S. Performance Standards Development Program*[:] *Initial Outline*. Washington, DC: American Wind Energy Association, 1980.

Pernoud, R. *Die Kreuzzuge in Augenseugenberichten*. Dusseldorf, Germany: Karl Rauch, 1961.

Phelps Light & Power Co., Rock Island, IL. "Phelps Pioneers the Way" [advertisement]. *Farm Light and Power Year Book* (New York) (1922): 27.

Potter, Andrey A. *Farm Motors*. New York: McGraw-Hill Book Company, 1917.

Powell, F. E. *Windmills and Wind Motors*[:] *How to Build and Run Them*. New York: Spon & Chamberlain, 1918.

Puthoff, Richard L. "100 KW Experimental Wind Turbine Generator Project." Paper presented at Second Workshop on Wind Energy Conversion Systems, Washington, DC, June 9–11, 1975. In *Proceedings of the Second Workshop on Wind Energy Conversion Systems*. Edited by Frank R. Eldridge. [Washington, DC]: The MITRE Corporation, 1975.

Putnam, Palmer Cosslett. *Power from the Wind*. New York: Van Nostrand Reinhold Company, 1948.

Quinn, William P. *The Saltworks of Historic Cape Cod: A Record of the Nineteenth Century Economic Boom in Barnstable County*. Orleans, MA: Parnassus Imprints, 1993.

Quistgaard, Therese. "The Experimental Windmills at Askov 1891–1903." In *Wind Power—The Danish Way*. Edited by Benny Christensen. Vejen, Denmark: The Poul la Cour Foundation, 2009.

Rader, Nancy, and Scott Hempling. *The Renewables Portfolio Standard*[:] *A Practical Guide*. Washington, DC: National Association of Regulatory Utility Commissioners, 2001.

Rankine, William John MacQuorn. *A Manual of the Steam Engine and Other Prime Movers*. London: Charles Griffin and Company, 1866.

Renzo, D. J., ed. *Wind Power*[:] *Recent Developments*. Park Ridge, N.J.: Noyes Data Corporation, 1979.

A Report on the Use of Windmills for the Generation of Electricity. Oxford University. Institute of Agricultural Engineering. Bulletin No. 1. Oxford, UK: The Clarendon Press, 1926. Available in the F. Hal Higgins Agricultural History Collection, University Library, University of California at Davis.

Rhoads-Weaver, H., and T. Forsyth. "Overcoming Technical and Market Barriers for Distributed Wind Applications: Reaching the Mainstream." Paper presented at the ASME 2006 International Solar Energy Conference, Denver, CO, July 8–13, 2006. In the *American Society of Mechanical Engineers 2006 International Solar Energy Conference Proceedings*.

New York: American Society of Mechanical Engineers, 2006.

Ridgeway, James. *The Last Play*[:] *The Struggle to Monopolize the World's Energy Resources*. New York: The New American Library, 1973.

Righter, Robert W. *Wind Energy in America*[:] *A History*. Norman: University of Oklahoma Press, 1996.

Rocky Mountain Institute, HOMER Energy, and CohnReznick Think Energy. *The Economics of Grid Defection*. Boulder, CO: Rocky Mountain Institute, 2014.

Roper, Stephen. *A Catechism of High Pressure or Non-Condensing Steam Engines*. Philadelphia: Edward Meeks, 1893.

Savino, Joseph M. "Introduction to Wind Energy Conversion Systems Technology." Paper presented at Second Workshop on Wind Energy Conversion Systems, Washington, DC, June 9–11, 1975. In *Proceedings of the Second Workshop on Wind Energy Conversion Systems*. Edited by Frank R. Eldridge. [Washington, DC]: The MITRE Corporation, 1975.

Savonius, Sigurd J. *The Wind-Rotor in Theory and Practice*. Helsingfors, Finland: Savonius & Co., [ca. 1925].

Schachle, Charles, and Robert L. Scheffler. "The Southern California Edison 3 MW Wind Turbine Generator (WTG) Demonstration Project." Paper presented at the American Wind Energy Association National Conference, Fall 78, Cape Cod, MA, September 25–27, 1978. In the *American Wind Energy Association National Conference, Fall 78 Proceedings*. Edited by Vaughn Nelson. Washington, DC: American Wind Energy Association, 1978.

Schaenzer, J. P. *Rural Electrification*. Milwaukee, WI: The Bruce Publishing Company, 1948.

Schaufelberger, M. *37 Years Experience with Elektro Windmills*. Winterthur, Switzerland: Elektro GmbH, [ca. 1985].

Schneider, Norman H. *Low Voltage Electric Lighting with the Storage Battery*. New York: Spon & Chamberlain, [ca. 1920].

Schroeder. "History of Electric Light." *Smithsonian Miscellaneous Collections* 76, no. 2 (August 15, 1923): i–xiii, 1–94.

Schurr, Sam H., Calvin C. Burwell, and Warren D. Devine Jr., and Sydney Sonenblum. *Electricity in the American Economy: Agent of Technological Progress*. Westport, CT: Greenwood Publishing, 1990.

Simmons, Daniel M. *Wind Power*. Park Ridge, NJ: Noyes Data Corporation, 1975.

Simonds, M. H., and A. Bodek. *Performance Test of a Savonius Rotor. Technical Report No. T10*[,] *January 1964*. Quebec, Canada: Brace Research Institute, McGill University, 1964.

Slattery, Harry. *Rural America Lights Up*. Washington, DC: National Home Library Foundation, 1940.

Smith, Beauchamp E. "Smith-Putnam Wind Turbine Experiment." Paper presented at the Wind Energy Conversion Systems Workshop, Washington, DC, June 11–13, 1973. In *Wind Energy Conversion Systems Workshop Proceedings*. Edited by Joseph M. Savino. Washington, DC: National Science Foundation, Research Applied to National Needs, and National Aeronautics and Space Administration, December 1973.

Soderholm, L. H. "Non-Interconnected Applications of Wind Energy." Paper presented at the Rural Electric Wind Energy Workshop, Boulder, CO, June 1–3, 1982. In the *Rural Electric Wind Energy Workshop Proceedings*. Washington, DC: The Energy and Environmental Cooperative Association, National Rural Electric Cooperative Association, 1982.

Solar Energy Industries Association. *US Solar Industry Year in Review 2008*. Washington, DC: Solar Energy Industries Association, 2009.

———. *US Solar Industry Year in Review 2009*. Washington, DC: Solar Energy Industries Association, 2010.

South, Peter, and Raj Rangi. "The Performance and Economics of the Vertical-Axis Wind Turbine Developed at the National Research Council, Ottawa, Canada." Paper presented at Annual Meeting of the Pacific Northwest Region of the American Society of Agricultural Engineers, Calgary, Canada, October 10–12, 1973. In the *Proceedings of the 1973 Meeting of the Pacific Northwest Region of the American Society of Agricultural Engineers*. St. Joseph, MI: American Society of Agricultural Engineers, 1973.

Spaulding, Allen P., Jr. "MP 1–200 Technical Description." Paper presented at the American Wind Energy Association National Conference, Fall 78, Cape Cod, MA, September 25–27, 1978. In the *American Wind Energy Association National Conference, Fall 78 Proceedings*. Edited by Vaughn Nelson. Washington, DC: American Wind Energy Association, 1978.

Sweeney, T. E. *The Princeton Windmill Program*[,] *AMS Report No. 1093*[,] *March, 1973*. Princeton, NJ: Princeton University, Department of Aerospace and Mechanical Sciences, 1973.

———. "Sailwing Windmill Technology." Paper presented at Second Workshop on Wind Energy Conversion Systems, Washington, DC, June 9–11, 1975. In *Proceedings of the Second Workshop on Wind Energy Conversion Systems*. Edited by Frank R. Eldridge. [Washington, DC]: The MITRE Corporation, 1975.

———, and W. B. Nixon. "An Introduction to the Princeton Sailwing Windmill." Paper presented at the Wind Energy Conversion Systems Workshop, Washington, DC, June 11–13, 1973. In *Wind Energy Conversion Systems Workshop Proceedings*. Edited by Joseph M. Savino. Washington, DC: National Science Foundation, Research Applied to National Needs, and National Aeronautics and Space Administration, December 1973.

Swisher, Randall, and Kevin Porter. "Renewable Policy Lessons from the US: The Need for Consistent and Stable Policies." In *Renewable Energy Policy and Politics: A Handbook for Decision-Making.* Edited by Karl Mallon. London: Earthscan, 2006.

Thomas, Ronald L. "Introduction to Large System Design." Paper presented at Second Workshop on Wind Energy Conversion Systems, Washington, DC, June 9–11, 1975. In *Proceedings of the Second Workshop on Wind Energy Conversion Systems.* Edited by Frank R. Eldridge. [Washington, DC]: The MITRE Corporation, 1975.

Thorndahl, Jytte. "Electricity and Wind Power for the Rural Areas 1903–1915." In *Wind Power—The Danish Way.* Edited by Benny Christensen. Vejen, Denmark: The Poul la Cour Foundation, 2009.

———. "Johannes Juul and the Birth of Modern Wind Turbines." In *Wind Power—The Danish Way.* Edited by Benny Christensen. Vejen, Denmark: The Poul la Cour Foundation, 2009.

———, and Benny Christensen. "Time for Survival and Development 1920–1945." In *Wind Power—The Danish Way.* Edited by Benny Christensen. Vejen, Denmark: The Poul la Cour Foundation, 2009.

Toepfer, Craig. *The Hybrid Electric Home.* Atglen, PA: Schiffer Publishing Ltd., 2010.

Torrey, Volta. *Wind-Catchers: American Windmills of Yesterday and Tomorrow.* Brattleboro, VT: Stephen Greene Press, 1976.

Tripp, Guy E. *Electric Development as an Aid to Agriculture.* New York: Knickerbocker Press, 1926.

US Department of Energy. *Home Wind Power.* Charlotte, VT: Garden Way Publishing, 1981.

Vance, W. "Vertical Axis Wind Rotors—Status and Potential." Paper presented at the Wind Energy Conversion Systems Workshop, Washington, DC, June 11–13, 1973. In *Wind Energy Conversion Systems Workshop Proceedings.* Edited by Joseph M. Savino. Washington, DC: National Science Foundation, Research Applied to National Needs, and National Aeronautics and Space Administration, December 1973.

Vargo, Donald J. *Wind Energy Developments in the 20th Century.* Cleveland, OH: Lewis Research Center, National Aeronautics and Space Administration, 1975.

Vick, Brian D., Byron Neal, and R. Nolan Clark. "Wind Powered Irrigation for Selected Crops in the Texas Panhandle and South Plains." Paper presented at Windpower 2001, Washington, DC, June 3–7, 2001. In *Windpower 2001 Proceedings.* Washington, DC: American Wind Energy Association, 2001.

———. Bryon A. Neal, and R. Nolan Clark. "Performance of a Small Wind Powered Water Pumping System."

Paper presented at Windpower 2008, Houston, TX, June 1–4, 2008. In *Windpower 2008 Proceedings.* Washington, DC: American Wind Energy Association, 2008.

———, and R. Nolan Clark. "Analysis of Wind Farm Energy Produced in the United States." Paper presented at Windpower 2007, Los Angeles, June 3–6, 2007. In *Windpower 2007 Proceedings.* Washington, DC: American Wind Energy Association, 2007.

———. "Performance and Economic Comparison of a Mechanical Windmill to a Wind-Electric Water Pumping System." Paper presented at 1997 ASAE (American Society of Agricultural Engineers) Annual International Meeting, Minneapolis, MN, August 10–13, 1997. In *Proceedings: 1997 ASAE Annual International Meeting.* St. Joseph, MI: American Society of Agricultural Engineers, 1997.

———. "Ten Years of Testing a 10 Kilowatt Wind-Electric System for Small Scale Irrigation." Paper presented at the 1998 ASAE (American Society of Agricultural Engineers) Annual International Meeting, Orlando, FL, July 12–16, 1998. In *Proceedings: 1998 ASAE Annual International Meeting.* St. Joseph, MI: American Society of Agricultural Engineers, 1998.

———. "Testing of a 2-Kilowatt Wind-Electric System for Water Pumping." Paper presented at Windpower 2000, Palm Springs, CA, April 30-May 4, 2000. In *Windpower 2000 Proceedings.* Washington, DC: American Wind Energy Association, 2000.

———, and Shitao Ling. "One and a Half Years of Field Testing a Wind-Electric System for Watering Cattle in the Texas Panhandle." Paper presented at Windpower 1999, Burlington, VT, June 20–23, 1999. In *Windpower 1999 Proceedings.* Washington, DC: American Wind Energy Association, 1999.

———, and Saiful Molla. "Performance of a 10 Kilowatt Wind-Electric Water Pumping System for Irrigating Crops." Paper presented at Windpower 1997, Austin, TX, June 15–18, 1997. In *Windpower 1997 Proceedings.* Washington, DC: American Wind Energy Association, 1997.

Vosburgh, Paul N. *Commercial Applications of Wind Power.* New York: Van Nostrand Reinhold Company, 1983.

———. "VAWTPOWER 185, Forecast Vertical Axis Wind Turbines." Paper presented at Wind Energy Expo '82 and National Conference, Amarillo, TX, October 24–27, 1982. In *Wind Energy Expo '82 and National Conference Proceedings.* Edited by Vaughn Nelson. Washington, DC: American Wind Energy Association, 1982.

Wegley, Harry L., and William T. Pennell. "Siting Small Wind Machines." Paper presented at the American Wind Energy Association National Conference, Fall 78. Cape Cod, MA, September 25–27, 1978. In the

American Wind Energy Association National Conference, Fall 78 Proceedings. Edited by Vaughn Nelson. Washington, DC: American Wind Energy Association, 1978.

Weis, Pat. "The Technical Information Dissemination Program for Wind Energy at SERI." Paper presented at the Panel on Information Dissemination for Wind Energy, Albuquerque, NM, August 2–3, 1979. In *Proceedings: Panel on Information Dissemination for Wind Energy.* Edited by Patricia Weis. Golden, CO: Solar Energy Research Institute, 1980.

"Wind Power Plants." *Farm Light and Power Year Book*[:] *Dealers' Catalog and Service.* New York: Farm Light and Power Co., Inc., 1922.

Wolfe, Ralph, and Peter Clegg. *Home Energy for the Eighties.* Charlotte, VT: Garden Way Publishing, 1979.

Wolff, Alfred R. *The Windmill as A Prime Mover.* 2nd ed. New York: John Wiley & Sons, 1900.

Wulff, Hans E. *The Traditional Crafts of Persia: Their Development, Technology, and Influence on Eastern and Western Civilization.* Cambridge, MA: MIT Press, 1966.

Articles

"3 MW Wind Turbine Slated for Southern California." *Wind Power Digest* no. 12 (Spring 1978): 26.

"The 11th Street Movement!" *Alternative Sources of Energy* no. 30 (February 1978): 31.

"$60 Million U.S. Program Helps Accelerate Use of Wind Power." *Canadian Renewable Energy News* (March 1979): 14.

"1978 Board of Directors." *AWEA Windletter* (Washington, DC) (June 1978): 4.

"The Acheval Wind Electronic Co.[:] A Profile." *Alternative Sources of Energy* no. 63 (September–October 1983): 44–46.

"AEA Introduces Amerenault Wind Turbine." *Wind Power Digest* no. 2 (Summer 1975): 31–32.

"Aerodynamic Wind Mills." *Scientific American* 140, no. 6 (June 1929): 525.

Ailworth, Erin, "The Race to Build a Wind Behemoth," *The Wall Street Journal* (weekend edition), August 25–26, 2018.

———, "Solar-Panel Makers See Daylight," *The Wall Street Journal* (weekend edition), May 12–13, 2018.

"All Day Service in Small Cities." *Popular Electricity* 2, no. 1 (May 1909): 23.

Allis-Chalmers Manufacturing Company, Milwaukee, WI. "A Complete Power Plant furnishing Light and Power to Small Communities or Large Farms" [advertisement]. *Farm Light and Power Year Book* (New York) (1922): 13.

Alsever, Jennifer, "A Nice Stiff Breeze, and a Nice Little Power Bill." *The New York Times*, April 1, 2007.

Altman, Alan. "'O$_2$ Powered Delight' Plans." *Alternate Sources of Energy* no. 14 (May 1974): 5.

"Amarillo Wind Expo '82[:] The Wind Industry Has Matured." *Wind Power Digest* no. 25 (Fall 1982): 48, 52.

"American Farm Light and Power Plants Purchased in India." *Farm Light and Power* (New York) 3, no. 7 (March 15, 1923): 18.

"The American Wind Energy Association," *The Mother Earth News* no. 30 (November 1974): 16.

"American Wind Energy Conference: April 10–11, 1975." *American Wind Energy Association Newsletter* (Detroit) (ca. March 1975), N.p.

"American Wind Turbine Revisited." *Wind Power Digest* no. 2 (Summer 1975): 13.

Anderson, Ann, "Plymouth Company Pioneers Hybrid Power Systems," *Plymouth Post* (Plymouth, MN), May 17, 1984.

"Anemometers and Recorders." In "The Solar Age Wind Products Supplement." *Solar Age Magazine* (1980).

"ANSI Approves AWEA as Standards-Making Body." *AWEA Wind Energy Weekly* (Washington, DC) 14, no. 639 (March 20, 1995): N.p.

"Are You Ready for State and County Fairs?" *Farm Light and Power* (New York) 2, no. 12 (August 15, 1922): 12.

Atchison, Chris, "Turbine Developer Feels the Wind at His Back," *The Globe and Mail* (Ontario, Canada), October 20, 2009.

"Atlantic Orient Plans Changes to Enertech E44 Wind Turbine." *AWEA Wind Energy Weekly* (Washington, DC) 10, no. 463 (1991): N.p.

"Atlantic Orient Plans Wind-Diesel Project in Alaska." *AWEA Wind Energy Weekly* (Washington, DC) 14, no. 637 (March 6, 1995): N.p.

Austvik, Ole Gunnar. The War over the Price of Oil: Oil and the Conflict on the Persian Gulf." *International Journal of Global Energy Issues* 5, nos. 2-3-4 (1993): 134–43.

"AWEA Asks Key Senators to Mull Production Incentives." *AWEA Wind Energy Weekly* (Washington, DC) 8, no. 348 (April 17, 1989): N.p.

"AWEA Board Developments." *AWEA Windletter* (Washington, DC) (June 1978): 7.

"AWEA Briefs Hill Staff on Wind and Developing World." *AWEA Wind Energy Weekly* (Washington, DC) 7, no. 310 (July 5, 1988): N.p.

"The AWEA Establishes an Office in Washington, D.C." *AWEA Windletter* (Washington, DC) (June 1978): 3.

"AWEA Joins Peers to Lobby for Tax Credit Extension." *World Wind* (Ottawa, Canada) 1, no. 2 (January 1985): 4.

"AWEA Members to Fund Legislative Push." *AWEA Wind Energy Weekly* (Washington, DC) 7, no. 327 (November 6, 1988): N.p.

"AWEA Reports to Members on Legislative Push." *AWEA Wind Energy Weekly* (Washington, DC) 8, no. 343 (March 12, 1989): N.p.

"AWEA Research." *AWEA Windletter* (Washington, DC) (February 1978): 10.

Bain, Don. "Small-Scale Wind and the Public Utility Regulatory Policies Act." *Wind Power Digest* no. 17 (Fall 1979): 5–6.

Baker, T. Lindsay. "Aermotor Windmill Towers." *Windmillers' Gazette* 11, no. 3 (Summer 1992): 2–14.

———. "Andrew J. Corcoran: Maker of America's Premiere Wooden-Wheel Windmills." *Windmillers' Gazette* 23, no. 2 (Spring 2004): 2–9.

———. "D. H. Bausman's Pennsylvania-Made Windmills." *Windmillers' Gazette* 6, no. 3 (Summer 1987): 6–7.

———. "'Every Farmer His Own Miller:' The Use of Power Windmills." *Windmillers' Gazette* 22, no. 4 (Autumn 2003): 2–7.

———. "Halladay & Wheeler's Patent Windmill." *Windmillers' Gazette* 6, no. 1 (Winter 1987): 8–10.

———. "Hard Times and Hard Feelings: The Untold Story of the Early 'Eclipse' Windmills," *Windmillers' Gazette* 19, no. 3 (Summer 2000): 2–9.

———. "How to Build a Wooden Windmill Tower." *Windmillers' Gazette* 6, no. 4 (Autumn 1987): 6–8.

———. "Iron Turbine." *Windmillers' Gazette* 7, no. 1 (Winter 1988): 2–4.

———. "Large-Diameter Halladay Standard Windmills." *Windmillers' Gazette* 30, no. 2 (Spring 2011): 2–6.

———. "An Overview of Horizontal Windmills." *Windmillers' Gazette* 18, no. 1 (Winter 1999): 2–7.

———. "Pioneer Metal Windmills of the Plains: The Kirkwood Iron Wind Engines." *Windmillers' Gazette* 14, no. 1 (Winter 1995): 2–6.

———. "Power Aermotor Windmills." *Windmillers' Gazette* 30, no. 3 (Summer 2011): 2–13.

———. "Power Windmills." *Windmillers' Gazette* 6, no. 1 (Winter 1987): 3–7.

———. "A Product History of the U. S. Wind Engine and Pump Company." *Windmillers' Gazette* 2, no. 1 (Winter 1983): 5–9.

———. "Steel Windmills versus Wooden: A War of Words." *Windmillers' Gazette* 7, no. 1 (Winter 1988): 8–10.

———. "'Turning Fair to the Wind'[:] An Occupational Vocabulary of Windmilling." *Windmillers' Gazette* 15, no. 4 (Autumn 1996): 2–8.

———. "Wind Electric News: The Papers of Oliver P. Fritchle." *Windmillers' Gazette* 10, no. 4 (Autumn 1991): 9–10.

———. "Windmill Museums as Destinations for Heritage Tourists." *Windmillers' Gazette* 31, no. 2 (Spring 2012): 2–9.

———. "Windmill Towers and Their Infinite Variety." *Windmillers' Gazette* 3, no. 1 (Winter 1984): 3–6.

———. "Windmills and Railroad Water Systems." *Windmillers' Gazette* 28, no. 2 (Spring 2009): 2–5.

———. "Windmills with Variable-Pitch Blades." *Windmillers' Gazette* 8, no. 4 (Autumn 1989): 2–7.

Balaskovitz, Andy, "Microgrid Boosters Hope Michigan 'Energy District' Will Spur More Interest," *Energy News Network* (St. Paul, MN), March 13, 2019.

Barkett, C. J. "Infringement—Exclusive Federal Court Jurisdiction over Patent Infringement Suits Does Not Preclude State Court from Exercising Jurisdiction over Due Process Takings and Conversion Claims Brought by Patent Holder against State." *The United States Law Week* 62 (October 26, 1993): 2241.

Barron, Rachel, "They Call the Small-Wind Firm Mariah," *Green Media Tech* (Mumbai, India), April 16, 2008.

Bates, Putnam A. "Farm Electric Lighting by Wind Power." *Scientific American* 107, no. 13 (September 28, 1912): 262.

"Battery, Switchboard and Resistance of a Wind-Driven Plant." *Farm Light and Power* (New York) 2, no. 12 (August 15, 1922): 32.

Beiter, Philipp, Joseph T. Rand, Joachim Seel, Eric Lantz, Patrick Gilman, and Ryan Wiser. "Expert Perspectives on the Wind Plant of the Future." *Wind Energy.* DOI: 10.1002/we.2735 (2022): 1-16. Accessed May 23, 2022. Onlinelibrary.wiley.com/doi/epdf/10.1002/we.2735

Bell, Benjamin. "Synchronous Wind Machines." *Solar Age* 8, no. 10 (October 1983): 36–41.

Belson, Ken, and David W. Dunlap, "Architects and Engineers Express Doubts about Bloomberg's Windmill Proposal," *The New York Times*, August 21, 2008.

Bergerson, Roger. "40 Years with Jacobs' Gems." *Public Power* 37, no. 5 (September–October 1979): 19.

Bergey, Karl H. "Wind Power Potential for the United States." *AWARE Magazine* (Madison, WI) no. 49 (October 1974): 2–5.

Bergey, Mike. "Wind Power in Oklahoma." *Wind Power Digest* no. 13 (Fall 1978): 8–10.

"Bergey Ships Nine Wind Pumping Systems to Indonesia." *AWEA Wind Energy Weekly* (Washington, DC) 14, no. 674 (November 27, 1995): N.p.

"Bergey Turbines in Brazil under World Bank Project." *AWEA Wind Energy Weekly* (Washington, DC) 15, no. 679 (January 8, 1996): N.p.

"Bergey Wind Turbines—Quality-Performance-Value" [advertisement]. *Home Power* (Ashland, OR) no. 57 (February-March 1997): 61.

"Best of What's New 2006[,] Home Tech[,] Skystream 3.7 Wind Power for Everyone," *Popular Science* 269, no. 6 (December 2006): 44.

"Best Inventions of 2006[,] The Home[,] Breezy Alternative," *Time* (2006). Accessed February 7, 2019. http://content.time.com/time/specials/packages/article/0,28804,1939342_1939395_1939670,00.html.

Betz, A. "The Maximum of the Theoretically Possible

Exploitation of Wind by Means of a Wind Motor." *Wind Engineering* 37, no. 4 (2013): 441–46.

Biello, David, "Where Did the Carter White House's Solar Panels Go?" *Scientific American*, August 6, 2010. Accessed December 24, 2017. http://wwwscientific american.com/article/carter-white-house-solar-panel-array/.

Biorn, Paul. "Recovering a Jacobs Wind Electric." *Wind Power Digest* 1, no. 3 (December 1975): 19–22.

Birkland, Sandy Jones. "Replica Russian Windmill Gifted to Fort Bliss: A Symbol of Peace Linking Two Nations." *Old Mill News* 40, no. 4 (Fall 2012): 15–17.

Block, Melissa, "Big Dreams for Small Wind Turbines," *National Public Radio*, August 27, 2009.

Bogart, Larry. "The Most Important Thing in the World." *Alternative Sources of Energy* no. 12 (October-November 1973): 15–16.

Bowman, Peyton G., III, Carl W. Ulrich, et al. "Report of the Committee on Cogeneration and Small Power Production Facilities." *Energy Law Journal* 4, no. 2 (1983): 279–88.

"Bridge Power," *San Francisco Chronicle*, September 10, 1983.

"Brilliant Street Lighting in Warren [, Ohio]." *Popular Electricity* 4, no. 5 (September 1911): 432–33.

Broad, William J., "Small, Cheap Windmills Outdo High Tech," *The New York Times*, August 14, 1984.

Buckwalter, Len. "Watts from the Wind." *Mechanix Illustrated* 71, no. 562 (March 1975): 40–41, 96.

Bumke, David. "Success Usually Doesn't Come Easy. People Who Wind with Wind Power." *Rodale's New Shelter* 4, no. 4 (April 1983): 44–52.

Buntjer, Julie, "Rural Landowners Air Opposition to Industrial Wind Farms," *Globe* (Worthington, MN), February 27, 2019.

Burnside, Diane. "Sound of the Wind." *Wind Power Digest* 1, no. 6 (September 1976): 28–30.

"Bush Signs Major Revision of Anti-Pollution Law," *The New York Times*, November 16, 1990.

California Electric Light Co., "The Brush Electric Light" [advertisement]. *The Pacific Rural Press* (September 9, 1882): 196.

"California May Allocate 15 Mil for State Wind Program." *Wind Power Digest* no. 11 (Winter 1977–1978): 29.

"California Utility to Test 3-MW Wind Turbine." *Wind Energy Report* (July 1978): 1–2.

Callahan, Elizabeth H. "Electricity in the Household[:] More Comfort in the Home." *Popular Electricity* 1, no. 2 (June 1908): 110–12.

Carlill, James. "On a New Form of Wind Power." Edited by Harold Cox. *The Edinburgh Review or Critical Journal* (Norwich, England) 228, no. 466 (October 1918): 343–54.

Carlin, P. W., A. S. Laxson, and E. B. Muljadi. "The History and State of the Art of Variable-Speed Wind Turbine Technology." *Wind Energy* 6 (2003): 129–59.

Carlisle, C. A., Jr. "How to Sell Wind Driven Light and Power Plants." *Farm Light and Power* (New York) 2, no. 8 (April 15, 1922): 28, 40.

Carter, Jay, Jr. Letter to the editor. *AWEA Windletter* (Washington, DC) (April 1980): 6.

Carter, Joe. "At Home with Martin Jopp[,] Wind Power Pioneer." *Wind Power Digest* no. 8 (Spring 1977): 6–8.

———. "Earl Rich's Big Windmill: Wind Power in N.H." *Wind Power Digest* no. 7 (December 1976): 27–30.

———. "New Machines . . . Wind Power Systems' Storm Master." *Wind Power Digest* no. 13 (Fall 1978): 34–35.

———. "Testing an Elektro-Gemini Wind System." *Wind Power Digest* 1, no. 6 (September 1976): 13–17.

———. "USDA's Wind Testing Program." *Wind Power Digest* no. 8 (Spring 1977): 38–40.

———. "VAWT Research at Sandia Labs." *Wind Power Digest* no. 9 (Summer 1977): 42–45.

———. "A Visit with Zephyr Wind Dynamo." *Wind Power Digest* no. 7 (December 1976): 14–17.

———, and Ted Finch. "Wind & Solar at East 11th." *Wind Power Digest* no. 9 (Summer 1977): 12–17.

"CDC Taking Charge of Jacobs Wind Electric." *Wind Energy Report* 6, no. 1 (January 14, 1983): 7.

"Centennial Improvements at Point Lookout, Md." *Lighthouse Service Bulletin* (Washington, DC) 4, no. 4 (April 1, 1930): 17.

Chapin, F. Stuart. "How We Waste Our Coal." *Popular Electricity Magazine* 6, no. 4 (August 1913): 366–67.

"Charging Windmills," *The Tribune* (Oakland, CA), September 11, 1983.

Chediak, Mark, and Prashant Gopal, "California Is First State to Mandate Solar Power," *Bloomberg* (in *The Frederick News-Post*, Frederick, MD), May 12, 2018.

Cheslow, Danielle, "Pittsburgh's Microgrids Technology Could Lead the Way for Green Energy," *NPR*, November 12, 2017.

J. G. Childs & Co., Ltd. "Electricity from the Wind" [advertisement]. *The Implement and Machinery Review* (London) 35, no. 410 (June 1, 1909): 248.

Church, Albert Cook. "The Padanaram Salt Works." *New England Magazine* 41, no. 2 (October 1909): 489–92.

"A City Electrical." *Popular Electricity* 1, no. 2 (June 1908): 77.

Clymer, Adam, "Gas Shortage Spurs Carter Decline in Poll," *The New York Times*, July 13, 1979.

Coffel, Steve. "Photovoltaics[:] Electricity from Sunshine." *Alternative Sources of Energy* no. 42 (March-April 1980): 3–7.

Cohn, Lisa. "Prospecting for Wind Pits Man Against Machine." *Windpower Monthly* (Knebel, Denmark) 15, no. 6 (June 1999): 42–43.

Cole, Cyndy. "Southwest Windpower gets $10M Investment," *Arizona Daily Sun*, April 7, 2009.

Comis, Don. "Wind Power Where You Want It." *Agricultural Research* (US Department of Agriculture, Washington, DC) 42, no. 6 (June 1994): 4–7.

"Comments by Charlie Wright." *Wind Power Digest* no. 2 (Summer 1975): 25–26.

"Competition Favors Renewables, says NIMO Vice President." *AWEA Wind Energy Weekly* (Washington, DC) 13, no. 594 (April 25, 1994): N.p.

Cook, Harold. "Water Stop." *Railroad Magazine* 66, no. 6 (October 1955): 12–23.

"Coping: Showing the Way to a Gas-Short Future." *Auto-Week* (May 14, 1979): 2.

Cornwall, Michael. "FloWind Begins Operation of Turbine in Washington State." *Sun*Up Energy News Digest* (Yucca Valley, CA) (September 1982): 34.

Cosgrove, R. C. "Developing the Farm Electric Industry." *Farm Light and Power* (New York) 3, no. 10 (June 15, 1923): 9–10, 29–30.

Cowan, Edward, "A Year of Costly Oil and Abrupt Changes," *The New York Times*, October 13, 1974.

"CPUC Restructuring Decision Includes Portfolio Standard[,] Order Calls for Partial Transition to Direct Access by 1998, Complete Phase-in by 2003." *AWEA Wind Energy Weekly* (Washington, DC) 14, no. 678 (December 25, 1995): N.p.

Crowley, C. A. "Wind-Driven Generator Charges Batteries." *Popular Mechanics* 70, no. 1 (July 1938): 146–51.

"Current from the Wind." *Popular Electricity* 4, no. 5 (September 1911): 429.

"DAF Darrius [sic.] Marketed." *Wind Power Digest* no. 5 (Summer 1976): 41.

Dare, Jerry. "Isolated Lighting Plants." *Everyday Engineering Magazine* (New York) 9, no. 2 (May 1920): 105–7.

"Darrieus-Silo Demonstration Project Now Underway." *Wind Power Digest* no. 14 (Winter 1978): 18–19.

Davidson, Ros. "Remote Area Potential Proving Attractive." *Wind Power Monthly* (Knebel, Denmark) 6, no. 9 (September 1990): 18.

Daysog, Rick, "A Planned Wind Farm Is Getting Opposition from an Unusual Source: Environmentalists," *Hawaii News Now* (Honolulu, HI), February 11, 2019.

DeKorne, James B. "The Answer Is Blowin' in the Wind." *The Mother Earth News* no. 24 (November 1973): 67–75.

"Design of 'Aerolite' Plant." *Farm Light and Power*[:] *Dealers' Service Book*. New York: Farm Light and Power Publishing Co., Inc., [ca. 1923].

"Designing a High Reliability Wind Machine—How One Company Did it." *Wind Power Digest* no. 25 (Fall 1982): 6–13.

"Details of Fritchle Attachable Unit." *Farm Light and Power*[:] *Dealers' Service Book*. New York: Farm Light and Power Publishing Co., [ca. 1923].

Dick, Everett. "Water: A Frontier Problem." *Nebraska History* 49, no. 3 (Autumn 1968): 215–45.

Diddlebock, Bob. "Small Wind Turbines[:] Blowing Hot." Special Section[:] Global Business/Small Business. *Time* 176, no. 2 (July 12, 2010): 10.

"DOE Contract Announced." *Wind Power Digest* no. 12 (Spring 1978): 28.

"DOE Contracts Awarded." *Wind Power Digest* no. 11 (Winter 1977–1978): 26–27.

"DOE Issues Three Contracts for 1-KW Systems." *AWEA Newsletter* (Washington, DC) (June 1978): 3.

"DOE Seeks Comment on Distributed Energy Generation Research Plans." *AWEA Wind Energy Weekly* (Washington, DC) 11, no. 499 (May 18, 1992): N.p.

Dolbear, A. E. "Moses Gerrish Farmer." *Proceedings of the American Academy of Arts and Sciences* 29 (May 1893—May 1894): 415–18.

Dole, F. L. "History and Development of the Windmill." *Export Implement Age* (Philadelphia) 14, no. 5 (August 1906): 27–28, 30.

"Domestic Electric Light Plant Driven by a Windmill." *Scientific American* 96, no. 22 (June 1, 1907): 448.

The Domestic Engineering Company, Dayton, OH. "Delco-Light[:] Electricity for Every Farm" [advertisement]. *The Saturday Evening Post* (Philadelphia) 189, no. 11 (September 9, 1916): 60–61.

Donn, Jeff, "Town Has Been Testbed for Wind Power," *Associated Press*, June 14, 2004.

Drees, Herman M. "Notes from the Editor." *Wind Technology Journal* 1, no. 1 (Spring 1977): 2.

Dunham, Bob. "Pinson[:] An Alternate Solution." *Wind Power Digest* no. 19 (Spring 1980): 50–52.

Dwyer, Jim. "Remembering a City Where the Smog Could Kill," *The New York Times*, February 28, 2017.

"The Early Work of Mr. Brush." *Gas Power* (St. Joseph, MI) 4, no. 6 (November 1906): 20.

Edwards, Frank B. "Windsteading into the Eighties." *Harrowsmith* (Ontario, Canada) 4, no. 22 (September 1979): 27–33, 75–79.

Eggleton, Claude W. "In Memorial . . . Marcellus Jacobs (1903–1985)." *Wind Power Digest* no. 27 (1985): 3.

Eisenberg, Anne, "Bringing Wind Turbines to Ordinary Rooftops," *The New York Times*, February 14, 2009.

Ekblaw, K. J. T. "Wind-Drive Power Plants." *The Farm Journal* (Philadelphia) 46, no. 8 (August 1922): 6–7.

Eldridge, Frank R. "A Preliminary Federal Commercialization Plan for WEC's." *AWEA Newsletter* (Bristol, IN) (Spring 1977): 14–19.

"Electric Farm Lighting." *Popular Electricity* 5, no. 5 (September 1912): 431.

"An Electric Laundry Washer." *Popular Electricity* 1, no. 2 (June 1908): 115.

"Electric Light the Safest in the Home." *Popular Electricity* 1, no. 10 (February 1909): 647.

"Electric Light Companies and the People." *Popular Electricity* 2, no. 9 (January 1910): 594.

"An Electric Light Plant Operated by a Wind Mill." *The Farm Implement News* (Chicago) 18, no. 3 (January 21, 1897): 46.

"Electric Lighting by Wind Power." *The Farm Implement News* (Chicago) 20, no. 17 (April 27, 1899): 16.

"The Electric Motor in the Home." *Popular Electricity* 1, no. 4 (August 1908): 250–51.

"Electric Signs Add Much to the Brilliance and Attractiveness of the Streets of Cities and Villages." *Popular Electricity* 1, no. 10 (February 1909): 619.

"An Electrical Fan the Year Around." *Popular Electricity* 3, no. 1 (May 1910): 53.

"Electrical Notes. Central Station Driven by Electric Power." *Scientific American Supplement* 56, no. 1459 (December 19, 1903): 23386.

"Electrical Notes. The Problem of Using Wind Power." *Scientific American Supplement* 57, no. 1469 (February 27, 1904): 23546.

"The Electrical Value of Wind Power." *Scientific American* 93, no. 21 (November 18, 1905): 394–95.

"Electricity, the Busiest Worker." *Popular Electricity* 2, no. 7 (November 1909): 462–64.

"Electricity—the Farm Hand." *Popular Electricity* 3, no. 12 (April 1911): 1056–61.

"Electricity Enhances Farm Values." *Popular Electricity* 4, no. 3 (July 1911): 230–31.

"Electricity from Wind," *Muskogee Phoenix* (Muskogee, Indian Territory), April 14, 1898.

"Electricity from the Wind." *Farm Mechanics* 17, no. 4 (August 1927): 25–27.

"Electricity in the Household[:] The Latest Electric Household Conveniences." *Popular Electricity* 1, no. 6 (October 1908): 382–84.

"Electricity in the Household[:] Sanitary Refrigeration for the Home." Popular Electricity 1, no. 7 (November 1908): 446–47.

"Electricity in the Household[:] Speaking of Wash Day." *Popular Electricity* 3, no. 2 (June 1910): 144–45.

"Electricity in the Household[:] The Wonders of 'Electric Shop.'" *Popular Electricity* 1, no. 12 (April 1909): 780–85.

"Electricity on the Farm," *Popular Electricity* 1, no. 4 (August 1908): 216–20.

"Electricity-Producing Windmills Generate Hope," *The Standard-Times* (New Bedford, MA), September 5, 1979.

"Electricity Supply for Small Communities." *Popular Electricity Magazine* 6, no. 3 (July 1913): 258.

Eller, Donnelle, "Neighbors in Eastern Iowa Fight to Bring Down Turbines—and Win," *Des Moines Register* (Des Moines, IA), November 21, 2018.

Ellis, Jonathan, and Cody Winchester, "Sales of Wind Turbines for Home Use Are Going Strong," *USA Today*, June 29, 2011.

"Engineering Notes. In a Lecture Recently Delivered at Copenhagen." *Scientific American Supplement* 46, no. 1185 (September 17, 1898): 18997.

Erdman, Jim. "Letters[:] 32 Volt Bulbs." *Alternative Sources of Energy* no. 21 (June 1976): 44.

"The Eustis Turbine Windmill." *Popular Electricity Magazine* 5, no. 10 (February 1913): 1033.

Evans, Michael. "Rocky Flats Small Wind System Tests[,] Manufacturers Prove Your Specs!" *Wind Power Digest* no. 24 (Summer 1982): 36–45.

Evans, Mike. "Interview with Rick Katzenberg." *Wind Power Digest* no. 13 (Fall 1978): 4–7.

———. "Understanding the P.U. R. P. A. Puzzle." *Wind Power Digest* no. 24 (Summer 1982): 16–17.

———. "Wind in Rural America." *Wind Power Digest* no. 25 (Fall 1982): 28–32.

Fairbanks, Morse & Co., Chicago. "What Our Selling Franchise Means to You" [advertisement]. *Farm Light and Power Year Book* (New York) (1922): 15.

"Fall Meeting Set[:} Warren, VT Chosen as Wind Exposition Site." *AWEA Newsletter* (Detroit) (September 1976): 3–4.

"Famous Wind Wheel Passes," *Cleveland Plain Dealer*, July 9, 1907.

"The Farmer's Light and Power." *Popular Electricity* 3, no. 5 (September 1910): 373–75.

Farrell, John, "Why Aren't Rural Electric Cooperatives Champions of Local Clean Power?" *Renewable Energy World.Com* (Tulsa, OK), December 3, 2014.

Farwell, Frank, "New Design Spurs a Resurgence of Windmills," *The New York Times*, February 3, 1980.

Fawcett, Waldon. "Electrical Invention and a Larger Patent Office." *Popular Electricity* 5, no. 7 (November 1912): 630–32.

Fawcet[t], Waldon. "The Pathfinders of the Wires." *Popular Electricity* 5, no. 5 (September 1912): 428–31.

"The Federal Wind Energy Program." *AWEA Newsletter* (Detroit) (September 1976): 10–11.

"The Federal Wind Program." *Wind Power Digest* no. 13 (Fall 1978): 40–43.

"Feedback: Marcellus Jacobs." *Wind Power Digest* no. 9 (Summer 1977): 5.

"FERC Deregulation Ignores Environmental Costs: AWEA." *AWEA Wind Energy Weekly* (Washington, DC) 7, no. 314 (July 30, 1988): N.p.

"FloWind Dedication." *Wind Power Digest* no. 24 (Summer 1982): 61.

"The Fuel Crisis—Nixon Acts." *Newsweek* (December 1973): 24–25.

Fleming, G. A. "The New Servant Girl." *Popular Electricity* 1, no. 2 (June 1908): 114.

Fletcher, Michael A., "Obama Leaves D. C. to Sign Stimulus

Bill[,] Renewable Energy a Focal Point in Denver as $787 Billion Effort is Made Law," *The Washington Post*, February 18, 2009.

Forrest, J. F. "A Practical Windmill Electric Plant." *Popular Electricity Magazine* 6, no. 1 (May 1913): 58–59.

Francis, Barbara. "Sandia Darrieus Technology Advances." *Wind Power Digest* no. 19 (Spring 1980): 27–29.

Frank, Ellen Perley. "Breaking the Energy Impasse[:] A Small Massachusetts Puts Windpower in the Marketplace." *New Roots* (Greenfield, MA) no. 7 (September–October 1979): 30–32.

Freeman, Harley. "Wind-Driven Power Plants You Can Build." *Popular Mechanics* 57, no. 6 (June 1932): 1043–45.

"From the Editor: Saddam Hussein's Timely Reminder." *Windpower Monthly* (Knebel, Denmark) 6, no. 9 (September 1990): 10.

"Functions and Methods of Trade Schools." *Popular Electricity* 4, no. 8 (December 1911): 699–704.

"The Future of the Windmill." *Scientific American* 108, no. 4 (April 5, 1913): 309.

Gailbrath, Kate, "Accessing the Value of Small Wind Turbines," *The New York Times*, September 3, 2008.

———, "Homeowners and Business Embracing Small Wind Turbines," *The New York Times*, October 2, 2011.

———, "Museum Shows History and Power of Wind Energy," *The New York Times*, July 30, 2011.

Gearino, Dan, "AEP Cancels Nations Largest Wind Farm: 3 Challenges Wind Catcher Faced," *Inside Climate News* (Brooklyn, NY), July 30, 2018.

"Generating Electricity with Windmills." *The Implement and Machinery Review* (London) 13, no. 150 (October 1, 1887): 9888.

"German Wind Power Plant." *Popular Electricity* 6, no. 7 (November 1913): 799.

Gery, Michael. "The New Alchemists: Creating a Gentle Science to Heal the Earth." *New Roots* (Greenfield, MA) (November–December 1979): 20–26.

Gillis, Chris. "Shake, Rattle and Roll." *American Shipper* (Jacksonville, FL) 52, no. 11 (November 2010): 58–62.

———. "Tapping Renewable Energies." *American Shipper* (Jacksonville, FL) 50, no. 6 (June 2008): 90–91.

Gillis, Christopher. "D. H. Bausman: A Pennsylvania Windmill Maker." *Windmillers' Gazette* 36, no. 1 (Winter 2017): 2–4.

———. "Sea Breezes to Salt." *Windmillers' Gazette* 32, no. 2 (Spring 2013): 9–10.

Gilmore, C. P. "What's New?[:] Blowing in the Wind." *Popular Science* 222, no. 4 (April 1983): 77.

Gipe, Paul. "Alcoa Soon to Dominate VAWT Market." *Wind Power Digest* no. 19 (Spring 1980): 20–25.

———. "Clowning Around at Dorny [sic.] Park." *Wind Power Digest* no. 15 (Spring 1979): 39–43.

———. "Dakota Sun & Wind[:] Making a Good Machine Better." *Wind Power Digest* no. 15 (Spring 1979): 13–14.

———. "The Mehrkams Re-visited[:] A 3 KW Unit This Time." *Wind Power Digest* no. 10 (Fall 1977): 13–17.

———. "Mehrkam's Windmills—He's Overpowering NASA." *Popular Science* 214, no. 4 (April 1979): 82–83.

———. "PURPA: A New Law Helps Make Small-Scale Power Production Profitable." *Sierra* 66 (November/December 1981): 52–55.

———. "USDA: Defining Wind's Role in Agriculture[,] Applied Wind Energy Research Continues." *Wind Power Digest* no. 18 (Winter 1979–1980): 14–19.

———. "VAWT's[:] A Brief Introduction." *Wind Power Digest* no. 19 (Spring 1980): 9–11.

———. "Wind and Utilities[:] Alcoa Markets VAWT's," *Wind Power Digest* no. 17 (Fall 1979): 29–32.

———. "Wind in America: The Changing Landscape." *Solar Age* 8, no. 10 (October 1983): 30–35.

———. "Wind Activists in the Commonwealth." *Wind Power Digest* no. 9 (Summer 1977): 32–35.

———. "Wind Catchers[:] Machines that Turn Wind into Electricity." *New Roots* (Greenfield, MA), no. 13 (Harvest 1981): 12–14.

———. "Wind Power in PA: Terry Mehrkam's Wind Machine." *Wind Power Digest* no. 9 (Summer 1977): 27–31.

———. "Windmills: Still Primed for Pumping Water." *Independent Energy* 19, no. 6 (July–August 1989): 62–64.

Gipe, Paul Bruce, "Tomorrow's Wind Turbines Being Built Today in Pennsylvania." *Alternative Sources of Energy*, no. 29 (December 1977): 5–10.

"Giromill Testing Underway." *Wind Power Digest* no. 21 (Winter 1980–1981): 10–12.

"Good Openings for Wind Engines." *The Implement and Machinery Review* (London) 32, no. 378 (October 2, 1906): 662–63.

Green, Wade, "The New Alchemists," *The New York Times*, August 8, 1976.

"Green Energy Technologies Introduces WindCube," *North American Windpower* (Southbury, CT), May 14, 2009.

Gross, Sharon P. "Energy Conservation—A New Factor in Certain Utility Rate Regulation." *Natural Resources Journal* 24 (Spring 1984): 487–96.

Grosvenor, Melville Bell. "Admiral of the Ends of the Earth." *The National Geographic Magazine* (Washington, DC) 112, no. 1 (July 1957): 36–48.

Gumpert, David. "The New Pioneers: Eliot and Sue Coleman find 'Homesteading' Is Satisfying Way of Life." *The Mother Earth News* no. 11 (September 1971): 38–41.

Hackleman, Michael. "More about the S-Rotor." *The Mother Earth News* no. 27 (May 1974): 39.

———. "The Savonius Super Rotor!" *The Mother Earth News* no. 26 (March 1974): 78–80.

Haislip, Alexander, "Startup Raises $2 Million for Small-Scale Wind Turbines," *Reuters*, May 12, 2009.

Hall, Lindsay, "Genoa Township Plans to Tear Down 3-Year-Old Wind Turbines," *Michigan Radio/National Public Radio*, July 16, 2013.

Hamilton, Roger. "Can We Harness the Wind?" *National Geographic* 148, no. 6 (December 1975): 812–28.

Hannaford, T. B. "To the Editor[:] Mr. Hannaford's Lighthouse." *New Zealand Herald* 27, no. 8232 (April 17, 1890).

"Harnessing the Wind." *New Hampshire Magazine* 8, no. 1 (October 1979): 7–8.

Harper, Ed. "Problems with a 6 KW Elektro." *Wind Power Digest* 1, no. 5 (June 1976): 11–14.

Hartwig, Jim. "PURPA Prevails." *Soft Energy Notes* (San Francisco) 5 (September–October 1982): 106.

Haugh, Brad, and Carol Hoerig. "Fortune!!![:] In Wind Energy, 40 Percent Tax Break, Mr. Hoyler." *Wind Power Digest* no. 25 (Fall 1982): 16–17.

Hayward, Charles B. "A Wind Electric Plant with Novel Features." *Farm Light and Power* (New York) 2, no. 12 (August 15, 1922): 20–21, 43.

"Hazardous Line Construction." *Popular Electricity* 2, no. 12 (April 1910): 782.

"He Waited 10 Years for It." *Rural Electrification News* 3, no. 1 (September 1937): 11–12.

Helmholz, George, and Larry Wheat. "22 Ft Wind Generator." *Alternative Sources of Energy* no. 18 (July 1975): 22–23.

Henry, Tom, "Wind Farms a Cash Cow for Communities, but Not Everyone's Sold," *The Blade* (Toledo, OH), April 7, 2019.

"Henry Clews' Miraculous Wind-Powered Homestead." *The Mother Earth News* no. 18 (November 1972): 25.

"Henry Ford Buys Aerolectric Plant." *Farm Light and Power* (New York) 4, no. 2 (October 15, 1923): 58.

Heronomous [sic.], William. "The U. Mass. Wind Furnace Program." *Wind Power Digest* 1, no. 7 (December 1976): 34–38.

Herts, Kenneth. "Energy Task Force[:] Rebuilding the Inner City with Affordable Sources of Energy." *Living Alternatives Magazine* 1, no. 6 (February 1980): 18–22.

Hirshberg, Gary, Earle Barnhart, Tyrone Cashman, and Bill Sheperdson. "Wind & Solar at New Alchemy." *Wind Power Digest* no. 11 (Winter 1977–1978): 16–25.

Hiscock, G. D. "Possibilities in Utilizing the Power of the Wind." *The Farm Implement News* (Chicago), 14, no. 9 (March 2, 1893): 19–20.

Hobart, F. G. "History of the 'Eclipse' Windmill and its Production." *Windmillers' Gazette* 1, no. 1 (Winter 1982): 6–9.

Hodges, Jeremy, "Wind Turbines Bigger Than Jumbo Jets Seen Growing Ever Larger," *Bloomberg*, January 24, 2019.

"A Holiday Is Given in Effort to Reduce Los Angeles Smog," *The New York Times*, July 27, 1973.

Holland, Elizabeth. "Jacobs Is Back in the Wind Machine Business." *Solar Age* 5, no. 2 (February 1980): 18.

"Home Wired for 20 Years[,] He Finally Gets Current." *Rural Electrification News* 2, no. 11 (July 1937): 13.

Hoose, Shoshana, "Wind Power Propels Family's Trade," *The Minneapolis Star*, January 7, 1982.

Hopson. Janet L. "They're Harvesting a New Cash Crop in California Hills." *Smithsonian* 13, no. 8 (November 1982): 122–27.

Hornig, Dana, "Energy Out of Thin Air: The Windmills of Marstons Mills," *The Register* (Cape Cod, MA), December 16, 1976.

"House Panel Explores PURPA Reform, Industry Structure[,] Karas Testifies That PURPA Has Not Yet Accomplished All of Its Goals." *AWEA Wind Energy Weekly* (Washington, DC) 15, no. 683 (February 5, 1996): N.p.

"How Are We Going to Sell the Farm Women?" *Farm Light and Power* (New York) 4, no. 2 (October 23, 1923): 51–52.

Howard, Lawrence M. "Power from the Winds." *Vermont Life* 10, no. 2 (Winter 1955–1956): 51–55.

"Hoyler's Response." *Wind Power Digest* no. 25 (Fall 1982): 17.

"Huge Windmill with One Blade Is Planned," *The New York Times*, April 8, 1986.

"An Inexpensive Farm Electric Plant." *Popular Electricity* 4, no. 8 (December 1911): 709.

"Inside a Small Wind-Turbine Beta Test," *Green Tech* (Boston, MA), July 22, 2009.

"An Interesting Wind Turbine Electric Plant Has Recently Been Installed at Buckenhill, Bromyard, by Messrs. J. G. Childs & Co., Ltd., of Willesden-green, London." *The Implement and Machinery Review* (London) 34, no. 416 (December 1, 1909): 982.

"Interview with Herman Drees[,] President—Pinson Energy Corp.," *Wind Power Digest* no. 7 (December 1976): 5–13.

"Interview with Ron Stimmel: AWEA Small Wind Advocate." *Small Wind Newsletter* (Interstate Renewable Energy Council, Inc., Latham, NY) no. 28 (July 2009): N.p.

Irwin, Bob. "The Gas Crunch of '79." *AutoWeek* (May 14, 1979): 2.

"J. Carter's Development of Fiberglass Technology." *Wind Power Digest*, no. 17 (Fall 1979): 40–44.

Jacobs, Ed. "Wind Generator has Variable-Pitched Blades

with Fixed-Blade Simplicity." *Popular Science* 220, no. 6 (June 1982): 114.

Jacobs, M. L. Letter to the editor. "Marcellus Jacobs on Raising a Tower." *Alternative Sources of Energy* no. 29 (December 1977): 11–13.

Jacobs, Mr. and Mrs. A. J. Jacobs, Dawson County, MT. Letter to Editor. "The Electricity Equipped Farm Home." *The Dakota Farmer* (Aberdeen, SD) 43, no. 5 (March 1, 1923): 1.

"Jacobs Unveils New Machine," *Wind Power Digest* no. 17 (Fall 1979): 18.

"Jacobs Wind Farm Being Built in Hawaii," *Wind Energy Report* 6, no. 1 (January 14, 1983): 7.

Jacobson, Arlo, "Air-Electric Machine Units Cut Weeds—Flab," *Des Moines Sunday Register*, December 6, 1970.

Janson, Donald, "Radiation Is Released in Accident at Nuclear Plant in Pennsylvania," *The New York Times*, March 29, 1979.

Johnson, J. R. "An Engineer's Perspective of Our Energy Dilemma." *Alternative Sources of Energy* no. 21 (June 1976): 9–12.

Johnson, Lee, and Marshall Merriam. "Small Groups, Big Windmills." *Wind Power Digest* no. 11 (Winter 1977–1978): 30–32.

Jopp, Martin. "Martin Answers." *Alternative Sources of Energy* no. 28 (October 1977): 46.

———. "Martin Answers[:] Questions and Answers on Windmills and Electricity." *Alternative Sources of Energy* no. 18 (July 1975): 21.

Jossi, Frank, "Nation's First Integrated Wind and Solar Project Takes Shape in Minnesota," *Energy News Update* (London), March 2, 2017.

Jureczko, M., M. Pawlak, and A. Mçżyk. "Optimisation of Wind Turbine Blades." *Journal of Materials Processing Technology* 167 (2005): 463–71.

Kahn, Debra, "San Francisco Plan Promotes Urban Wind Power," *The New York Times*, September 30, 2009.

Kanellos, Michael, "Big Boost for Small Wind: GE Invests in Southwest Windpower," *Greentech Media* (Boston, MA), April 6, 2009.

Katzenberg, Richard. "President's Report." *AWEA Newsletter* (Bristol, IN) (Summer 1977): 5–6.

———. "Quarterly Report." *AWEA Newsletter* (Bristol, IN) (Spring 1977): 23.

———. "Where Are the Dallas Cowgirls When We Need Them?" *Wind Power Digest* no. 16 (Summer 1979): 5–7.

Kavlak, Goksin, James McNerney, and Jessika E. Trancik. "Evaluating the Causes of Cost Reduction in Photovoltaic Modules." *Energy Policy* 123 (2018): 700–10.

Kay, Julie, "Adorno Lawsuit Was under Radar until Paychecks Bounced," *Daily Business Review* (Miami-Dade, FL), December 10, 2010.

"The Kedco 1200." *Wind Power Digest* no. 2 (Summer 1975): 14.

Keily, William. "Electricity in the Household[:] Is Electricity Light Too Dear for Modest Purses?" *Popular Electricity* 2, no. 1 (May 1909): 42–45.

Kellner, Thomas, "A Towering Achievement: This Summer in Holland, GE Will Build the World's Largest Wind Turbine," *GE Reports*, January 18, 2019.

"Kenetech Board Struggles to Resolve Financial Troubles." *AWEA Wind Energy Weekly* (Washington, DC) 14, no. 676 (December 11, 1995): N.p.

Keyes, Ralph. "Learning to Love the Energy Crisis." *Newsweek* (December 3, 1973): 17.

Kidd, Stephen. "Windmills and Other Things." *Princeton Alumni Weekly* 73, no. 24 (April 24, 1974): 12–13.

———, and Doug Garr. "Can We Harness Pollution-Free Electric Power from Windmills?" *Popular Science* 201, no. 5 (November 1972): 70–72.

Kihss, Peter, "Worried Drivers Swamp Stations Selling Gasoline," *The New York Times*, February 5, 1974.

Kilduff, Paul, "Homeowners Can Harness the Wind," *The Chronicle* (San Francisco, CA), December 20, 2008.

Kimsey, Carolyn. "The Plowboy Interview: Dr. Ralph Borsodi." *The Mother Earth News* no. 26 (March 1974): 6–13.

Klassen, Victor. "The Energy Crisis in Historical Perspective." *Alternative Sources of Energy* no. 12 (October–November 1973): 21–22.

Klinkenberg, Karen. "Radisson Farm & Its Influence[:] Historical Notes on the Farm, the Hotel and the Road." *Blaine Historical Society* (Blaine, MN) (December 2010): 1–2. Accessed May 13, 2017. http://www.blainehistory.org/Radisson_Farm/Radisson_Farm_Article.pdf.

Kloeffler, R. G., and E. L. Sitz. "Electric Energy from Winds." *Kansas State College Bulletin* (Manhattan, KS) 30, no. 9 (September 1, 1946): 1–32.

Knox, Keith R. "Strategies & Warnings for Wind Generator Buyers." *Wind Power Digest* no. 24 (Summer 1982): 54–56.

Kolbe, Harry. "Build Your Own Budget Windcharger." *Mechanix Illustrated* 75, no. 597 (February 1978): 57–60.

LaFollette, Douglas. "Guest Editorial[:] The Economic Myth of Nuclear Power." *Alternative Sources of Energy* no. 30 (February 1978): 2–3.

Laitner, Skip. "The Case against Nuclear Power: An Overview[,] Part I." *Alternative Sources of Energy* no. 23 (December 1976): 22–24.

Lammers, Dick, "Federal Tax Credit Boosts Home Wind Turbine Market," *Associated Press*, January 25, 2009.

LaMonica, Martin, "Inside a Small Wind-Turbine Beta Test," *Green Tech—CNET News*, July 22, 2009.

Lee, Byran. "Highlights of the Clean Air Act Amendments of 1990." *Journal of the Air & Waste Management Association* (Pittsburgh, PA) 41, no. 1 (January 1991): 16–19.

Lee, Patrick, "Impact of the Gulf War: Crude Plunges; Gasoline Prices to Dealers Cut: Energy: The Movement Defies Predictions," *Los Angeles Times*, January 18, 1991.

"Legislative Alert[:] Make Your Voice Heard on the Renewable Energy Research Budget!" *AWEA Wind Energy Weekly* (Washington, DC) 15, no. 685 (February 19, 1996): N.p.

Lewis, I. N., Lt. "Generating Electricity by Wind Mills." *The Farm Implement News* (Chicago) 16, no. 34 (August 22, 1895): 20–21, 24.

"The Lewis Electric Car Lighting System." *The Electrical Engineer* (New York) 17, no. 298 (January 17, 1894): 43–44.

"The Lewis System of Electric Lighting by Windmills." *The Electrical Engineer* (New York) 17, no. 300 (January 31, 1894): 86–87.

"The Lewis Train Electric Lighting System." *The Electrical World* (New York) 24, no. 4 (July 28, 1894): 85–87.

"The Lighting of Show Windows." *Popular Electricity* 1, no. 12 (April 1909): 735–38.

"Lights Enough to Encircle the Globe." *Popular Electricity* 1, no. 2 (June 1908): 91.

Lillian, Betsy, "DOE: U. S. Installed More Than 50 MW of Distributed Wind in 2018," *North American Windpower* (Southbury, CT), August 27, 2019.

———, "GE, Juhl Energy Build Wind-Solar Hybrid Project in Minnesota," *North American Windpower* (Southbury, CT), November 21, 2018.

———, "GE Haliade-X 12 MW Giant to Be Tested in Rotterdam," *North American Windpower* (Southbury, CT), January 21, 2019.

———, "Microgrid Enables California Solar Developer to Go Off the Grid," *Solar Industry* (Southbury, CT), February 28, 2019.

———, "National Lab: Costs Continue to Drop for Residential, Commercial PV," *Solar Industry* (Southbury, CT), December 20, 2018.

———, "Smithfield Foods Adding Wind Turbines to Colorado Hog Farms," *North American Windpower* (Southbury, CT), March 20, 2019.

———, "Two Worlds Collide: GE, Juhl Energy Building Hybrid Wind-Solar Project," *North American Windpower* (Southbury, CT), February 22, 2017.

———, "Vestas Wind Installation to Be U.S.' Tallest," *North American Windpower* (Southbury, CT), December 21, 2018.

Lindsley, E. F. "33 Windmills You Can Buy Now!" *Popular Science* 221, no. 1 (July 1982): 52–55.

———. "Advanced Design Puts New Twists in Windmill Generator." *Popular Science* 215, no. 4 (October 1979): 84–85.

———. "New Inverter Gives Wind Power without Batteries," *Popular Science* 207, no. 4 (October 1975): 50–52.

———. "Wind Power: How New Technology Is Harnessing an Age-Old Energy Source." *Popular Science* 205, no. 1 (July 1974): 54–59, 124–25.

Ling, Katherine, "Small Turbines Get Second Wind from Climate Regs," *E&E News* (Washington, DC), August 17, 2015.

"Linking the Links of the Long Distance 'Phone." *Popular Electricity Magazine* 5, no. 7 (November 1912): 670–72.

"A Little Motor for the Home." *Popular Electricity* 3, no. 7 (November 1910): 639–40.

Lo, Chris. "Vestas's Multi-Rotor Wind Turbine: Are Four Rotors Better Than One?" *Power Technology* (October 17, 2016). Accessed February 10, 2018. http://www.power-technology.com/features/featurevestass-multi-rotor-wind-turbine-are-four-rotors-better-than-one-5001819/.

Lokter, Michael. "Making the Most of Federal Tax Laws: A New Way to Look at WECS Development." *Alternative Sources of Energy* no. 63 (September–October 1983): 38–43.

Lombardi, Candice, "Home Depot's Latest Small Wind Deal," *CNET*, June 22, 2011.

"Lone Star Wind[:] An Overview of Current Wind Energy Activities in Texas." *Wind Power Digest* no. 17 (Fall 1979): 36–40.

"Looking in on Altamont's Wind Farms . . . by Car or on a Tour." *Sunset* 174 (April 1985): 78.

Love, Shayla, "The Almost Forgotten Story of the 1970s East Village Windmill," *The Gothamist*, September 29, 2014. Accessed September 30, 2014. http://gothamist.com/2014/09/29/east_village_windmill_nyc.php#photo-1.

Lowstutter, William R., Jr. "Wind Supplies Much of Cuttyhunk Island's Electric Power." *Electrical Consultant* 59, no. 5 (September–October 1979): 26–30.

Lundquist, Enoch. "Selling Electrical Appliances to the Farmer." *Farm Light and Power* (New York) 3, no. 11 (July 15, 1923): 19, 29, 31.

Lydersen, Kari, "In Wisconsin, Many Oppose Transmission Line to Bring Western Wind Power," *Energy News Network* (St. Paul, MN), January 22, 2019.

McDermott, Jeanne. "Harnessing Urban Winds[:] The Bronx Frontier Organizes a Unique Wind-Assisted Composting Project in the Big Apple." *Wind Power Digest* no. 18 (Winter 1979–1980): 7–12.

———. "How to Make a Wind-Site Analysis." *Popular Science* 217, no. 1 (July 1980): 100–02.

"Making Electricity by Wind Power." *Popular Electricity* 2, no. 9 (January 1910): 583–84.

Manning, Roger S. "The Windmills in California." *Journal of the West* 14, no. 3 (July 1975): 33–39.

Marier, Donald. "The Bergey Windpower Corp.: A Profile." *Alternative Sources of Energy* no. 61 (May-June 1983): 20–21.

———. "Conference Report: The CAL-WEA Annual Meeting[,] Setting the Record Straight on Jacobs." *Alternative Sources of Energy* no. 61 (May–June 1983): 31.

———. "The Consumer Wind Market: What Will the Future Hold," *Alternative Sources of Energy* no. 61 (May–June 1983): 4.

———. "Marcellus Jacobs: 1903–1985." *Alternative Sources of Energy* no. 75 (September–October 1985): 6.

———. "Reaching Up, Reaching Out." In "Wind Energy & Wood Heating." Special issue, *Alternative Sources of Energy* no. 41 (January–February 1980): 11–14.

———. "Some Notes on Windmills." *Alternative Sources of Energy* no. 12 (October–November 1973): 2–3.

———. "Some Notes on Windmills." *Alternative Sources of Energy* no. 13 (February 1974): 28.

Martin, Hannah, "Buckminster Fuller's Geodesic Dome and Other Forward-Looking Architecture," *Architectural Digest*, February 12, 2016. Accessed September 9, 2018. https://www.architecturaldigest.com/gallery/buckminster-fuller-architecture/all.

Maule, Pete. "Oregon Anemometer Loan Program." *Wind Power Digest* no. 20 (Summer 1980): 36–41.

Mayer, Don. "The Development of the AWEA." *AWEA Windletter* (Washington, DC) (February 1978): 1, 6.

Maynard, Joyce, "519 East 11th St.: Neighbors Rebuild Hopes," *The New York Times*, July 8, 1976.

"Membership Application." *American Wind Energy Association Newsletter* (Detroit) (June 1976), N.p.

Merriam, Marshall. "The Gedser Mill Re-vitalized." *Wind Power Digest* no. 11 (Winter 1977–1978): 34–38.

Meter, Ken. "Interview with Martin Jopp." *Alternative Sources of Energy* no. 28 (October 1977): 27–31.

Meyer, Hans. "Wind Generators: Here's an Advanced Design You Can Build." *Popular Science* 201, no. 5 (November 1972): 103–5, 142.

Miller, Warren H. "Fifty Kilowatts of Water Power." *Popular Electricity* 3, no. 7 (November 1910): 618–21.

Miner, Colin, "Standards for Small-Scale Wind Power," *The New York Times*, August 28, 2009.

"Model of Windmill Electric Plant." *Popular Electricity Magazine* 6, no. 1 (May 1913): 57.

Mohrbacher, Bill. "History of Model Engines." *Model Aviation* (February 2012). Accessed April 23, 2016. http://www.modelaviation.com/enginehistory.

"More on Small Wind Generators." *AWEA Windletter* (Washington, DC) (February 1978): 9.

Moretti, Peter M., and Louis V. Divone. "Modern Windmills." *Scientific American* 254, no. 6 (June 1986): 110–18.

"A Motor for the Sewing Machine." *Popular Electricity* 1, no. 2 (June 1908): 115.

"Mountain-Top Windmill to Feed Vermont Electric Lines." *Popular Science* 139, no. 1 (July 1941): 115–17.

"Mr. Brush's Windmill Dynamo," *Cleveland Plain Dealer*, July 11, 1887.

"Mr. Brush's Windmill Dynamo." *Scientific American* 63, no. 25 (December 20, 1890): 389.

"Mr. J. Wallis Titt's New Wind Engine at Boyle Hall: A Remarkably Economical and Automatic Power." *The Implement and Machinery Review* (London) 24, no. 288 (April 1, 1899): 23879–80.

"Mr. John Wallis Titt." *The Implement and Machinery Review* (London) 21, no. 244 (August 1, 1895): 1908.

Mullendore, Daniel. "Letter to the Members[,] April 15, 1975." *American Wind Energy Association Newsletter* (Detroit) (August 1975): 13–14.

Munoz, Sara Schaeffer, "A Novel Way to Reduce Home Energy Bills," *The Wall Street Journal*, August 15, 2006.

"NAE, Micon to Install Largest Single Turbine in Midwest." *AWEA Wind Energy Weekly* (Washington, DC) 14, no. 678 (December 25, 1995): N.p.

National Association of Regulatory Utility Commissioners Subcommittee on Renewable Energy. "State Activities[:] Texas." *State Renewable Energy News* (Washington, DC) 8. No. 2 (Summer 1999): 4.

"National Renewable Energy Laboratory to Trim Staff." AWEA Wind Energy Weekly (Washington, DC) 14, no. 673 (November 20, 1995): N.p.

Naughton, Brian. "Wind Plant Optimization[:] Sandia Wake Imaging System Successfully Deployed at the SWiFT Facility." *Wind & Water Power Newsletter* (Sandia National Laboratories, Albuquerque, NM) (July–August 2015). Accessed February 3, 2018. http://content.govdelivery.com/accounts/USDOESN LEC/bulletins/114227e.

"New Bergey Unit Coming." *Wind Power Digest* no. 25 (Fall 1982): 33.

"New Center Formed[,] DOE Merges SERI and Rocky Flats." *World Wind* (Ottawa, Canada) 1, no. 2 (January 1985): 1.

"New Dawn in the City." *Wind Power Digest* no. 20 (Summer 1980): 18.

"New Design for Windmill on Display," *Osceola Sun* (Kissimmee, FL), February 20, 1974.

"The New Highway Guerillas." *Time* 102, no. 25 (December 17, 1973): 33.

"A New Windmill Electricity Generating Plant." *The Implement and Machinery Review* (London) 51, no. 601 (May 1, 1925): 70.

"New Schemes for Harnessing the Winds." *Popular Science* 135, no. 2 (August 1939): 100–01.

"New U.S. Administration May Bring Changes." *AWEA Wind Energy Weekly* (Washington, DC) 7, no. 328 (November 13, 1988): N.p.

"NH Utility Purchasing Experimental VAWT." *Wind Energy Report* (July 1978): 5.

Nichols, Leander. "Wind-Driven Alternators." *Wind Power Digest* no. 9 (Spring 1977): 34–37.

"Nixon Approves Limit of 55 M.P.H.," *The New York Times*, January 3, 1974.

Norberg, Bob. "Fort Ross Shows Off New Russian-Built Windmill," *Press Democrat* (Santa Rosa, CA), October 20, 2012.

"Nordtank, Micon Proceed with Merger Proposal." *AWEA Wind Energy Weekly* (Washington, DC) 16, no. 751 (June 9, 1997): 2.

"Northern Power Advanced Turbine Would Cut Power Costs by 46%." *AWEA Wind Energy Weekly* (Washington, DC) 10, no. 464 (1991): N.p.

"Northern Power to Design Large-Scale Polar Turbine." *AWEA Wind Energy Weekly* (Washington, DC) 14, no. 634 (February 13, 1995): N.p.

O'Gara, P. J. "How a Farmer Built His Own Electric Power Plant." *Popular Electricity* 1, no. 7 (November 1908): 432–36.

O'Leary, Melissa, "Experimental Turbine to Bring Wind Energy to the Masses," *Composites Manufacturing* (Arlington, VA), March 8, 2019.

O'Shea, Allan. "Memo from the President." *American Wind Energy Association Newsletter* (Detroit) (August 1975): 2.

"Obituaries[:] Elliot John Bayly." *Steamboat Pilot & Today* (Steamboat Springs, CO), January 28, 2012.

Ogden, Derek, and Anne Burke. "The Windmill at Flowerdew Hundred." *Old Mill News* 6, no. 1 (January 1978): 4–6.

"Ordinary Windmill Will Not Do." *Farm Light and Power Year Book*[:] *Dealers' Catalog and Service*. New York: Farm Light and Power Publishing Co., 1922.

Paris, Ellen. "The Great Windmill Tax Dodge." *Forbes* 133 (March 12, 1984): 40.

———. "Palm Springs and the Wind People." *Forbes* 135 (June 3, 1985): 170–01.

Parsons, Sarah, "A Personal Turbine Makes Your Rooftop into a Wind Farm," *Popular Science*, September 16, 2011. Accessed September 19, 2019. https://www.popsci.com/gadgets/article/2011–09/personal-turbine-makes-your-rooftop-wind-farm/.

"A Pennsylvania Farm Electrified." *Popular Electricity* 5, no. 2 (June 1912): 138–39.

Phelps, David, and Steve Gross, "Against the Wind: CDC's $90 Million Failure," *Minneapolis Star and Tribune*, December 9, 1986.

Piacente, Steve, "Tide Gate Could Improve Bay Water Quality," *Tampa Tribune* (Tampa, FL), May 22, 1980.

"Plans for '83[:] Rocky Flats." *Wind Power Digest* no. 25 (Fall 1982): 27–28.

Podevin, Mary. "Promoting Large-Scale Wind[:] A Conversation with Wind Farm Entrepreneur Alvin Duksin." *Wind Power Digest* no. 18 (Winter 1979–1980): 40–41.

Pope, Franklin L. "An Isolated Electric-Lighting Plant Has Recently Been Installed in London Which Is Being Successfully Operated by an American Windmill." *Engineering Magazine* 3, no. 3 (June 1892): 399–400.

Porcello, Samuel. "Windmills and Railroads: A Successful Partnership." *Windmillers' Gazette* 35, no. 3 (Summer 2016): 6–7.

"Power!" *The Mother Earth News* no. 6 (November 1970): 60–61.

Priest, Tyler. "The Dilemmas of Oil Empire." *Journal of American History* 99, no. 1 (June 2012): 236–51.

"Primus Wind Power Acquires the AIR Wind Turbine Line from Southwest Windpower," *Cruising Outpost Magazine*, August 29, 2013.

Prince, Mary, and Woody Stoddard. "University of Massachusetts Solar Wind House." *Alternative Sources of Energy*, no. 26 (June 1977): 26–27.

"Professor Blyth on Wind Power for Electric Lighting." *The Electrician* (London) 21, no. 522 (May 18, 1888): 38.

Pugh, Robert C. "Wind Power Is Poised for the Future." *North American Windpower* (Southbury, CT) 14, no. 9 (October 2017): 30–32.

Pullen, John J. "Will It Pay You to Put up a Windmill?" *Country Journal* 7, no. 7 (July 1980): 72–81.

Raasch, Chuck, "Wind Power on the Brink of Revival," *USA Today*, April 5, 1983.

Rader, Nancy A., and Richard B. Norgaard. "Efficiency and Sustainability in Restructured Electric Markets: The Renewables Portfolio Standard." *The Electricity Journal* 9, no. 6 (July 1996): 37–49.

Randolph, William. "Power to the People (Wincharger Style)." *Wind Power Digest* no. 8 (Spring 1977): 10–14.

Rayasam, Renuka. "For Building This Business, the Answer's in the Wind." *U.S. News & World Report* (May 28, 2007): 64.

"Restructuring Good for Environment, Utility Executive Contends." *AWEA Wind Energy Weekly* (Washington, DC) 15, no. 692 (April 8, 1996): N.p.

"Review of Hardware." *Wind Power Digest* no. 6 (September 1976): 44–46.

Richtmyer, Richard, "NY Testing Wind-Energy atop Albany Skyscraper," *Associated Press*, February 2, 2009.

Righter, Robert W. "Reaping the Wind[:] The Jacobs Brothers, Montana's Pioneer 'Windsmiths,'" *Montana*[:] *The Magazine of Western History* 46, no. 4 (Winter 1996): 38–49.

Roberts, Nancy L., "Wind Power[:] Time Has Come, Again, for This Energy Pioneer," *The Dispatch* (Minneapolis), May 4, 1981.

"Rocky Flats: Testing Small-Scale Wind Systems." *Wind Power Digest*, no. 7 (December 1976): 39–41.

Rogier, Etienne. "The First Wind Generator in France, 1887." *Windmillers' Gazette* 38, no. 2 (Spring 2019): 4–5.

———. "Georges Darrieus and the 'Egg-Beater' Vertical-

Axis Wind Turbines." *Windmillers' Gazette* 23, no. 1 (Winter 2004): 12–14.

Rogier, Etienne, and T. Lindsay Baker. "The Polar Wind Machines: The Wind Generator on the *Fram* during Nansen's 1893–96 Arctic Expedition." *Windmillers' Gazette* 17, no. 2 (Spring 1998): 7–11.

Roguski, Randy, "Demand for Small-Scale Wind Turbines Growing among Ohio Businesses," *The Plain Dealer* (Cleveland, OH), May 2, 2009.

Rounds, Tim. "PCR Employs Windpower," *The Clock* (Plymouth State College, Plymouth, NH), November 30, 1978.

Royden, Amy. "US Climate Change Policy under President Clinton: A Look Back." *Golden Gate University Law Review* 32, no. 4 (January 2002): 415–78.

Ruess, Henry S., Letter to Editor, *The New York Times*, June 27, 1976.

Sagrillo, Mick. "Home-Built Wind Generators. *AWEA WindLetter* (Washington, DC) 21, no. 5 (May 1994): 4.

———. "How It All Began." *Home Power* no. 27 (February–March 1992): 14–17.

———. "So You Want to Build a Wind Generator." *Home Power* no. 17 (June–July 1990): 28–30.

Sandru, Ovidiu, "$4,500 WindTronics Ultra-Efficient Low Speed Wind Turbine Available This Fall," *The Green Optimistic*, June 9, 2009.

Schachle, Charles, and Robert L. Scheffler. "Southern California Edison's 3MW Wind Demonstration Project." *Wind Power Digest* no. 16 (Summer 1979): 32–34.

Schefter, Jim. "New Harvest of Energy from Wind Farms." *Popular Science* 222, no. 1 (January 1983): 58–61.

Schmidt, Mitchell, "Wind Turbines Haven't Been Universally Welcomed by Everyone in Iowa," *The Gazette* (Cedar Rapids, IA), February 23, 2019.

Scott, Mike, "Battery Energy Storage Is a $620 Billion Opportunity as Cost Continue to Crash," *Forbes*, November 9, 2018. Accessed September 1, 2019. https://www.forbes.com/sites/mikescott/2018/11/09/battery-energy-storage-is-a-1-trillion-opportunity-as-costs-continue-to-crash/#6edb23d24684.

Searles, Edmund. "What Electricity Offers as a Life Work." *Popular Electricity* 3, no. 10 (February 1911): N.p.

Sencenbaugh, Jim. "I Built a Wind Charger for $400!" *The Mother Earth News* no. 20 (March 1973): 32–36.

Shepherdson, William. "Dynergy[:] One Entrepreneur's Experience in Wind." *Wind Power Digest* no. 19 (Spring 1980): 13–18.

———. "W. T. G. Energy Systems: 200 KW for Cuttyhunk Island." *Wind Power Digest* no. 10 (Fall 1977): 6–11.

Sherwood, Larry. "Interview: Mike Bergey of Bergey Windpower." *Small Wind Energy News* (Interstate Renewable Energy Council, Latham, NY), November

5, 2013. Accessed November 12, 2016. http://www.ireusa.org/2013/11/interview-mike-bergey-of-bergey-windpower/.

———. "What Happened to Southwest Wind Power?" *Small Wind Energy News* (Interstate Renewable Energy Council, Latham, NY), May 3, 2013. Accessed November 27, 2017. http://www.irecusa.org/2013/05/what-happened-to-southwest-wind-power/.

Shumaker, Varney V. "The House without a Chimney." *Popular Electricity* 3, no. 10 (February 1911): 926–28.

"Slash R&D? 'Historic Blunder,' Say DOE Executives[,] Renewable R&D Vital to Nation's Future: DOE Officials." *AWEA Wind Energy Weekly* (Washington, DC) 15, no. 693 (April 15, 1996), N.p.

"Small Is Beautiful." *New Scientist* 73, no. 1042 (March 10, 1977): 574.

"Small-Scale Wind Turbines Catching On," *The Associated Press*, May 17, 2009.

"Small Wind Systems 1979[,] SWECS Representatives Meet in Boulder." *Wind Power Digest* no. 15 (Spring 1979): 25–35.

"Small Wind Systems Continue to Find Markets." *Alternative Sources of Energy* no. 71 (January–February 1985): 55–56.

Smeaton, J. "An Experimental Enquiry Concerning the Natural Powers of Water and Wind to Turn Mills, and Other Machines, Depending on a Circular Motion." *Philosophical Transactions of the Royal Society* 51 (1759): 100–74.

Smith, Bill, "Happily, This Windmill Does More Than Spin—It Works for a Living," *Sunday Cape Cod Times* (Hyannis, MA), May 22, 1977.

Smith, Ken. ".75 KW Windcycle." *Alternative Sources of Energy* no. 14 (May 1974): 2–4.

Sønder, Niels. "Darrieus Windmills: Past, Present—and Future?" *Windpower Monthly* (Knebel, Denmark) 1, no. 12 (December 1985): 18–19.

Spear, Kevin, "Lake County Man's Wind Turbine Could Help Feed Florida's Power Grid," *Orlando Sentinel* (Orlando, FL), September 29, 2009.

"A Special Report[:] Section 2[,] P.U. R. P. A. and Small Systems." *Wind Power Digest* no. 22 (Spring 1981): 13–23.

"A Special Report: U.S. Wind Energy after Seven Years." *AWEA Wind Energy Weekly* (Washington, DC) 7, no. 300 (April 24, 1988), N.p.

"This Special Issue of Alternative Sources of Energy Magazine Is Dedicated to the Memory of Martin Jopp." In "Wind/Hydropower." Special issue, *Alternative Sources of Energy*, no. 46 (November–December 1980): 5.

"The Squeeze Hits the Purse." *Newsweek* (December 3, 1973): 92.

"Staff Notes." *Alternative Sources of Energy*, no. 28 (October 1977): 3.

Starks, Carolyn, "More Turn to Wind Turbines," *Chicago Tribune*, April 24, 2009.

"Startup Green Energy Tech Installs First Small-Wind Concentrators," *Reuters*, June 3, 2009.

"At State College Wind Power used for Radio Station," *The Herald* (Provo, UT), December 7, 1978.

Steenrod, A. H. "Fritchle Wind Electric System for Windmills." *Farm Light and Power* (New York) 2, no. 10 (June 15, 1922): 29–30, 38.

Stepler, Richard. "Eggbeater Windmill Is Self-Starting, Cheaper to Build." *Popular Science* 206, no. 5 (May 1975): 74–76.

Stoddard, Woody. "Appreciation[:] The Life and Work of Bill Heronemus, Wind Engineering Pioneer." *Wind Engineering* 26, no. 5 (2002): 335–41.

———. "Commentary and Review[:] American Wind Energy Association Annual Meeting and Exposition." *Alternative Sources of Energy*, no. 23 (December 1976): 31.

Stoeffler, David, "WPL Hopes to Get Boost from Wind," *Wisconsin State Journal* (Madison, WI), June 17, 1982.

Stoiaken, Larry N. "1984 SWECS Shakeouts." *Alternative Sources of Energy* no. 73 (May–June 1985): 57.

———. "Going International[:] Bergey Approached by Japanese and Chinese for License Production of BWC 1000." *Alternative Sources of Energy* no. 63 (September–October 1983): 25.

———. "Jacobs Finalizes Dyna Technology Acquisition: Unveils New Name—Earth Energy Systems Inc." *Alternative Sources of Energy* no. 72 (March–April 1985): 53.

———. "The Small Wind Energy Conversion System Market: Will 1984 Be 'The Year of the SWECS'?" *Alternative Sources of Energy* no. 63 (September–October 1983): 10–17, 20–23.

"Storing Wind Power." *Scientific American* 49, no. 15 (October 13, 1883): 229.

"The Story of an Electric Farm." *Popular Electricity* 4, no. 4 (August 1911): 289–97.

"Successful Electric Wind Mill Plant." *The Farm Implement News* (Chicago) 16, no. 11 (March 14, 1895): 17.

Swanson, Glen E. "The Great Eclipse of 1878 and Thomas Edison's Wind Turbine." *Windmillers' Gazette* 36, no. 3 (Summer 2017): 6–8.

"Swisher Finds NARUC Meeting 'Encouraging' for Wind." *AWEA Wind Energy Weekly* (Washington, DC) 14, no. 647 (May 22, 1995): N.p.

"In Sync[,] Up and Coming in Broadcast Technology[,] It's a Breeze." *Broadcasting* (Washington, DC) 48 (January 8, 1979): 42.

Szostak, Joe. "DAF" *Wind Power Digest* no. 19 (Spring 1980): 30–32.

———, Brian Toller, and Doug Whiteway. "NRC[:] Rebirth of the Darrieus Rotor." *Wind Power Digest* no. 19 (Spring 1980): 33–35.

"A Talk with Jim Sencenbaugh." *Wind Power Digest* 1, no. 6 (September 1976): 4–9.

Tan, Lot. "Small, Cheaper Turbine Might Work for Homes," *The Columbus Dispatch* (Columbus, OH), September 14, 2008.

Tanner, A. M. "The Electrical Utilization of Water and Wind Power First Proposed by Nollet in the Year 1840." *The Electrical World* (New York) 19, no. 15 (April 9, 1892): 242.

Tennyson, George. "Test and Product Standards for WECS." *AWEA Newsletter* (Bristol, IN) (Summer 1977): 4.

"Texas Passes Law for Big Renewable Energy Portfolio." *Windpower Monthly* (London), July 1, 1999. Accessed March 18, 2018. http://www.windpowermonthly.com/article/955103/texas-passes-law-big-renewable-energy-portfolio.

"This Month in Physics History[:] April 25, 1954: Bell Labs Demonstrates the First Practical Silicon Solar Cell." *APS News* (American Physical Society, College Park, MD) 18, no. 4 (April 2009): 2.

Thomas, Ken, "Wind Energy Groups Seeking Economic Stimulus Aid," *Associated Press*, January 30, 2009.

Thomas, Robert Mcg., "11th St. Tenants with Windmill and Con Edison," *The New York Times*, November 13, 1976.

Thomson, Sir William. "Address to the Mathematical and Physical Science Section of the British Association." *The Chemical News and Journal of Physical Science* (London) 44, no. 1138 (September 16, 1881): 135–37.

"'Thousand Cuts' Cause Drop in Kenetech's Earnings." *AWEA Wind Energy Weekly* (Washington, DC) 14, no. 659 (August 14, 1995): N.p.

"Three-Machine Windfarm Starts Up, Then Shuts Down for Repairs." *Solar Times* 3, no. 7 (July 1981): 1.

"Three Years of REA." *Rural Electrification News* 3, no. 9 (May 1938): 3–5, 18.

Tievsky, Charles A. "Full Electric Utility Ownership of PURPA Qualifying Cogeneration Facilities: Trouble Down the Line?" *Washington University Journal of Urban and Contemporary Law* 27 (January 1984): 321–58.

"To Catch the Wind." *Rockwell News, Rocky Flats Plant* (Boulder, CO) 7, no. 8 (June 20, 1980): 3–4.

Tocco, Peter. "The Night They Turned the Lights on in Wabash." *Indiana Magazine of History* 95, no. 4 (December 1999): 350–63.

"Tom Chalk's American Wind Turbine." *Wind Power Digest* no. 21 (Winter 1980–1981): 19.

Torinus, Jr., John, "Windworks Finds Home with Utility," *Milwaukee Sentinel*, September 27, 1983.

Trenka, Andrew R. Letter to the Editor. "Update: Rocky Flats/SERI Merger." *Wind Power Digest* no. 27 (1985): 5, 7.

Trenner, Patricia. "10 All-Time Great Pilots." *Air & Space Magazine* 17, no. 6 (March 2003): 67–72.

Tronche, John-Laurent, "Fort Worth Energy Firm Prepares First Turbine Prototype," *Fort Worth Business Press* (Fort Worth, TX), January 19, 2009.

"Tustin's Improved Adjustable Windmill." *Mining and Scientific Press* (San Francisco) 16, no. 16 (April 18, 1868): 241.

Unwin, Jack, "The Top 10 Countries in the World by Wind Energy Capacity," *Power Technology* (London), March 14, 2019.

"US Market for Small Wind Turbines Grew 78% in 2008," *Reuters*, June 1, 2009.

"US Renewable Energy Industry Reacts to Stimulus Package Passage," *RenewableEnergyWorld.com*, February 18, 2009.

"The Use of Wind Power." *The Electrical Engineer* (London) 13, no. 2 (January 12, 1894): 36.

"Utilities Upbeat about Windpower Prospects." *AWEA Wind Energy Weekly* (Washington, DC) 13, no. 597 (May 23, 1994): N.p.

Valley Industries, Inc. "Giromill Begins Performance Tests." *Pipeline* (St. Louis, MO) (June 1980): 1.

Van Vlissingen, Arthur. "Riches from the Wind[:] How Two Boy Mechanics Started a Brand-New Industry." *Popular Science Monthly* 132, no. 5 (May 1938): 54–55, 116–17.

Vaughn, Martin, "Tax Report[:] Taking Credit for Energy Efficiency," *The Wall Street Journal*, December 31, 2008.

Velasquez, Manziel E., Stuart Laval, Prateek Pandey, and David Blood. "Partners Tackle How to Achieve Seamless Transfer in a Microgrid." *Solar Industry* (Southbury, CT) 10, no. 1 (February 2017): 16–19.

Vick, Brian, and Sylvia Broneske. "Effect of Blade Flutter and Electrical Loading on Small Wind Turbine Noise." *Renewable Energy* 50 (2013): 1044–52.

Vick, Brian D., and Tim A. Moss. "Adding Concentrated Solar Power Plants to Wind Farms to Achieve a Good Utility Electrical Load Match." *Solar Energy* 92 (2013): 298–312.

Vick, Brian D., and Byron A. Neal. "Analysis of Off-Grid Hybrid Wind Turbine/Solar PV Water Pumping Systems." *Solar Energy* 86 (2012): 1197–207.

Voaden, Grant H. "The Smith-Putnam Wind Turbine . . . A Step in Aero-Electric Power Research." *Turbine Topics* (York, PA) 1, no. 3 (June 1943): 3–8.

W. O. A. "The Storage of Wind Power." *Scientific American* 49, no. 2 (July 14, 1883): 17.

Wade, Nicolas. "The New Alchemy Institute: Search for an Alternative Agriculture." *Science* 187, no. 4178 (February 28, 1975): 727–29.

Walin, Mike. "Value of Handling a Diversified Line." *Farm Light and Power* (New York) 4, no. 6 (February 15, 1924): 193–94, 213–15.

Walker, James, "Tide Gate Eyed to Ease Pollution," *Tampa Tribune* (Tampa, FL), April 5, 1975.

Walter, Helen. "Hannan Bros Ltd." *The Windmill Journal* (Morawa, Australia) 14, no. 3 (September 2015): 9–11.

——. "Saunders' Break-of-Gauge and Engineering Company Ltd, Saunders Engineering Co. Ltd & Speedy Windmill and Pump Company," *The Windmill Journal* (Morawa, Australia) 11, no. 4 (December 2012): 4–9.

Walters, Samuel. "Power from Wind." *Mechanical Engineering* 96, no. 4 (April 1974): 55.

Ward, Mitchell. "The PTC and Wind Energy: Restructuring the Production Tax Credit as a More Effective Incentive." *Houston Business and Tax Law Journal* 11, no. 2 (2011): 455–88.

"The Wasted Wind." *Chambers's Journal* 1, no. 6 (December 1897–November 1898): 247–50.

Watkins, John R. "A Common Crystal." *The Strand Magazine* (London) 17, no. 98 (February 1899): 174–78.

Weichelt, Steve. "The Plowboy Interview[:] Marcellus Jacobs." *The Mother Earth News* no. 24 (November 1973): 52–58.

Well, Josh. "A New Spin on Wind." *Orion* 24, no. 6 (November–December 2005): 60–67.

——, "The Wind Farmers of East 11th Street," *The New York Times*, August 3, 2008.

Welsh, Melinda. "Congress Not Likely to Take Decision on Wind in '85." *Windpower Monthly* (Knebel, Denmark) 1, no. 9 (September 1985): 8.

——. "Future for Tax Credits Darkens." *Windpower Monthly* (Knebel, Denmark) 1, no. 7 (July 1985): 4.

——. "Tax Credits Update." *Windpower Monthly* (Knebel, Denmark) 1, no. 6 (June 1985): 14.

Werst, Mike. "Herbert E. Bucklen Corporation and Its Propeller-Driven Wind Plants." *Windmillers' Gazette* 37, no. 1 (Winter 2018): 2–4.

——. "Pre-Rural Electrification Wind Generators— Speed Governing Mechanisms." *Windmillers' Gazette* 35, no. 1 (Winter 2016): 2–4.

Werst, Mike, and John Killam. "Air-Electric Machine Company: The Early Years." *Windmillers' Gazette* 38, no. 1 (Winter 2019): 2–5.

Western Electric Company, New York. "Western Electric Power & Light" [advertisement]. *Farm Light and Power Year Book* (New York) (1922): 35.

Westinghouse Electric & Manufacturing Company, East Pittsburgh, PA. "Everything Electrical for the Farm" [advertisement]. *Farm Light and Power Year Book* (New York) (1922): 7–11.

Whitehead, Dan. "Living with a Wind Machine." *Home Power* no. 57 (February-March 1997): 18–21.

Whitehead, Daniel. "Alternative Energy . . . or Just Plain Crazy." *Home Power* no. 53 (June–July 1996): 6–10.

Whites-Koditschek, Sarah, "Wisconsin Wind Turbine Project Pits Brother against Brother, Clean Energy against Rural Vistas," *Milwaukee Journal Sentinel*, April 16, 2019.

"Who's Mill Is This?" *Wind Power Digest* no. 25 (Fall 1982): 16.

"Why the Farmer Buys a Plant." *Farm Light and Power* (New York) 3, no. 8 (April 14, 1923): 24.

Widhalm, Shelley, "Rooftop Wind Turbine Seen as Green Energy Starter Kit," *Loveland Reporter-Herald* (Loveland, CO), May 16, 2009.

Wilbur, John B. "The Smith-Putnam Wind Turbine Project." *Journal of the Boston Society of Civil Engineers* 29 (July 1942): 211–28.

Wilford, John Noble, "Nation's Energy Crisis: Is Unbridled Growth Indispensable to the Good Life?" *The New York Times*, July 8, 1971.

Willson, Elizabeth. "Fayette Manufacturing: Is Wind Power Gone with the Tax Credits?" *High Technology* 6, no. 1 (January 1986): 17.

Wilson, Alex. "The Folly of Building-Integrated Wind." *Building Green* 18, no. 5 (April 29, 2009): N.p. Accessed February 11, 2019. http://www.buildinggreen.com/feature/folly-building-integrated-wind.

"Wind and Electricity: Extensive Improvements at Boyle Hall, West Ardsley." *The Implement and Machinery Review* (London) 24, no. 287 (March 2, 1899): 23774.

"Wind Can Help Avert Greenhouse Effect, AWEA Says." *AWEA Wind Energy Weekly* (Washington, DC) 7, no. 311 (July 11, 1988): N.p.

"Wind Conversion," *Daily Globe* (Dodge City, KS), October 27, 1982.

"Wind-Driven Electric Generator at Kalae Light Station, Hawaii." *Lighthouse Service Bulletin* (Washington, DC) 3, no. 1 (January 2, 1930): 6.

"Wind-Driven Electric Plant at Arecibo Light Station." *Lighthouse Service Bulletin* (Washington, DC) 3, no. 54 (June 1, 1928): 247.

"Wind Electric Company[,] Minneapolis, Minn." *Farm Light and Power*[:] *Dealers' Service Book*. New York: Farm Light and Power Publishing Co., [ca. 1923].

Wind Energy Society of America. "Annual Meeting of the Wind Energy Society of America, August 2, 1975." *Wind Energy Society of America Newsletter* (Pasadena, CA), no. 8 ([August] 1975): 1.

———. "WESA Status." *Wind Energy Society of America Newsletter* (Pasadena, CA), no. 10 ([August] 1976): 1.

"Wind Journal Arrives." *AWEA Newsletter* (Bristol, IN) (Spring 1977): 13.

"The Wind Mill Industry." *The Farm Implement News* (Chicago) 15, no. 49 (December 6, 1894): 20.

"Wind Power for Electric Lighting." *The Farm Implement News* (Chicago) 12, no. 2 (February 1890): 30–31.

"Wind Power Electric Plants." *Popular Electricity* 2, no. 7 (November 1909): 451.

"Windependence." In "Wind/Hydropower," edited by Don Marier. Supplement, *Alternative Sources of Energy* no. 46 (November–December 1980): 50–51.

"Windmill Dynamo Created," *Lubbock Avalanche-Journal* (Lubbock, TX), March 12, 1975.

"The Windmill Electric Lighting Plant at Marblehead Neck, Mass." *The Electrical Engineer* (New York) 18, no. 342 (November 21, 1894): 412–13.

"Windmill Farm Lighting Plant." *Popular Electricity Magazine* 5, no. 7 (November 1912): 655–56.

"Windmill Generation Plants. A Detailed Report on Their Practicability and Economy." *The Implement and Machinery Review* (London) 51, no. 611 (March 1, 1926): 1220–24.

Windmill Light & Power Co., Walpole, MA. "Do Your Own Electric Lighting" [advertisement]. *Munsey's Magazine* (New York) 17, no. 4 (July 1897): N.p.

"Windmill for Producing Electric Light at Cape de la Heve." *Scientific American Supplement* 28, no. 709 (August 3, 1889): 11326–27.

"Windmill Supplies Light." *Popular Science* 103, no. 6 (December 1923): 69.

"The Windmill's Electric Generating Services." *The Implement and Machinery Review* (London) 24, no. 282 (October 1, 1898): 23270.

Windmills for Electric Lighting." *The Electrical World* (New York) 23, no. 5 (February 3, 1894): 157–58.

"Windmills Protect Pipelines from Corrosion." *Agricultural Engineering* 18, no. 7 (July 1937): 295.

"Windpower '85 a Success[,] Over 650 Go to San Francisco." *Alternative Sources of Energy* no. 76 (November–December 1985): 50.

"Windworks/Wisconsin Power Merger." *Wind Power Digest* no. 24 (Summer 1982): 18–19.

Wisconsin Power & Light Company, "The Winds of Change Blow Strong and Sure for Windworks." *Concepts* (Madison, WI) 9, no. 4 (Fall 1983: 2–9.

Wiser, Mike. "Growing Tall Turbines," *Iowa Farmer Today* (Cedar Rapids, IA), November 18, 2011.

Wiser, Ryan, and Ole Langniss. "The Renewable Portfolio Standard in Texas: An Early Assessment." *Energy Policy* 31, no. 6 (May 2003): 527–35.

"With a Little Coaxing, the Wind Can Be and Is Made Very Serviceable as a Power Generator." *The Implement and Machinery Review* (London) 35, no. 411 (July 2, 1909): 375.

Woelke, Judith. "Minutes." *AWEA Newsletter* (Detroit) (April 1976): 3–6.

[Wolff, A. R.]. "Are Windmills Expensive Prime Movers?" *The American Engineer* (Chicago) 2, no. 11 (November 1881): 209.

Wolf[f], Alfred B. [sic.] "Windmills for Generating Electricity." *The Engineer* (London) 65 (February 3, 1888): 88.

Wolfson, Karen, "Tide Gates Ready to Flush Stagnant Waterways," *Tampa Tribune* (Tampa, FL), December 23, 1982.

Wood, Karen, "Wind Turbine Blades: Glass vs. Carbon Fiber," *CompositesWorld*, May 31, 2012. Accessed January 14, 2018. https://www.compositesworld.com/articles/wind-turbine-blades-glass-vs-carbon-fiber.

"Woodmanse Mfg. Co.[,] Freeport, Ill." *Farm Light and Power*[:] *Dealers' Service Book*. New York: Farm Light and Power Publishing Co.,[ca. 1923].

"World's First Windfarm." *Farm Show Magazine* 5, no. 5 (May 1981): 18.

Worts, George F. "Running a Farm by the Power of a Brook." *Popular Electricity* 5, no. 10 (February 1913): 1058–59.

Zayas, Jose. "DOE's Competitiveness Improvement Projects Are Delivering Substantial Cost Reductions in the U.S. Distributed Wind Industry." *North American Clean Energy* (July–August 2017). Accessed January 31, 2018. http://www.nacleanenergy.com/articles/27632/doe-s-competitiveness-improvement-projects-are-delivering-substantial-cost-reductions-in-the-u-s-distributed-wind-industry.

"Zephyr Unveils 7.5 KW Generator." *Wind Power Digest* no. 2 (Summer 1975): 29.

Ziter, Brett G. "Electric Wind Pumping for Meeting Off-Grid Community Water Demands." *Guelph Engineering Journal* (Ontario, Canada) 2 (2009): 14–23.

Trade Literature

Aermotor Company, Chicago. *Electric Aermotor*. Publication E. A. I. Chicago: Aermotor Company, [ca. 1920]. Folder. Available in *Windmillers' Gazette* research files, Rio Vista, TX.

Aerodyne Corporation, Minneapolis. *Electricity from Wind!* Minneapolis: Aerodyne Corporation, [ca. 1930]. Folder. In possession of Christopher Gillis.

American Energy Alternatives Inc., Boulder, CO. *Amerenalt Series Energy Conversion Systems*. Boulder, CO: American Energy Alternatives Inc. 1975. Brochure. Available in the *Windmillers' Gazette* research files, Rio Vista, TX.

American Wind Turbine, St. Cloud, FL. *Dear Wind Power Enthusiast*. St. Cloud, FL: American Wind Turbine, [ca. 1975]. Handbill. 3 leaves. Available in the *Windmillers' Gazette* research files, Rio Vista, TX.

Bergey Windpower Company, Norman, OK. *Annual Energy Output Curve*[,] *Revision 2-3-82*. Norman, OK: Bergey Windpower Company, 1982. Handbill. Available in *Windmillers' Gazette* research files, Rio Vista, TX.

———. *Bergey Excel 15*[:] *Advanced Technology—Superior Economics*. Norman, OK: Bergey Windpower, [ca. 2019]. Handbill. In possession of Christopher Gillis.

———. *Bergey Windpower*. Norman, OK: Bergey Windpower Company, [ca. 1982]. Handbill. Available in *Windmillers' Gazette* research files, Rio Vista, TX.

———. *Bergey Windpower Powersync Inverter Data Sheets*. Norman, OK: Bergey Windpower Company, 1982. 3 leaves. Available in *Windmillers' Gazette* research files, Rio Vista, TX.

———. *BWC 1000*[,] *Bergey Windpower*. Norman, OK: Bergey WIndpower Company, [ca. 1982]. Folder. Available in *Windmillers' Gazette* research files, Rio Vista, TX.

———. *BWC 1500 Wind Turbine System*. Norman, OK: Bergey Windpower, 1994. Handbill. In possession of Christopher Gillis.

———. *BWC EXCEL*[,] *Bergey Windpower Announces the BWC EXCEL*. Norman, OK: Bergey Windpower Company, 1983. Handbill. In possession of Christopher Gillis.

———. *BWC EXCEL Series*[,] *Bergey Wind Energy Systems*. Norman, OK: Bergey Windpower Company, 1983. Handbill. In possession of Christopher Gillis.

———. *BWC XL.1*[,] *1 KW Class Wind Turbine*. Norman, OK: Bergey Windpower, [ca. 2000]. Handbill. In possession of Christopher Gillis.

———. *BWC XL.50*[,] *50 KW Class Wind Turbine*. Norman, OK: Bergey Windpower, [ca. 2000], Handbill. In possession of Christopher Gillis.

———. *Windpower! BWC 1000 Series*. Norman, OK: Bergey Windpower Company, [ca. 1983]. Brochure. In possession of Christopher Gillis.

———. *Windpower! BWC EXCEL Series*. Norman, OK: Bergey Windpower Company, [ca. 1983]. Brochure. In possession of Christopher Gillis.

Bonus Energy A/S, Brande, Denmark. *Bonus Energy A/S*. Brande, Denmark: Bonus Energy, [ca. 2000]. Handbill. Available in the files of Warren S. Bollmeier, II, Kaneohe, HI.

Carter Wind Systems, Inc., Burkburnett, TX. *Carter Model 250*. Burkburnett, TX: Carter Wind Systems, Inc., [ca. 1983]. Brochure. Available in the files of Bob McBroom, Holton, KS.

———. *Model 200 Generator Specifications*. Burkburnett, TX: Carter Wind Systems, Inc., [ca. 1983]. Handbill. Available in the files of Bob McBroom, Holton, KS.

———. *Carter Wind Systems Model 25 Generator—July 1983*. Burkburnett, TX: Carter Wind Systems, Inc., 1983. Handbill. Available in the files of Bob McBroom, Holton, KS.

Chalk Wind Systems, St. Cloud, FL. *Modern Technology Comes to an Old Idea*. St. Cloud, FL: Chalk Wind Systems, [ca. 1979]. Folder. Available in the *Windmillers' Gazette* research files, Rio Vista, TX.

Charles J. Jager Company. *Illustrated Catalogue of Windmills, Tanks and Pumps as Applied to Water Supply Systems, also Windmills Adapted for Power*. Boston: Charles J. Jager Company, [ca. 1895]. Available in the Library and Archives, Museum of the Great Plains, Lawton, OK.

Chava Wind, Homestead, FL. *Chava Windleaf 2500*[:] *Chava Wind—The Art of Wind Power*. Homestead, FL: Chava Wind, [ca. 2019]. Brochure. In possession of Christopher Gillis.

Clean Energy Products, Seattle, WA. *Wind Power Systems and Components*. Seattle, WA: Clean Energy Products, 1979. Brochure. In possession of Christopher Gillis.

Coulson Wind Electric, Polk City, IA. *New Parts and Price List as of April 1, 1977*. Polk City, IA: Coulson Wind Electric, 1977. Typewritten handbill. In possession of Christopher Gillis.

Coyne Electrical School, Chicago. *Electricity*[,] *Gateway to Opportunity*[,] *Coyne Electrical School*[,] *Established 1899*[,] *Chicago*[,] *World's Greatest Electrical Center*. Chicago: Coyne Electrical School, [ca. 1940]. Brochure. In possession of Christopher Gillis.

———. *I Will Include An . . . Extra Radio Course*. Chicago: Coyne Electrical School, [ca. 1940]. Folder. In possession of Christopher Gillis.

Dakota Wind & Sun Ltd., Aberdeen, SD. *Common Questions and General Information*[,] *April 1978*. Aberdeen, SD: Dakota Wind & Sun Ltd., 1978. Folder. Available in the *Windmillers' Gazette* research files, Rio Vista, TX.

———. *Common Questions and General Information*[,] *September 1978*. Aberdeen, SD: Dakota Wind & Sun Ltd., 1978. Folder. Available in the *Windmillers' Gazette* research files, Rio Vista, TX.

———. *Greetings from the Factory. February 19, 1979*. Aberdeen, SD: Dakota Wind & Sun Ltd., 1979. 5 leaves. Available in the *Windmillers' Gazette* research files, Rio Vista, TX.

———. *Greetings from the Factory! April 4, 1979*. Aberdeen, SD: Dakota Wind & Sun Ltd., 1979. 7 leaves. Available in the *Windmillers' Gazette* research files, Rio Vista, TX.

———. *Greetings from the Factory! May 18, 1979*. Aberdeen, SD: Dakota Wind & Sun Ltd., 1979. 7 leaves. Available in the *Windmillers' Gazette* research files, Rio Vista, TX.

Delco-Light Company, Dayton, OH. *The Delco-Light Story*. Dayton, OH: Delco-Light Company, 1922. In possession of Christopher Gillis.

Dominion Aluminum Fabricating Limited, Mississauga, Canada. *Wind Turbines*. Mississauga, Canada: Dominion Aluminum Fabricating Limited, 1975. Folder. Available in the *Windmillers' Gazette* research files, Rio Vista, TX.

Ducted Turbines International, Potsdam, NY. *DTI*[:] *Design Technology*. Potsdam, NY: Ducted Turbines International, 2019. Handbill. In possession of Christopher Gillis.

Earth Energy Systems Inc., Eden Prairie, MN. *Bringing Energy to the World*. Eden Prairie, MN: Earth Energy Systems, 1985. Brochure. Available in the files of Craig Toepfer, Chelsea, MI.

———. *Class A Hybrid Power Plant*[,] *Wind/Solar/Engine/Battery*[:] *Energy-Minder Theory of Operation*. Eden Prairie, MN: Earth Energy Systems Inc., [ca. 1985]. Handout. Available in the files of Craig Toepfer, Chelsea, MI.

———. *Hybrid Power Plant*[:] *Class A up to 10 KW Continuous*. Eden Prairie, MN: Earth Energy Systems Inc., [ca. 1985]. Brochure. Available in the files of Craig Toepfer, Chelsea, MI.

———. *Hybrid Power Plant*[:] *Wind/Solar/Engine/Battery*[:] *Wincharger 1000 Theory of Operation*. Eden Prairie, MN: Earth Energy Systems Inc., [ca. 1985]. Handout. Available in the files of Craig Toepfer, Chelsea, MI.

The Electric Storage Battery Company, Philadelphia. *The Exide-Hyray Battery*. Philadelphia: The Electric Storage Battery Company, 1921. Brochure with price sheets. In possession of Christopher Gillis.

Energy Development Company, Hamburg, PA. *Energy of the Future Available Today*. Hamburg, PA: Energy Development Company, [ca. 1980]. Brochure. Available in the files of Bob McBroom, Holton, KS.

Energy Sciences Incorporated, Boulder, CO. *The New Generation in Wind Technology*. Boulder, CO: Energy Sciences Incorporated, [ca. 1982]. Brochure. In possession of Christopher Gillis.

Enertech, Norwich, VT. *Enertech 1500*[,] *A Beautiful Way to Save Electricity*. Norwich, VT: Enertech, 1979. Brochure. Available in *Windmillers' Gazette* research files, Rio Vista, TX.

———. *Enertech 1800*[,] *A Beautiful Way to Save Electricity*. Norwich, VT: Enertech, 1981. Brochure. Available in *Windmillers' Gazette* research files, Rio Vista, TX.

———. *Enertech E44*. Norwich, VT: Enertech, [ca. 1982]. Brochure. In possession of Christopher Gillis.

———. *Enertech E44*. Norwich, VT: Enertech, [ca. 1986]. Brochure. Available in *Windmillers' Gazette* research files, Rio Vista, TX.

Etablissement des Aéromoteurs Cyclone, Margny-lès-Compiègne, France. *Aéromoteurs "Cyclone" pour L'Élévation de L'Eau et al. Production de L'Électricité*[,] *Pompes*[,] *Pompes a Bras—Pompes a Moteurs*[,] *Pompes D'Irrigation—Réservoirs*. Margny-les-Compiègne, France: Étabits des Aéromoteurs "Cyclone," [ca. 1928]. Brochure. Available in research files of Etienne Rogier, Toulouse, France.

Fayette Manufacturing Corporation, Fayette, CA. *Fayette Manufacturing Corporation*. Fayette, CA: Fayette Manufacturing Corporation, [ca. 1985]. Brochure. Available in the files of Earle Rich, Mont Vernon, NH.

FloWind Corporation, Kent, WA. *FloWind Corporation*[:] *An Integrated Wind Energy Company*. Kent, WA: FloWind Corporation, [ca. 1985]. Folder. In possession of Christopher Gillis.

———. *FloWind Model 120 Vertical Axis Wind Turbine*.

Kent, WA: FloWind Corporation, [ca. 1985]. Folder. In possession of Christopher Gillis.

———. *FloWind Model 170 Vertical Axis Wind Turbine.* Kent, WA: FloWind Corporation, [ca. 1985]. Folder. In possession of Christopher Gillis.

Grumman Energy Systems, Ronkonkoma, NY. *Grumman Windstream 25.* Ronkonkoma, NY: Grumman Energy Systems, 1976. Handbill. Available in the files of Bob McBroom, Holton, KS.

Herbert E. Bucklen Corporation, Elkhart, IN. *HEBCO All Service Wind Electrics Price List 101-F[,] Aug. 5, 1928.* Elkhart, IN: Herbert E. Bucklen Corporation, 1928. Folder. Available in *Windmillers' Gazette* research files, Rio Vista, TX.

———. *Light and Power at the Cost of Pumping Water by Wind-Mill.* Elkhart, IN: Herbert E. Bucklen Corporation, [ca. 1927]. Brochure. In possession of Christopher Gillis.

———. *Modern Conveniences Are Yours[,] Just Snap on the Switch.* Elkhart, IN: Herbert E. Bucklen Corporation, [ca. 1932]. Brochure. In possession of Christopher Gillis.

———. *What the United States Government and Others Think and Say about HEBCO Wind-Electrics.* Publication Form 129-A. Elkhart, IN: Herbert E. Bucklen Corporation, [ca. 1927]. Circular. In possession of Christopher Gillis.

HOMER Energy, Boulder, CO. *HOMER Grid[:] Intelligently Reduce Your Peak Power.* Boulder, CO: HOMER Energy, 2019. Handbill. In possession of Christopher Gillis.

———. *HOMER Pro[:] The Global Standard for Optimizing Microgrid Design.* Boulder, CO: HOMER Energy, 2019. Handbill. In possession of Christopher Gillis.

Honeywell International, Inc., Ontario, Canada. *Honeywell Wind Turbine Model WT6500 Owner's Manual.* Ontario, Canada: Honeywell International Inc. [ca. 2011]. In possession of Christopher Gillis.

Independent Energy Systems, Inc., Fairview, PA. *IES Independent Energy Systems, Inc. Catalog #216.* Fairview, PA: Independent Energy Systems, Inc., 1979. Brochure. Available in the *Windmillers' Gazette* research files, Rio Vista, TX.

———. *IES Independent Energy Systems Inc. Integrated Utility Free Energy Systems.* Fairview, PA: Independent Energy Systems, Inc., [ca. 1977]. Brochure. Available in the *Windmillers' Gazette* research files, Rio Vista, TX.

Jacobs Wind Electric Company, Fort Myers, FL. *Important Features of the New Jacobs 8 KVA Wind Energy System.* Fort Myers, FL: Jacobs Wind Electric Company, 1979. Handbill. Available in the files of Paul R. Jacobs, Corcoran, MN.

———. *New Jacobs Wind Energy System Summary Information.* Fort Myers, FL: Jacobs Wind Electric Com-

pany, 1979. Handbill. Available in the files of Paul R. Jacobs, Corcoran, MN.

Jacobs Wind Electric Company, Minneapolis, MN. *Electricity from the Air[:] The Super Automatic Wind Electric Plant with Fly Ball Governor Control.* Minneapolis: Jacobs Wind Electric Company, [ca. 1940]. Brochure. In possession of Christopher Gillis.

———. *Introduction to Wind Electric Generation[:] Background (1920–1970).* Minneapolis, MN: Jacobs Wind Electric Company, 1980. Handbill. In possession of Christopher Gillis.

———. *Jacobs[:] The Standard of Comparison.* Minneapolis, MN: Jacobs Wind Electric Company, [ca. 1935]. Brochure. In possession of Christopher Gillis.

———. *Jacobs Air Way[,] The New Model 35.* Minneapolis, MN: Jacobs Wind Electric Company, [ca. 1949]. Brochure. In possession of Christopher Gillis.

———. *The Jacobs Super Automatic Wind Electric Plant.* Minneapolis, MN: Jacobs Wind Electric Company, [ca. 1940]. Handbill. In possession of Christopher Gillis.

———. *The Jacobs Twin Motor Electric.* Minneapolis, MN: Jacobs Wind Electric Company, 1938. Brochure. In possession of Christopher Gillis.

———. *Jacobs Wind Electric[,] Pipe Line Cathodic Protection.* Minneapolis, MN: Jacobs Wind Electric Company, [ca. 1950]. Handbill. Available in the files of Paul R. Jacobs, Corcoran, MN.

———. *Navy Special Jacobs Engine Electric Plant.* Minneapolis, MN: Jacobs Wind Electric Company, [ca. 1943]. Handbill. Available in the files of Paul R. Jacobs, Corcoran, MN.

———. *New 4th Generation of Jacobs Wind Systems.* Minneapolis, MN: Jacobs Wind Electric Company, 2009. Handbill. In possession of Christopher Gillis.

———. *Pipeline Cathodic Protection Plants.* Minneapolis, MN: Jacobs Wind Electric Company, [ca. 1950]. Handbill. Available in the files of Paul R. Jacobs, Corcoran, MN.

———. *Pipe Line Protection with the New Jacobs System.* Minneapolis, MN: Jacobs Wind Electric Company, [ca. 1950]. Handbill. Available in the files of Paul R. Jacobs, Corcoran, MN.

———. *Twin Motor Electric[,] Automatic Proven Speed Control, Charging Control, Voltage Control[,] Price List[,] Effective February 1, 1938.* Minneapolis, MN: Jacobs Wind Electric Company, 1938. Handbill. In possession of Christopher Gillis.

———. *With Byrd in Little America.* Minneapolis, MN: Jacobs Wind Electric Company, [ca. 1940]. Handbill. In possession of Christopher Gillis.

James Alston & Sons Pty. Ltd., South Melbourne, Australia. *"Alston" Patent Electric Generating Windmill.* South Melbourne, Australia: James Alston & Sons Pty.

Ltd., [ca. 1931]. Brochure. Available in *The Windmill Journal* research files, Morawa, Australia.

Jay Carter Enterprises, Inc., Burkburnett, TX. *Carter Wind Generator Model 25 General Description*. Burkburnett, TX: Jay Carter Enterprises, Inc., [ca. 1981]. Brochure. Available in the files of Bob McBroom, Holton, KS.

Kaman Aerospace Corporation, Bloomfield, CT. *Kaman Wind Energy Systems*. Bloomfield, CT: Kaman Aerospace Corporation, 1980. Available in the *Windmillers' Gazette* research files, Rio Vista, TX.

Kedco, Inc. Inglewood, CA. *Harness Electricity from the Wind*[,] *Kedco 1200*. Inglewood, CA: Kedco, Inc., [ca. 1975]. Handbill. Available in the *Windmillers' Gazette* research files, Rio Vista, TX.

Klinsick Mechanical Shop, Optima, OK. *Price List*. Optima, OK: Klinsick Mechanical Shop, [ca. 1935]. Loose-leaf page. Available in the files of Norman Marks, Geneva, NE.

———. *The Wind is Free*[,] *Why Not Use It!* Optima, OK: Klinsick Mechanical Shop, [ca. 1935]. Brochure. Available in the files of Norman Marks, Geneva, NE.

Kohler Company, Kohler, WI. *The Principle and the Proof*. Kohler, WI: Kohler Company, 1924. In possession of Christopher Gillis.

———. *"And It's the Thriftiest Thing on the Place."* Kohler, WI: Kohler Company, 1925. Brochure. In possession of Christopher Gillis.

LeJay Manufacturing Company, Minneapolis, MN. *LeJay 6 and 32 Volt Slow Speed Wind Plants*. Minneapolis: LeJay Manufacturing Company, [ca. 1935]. Brochure. In possession of Christopher Gillis.

———. *LeJay 6 and 32 Volt Slow Speed Wind Plants Catalog No. 12*. Minneapolis, MN: LeJay Manufacturing Company, [ca. 1935]. Brochure. In possession of Christopher Gillis.

———. *LeJay Slow Speed Wind Plants Catalog No. 15*. Minneapolis, MN: LeJay Manufacturing Company, 1936. Brochure. In possession of Christopher Gillis.

———. *LeJay Slow Speed Wind Plants Catalog No. 16*. Minneapolis, MN: LeJay Manufacturing Company, 1936. Brochure. In possession of Christopher Gillis.

———. *LeJay Slow Speed Wind Plants Catalog No. 17*. Minneapolis, MN: LeJay Manufacturing Company, 1936. Brochure. In possession of Christopher Gillis.

———. *LeJay Slow Speed Wind Plants Catalog No. 18A*. Minneapolis, MN: LeJay Manufacturing Company, [ca. 1937]. Brochure. In possession of Christopher Gillis.

Manning Engineering and Sales Company, Elkhart, IN. *Wind Driven Electric Light, Power and Irrigation Plants*. Elkhart, IN: Manning Engineering & Sales Company, [ca. 1923]. Folder. Available in the Kregel Windmill Company Papers, Nebraska State Historical Society, Lincoln, Nebraska.

Montgomery Ward & Company, Chicago. *Wards Powerlite Electric Light and Power Plants*. Publication 2–15–38. Chicago: Montgomery Ward & Company, 1938. In possession of Christopher Gillis.

National Batteries, St. Paul, MN. *Isolated Lighting Plant Batteries*. St. Paul, MN: National Batteries, [ca. 1930]. Brochure. In possession of Christopher Gillis.

National Manufacturing Company, Lincoln, NE. *Hang This Card Up on the Wall Near Your Light Plant, Storage Batteries on in Work Shop*. Lincoln, NE: National Manufacturing Company, [ca. 1935]. Handbill with cover letter. In possession of Christopher Gillis.

Natural Power Tower Inc., New Boston, NH. *About Natural Power Inc.* New Boston, NH: Natural Power Inc., 1977. Handbill. Available in *Windmillers' Gazette* research files, Rio Vista, TX.

———. *Anemometer—Recorder System*. New Boston, NH: Natural Products Inc., 1976. Handbill. In possession of Christopher Gillis.

———. *Data Accumulator*. New Boston, NH: Natural Power Inc., 1976. Handbill. In possession of Christopher Gillis.

———. *Octahedron Module Tower*. New Boston, NH: Natural Power Inc., 1976. Handbill. In possession of Christopher Gillis.

———. *Octahedron Module Tower*. New Boston, NH: Natural Power Inc., 1977. Handbill. In possession of Christopher Gillis.

———. *Price List*. New Boston, NH: Natural Power Inc., 1976. Handbill. In possession of Christopher Gillis.

———. *Specialists in Instrumentation and Controls for Alternate Energy Industry*. New Boston, NH: Natural Power Inc., 1976. Handbill. In possession of Christopher Gillis.

———. *Tachometer*. New Boston, NH: Natural Power Inc., 1976. Handbill. In possession of Christopher Gillis.

———. *Wind and Solar Energy Products*. New Boston, NH: Natural Products Inc., 1977. Brochure. Available in the files of Richard Katzenberg, Amherst, NH.

———. *Wind and Solar Energy Products*. New Boston, NH: Natural Products Inc., 1984. Catalog. In possession of Christopher Gillis.

———. *Wind Data Compiler*. New Boston, NH: Natural Power Inc., 1976. Handbill. In possession of Christopher Gillis.

———. *Wind Energy Monitor*. New Boston, NH: Natural Power Inc., 1976. Handbill. In possession of Christopher Gillis.

———. *Wind Survey Techniques Using Natural Power Inc. Recording Anemometers*. New Boston, NH: Natural Power Inc., 1976. Brochure. In possession of Christopher Gillis.

Nordtank Energy Group A/S, Balle, Denmark. *The 1.4 MW Wind Turbine of Tomorrow*. Balle, Denmark: Nordtank Energy Group, [ca. 1997]. Brochure. Available in the files of Warren S. Bollmeier, II, Kaneohe, HI.

North Wind Power Company, Moretown, VT. *North Wind Power Co.* Moretown, VT: North Wind Power Company, [ca. 1982]. Brochure. In possession of Christopher Gillis.

———. *Reliable Power from the Wind.* Moretown, VT: North Wind Power Company, 1981. Brochure. In files of Christopher Gillis.

North Wind Power Company, Warren, VT. *The 2KW High Reliability Wind Turbine Generator.* Warren, VT: North Wind Power Company, 1979. Handbill, 5 leaves. Available in the *Windmillers' Gazette* research files, Rio Vista, TX.

———. *North Wind Power Co*[.] Warren, VT: North Wind Power Company, [ca. 1979]. Brochure. In possession of Christopher Gillis.

———. *North Wind Power Co., Inc.* Warren, VT: North Wind Power Company, [ca. 1980]. Brochure. In possession of Christopher Gillis.

———. *Reliable Power from the Wind.* Warren, VT: North Wind Power Company, [ca. 1980]. Brochure. In possession of Christopher Gillis.

NRG Systems, Charlotte, VT. *NRG Systems Logger 9000*[,] *an Advanced Real-Time Serial Wind Data Logger.* Charlotte, VT: NRG Systems, 1987. Folder. Available in the *Windmillers' Gazette* research files, Rio Vista, TX.

Parker-McCrory Manufacturing Company, Kansas City, MO. *The Modern Fencing Method*[,] *Parmak Electric Fencer with the Amazing Flux Diverter and Dry Weather Intensifer.* Kansas City, MO: Parker-McCrory Manufacturing Company, 1939. Brochure. In possession of Christopher Gillis.

Parris-Dunn Associates, Clarinda, IA. *Manual of Arms for the Victory TraineRifle.* Clarinda, IA: Parris-Dunn Associates, [ca. 1943]. Brochure. In possession of Christopher Gillis.

Parris-Dunn Corporation, Clarinda, IA. *32 Volt Free Lite*[:] *World's Most Complete Line of Direct-Drive Wind Electric Power Plants.* Clarinda, IA: Parris Dunn Corporation, [ca. 1940]. Brochure. In possession of Christopher Gillis.

———. *Hy-Tower*[,] *The World's Best 6-Volt Charger*[,] *Free Power for Radio and Lights with the Hy-Tower.* Clarinda, IA: Parris-Dunn Corporation, [ca. 1935]. Brochure. In possession of Christopher Gillis.

———. *Instruction Manual for Model 206—Six Volt* [and] *Model 212—Twelve Volt*[,] *Direct-Drive*[,] *Slip-the-Wind Power Plants.* Clarinda, IA: Parris-Dunn Corporation, [ca. 1935]. Available in the files of Mike Werst, Manor, TX.

———. *Direct Drive 6–12–32–110 Volts Wind Electric Light & Power Plants.* Clarinda, IA: Parris-Dunn Cor-

poration, [ca. 1940]. Handbill. Available in the files of Mike Werst, Manor, TX.

———. *Power for Radio and Lights with the Hy-Tower.* Clarinda, IA: Parris-Dunn Corporation, [ca. 1940]. Brochure. In possession of Christopher Gillis.

Peerless Battery Manufacturing, Lincoln, NE. *Let the Wind Do Your Work*[,] *the System Without a Peer*[,] *Peerless Battery Manufacturing Company*[,] *Lincoln*[,] *Nebraska.* Lincoln, NE: Peerless Battery Manufacturing, [ca. 1920]. Folder. Available in the Kregel Windmill Company Papers, Nebraska State Historical Society, Lincoln, Nebraska.

Perkins Corporation, Mishawaka, IN. *Aerolectric Light and Power from the Wind*[,] *Electricity*[,] *a City Luxury for Every Farm.* Form 1001. Mishawaka, IN: Perkins Corporation, [ca. 1922]. Folder. Available in *Windmillers' Gazette* research files, Rio Vista, TX.

———. *Electricity from the Wind.* Mishawaka, IN: Perkins Corporation, [ca. 1922]. Brochure. Available in *Windmillers' Gazette* research files, Rio Vista, TX.

———. *Electricity from the Wind.* Mishawaka, IN: Perkins Corporation, 1923. Folder with front printed as letterhead stationary. Available in *Windmillers' Gazette* research files, Rio Vista, TX.

Perkins Corporation, South Bend, IN. *The Guiding Star of the Air Mail Service.* South Bend, IN: Perkins Corporation, [ca. 1926]. Folder. In possession of Christopher Gillis.

———. *The New Perkins Aerolectric.* South Bend, IN: Perkins Corporation, [ca. 1924]. Folder. Sent with letter dated December 9, 1924. Available in *Windmillers' Gazette* research files, Rio Vista, TX.

Pinson Energy Corp., Marstons Mills, MA. *The Cycloturbine Vertical Axis Wind Turbine.* Marstons Mills, MA: Pinson Energy Corp., 1977. Brochure. In possession of Christopher Gillis.

———. *Power from the Wind, A Renewable Energy Source.* Marstons Mills, MA: Pinson Energy Corp., [ca. 1980]. Folder. In possession of Christopher Gillis.

Primus Wind Power, Lakewood, CO. *Air Completes Any Off Grid System.* Lakewood, CO: Primus Windpower, [ca. 2018]. Brochure. In possession of Christopher Gillis.

———. *Air X Owner's Manual*[:] *Installation * Operation * Maintenance.* Lakewood, CO: Primus Wind Power, 2013. Manual. In possession of Christopher Gillis.

Renewable Energy Ventures, Inc., Encino, CA. *Renewable Energy Ventures, Inc.*[,] *a Subsidiary of Earth Energy Systems, Inc.* Encino, CA: Renewable Energy Ventures, Inc., 1985. Brochure. Available in the files of Paul R. Jacobs, Corcoran, MN.

Rohn Manufacturing Company, Peoria, IL. *"This Is Your Line"*[,] *TV and Communication Towers.* Peoria, IL: Rohn Manufacturing Company, 1959. Brochure. In possession of Christopher Gillis.

Ruralite Engineering Company, Sioux City, IA. *Plenty of Dependable Electric Light and Power*. Sioux City, IA: Ruralite Engineering Company, [ca. 1938]. Folder. Available in the *Windmillers' Gazette* research files, Rio Vista, TX.

———. *Ruralite Wind Chargers 1938 Models*. Sioux City, IA: Ruralite Engineering Company, 1938. Brochure. In possession of Christopher Gillis.

Sears, Roebuck and Company, Atlanta. *The Latest Merchandise for Spring and Summer 1939*. Atlanta: Sears, Roebuck and Company, 1939. In possession of Christopher Gillis.

Sheldon & Sartor Wind Electric Manufacturing Company, Nehawka, NE. *Specially Designed Wind Electric Propeller Hubs Made to Fit Our Aluminum Alloy Propellers*. Nehawka, NE: Sheldon & Sartor Wind Electric Manufacturing Company, [ca. 1935]. Handbill. In possession of Christopher Gillis.

———. *Wind Electrics and Aluminum Propellers!* Nehawka, NE: Sheldon & Sartor Wind Electric Manufacturing Company, [ca. 1935]. Handbill. In possession of Christopher Gillis.

Solar Wind Company, Holden, ME. *Solar Wind Company*[,] *System and Component List*[,] *January 1974*. East Holden, ME: Solar Wind Company, 1974. Brochure. In possession of Christopher Gillis.

———. *Wind Generators Currently in Production Throughout the World*. East Holden, ME: Solar Wind Company, 1974. Typewritten letter. In possession of Christopher Gillis.

Southwest Windpower, Flagstaff, AZ. *Air Power*[,] *New Air 403*[,] *an All New Design that Once Again Redefines How the World Looks at Wind . . .* Flagstaff, AZ: Southwest Windpower, [ca. 1998]. Brochure. In possession of Christopher Gillis.

———. *Skystream 3.7 by Southwest Windpower*. Flagstaff, AZ: Southwest Winpower, Inc., [ca. 2009]. Brochure. In possession of Christopher Gillis.

———. *Skystream 3.7 Owner's Manual*. Flagstaff, AZ: Southwest Windpower, Inc., 2006. In possession of Christopher Gillis.

Star Wind Turbines, East Dorset, VT. *Star Wind Turbines LLC*. East Dorset, VT: Star Wind Turbines, [ca. 2018]. Brochure. In possession of Christopher Gillis.

Syverson Consulting, Mankato, MN. *Dunlite*[,] *Brushless*[,] *Wind Driven Power Plants*. North Mankato, MN: Syverson Consulting, 1975. Brochure. Available in the files of Bob McBroom, Holton, KS.

Twiford Corporation, Chicago Heights, IL, *Twiford Wind Motors with Electrical & Irrigation Equipment Especially Designed for Farm Lighting, Water Systems, and the Operation of All Household Electrical Appliances. Also for Power Driven Machinery Such as Feed Grinders, Corn Shellers, Circular Saws, Drilling Machines, Iron and Wood Lathers, Etc.* Chicago Heights, IL: Twiford Corporation, [ca. 1924]. Brochure. Available in the *Windmillers' Gazette* research files, Rio Vista, TX.

Unarco-Rohn, Division of Unarco Industries, Inc., Peoria, IL. *Rohn*. Peoria, IL: Unarco-Rohn, Division of Unarco Industries, Inc., [ca. 1978]. Handbill. In possession of Christopher Gillis.

———. *Rohn 25G Tower*. Peoria, IL: Unarco-Rohn, Division of Unarco Industries, Inc., 1978. Brochure. In possession of Christopher Gillis.

———. *Rohn No. 45G Communication Tower*. Peoria, IL: Unarco-Rohn, Division of Unarco Industries, Inc., 1979. Brochure. In possession of Christopher Gillis.

———. *Rohn SSV Self-Supporting Communication Towers*. Peoria, IL: Unarco-Rohn, Division of Unarco Industries, Inc., 1978. Brochure. In possession of Christopher Gillis.

Universal Battery Company, Chicago. *The New Universal Guide for Lengthening Battery Life*. Chicago: Universal Battery Company, 1932. Brochure. In possession of Christopher Gillis.

Utility Wind Interest Group, Palo Alto, CA. *The Evolving Wind Turbine*. Palo Alto, CA: Utility Wind Interest Group, 1993. Brochure. Available in the files of Warren S. Bollmeier, II, Kaneohe, HI.

Vestas Wind System A/S, Lem, Denmark. *1.65 MW*[:] *Vestas V66—1.65 MW Pitchregulated* [sic.] *Wind Turbine with OptiSlip and OptiTip*. Lem, Denmark: Vestas Wind System, [ca. 1997]. Brochure. Available in the files of Warren S. Bollmeier, II, Kaneohe, HI.

Whirlwind Power Company, Denver, CO. *All New for 1982! Economical, Reliable Wind Electric Systems for Home, Business and Remote Power Needs. Whirlwind Series 3000*[,] *3 KW Wind Generators*. Denver, CO: Whirlwind Power Company, 1982. Handbill. Available in the *Windmillers' Gazette* research files, Rio Vista, TX.

———. *Finally, a Cost-Effective System*. Denver, CO: Whirlwind Power Company, [ca. 1980]. Folder. Available in the *Windmillers' Gazette* research files, Rio Vista, TX.

Whirlwind Power Company, Duluth, MN. *WhirlWind Power Company*[,] *Catalog of Wind-Powered Generators and Accessories*. Duluth, MN: Whirlwind Power Company, [ca. 1983]. Brochure. Available in the *Windmillers' Gazette* research files, Rio Vista, TX.

———. *Whirlwind Power Company*[,] *Wind Powered Generators*[,] *2 Kilowatt—Series W2*[,] *4 Kilowatt—Series W4*[,] *10 Kilowatt—Series W10*. Duluth, MN: Whirlwind Power Company, [ca. 1987]. Brochure. Available in the *Windmillers' Gazette* research files, Rio Vista, TX.

Wincharger Corporation, Sioux City, IA. *Enjoy Your Own Electric Light & Power*[,] *Pay Only 10¢ a Year*

Power Operation Cost. Form No. 839. Sioux City, IA: Wincharger Corporation, [ca. 1938]. Brochure. In possession of Christopher Gillis.

———. *Every Farm Home Can Now Enjoy "Big City" Radio Reception.* Sioux City, IA: Wincharger Corporation, [ca. 1937]. Brochure. In possession of Christopher Gillis.

———. *Free Electricity for Radio and Lights*[,] *Brought to the Farm by Wincharger.* Sioux City, IA: Wincharger Corporation, [ca. 1935]. Brochure. In possession of Christopher Gillis.

———. *Giant 32 Volt Wincharger*[,] *Factory Direct to You*[,] *$69.95.* Sioux City, IA: Wincharger Corporation, [ca. 1936]. Brochure. In possession of Christopher Gillis.

———. *Here's That Popular, Low-Priced Steel Antenna Tower Made by Wincharger.* Sioux City, IA: Wincharger Corporation, [ca. 1953]. Brochure. In possession of Christopher Gillis.

———. *Light Your Farm for 50¢ a Year*[,] *Power Operating Cost.* Sioux City, IA: Wincharger Corporation, [ca. 1937]. Brochure. In possession of Christopher Gillis.

———. *New and Stronger Wincharger Tower.* Sioux City, IA: Wincharger Corporation, [ca. 1950]. Handbill. In possession of Christopher Gillis.

———. *Parts Price List for 6-Volt Winchargers*[,] *Covers Models '36, '37, '38.* Sioux City, IA: Wincharger Corporation, 1938. Brochure. In possession of Christopher Gillis.

———. *Self-Supporting Wincharger Tower.* Sioux City, IA: Wincharger Corporation, [ca. 1950]. Handbill. In possession of Christopher Gillis.

———. *Souvenir Booklet of the Army-Navy "E" Presentation to the Employees and Management of Wincharger Corporation.* Sioux City, IA: Wincharger Corporation, 1943. Brochure. In possession of Christopher Gillis.

———. *There's Power in the Air.* Sioux City, IA: Wincharger Corporation, [ca. 1937]. Folder. In possession of Christopher Gillis.

Winco, Division of Dyna Technology, Inc., Sioux City, IA. *Wincharger*[:] *12 Volt Wind Electric Battery Charger.* Sioux City, IA: Winco, Division of Dyna Technology, Inc., [ca. 1975]. Handbill. In possession of Christopher Gillis.

Wind-Power Light Company, Newton, IA. *End Your Battery Troubles with the Battery Made for Submarines and Supertrains*[:] *Get a Practically New Edison All Steel Alkaline Battery and . . . You're Through!* Newton, IA: Wind-Power Light Company, [ca. 1935]. Brochure. In possession of Christopher Gillis.

Wind Turbine Industries Corp., Prior Lake, MN. *Jacobs Wind Energy Systems 10–17.5 KW Wind Turbines.* Prior Lake, MN: Wind Turbine Industries Corp., 1986. Brochure. Available in the files of Dan Whitehead, St. Johns. FL.

Windependence Electric Co., Ann Arbor, MI. *Windependence Electric Co.* Ann Arbor, MI: Windependence Electric Co., [ca. 1976]. Brochure. Available in the files of Craig Toepfer, Chelsea, MI.

Windworks, Inc., Mukwonago, WI. *Gemini Synchronous Inverter Systems.* Mukwonago, WI: Windworks, Inc. Brochure. Available in the files of Craig Toepfer, Chelsea, MI.

———. *Windworks.* Mukwonago, WI: Windworks, Inc., [ca. 1980]. Brochure. Available in the files of Craig Toepfer, Chelsea, MI.

———. *Windworks—An Energy System Corporation • R&D • Product Engineering • Manufacturing.* Mukwonago, WI: Windworks, Inc., [ca. 1980]. Brochure. In possession of Christopher Gillis.

Woodmanse Manufacturing Company, Freeport, IL. *The Fritchle Wind-Electric.* Freeport, IL: Woodmanse Manufacturing Company, [ca. 1922]. Brochure. Available in *Windmillers' Gazette* research files, Rio Vista, TX.

———. *Light, Power & Water from the Wind.* Freeport, IL: Woodmanse Manufacturing Company, [ca. 1922]. Brochure. Available in *Windmillers' Gazette* research files, Rio Vista, TX.

———. *Price List No. 1 Covering The Fritchle Wind-Electric System.* Freeport, IL: Woodmanse Manufacturing Company, 1922. Folder. Available in *Windmillers' Gazette* research files, Rio Vista, TX.

WTG Energy Systems Inc., Buffalo, NY. *MP-20 Detailed Schematic*[,] *WTG Energy Systems Inc.* Buffalo, NY: WTG Energy Systems Inc., [ca. 1980]. Handbill. 4 leaves. Available in *Windmillers' Gazette* research files, Rio Vista, TX.

WTG Energy Systems Inc., Buffalo, NY. *MP 1–200 General Description.* Buffalo, NY: WTG Energy Systems Incorporated, [ca, 1979]. Handbill. Available in the files Earle Rich, Mont Vernon, NH.

Zephyr Wind Dynamo Company, Brunswick, ME. *Zephyr Wind Dynamo Company.* Brunswick, ME: Zephyr Wind Dynamo Company, [ca. 1976]. Brochure. Available in the files of Earle Rich, Mont Vernon, NH.

U.S. and Foreign Government Documents

Anderson, John W., Robert V. Brulle, Edwin B. Birchfield, and William D. Duwe. *Development of a 40 KW Giromill*[,] *Phase I*[,] *Volume II—Design and Analysis.* Prepared for Rockwell International Corporation, Energy Systems Group, Rocky Flats Plant, Wind Systems Program, Golden, CO, as part of the US Department of Energy, Division of Distributed Solar Technology, Federal Wind Energy Program. St. Louis, MO: McDonnell Aircraft Company, 1979.

Bolle, Thomas G. *Financial Problems Facing the Manufacturers of Small Wind Energy Conversion Systems*[,]

Final Report[,] *November 1979*. Washington, DC: JDB & Company, 1979.

Bollmeier, W. S., C. P. Butterfield, R. P. Dingo, D. M. Dodge, A. C. Hansen, D. C. Shepherd, and J. L. Tangler. *Small Wind Systems Technology Assessment*[:] *State of the Art and Near Term Goals*[,] *February 1980*. Golden, CO: Rockwell International, 1980.

Brulle, R. V. *Feasibility Investigation of the Giromill for Generation of Electrical Power*[,] *Volume I—Executive Summary*[,] *Final Report for the Period April 1975—April 1976*. Prepared for the US Energy Research and Development Administration, Division of Solar Energy. St. Louis: McDonnell Aircraft Company, 1977.

———. *Feasibility Investigation of the Giromill for Generation of Electrical Power*[,] *Volume II—Technical Discussion*[,] *Final Report for the Period April 1975—April 1976*. Prepared for the US Energy Research and Development Administration, Division of Solar Energy. St. Louis: McDonnell Aircraft Company, 1977.

Day, P. C. "The Winds of the United States and Their Economic Uses." *Yearbook of the United States Department of Agriculture*. Washington, DC: US Department of Agriculture, 1911.

Dodge, D. M., and W. S. Bollmeier, II. *Wind-Hybrid Systems Tests, US Department of Energy, National Renewable Energy Laboratory, Cooperative Field Test Program for Wind Systems*[,] *Final Report*. Golden CO: National Renewable Energy Laboratory, 1992.

Edwards, Jennifer L., Ryan Wiser, Mark Bolinger, and Trudy Forsyth. *Evaluating State Markets for Residential Wind Systems: Results from an Economic and Policy Analysis Tool*[,] *Prepared for the Wind & Hydropower Technologies Program*[,] *Paper LBNL-56344*[,] *December 2004*. Berkeley, CA: Ernest Orlando Lawrence Berkeley National Laboratory, 2004.

Energy Task Force. *Windmill Power for City People*[:] *A Documentation of the First Urban Wind Energy System*. Washington, DC: US Government Printing Office, 1977.

Fu, Ran, David Feldman, and Robert Margolis. *U. S. Solar Photovoltaic System Cost Benchmark: Q1 2018*. Golden, CO: National Renewable Energy Laboratory, 2018.

Huskey, A., and T. Forsyth. *NREL Small Wind Turbine Test Project: Mariah Power's Windspire Wind Turbine Test Chronology*[,] *Technical Report*[,] *NREL/TP-500–45552*[,] *June 2009*. Golden, CO: National Renewable Energy Laboratory, 2009.

Moran, W. A. *Giromill Wind Tunnel Test and Analysis*[,] *Volume 2. Technical Discussion*[,] *Final Report for the Period July 1976—October 1977*. Prepared for the US Energy Research and Development Administration, Division of Solar Energy. St. Louis: McDonnell Aircraft Company, 1977.

Murphy, Edward Charles. *Windmills for Irrigation*. US Department of the Interior, Geological Survey, Water-Supply and Irrigation Paper No. 8. Washington, DC: Government Printing Office, 1897.

National Academy of Sciences. *Establishment of a Solar Energy Research Institute*. Washington, DC: National Academy of Sciences, 1975.

National Renewable Energy Laboratory. *National Renewable Energy Laboratory*[:] *Ten-Year Site Plan*[,] *FY2007-FY2018*. Washington, DC: Office of Energy Efficiency and Renewable Energy, US Department of Energy, 2006.

National Research Council Canada, National Aeronautical Establishment. *Laboratory Technical Report*[,] *LTR-LA-74*[,] *Preliminary Tests of a High Speed Vertical Axis Windmill Model*. By P. South and R. S. Rangi. Ottawa, Canada: National Research Council Canada, 1971.

Noll, R. B., N D. Ham, H. M. Drees, and L. B. Nichol. *ASI/Pinson 1 Kilowatt High Reliability Wind System Development*[,] *Phase I—Design and Analysis*[,] *Technical Report*[,] *March 1982*. Prepared for Rockwell International Corporation, Energy Systems Group, Rocky Flats Plant, Wind Systems Program, Golden, CO, as part of the US Department of Energy, Wind Technology Division, Federal Wind Energy Program. Burlington, MA: Aerospace Systems Inc., 1982.

Orrell, A. C., L. T. Flowers, M. N. Gagne, B. H. Pro, H. E. Rhoads-Weaver, J. O. Jenkins, K. M. Sahl, and R. E. Baranowski. *US Department of Energy*[,] *Energy Efficiency & Renewable Energy*[,] *Wind Program*[,] *2012 Market Report on U. S. Wind Technologies in Distributed Applications*[,] *August 2013*. Richland, WA: Pacific Northwest Laboratory, 2013.

Orrell, Alice C., Nikolas F. Foster, Scott L. Morris, and Juliet S. Homer. *2016 Distributed Wind Market Report*. Richland, WA: Pacific Northwest National Laboratory, 2017.

Rockwell International, Rocky Flats Wind Systems Program, Golden, CO. *Commercially Available Small Wind Systems and Equipment*[,] *March 31, 1981*[,] *A Checklist Prepared by the Rocky Flats Wind Systems Program*. Golden, CO: Rockwell International, 1981.

Schefter, James L. *Capturing Energy from the Wind*. Washington, DC: Scientific and Technical Information Branch, National Aeronautics and Space Administration, 1982.

Sutherland, Herbert J., Dale E. Berg, and Thomas D. Ashwill. *Sandia Report*[:] *A Retrospective of VAWT Technology*. Albuquerque, NM: Sandia Laboratories, 2012.

Tangler, James. *Advanced Wind Turbine Design Studies*[,] *Advanced Conceptual Study*[,] *Final Report*[,] *Atlantic Orient Corporation*[,] *Norwich, Vermont*. Prepared for the US Department of Energy, National Renewable Energy Laboratory. Golden CO: National Renewable Energy Laboratory, 1994.

US Congress. House. *Production of Power by Means of Wind-Driven Generator: Hearing before the Committee on Interior and Insular Affairs*. HR 4286. 82nd Cong., 1st sess., September 19, 1951.

———. Senator Metcalf of Montana. Speaking on Wind Power, on August 7, 1974. *Congressional Record*. 93rd Cong., 2nd sess., 1974. Vol. 120, pt. 20.

US Department of Agriculture, Agricultural Research Service. *Agriculture Information Bulletin Number 446[,] Wind Power Research for Agriculture*. Washington, DC: US Department of Agriculture, 1981.

———. *Development of Rural and Remote Applications of Wind-Generated Energy[,] Summary of Expected Requests for Proposal[,] July 1977–June 1978*. Beltsville, MD: US Department of Agriculture, 1977.

———, Southern Great Plains Research Center. *Irrigation Pumping with Wind Energy*. Bushland, TX: US Department of Agriculture, 1978.

US Department of Energy. *20% Wind Energy by 2030[:] Increasing Wind Energy's Contribution to U. S. Electricity Supply[,] DOE/GO-102008–2567[,] May 2008*. Washington, DC: US Department of Energy, 2008.

———. "Alaska to Harness Arctic Winds for Village Power." *Wind Power Today* (1997): 24–29, 31.

———. "A New Generation of Wind Turbines on the Horizon." *Wind Power Today* (1997): 12–21.

———. "Technology Key to Adoption of Wind Energy in Alaska." *Wind Power Today* (1997): 30.

———. *2016 Distributed Wind Market Report*. Richland, WA: Pacific Northwest National Laboratory, 2017.

———. *2017 Distributed Wind Market Report*. Richland, WA: Pacific Northwest National Laboratory, 2018.

———. *2018 Distributed Wind Market Report*. Richland, WA: Pacific Northwest National Laboratory, 2019.

———. *Distributed Wind Competitiveness Improvement Project*. Golden, CO: National Renewable Energy Laboratory, 2019. Handbill. In possession of Christopher Gillis.

———. *Wind Energy Systems Program Summary[,] Fiscal Years 1981 and 1982*. Washington, DC: US Department of Energy, 1983.

———. *Workshop Report: Wind Innovations for Rural Economic Development (WIRED)[,] December 2018*. Richland, WA: Pacific Northwest National Laboratory, 2018.

US Department of Energy, Division of Solar Technology. *Federal Wind Energy Program[,] Program Summary[,] January 1978*. Washington, DC: US Department of Energy, 1978.

US Department of Energy, Office of Energy Efficiency and Renewable Energy. *2015 Distributed Wind Market Report*. Richland, WA: Pacific Northwest National Laboratory, 2016.

US Department of Energy, Wind Energy Technology Division. *Five Year Research Plan 1985–1990[,] Wind Energy Technology: Generating Power from the Wind[,] January 1985*. Washington, DC: US Department of Energy, 1985.

US Department of Energy, Wind Systems Branch, Division of Solar Technology, Washington, DC. *A guide to Commercially Available Wind Machines[,] April 3, 1978*. Washington, DC: US Department of Energy, 1978.

———. *Technical and Management Support for the Development of Wind Systems for Farm, Remote and Rural Use[,] Annual Report for the Period October 1976-September 1977[,] Wind Systems Program[,] Rocky Flats Plant[,] Rockwell International[,] October 1977*. Contract E(20–2) 3533. Washington, DC: US Department of Energy, 1977.

US Energy Information Administration. *Total Energy[,] Annual Energy Review[,] Average Retail Prices of Electricity, 1960–2011*. September 2012. Accessed July 25, 2017. https://www.eia.gov/totalenergy/data/annual/showtext.php?t=ptb0810.

US Internal Revenue Service. *Property Qualifying for the Energy Credit under Section 48[,] Notice 2015–4*. Washington, DC: US Internal Revenue Service, 2015. Accessed July 14, 2019. https://www.irs.gov/pub/irs-drop/n-15–04.pdf.

US Office of Consumer Affairs. *People Power: What Communities are Doing to Counter Inflation*. Report No. ED193164. Edited by Mary S. Gordon. Washington, DC: US Office of Consumer Affairs, 1980.

Wegley, Harry L., James V. Ramsdell, Montie M. Orgill, and Ron L. Drake. *A Siting Handbook for Small Wind Energy Conversion Systems[,] PNL-2521 Rev 1[,] March 1980*. Prepared for the US Department of Energy, Washington, DC, as part of the Pacific Northwest Laboratory, Battelle Memorial Institute. Richland, WA: Pacific Northwest Laboratory, 1980.

Zayas, Jose R., and Wesley D. Johnson. *Sandia Report[:] 3X-100 Blade Field Test*. Albuquerque, NM: Sandia National Laboratories, 2008.

Index

APRS World, 224
Archer, William R. Jr., 90. *See also* production tax credit
ARCO Solar, 144. *See also* solar photovoltaic systems
Arndt, Ernest A., 43
Ashe, William, 72
Atlantic Orient Corporation, 85, 91, 125, 200, 205. *See also* wind farms
Autocon, 138, 141. *See also* Control Data Corporation; Jacobs Wind Electric Company

Barnes, Bill, 72. *See also* Windlite Alaska
Bass, Charles and Mabel, 49–50. *See also* Rural Electrification Administration
Bass, Delwin, 50. *See also* Rural Electrification Administration
Bates, Putnam, A., 27
Batesole, William, 77. *See also* American Wind Energy Association; Kaman Aerospace Corporation
Battelle Pacific Northwest Laboratories, 75, 118
batteries: glass, 7, 41; lead-acid, 62; storage systems, 16–18, 23–24, 144, 220–22
Baucus, Max, 88. *See also* American Wind Energy Association
Baylor University, 125. *See also* Southwest Windpower
Bayly, Elliott, 67. *See also* Wind Power Technologies
Becquerel, Edmond, 214
Bell Laboratories, 214. *See also* solar photovoltaic systems
Bergey, Karl Halteman, Jr.: 1920s and 1930s, 150–51; aviation influence, 150–52; Brown Junior engines, 151; education, 151–52; North American Rockwell, 152; Piper Development Center, 152; University of Oklahoma Department of Aerospace, Mechanical and Nuclear Engineering, 152–54; Urban Car, 152–53; wind farm study, 161; wind generator studies and testing, 153–56
Bergey, Michael L. S., 79–80, 85, 92, 93–94, 95, 98, 152–168, 213. *See also* American Wind Energy Association; Distributed Wind Energy Association

Bergey Windpower Company: aluminum blades, 156–57; BWC 1000, 157–59; BWC 1000-S, 158; BWC 1500, 163–64; BWC Excel, 159, 161, 164; BCW Excel 6, 166; BWC Excel 10, 92, 161, 166, 168, 213, 216, 218, 219; BWC Excel 15, 127–28, 168, 219; BWC XL.1, 164–65; BWC XL.50, 166; China manufacturing, 97, 165; fiberglass blades, 159, 176; exports, 159, 164–65; hybrid systems, 159, 164; leasing, 219–220 Norman, Oklahoma, 80, 85, 155–168, 171, 198, 202, 203, 205, 208; Powerflex, 157; Powersync inverter, 159; water-pumping wind turbines, 85, 91, 123–24, 125, 162–64
Betz, Albert, 172
Betz Limit, 172. *See also* wind turbine rotors
Biorn, Paul, 62
Bipartisan Budget Act of 2018, 98
Blake, Steve, 74, 77. *See also* American Wind Energy Association; Sunflower Power
Blittersdorf, David and Jan, 93–94, 197. *See also* Distributed Wind Energy Association; NRG Systems; wind measurement technology
Bloomberg, Michael R., 211
Blyth, James, 9–10
Bode, Denise, 95. *See also* American Wind Energy Association
Boeing, 81, 102, 119–120, 152, 161. *See also* Federal Wind Energy Program; National Aeronautics and Space Administration
Bogert, Frank, 85. *See also* wind farms
Bollmeier, Warren, 199–200
Bonus Energy, 90. *See also* wind farms
Borsodi, Ralph, 53, 55
Branstad, Terry, 88. *See also* Renewable Portfolio Standard
Braun, Tom, 137. *See also* Jacobs Wind Electric Company
Braun Engineering, 137. *See also* Jacobs Wind Electric Company
Brewer, Roy, 72. *See also* Winpower Manufacturing Company
British Wind Energy Association, 93. *See also* small wind turbine performance standards
Bronx Frontier Development Corporation, 179
Brown Hannaford, Thomas, 11

Brulle, Robert V., 188. *See also* Giromill; McDonnell-Douglas Aircraft Corporation; Valley Industries Aermotor Division
Brush, Charles F., 11–13
Brush Electric Company, 11
Buck, Foster, 45. *See also* North Dakota Agricultural College
Bucklen, Herbert E., 34–36
Bucklen-Perkins Aerolectric, 42–43
Buckwalter, Len, 58
Burnham, John, 5
Bush, George H. W., 85–86, 121, 199
Bush, George W., 89, 122
Bushe, Allen, 157, 162. *See also* Bergey Windpower Company
Bushland, Texas, 81, 116–18, 123–25, 162–64. *See also* US Department of Agriculture
Byrd, Richard, 39

C. H. Charlson Manufacturing Company, 42
Cadbury, George, 19–20. *See also* wind electric generators
Cadillac, Michigan, 22
Caldwell, Frank W., 32
California Public Utilities Commission, 89. *See also* Renewable Portfolio Standard
California Wind Energy Association, 88–89, 143. *See also* Renewable Portfolio Standard
Calley, David, 85, 203, 217. *See also* Southwest Windpower
Canadian National Research Council, 100. *See also* vertical-axis wind turbine
Canadian Wind Energy Association, 93. *See also* small wind turbine performance standards
Candee, Harry T., 24
Cannon Wind Eagle Corporation, 122, 203. *See also* US Department of Energy
Carter, Jay Jr., 77. *See also* Carter Enterprises Inc.
Carter, Jay Sr., 182. *See also* Carter Enterprises Inc.
Carter, Jimmy, 76, 119, 120
Carter Enterprises Inc., 77, 118, 177, 182. *See also* Rockwell International; Rocky Flats
Carter Wind Energy (Systems), 158, 171, 191, 223.

cathodic protection, 46. *See also* Jacobs Wind Electric Company; Wincharger Corporation; Wind Power Light Company

Central Vermont Public Service Corporation, 49, 190–91

Chalk, Thomas O., 112, 175. *See also* American Wind Turbine Company

Challenge Company, 40. *See also* towers

Chapin, Daryl M., 214

Charles E. Miller Company, 42

Charles J. Jager Company, 19

Chava Wind, 223. *See also* vertical-axis wind turbine

China, 2, 97. *See also* horizontal windmill

Christianson, Mary, 70

Clark, R. Nolan, 123, 125. *See also* Agricultural Research Service; Bushland, Texas; Conservation and Production Research Laboratory; US Department of Agriculture

Clark, Ramsey, 70

Clarkson College (now Clarkson University), 186, 223. *See also* vertical-axis wind turbine

Clean Air Act of 1990, 199

Clean Renewable Energy Bonds, 206

Clews, Henry and Rita, 56–58, 170. *See also* Solar Wind Company

Climate Change Action Plan, 122

Climate Change Technology Initiative, 122

Clinton, William J., 122

Cohn, Ken, 197. *See also* Second Wind Inc.; wind measurement technology

Coleman, Clint (Jito), 200. *See also* Northern Power Systems

Coleman, Eliot and Sue, 53–54

Colorado State University, 117. *See also* Agricultural Research Service; US Department of Agriculture

Columbus Windmill Company, 6

Commercially Available Small Wind Systems and Equipment, 113. *See also* Rockwell International; Rocky Flats

Commonwealth Edison Company of Chicago, 21, 22. *See also* electric power

community wind, 206–207, 221

Compagnie Electro-Mécanique, 49

Conservation and Production Research Laboratory, 118, 123. *See also* Agricultural Research Service; Bushland, Texas; US Department of Agriculture; wind electric generators

Conservation Tools and Technology, 72

Consolidated Edison (ConEd) (New York), 69–70

Consumer's Guide to Commercially Available Wind Turbine Generators, 76. *See also* Rockwell International; Rocky Flats

Continental Windmill Company, 5

Control Data Capital Corporation (CDCC), 137–148. *See also* Jacobs Wind Electric Company

Control Data Corporation (CDC), 137–148. *See also* Jacobs Wind Electric Company

Cook, Jerry, 72

Cook Brothers, 30. *See also* horizontal windmills

Cooperative Field Test Program, 120. *See also* Rocky Flats; Solar Energy Research Institute

Corcoran, Andrew J., 18–19

Cornell University, 117. *See also* Agricultural Research Service; US Department of Agriculture

Coulson, Richard, 63, 67

Coulson Wind Electric, 63, 67

Council for Renewable Energy Education, 86

Creative Electronics, 72

Crosiar, Eli, 23–24

Crotched Mountain (New Hampshire), 161. *See also* US Windpower; wind farms

Crude Oil Windfall Profits Act of 1980, 79, 82

Cuttyhunk, Massachusetts, 180, 182

Cycloturbine, 77, 103, 109, 154, 187–88. *See also* vertical-axis wind turbine

D. H. Bausman, 30. *See also* horizontal windmills

D. W. Onan and Sons, 48

Dakota Farmer, 37

Dakota Wind & Sun, 171

Danish Wind Electric Society, 15–16

Danish Wind Energy Association, 26

Darr, Robert W., 138. *See also* Control Data Corporation; Jacobs Wind Electric Company

Darrieus, Georges Jean Marie, 32, 183

Data General Nova computer, 104. *See also* Rockwell International; Rocky Flats

Day, Jason, 223

Day, P. C., 27

De Goyon, Charles Marie Michel, 10–11

Deerborn, Henry, 68

DeKorne, James, 62

Delatron Systems Corporation, 72

Delco-Light Plant, 24–25, 37

Delp, Bill, 72

Det Norske Veritas, 212–13

Deukmejian, George, 85. *See also* wind farms

Development, Planning and Research Association (Manhattan, Kansas), 116. *See also* Bushland, Texas; US Department of Agriculture

DiFrancisco, Lisa, 94. *See also* Distributed Wind Energy Association

The Dispatch (Minneapolis, Minnesota), 142. See also Jacobs Wind Electric Company

distributed wind, 94, 96–97, 123, 167, 221–22. *See also* residential wind systems

Distributed Wind Competitiveness Improvement Project (CIP), 126–28, 168, 218–19. *See also* National Renewable Energy Laboratory; US Department of Energy; Wind Energy Technologies Office

Distributed Wind Energy Association (DWEA), 94–98, 167, 211.

Distributed Wind Vision—2015–2030, 96–97. *See also* Distributed Wind Energy Association

Divone, Louis, 100

Dodge, Darrell, 114. *See also* Federal Wind Energy Program

Dominion Aluminum Fabricating (DAF) Limited (Canada), 117, 185–86, 198. *See also* vertical-axis wind turbine

Dorney Amusement Park (Pennsylvania), 179

Drees, Herman M., 74, 75, 77, 154, 187–88. *See also* vertical-axis wind turbine

Droke, Farrel, 157, 162. *See also* Bergey Windpower Company

Ducted Turbines International, 223. *See also* Clarkson College

propellers: aircraft, 32–34, 172; aluminum, 42; governor, 44; Sitka spruce, 37; wood, 32–36, 172; World War I, 32, 172

Proven Energy (Scotland, United Kingdom), 208

Public Citizen, 89. *See also* Renewable Portfolio Standard

Public Power, 135

Public Utility Regulatory Policies Act (PURPA), 70, 78–79, 80, 81–82, 90, 144, 149, 170–71, 190, 202

Putnam, Palmer Cosslett, 49, 190–91

Putt, Harlie O., 35

R. M. Young Propvane, 103, *See also* wind measurement technology

R. Lynette & Associates, 200. *See also* wind farms

Radner, Nancy A., 88–89. *See also* Renewable Portfolio Standard

Randolph, William, 66

Rangi, Raj, 183. *See also* vertical-axis wind turbine

Reagan, Ronald, 81, 82–83, 85, 119, 120, 147, 162, 171, 172, 203

Real Gas and Electric Company, 72, 73, 137, 196. *See also* Jacobs Wind Electric Company

Rendell, Edward G., 92

Renewable Energy Venture (REV), 82, 143–44, 147–48. *See also* Control Data Corporation; Jacobs Wind Electric Company; wind farms

Renewable Portfolio Standard (RPS), 88–90, 200, 206

Research Applied to National Needs program, 100. *See also* National Science Foundation

Residential Renewable Energy Tax Credit, 128

residential wind systems, 79, 80, 81–82, 92–93, 94–95, 96–97, 98, 100, 108,123, 126–28, 136–37, 156–162, 165–66, 167, 170–72, 174, 177, 188, 202–203, 204–206, 207–209, 212–14, 216, 218–19, 221–23, 225

Reuss. Henry S., 189

REV Wind Power Partners, 147–48. *See also* Jacobs Wind Electric Company; Paul R. Jacobs; wind farms

Rich, Earle, 58. *See also* American Wind Energy Association; Natural Power Company

Righter, Robert W., 84, 135. See also Marcellus L. Jacobs

Risø National Laboratory (Denmark), 81

Ritter, Lloyd, 94. *See also* Distributed Wind Energy Association

Rock Concrete, 219. *See also* Distributed Wind Competitiveness Improvement Project (CIP); residential wind systems

Rocky Flats (Boulder, Colorado), 74, 75, 76–77, 79–80, 99–100, 102–116, 118, 119–120, 121, 135–36, 179, 188. *See also* Federal Wind Energy Program; US Energy Research and Development Administration

Rocky Flats Standard Workshop (Boulder, Colorado), 76

Rockwell International, 75, 76–77, 99–100, 103, 112, 113, 115, 136. *See also* US Energy Research and Development Administration

Rohn, Dwight R., 196. *See also* Rohn Tower Company

Rohn Tower Company (also Rohn Manufacturing and Unarco-Rohn), 69, 72, 74, 103, 137, 141, 195–96

Rollins, John, 186. *See also* vertical-axis wind turbine

rooftop wind turbine, 93, 209, 211–13

Roosevelt, Franklin D., 47. *See also* Great Depression and Rural Electrification Administration

Royal Technical Institute (Dresden, Germany), 26

Ruff, Hagen, 223. *See also* vertical-axis wind turbine

Rural Electrification Act, 47, 61. *See also* Rural Electrification Administration

Rural Electrification Administration (REA), 47, 49–51, 129, 131, 188

Rural Electric Wind Energy Workshop of 1982, 158

Rural Energy Producers Electric Power Cooperative (REPCO), 149. *See also* Jacobs Wind Electric Company; Paul R. Jacobs

Rural Energy for America Program (REAP), 125, 207. *See also* US Department of Agriculture

Ruralite Engineering Company, 43

S. Morgan Smith Company, 190–91, 201. *See also* wind turbine generators

Sagrillo, Mick, 204. *See also* wind electric generators

Sailwing, 173–74

Salter, Ed, 182–83

San Gorgonio Pass (California), 83–84, 144, 182, 186, 199, 205. *See also* wind farms

Sandia National Laboratories, 116, 118, 119, 125–26, 154, 183–84, 185. *See also* vertical-axis wind turbine

Sandia Scaled Wind Farm Technology (SWiFT) facility, 126. *See also*, Sandia National Laboratories

Sass, Walter, 197. *See also* wind measurement technology

Saturday Evening Post, 25

Saunders Engineering Co. Ltd., 43. See *also* wind electric generators

Savino, Joseph, 135. *See also* Marcellus L. Jacobs; National Aeronautics and Space Administration

Savonius, Sigurd J., 32, 61, 183

Saylor, John, 72, 74, 77

Schachle & Sons, 182. *See also* wind farms.

Schaefer, Daniel, 89. *See also* Renewable Portfolio Standard

Schmidt, Robert D., 142, 147. *See also* Jacobs Wind Electric Company

Schoenball, Walter, 72

Scholes, Daniel, 28

Schulte, Kevin, 94. *See also* Distributed Wind Energy Association

Scientific American, 13, 15, 19, 27

Sears, Roebuck and Company, 43

Seaverson, Lou, 107. *See also* Rockwell International; Rocky Flats

Second Wind Inc., 86, 197. *See also* wind measurement technology

Sellers, David, 63. *See also* North Wind Power Company

Sencenbaugh, Jim, 59–61, 70, 74, 158. *See also* Sencenbaugh Wind Electric Company

Sensenbaugh Wind Electric Company, 59–61, 103, 176, 196

Sheldon & Sartor Wind Electric Manufacturing Company, 42. *See also* propellers

Sherwin, Robert, 83, 85. *See also* American Wind Energy Association; Atlantic Orient Corporation; Enertech Corporation

Shulins, Paul, 67–68

Shuttleworth, John, 55. *See also Mother Earth News*

Simms, David, 70

Other Books in the Tarleton State University Southwestern Studies in the Humanities Series

In the Deep Heart's Core: Reflections on Life, Letters, and Texas
Craig E. Clifford

Cannibals and Condos: Texans and Texas along the Gulf Coast
Robert L. Maril

Rough and Rowdy Ways: The Life and Hard Times of Edward Anderson
Patrick Bennett

Larry McMurtry and the Victorian Novel
Roger W. Jones

Nueva Granada: Paul Horgan and the Southwest
Robert F. Gish

Beautiful Swift Fox: Erna Fergusson and the Modern Southwest
Robert F. Gish

Texas Women Writers: A Tradition of Their Own
Sylvia A. Grider and Lou H. Rodenberger

Riding the Wind and Other Tales
James Hoggard

State of Mind: Texas Literature and Culture
Tom Pilkington

Culture in the American Southwest: The Earth, the Sky, the People
Keith L. Bryant

Donald Barthelme: The Genesis of a Cool Sound
Helen M. Barthelme

West of the American Dream: An Encounter with Texas
Paul Christensen

San Antonio on Parade: Six Historic Festivals
Judith B. Sobré